Technics and Architecture

Technics and Architecture

The Development of Materials and Systems for Buildings

Cecil D. Elliott

The MIT Press **Cambridge, Massachusetts** **London, England**

First MIT Press paperback edition,
1994

This book was set in 9.5 point Janson
by DEKR Corporation and was
printed and bound in the United
States of America.

Library of Congress Cataloging-in-
Publication Data
Elliott, Cecil D.
Technics and architecture : the
development of materials and
systems for buildings / Cecil D.
Elliott.
p. cm.
Includes bibliographical refer-
ences and index.
ISBN 0-262-05045-5 (HB)
ISBN 0-262-55024-5 (PB)
1. Building—History. 2. Build-
ing materials—History. I. Title.
TH18.E45 1992
690′.09—dc20 91-28298
CIP

Contents

Preface

This study gathers together stories of the production of building materials and the development of building equipment and other systems. These are the media in which architects work. They constitute the array of possibilities from which architects, builders, and investors make the choices that largely define the nature of a building, and because they are on the whole pragmatic departures, successful new materials and systems soon become accepted as necessities, elevating the standards of performance by which buildings are to be judged.

The accounts that follow focus on the period of the Industrial Revolution and the times that came after, with introductory descriptions of earlier events. Much of the narrative centers on England, where mechanization first flourished. Later, in the nineteenth century, leadership shifted to the United States, where the force of expansion generated opportunities for experimentation. Emphasis has been given to the general adoption of systems and materials, and "firsts" have been neglected because such claims are as a rule questionable.

The text has been divided into two parts. The first groups seven categories of building materials in an order that is roughly chronological. The second discusses nine systems of building functions, including five developed to improve comfort and convenience and four that embodied the technical and scientific knowledge of the period when they were developed. In some cases different chapters may share a subject, treating separate aspects. For instance, terra-cotta fireproofing is discussed both in the chapter devoted to the manufacture of that material and in the chapter dealing with problems of fire protection. Structural engineering related to a specific material (iron, steel, or reinforced concrete) is treated as an aspect of that material's development and use, and general structural principles are considered in a chapter that reviews the foundations of structural theory and practice.

Each of these stories has its own character. In some cases, governmental and economic influences determined the outcome; in others, scientific discoveries or the growth of other industries may have controlled events. The distinctive characteristics of each story have been stressed, relinquishing a uniformity of treatment for the emphasis of individual features.

Beyond the matters discussed in this volume, there remain many subjects related to the construction industry, laborers in building crafts, the use of machines in construction, and other aspects of building and architecture. It is hoped that this volume may encourage scholars' study of such topics.

Acknowledgments

Research for this book would have been impossible without use of the Interlibrary Loan system through which academic and public libraries provided copies of publications. I am grateful to Deborah Sayler, Mary Jo Stanislao, and Lorretax Mindt of the Interlibrary Loan Service, North Dakota State University Library, for the diligence and patience with which they pursued my requests. Susan Wee of the Architecture Library, North Dakota State University, kindly searched out the answers to my many phone requests.

Local and state historical organizations (too many to mention individually) delved into their files and provided much of the information that expanded my understanding of places and people.

In preparation of the manuscript, invaluable assistance was given by Ronald L. M. Ramsay, who generously shared with me his knowledge of research techniques and material from his own files; Frances Fisher, who gave encouragement and commented on several chapters; Dennis C. Colliton, who commented on the chapters about elevators and brickmaking; Joel B. Goldsteen, who commented on the chapter on cements and made suggestions about organization; Bernard A. Nagengast, who gave expert advice regarding the chapters on heating and air conditioning; Earl E. Stewart, who commented on the chapter related to acoustics; and George E. LaPalm, who commented on the chapter on structural engineering. Errors of fact, interpretation, and omission remain the author's responsibility.

I am also indebted to students of the Department of Architecture and Landscape Architecture, North Dakota State University, with whom I had the opportunity to discuss these subjects in seminars.

Introduction

The architecture of the Industrial Revolution, from the middle years of the eighteenth century to the last of the nineteenth, is best identified as Romantic. A variety of terms have been coined to mark subdivisions of this period's stylistic attitudes, using the term "Revival" in some cases and the prefix "Neo-" in others, but always adding the designation of a previous style of architecture. (Art Nouveau, at the very end of the nineteenth century, is perhaps the sole exception to this tendency.) From the start of the Renaissance to World War II, the dominant stylistic theme in architecture was a reliance on the past, the extent of historical accuracy varying according to the patronage, physical requirements, and architectural talent under which a project was executed. Until the middle of the eighteenth century, European designers, both Renaissance and Baroque, adapted ancient motifs with considerable freedom, and their styles themselves later became models for the eclecticism of the nineteenth century. Each of the Romantic movements—revisitations of the Greek, Roman, Romanesque, Gothic, and even more exotic styles—responded to current ideals. Newly established representative governments (for the Industrial Revolution was also a period of political revolution and theorizing) echoed the forms of Greece and Rome in their buildings; and religious groups, according to their outlooks, chose between the perceived rationality of classical styles and the emotionalism of Gothic.

The concurrence of the Industrial Revolution and Romanticism is understandable if one accepts the historical realities. The innovations of industrialization, so stunningly compressed and dramatized in history books, were slow to be accepted and operated throughout any industry or any country. During his lifetime the average citizen of the western world in the eighteenth and nineteenth centuries had contact with only a few significant technological advances, and these were accepted as merely sensible solutions to practical problems, not part of a juggernaut of industrial development. For instance in 1880, some four generations after the first steamboat, about three-quarters of the world's shipping capacity was still under sail. Many of the innovations that proved to be most influential were at first viewed as curious experiments, having questionable value under realistic conditions.

The reality of Romanticism is similar in that it was to some extent a part of the literature, painting, music, and architecture of all periods, and its dominance during much of the eighteenth and nineteenth centuries developed gradually in the company of considerable social and political change. While classicism had espoused the cause of simplicity and the evolution of acceptable forms of expression, Romanticism encouraged individualism, imagination, and emotion, accepting even fear and morbidity as beneficial experiences. Such preferences fostered an appreciation of variety and change, and the drama of invention was to the Romantic often more attractive than its results.

Our histories of architecture correctly stress the importance of milestone buildings that were far in advance of the normal course of design and construction; reading them, one might almost be persuaded that the innovations of Louis Sullivan or H. P. Berlage were quickly, enthusiastically, and widely accepted as models by the architectural profession and its clients. In truth, new ideas such as these prospered among narrow ranges of patronage and professional practice. The rate at which technological advances were adopted was far more rapid for factories, office buildings, department stores, and other types of buildings that were blessedly free of professional or public preconceptions about architectural style than it was for the more traditional types of buildings. Later in the twentieth century the impetus toward nonhistorical designs and the incorporation of technological advances was severely limited by the twenty-year hiatus of the Great Depression, World War II, and the years of painful postwar recuperation.

It would be convenient, but only somewhat accurate, to attribute twentieth-century changes in architecture to the influence of technology as it was applied to buildings. Architecture is a complex art having many masters. A building is at the same time an object, an investment, and a cultural and personal expression of beliefs. Any change in the way buildings are built or the way they look must be tested against a variety of standards, their relative importance being somewhat different for every project. This truism explains why certain technological aspects of architecture have been readily adopted and others have been long delayed. For instance, elevators were a vital factor in the economic and social changes related to the great sweep of urbanization, and therefore elevator technology was immediately accepted and quickly developed. No similar urge spurred the development of a more rational system of plumbing and waste handling.

Often the mood of historicism was superficial. The choice of a Greek temple as model did not necessarily rule out grafting a dome or spire onto a design, and with up-to-date construction techniques Gothic naves often acquired proportions that bore limited relationship to those of the buildings they emulated. New technologies intruded gently because historical accuracy and consistency, not prime criteria of Romanticism, were generally less valued than the evocation of appropriate sentiments.

Nearly all buildings are singular designs, though with pronounced similarities to other buildings of the same purpose, place, and period. The individuality of projects has effectively limited the extent to which industrialization has been applicable to architectural construction. In the early eighteenth century, the process of assembling a building might well begin on the site with preparation of the clay and kiln that would produce brick for the walls. Timbers for trusses would be cut, mortised, and drilled by the workmen who would set them in place. Today such elements of the construction will have been largely prepared before they are brought to the site for assembly into the final structure. Although craftsmen still work to fit and fasten in place the parts of a building, the nature of their work has gradually changed as industrialization and the concomitant standardization have increased the relative completeness of elements that are brought to the site of construction. At the same time, the Industrial Revolution and the revolution in transportation that accompanied it reduced the architectural significance of a region's own materials. Local differences in the prices of materials still exerted appreciable influence on selections, but their significance diminished as decades passed. Climatic differences have become less important, because equipment and materials have been developed to ameliorate the effect of external conditions on buildings' interiors, and cultural and economic exchanges between nations and places have acted to reduce the differences between buildings in different settings. Good or bad, office buildings throughout the world are much alike, government buildings have been similar since the Classical Re-revival at the start of this century, and even residences are moving toward greater similarity.

The modern movement in architecture, which blossomed in the 1920s and bore fruit after World War II, has been on the whole more eager to utilize new materials and incorporate new systems than were the Romantic historically based fashions that came before. In the chapters that follow, the reader will note that in the majority of cases the most advantageous improvements in the manufacture of materials and in the quality of systems for buildings were achieved in the last quarter of the nineteenth century and the first decade of the twentieth century. Since that time the acceptance and application of those advances have increased, and in many cases the need for further fundamental development has become apparent. At present, architects work with a relatively stable palette of technological development, seeking viable responses to social and economic functions and awaiting a revival of inventive enthusiasm that may improve the art's technological capabilities.

Technics and Architecture

Materials

The materials that have been used to construct buildings during the last three centuries were already present in ancient and medieval times. The change, in essence, has been largely in the development of mechanized methods for manufacturing them and in the networks of commerce that have made raw materials and finished products available. Early technological advances improved the tools and methods used in the preparation of building materials, but such changes often resulted (as with early advances in glassmaking and stone quarrying) in little more than transferring traditional handwork procedures to the operation of machinery. The more startling advances in the western world's material accomplishments were rooted in the enlargement of scale, which extended markets and challenged industrialists with a scale of operation that had previously been available only to some political leaders. Industrial "empires" and competition (so fierce that terms of warfare were often used to describe it) in the late nineteenth century found fitting reward in the medals presented to competing manufacturers at the international expositions of that period.

Although the use of a material in building construction provided the major market for few industries, this application profited from the advances for which other uses provided the stimulus. In addition to its fostering the expansion of industrial activity, transportation required mammoth constructions. By the end of the eighteenth century the tonnage of European shipping was about five times as great as it had been two centuries before. To attend these vessels, it was necessary to expand and improve harbor facilities. The establishment of railroad lines required construction of bridges and viaducts along routes of remarkable length. These developments, along with the growth of cities between which goods were carried, stimulated the production of many of the materials needed for building construction. Such activity was by no means continuous or untroubled. In England the two waves of railroad construction were divided by six years of desperately hard times, the so-called "Hungry Forties," which were reflected in a reduction in brick production (then a fair indicator of building activity) of almost one-fifth. But this was much smaller than the one-third reduction of brick production that occurred at the beginning of the Napoleonic wars, when construction workers joined the military forces and the prices of building materials rose sharply.

Although construction itself lagged far behind factory industries in its mechanization and industrialization, building activity thrived. Between 1875 and 1907 in Germany, employment in construction grew from 10 to 16 percent of the total national work force. The massive movement to cities increased population densities there, and merchants and manufacturers became the dominant patrons of building construction. In England, public expenditures and the interest rate on government bonds decreased radically around 1820 and remained relatively low during the remainder of the century. During that period construction work and architectural commissions rose. It appears that buildings, whether commercial, industrial, or residential in function, came to be recognized as alternative investments. Frugality and the demand for more intensive and profitable use of sites, at the same time, changed the character of buildings and the materials of which they were made. On one hand, new and improved materials made it possible for buildings to satisfy needs that were new to the period, such as larger spans and taller structures. On the other hand, for more traditional functions it was possible to make these materials conform to the standards of propriety and taste to which bourgeois patronage aspired. This meant that architecture was soon forced to contend with opposing forces: the material nature of the media in which it worked and the visual expectations of the art, whether the latter were founded on historicism or cubistic abstraction.

As animal power gave way to water power and steam power, the simple procedures of local manufacture for a regional market were necessarily superseded by those of larger work forces, sufficient to produce enough goods to justify the required investment in equipment. At one extreme, the tightly knit work groups of glasshouses isolated in forest land became blowers' teams, reporting to the factory when a batch of molten glass was ready for their work to commence; under full mechanization, glassworkers became machine operators and supervisors in large factories. At the same time the casually assembled crews brought together for such seasonal operations as brick kilns became regularly employed work forces, once mechanized preparation of the clay permitted more continuous operation of brickyards.

For those materials manufactured with the use of heat, a critical factor was the economical utilization of fuel, which was often the principal determinant of price. Wood and charcoal, the most ancient fuels, became too costly as forests were devastated for use in buildings and ships. The specific locations of coal, petroleum, and natural gas required the location of factories where fuel was to be found or at points to which it could easily be brought. Nevertheless, profitable manufacturing required the most efficient possible use of fuels. This problem was common to all industries using furnaces and kilns. Efforts were made to improve processes by eliminating the intermittent heating of furnace masses and utilizing the heat of escaping gases. Such improvements were usually expensive to institute and their importance varied among industries according to the relative significance of fuel costs, the level of competition present, and economic trends. In many cases, the adoption of new fuel-saving methods was delayed because of the reluctance of factory owners to make the large investments that were required. In most industries it appears that the effects of competition were dulled in the late nineteenth century by localization of markets, the imposition of tariffs, and price-fixing manipulations by groups of manufacturers. Scientific knowledge became increasingly important as the factories producing building materials grew in size and international competition intensified. Mixtures and methods that were part of a craft's heritage and innovations that had been discovered by chance gave way to scientific analysis. The development of metering and testing equipment permitted an increased accuracy in controlling production and gauging the quality of the material produced, and advances in the scientific understanding of heat permitted more efficient use of fuel.

Architects' use of materials during this period followed three patterns. Industrialization of the production of some materials, such as masonry, wood, and glass, did little to alter their nature and quality but increased their availability. For others, terracotta and concrete, their early use as substitutes for traditional materials soon waned. The most dramatic influences were structural materials, iron, steel, and reinforced concrete, which permitted the development of new building forms responding to changed needs.

1 Wood

1777 Circular saw patented (Britain)

1832 Balloon framing introduced in Chicago

c. 1850 Circular saw first employed in processing lumber

1860s Great Lakes area becomes center of the U.S. lumber industry

1867 Rotary veneer lathe developed

1870s Double-edged axe introduced

Peak of investigations of chemical preservatives by U.S. railroad interests

1880s Saws with raker teeth used in felling

1883 Completion of railroad to Puget Sound opens forests of northwestern U.S.

1884 Production of three-ply chair seats in Estonia

1890s "Hot ponds" make year-round operation possible for sawmills in North America

c. 1900 Resawing introduced

1905 Softwood plywood displayed at Lewis and Clark Expedition Centennial

1933 Synthetic resin glues produced in Germany

European life until the nineteenth century was largely based on the use of wood and the exploitation of the continent's vast forest lands. Wood provided fuel for industrial and household purposes, raw material from which tools and utensils were fashioned, and timbers from which carts, ships, and buildings were assembled. As the European population grew and spread toward the deep woodlands of the north, trees were felled in increasing numbers. Wood and charcoal had provided heat for Greek smelting and other crafts, and charcoal braziers heated Greek buildings when the sun's rays were not sufficient. Greek ships of trade or war required much timber, and when the Delian League was established in the fifth century B.C. to protect Greek interests in the Aegean Sea, the city-states were pledged to contribute a fleet of about three hundred warships. By that time so many trees had already been cut that the Greek landscape had been altered significantly. Indeed, the Greek city-states had introduced restrictions designating those processes for which wood or charcoal might be used as fuel. Several centuries later, Rome was confronted with a similar problem. Forests near the city were so depleted that it was necessary to obtain wood from almost a thousand miles away. By the fourth century A.D. the needs of Rome were served by a fleet of "wood ships," which brought timber from France and North Africa to the port of Ostia, from where it was moved upriver to Rome.

Although the amount of wood used in the construction of buildings was certainly less than that used as fuel and probably less than that used in ship building, deforestation made lumber more expensive, especially for the heavy timbers needed in roof construction. In the sixth century B.C., Greek temples, which had previously been of wood with terra-cotta surfacing attached, began to be built of stone, except for the roof construction and ceilings, which were still made of wood.[1] In later centuries, Roman settlements were constructed in northern Europe, where the forests still grew dense and wood was plentiful. Remains of the structures testify that there was a plentiful supply of seasoned timber, and the skill with which it was assembled is evidence of the efficient organization of the Roman army and the variety of iron tools that were available at that time.[2]

Accounts of medieval building projects often begin with descriptions of bands of carpenters going into a forest to fell the oaks that would be used. This meant that the lumber might not be properly cured by the time it was used, with the accompanying probability of warpage, but the slow progress of work on large projects in that period often provided for them a degree of curing during construction. In the fifteenth century, average sizes of wood framing in England would be about 13 by 11 inches for principal posts and 10 by 7 inches for joists spaced with around 8 inches between them (fig. 1.1).[3] Improving craftsmanship and rising costs of lumber reduced those formidable dimensions so that after the Great Fire of London (1666) joists were more likely to be 7 by 3 inches, less than half the cross-sectional area of earlier times.

When Abbot Suger required 12 beams of extraordinary size for one of his projects, his carpenters and those he consulted in Paris knew of no French forests in which timbers so large could be found, and the abbot and his aides are said to have searched the woods until suitable trees were discovered.[4] In England an exhaustive inquiry was required to locate the long timbers required for construction of the lantern at Ely Cathedral, for by that time, the middle of the fourteenth century, lengths greater than 30 feet were rarely available.[5] In early times structures of appreciable size were built by the crown, the greater nobles, or the church, and each of these classes of patrons usually possessed broad forests from which trees could be selected for the projects that they might undertake. As building came to be a more widespread activity and English forests were more rapidly depleted, there arose a trade in timber, shipping fir and oak from the Baltic ports to England. This commerce brought timber principally from the Hanseatic ports of Danzig, Riga, and Memel and exchanged it for English cloth. Oak was the wood most highly prized for English ships and buildings, whether built for government or for trade, and the forests upriver from Baltic ports held rich stands of oak, as well as fir and pine. The forests of Sweden and Norway were an additional source of timber, a source that grew to be so important that in the seventeenth century it was said that "the Norwegians warmed themselves by the [Great] fire of London."[6] From German forests, logs were floated down the Rhine to be shipped from Dutch ports.

To prepare lumber for use in construction, medieval carpenters (for in most areas carpenters undertook the work from felling the trees to completing the joinery) rough-hewed the logs with an adze or split the wood with iron wedges and mallets. It was usually after the squared logs were removed from the forest that pit sawing took place. For sawing, a log or a riven half-log was raised on trestles or laid on the ground above a pit. The line to be followed in sawing was marked by a chalked cord and a pair of sawyers, one standing atop the log and another beneath, used a two-handled saw or a frame saw in which the blade was stretched within a wooden rectangle (fig. 1.2). The saw

1.1 As in this fifteenth-century house in central France, when medieval buildings were made of wood, massive timbers occupied much of the exterior wall surface, leaving small interstitial areas to be filled with masonry or other materials. Connections, which were mortised and pegged, required that the timbers be large. (*Architectural Record*, April 1900.)

1.2 In this view of an eighteenth-century lumberyard, a frame saw is used to divide a plank laid across trestles. Most of the energy for sawing in this manner was provided by the lower sawyer, and the worker above was responsible for guiding the cut along a chalked line. (Diderot, *Encyclopédie*, 1762–1777, "Menuiserie," plate 1.)

cut as it was pulled downward by the worker who stood, covered with sawdust, in the bottom of the pit. Pit sawing, or similar methods, is still practiced today by residents of some of the least developed regions of the world.

1.3 A sixteenth-century design for a hand-cranked gang saw. In this period, drawings of mechanical inventions were often published to demonstrate the principles involved; the inventions may never have been actually constructed and operated. (*Iconographic Encyclopedia*, 1889, vol. 6.)

The work of the pit sawyer was easily mechanized. During the sixteenth century, sawmills powered by water or wind were erected in Norway, Holland, and some Baltic areas.[7] An estimate at the end of the eighteenth century said that "one mill, attended by one man . . . will saw more than twenty men with whip saws, and much more exactly," but it is difficult to estimate the financial advantages because one must consider the relative wages of the different sorts of workers and the amount saved by not shipping wood that would become useless sawdust. Apparently, English pit sawyers were apprehensive about changes that might result from mechanization. Water-powered mills for other purposes were built in England from the thirteenth century, but there were few sawmills, although many travelers gave detailed accounts of those they had seen on the Continent. In the sixteenth century, an increase of the English birth rate caused a population explosion, and unemployment among unskilled workers accompanied migration from rural districts into towns. In the early seventeenth century, a young traveler wrote his father an enthusiastic description of a sawmill he had seen in Germany but closed with a comment that he "would recommend the use of mills to saw timber in England, were it not that it would hinder the employment of poor men."[8] The public's fear of the social results of technological change persisted. As late as the 1760s, when a prosperous timber merchant built a wind-driven sawmill, a mob of irate pit sawyers attacked and pulled it down.[9]

The earliest sawmills of note were based on a turning source of power (whether from wind or water), a crank that converted that rotary motion to a linear motion, a rocker arm attached to the vertical frame that held the sawblade, guides in which the frame moved, and perhaps a spring system (sometimes formed by a bent piece of wood) that would return the frame after the completion of each stroke (fig. 1.3). In addition, a system of weights or rollers was required to move the timber forward and maintain its contact with the saw blades. The frames of these saws were heavy, which limited the speed with which they could be operated. In nineteenth-century Mississippi, 120 strokes per minute was the greatest speed at which a sash (or frame) sawmill could run. This allowed the mill to saw between 3,000 and 5,000 feet daily, with a kerf somewhat greater than a half-inch.

The muley saw—its name derived from the German word for "mill"—was a later mechanism using vertical reciprocating action. Dispensing with the heavy frame of previous saws, the blade of the muley saw was mounted at top and bottom in a manner that allowed adjustment according to the size of the logs being sawed (fig. 1.4).[10] The greatest advantage of the muley saw was the smoothness of its cut, its kerf being somewhat narrower than that of previous saws. (The width of saw kerf, which set the amount of sawdust produced and hence largely determined how much wood would be wasted, mattered little in the early spendthrift period of a forest's exploitation, but as the wealth of wood dwindled lumbermen became acutely concerned about narrowing kerfs.) The principal disadvantage of the muley saw was its speed, which was no greater than that of the saws introduced earlier.[11]

The gang saw resulted from placing several blades together, so that a series of boards could be cut with one pass of a log through the saw (fig. 1.5). It was the first of a series of devices that accelerated the work of sawmills and consequently required the invention of machines that would assist in tending the saws. Lumber moved so quickly through the new saws that it was no longer sensible for workmen to bring logs to the saw, turn logs on the saw, move timbers from one saw to another, and carry finished boards to the drying shed without the assistance of machines devised for those purposes. The demands of speed were particularly felt after the advent of the circular saw.

Although a British patent was awarded as early as 1777 and it was introduced into the United States in 1814, until the middle of the nineteenth century the circular saw was used principally for cutting veneer. At high speeds the centrifugal forces within a spinning blade and its expansion from heat while sawing could cause vibration and curvature of the blade. The need to mechanically flatten circular blades in their manufacture and during their use limited the hardness of the metal from which they could be made, and the use of soft metal led to a thicker blade and teeth that were quickly dulled.[12] Strains within the blades were somewhat relieved by providing radial slots toward the center of blades, and the effect of centrifugal pressures was reduced by tapering the thickness of blades (fig. 1.6). The greatest improvement of the circular saw resulted from inserting teeth of hard steel around the edges of a disc of softer steel. The maximum depth of cut for a circular saw was slightly less than half the blade's diameter, and increasing the diameter beyond certain limits would, of course, increase the saw's vibration and thus widen its

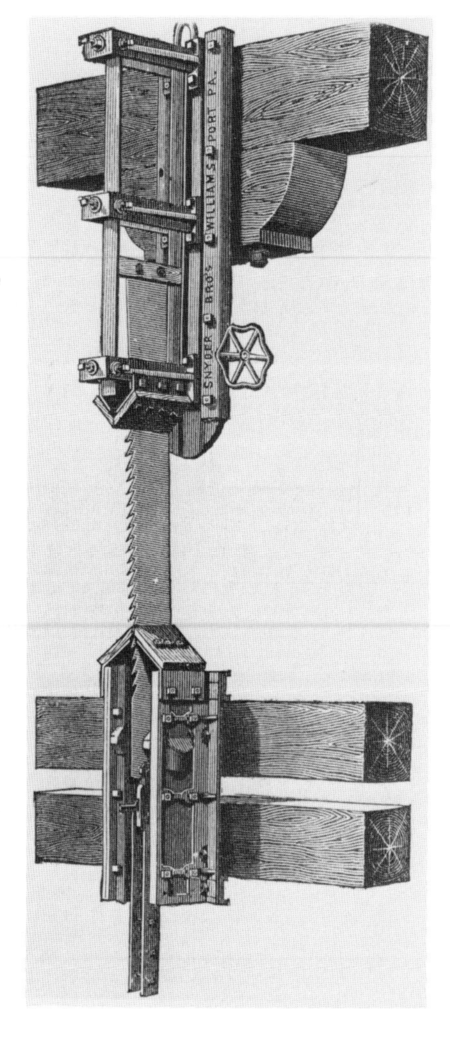

1.4 In the muley saw, variable lengths of the blade could be exposed between the adjustable muley heads. This ability to adapt the saw to the dimensions of logs resulted in a narrower kerf and consequent savings. (*Appleton's Cyclopedia*, 1880, 2:708.)

1.5 This gang saw of the 1880s includes a mechanism that moves the log forward at the same moment that the blades cut on their downward stroke. The bar below (*E*), which pulled the frame of blades down, was called the "pitman," an echo of the days of pit sawyers. (*Appleton's Cyclopedia*, 1880, 2:709.)

1.6 To avoid cupping of circular saw blades, slots could widen if the rim of the blades became hot and expanded or narrow if the center became hot. Inserted teeth were shaped of metal harder than that of the blade. The planer-tooth (above) was said to produce smooth-sided kerfs; the clipper-saw (below) was particularly suited for use on thin saw blades. (*Appleton's Cyclopedia*, 1880, 2:700.)

kerf. By the middle of the nineteenth century, the practical maximum diameter for a circular saw was usually considered to be around 5½ feet, which would saw a log little larger than 2½ feet in diameter. A solution to this problem was found by mounting a second and smaller circular blade above the first, increasing the total cut by about half this smaller saw's diameter (fig. 1.7). Whatever the difficulties inherent in the circular saw, it was fast. Other saws might average cutting as much as 5,000 feet daily (gang saws affording the pronounced advantage of their multiple blades), but in the 1850s circular saws could be expected to cut at least 1,000 feet per hour. Before many years had gone by that number quadrupled.[13]

Circular and muley saws had kerfs of 5/16 of an inch at best, but usually cut a width of 3/8 of an inch, the same dimension as the kerf of vertical frame saws (fig. 1.8). A kerf of that width resulted in a loss of more than a third of the wood after a log was squared, the tree's taper was accounted for, and the remaining timber was sawed into boards one inch thick.[14] Vast piles of waste accumulated at mills with little possible use for them. Sawdust could be dumped into a stream, but it quickly accumulated and interfered with floating logs to mills downstream. A fraction of the sawdust might be sold for use in filling ice houses or as an agricultural fertilizer, and a small portion of the bark was needed by tanneries. Steam power for a sawmill could be produced with little cost by burning a sawmill's plentiful waste, trimmings, bark, and sawdust. In fact, fuel was so plentiful that in the 1890s mills in the northern parts of the United States instituted year-round operation, a result of the introduction of the "hot pond," the pool in which logs were kept being heated by a pipe carrying steam from the sawmill's boiler. Still, enough waste remained to cause a constant danger of fire, a risk always confronting lumbermen.

Although the band saw was invented and patented quite early in the nineteenth century, it was not useful until a blade could be manufactured that would run at high speeds without snapping. After the Civil War, the rising quality of metal and workmanship in the United States permitted manufacture of fine-toothed band saw blades of well-tempered metal. Blades as long as 60 feet and about 15 inches wide were looped over wheels above and below, and there was virtually no limit to the size of log that could be taken by a band saw. The narrow kerf of these blades produced, in most cases, less than half the amount of sawdust that came from other saws. Until the start of the twentieth century, much lumber was sawn at the start to the size at which it was to be used. Resawing, a method of shaping large timbers from a log and sawing them into boards and planks after shipping and as needed, was common by the end of World War I. For the second sawing, bandsaws mounted as multiple blades proved to be useful because of the accuracy and speed of their cut.[15]

It should be understood that at no time did one type of saw completely supplant the preceding types. At all times the larger share of sawmills were small plants, perhaps the wintertime occupation of a farm family. In such situations, investment in labor-saving equipment was seldom logical, for there was no serious shortage of man-hours for the work. Different species of trees and the probable uses for their wood often governed the choice of machinery. Only a few decades ago it was reported that the vertical frame saw (first to follow the pit-sawyer) was finding new favor in the United States and Canada, because it was well suited for sawing small and medium-sized logs.[16] One of the criteria for the planning of many mills was the simplicity with which the machinery might be dismantled when an area of forest land had been depleted and the time had come to relocate the saw at a place that offered a fresh stand of trees.

Machines for finishing lumber were a popular area of invention, so popular indeed that the case of the Woodworth planer, which combined feed rollers and cylindrical rotating cutters, became a *cause célèbre* of U.S. patent law. Woodworth's son in 1842 applied for an extension of his father's 1828 patent on the grounds that his father had been forced to sell off rights to the patent after workmen's protest demonstrations closed his initial display of the machine in New York (fig. 1.9). In 1836 a law had been enacted

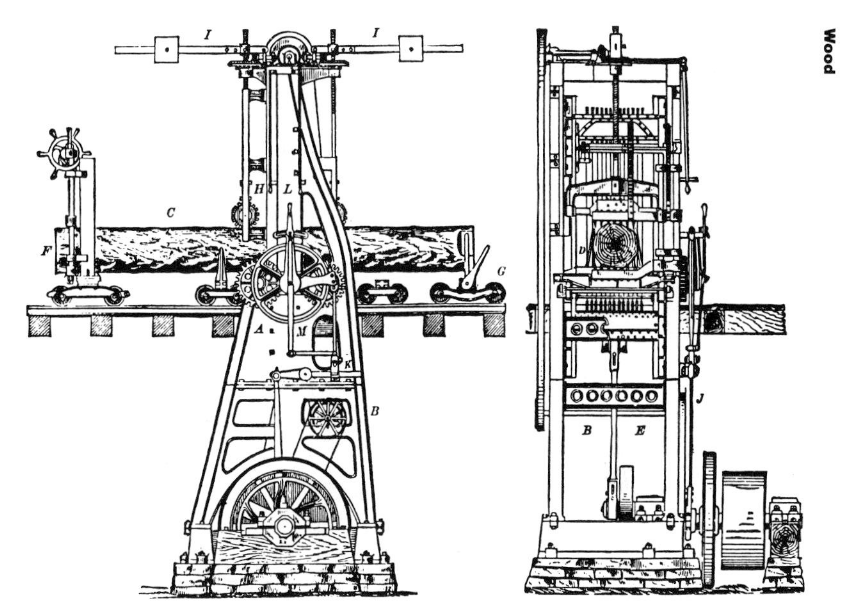

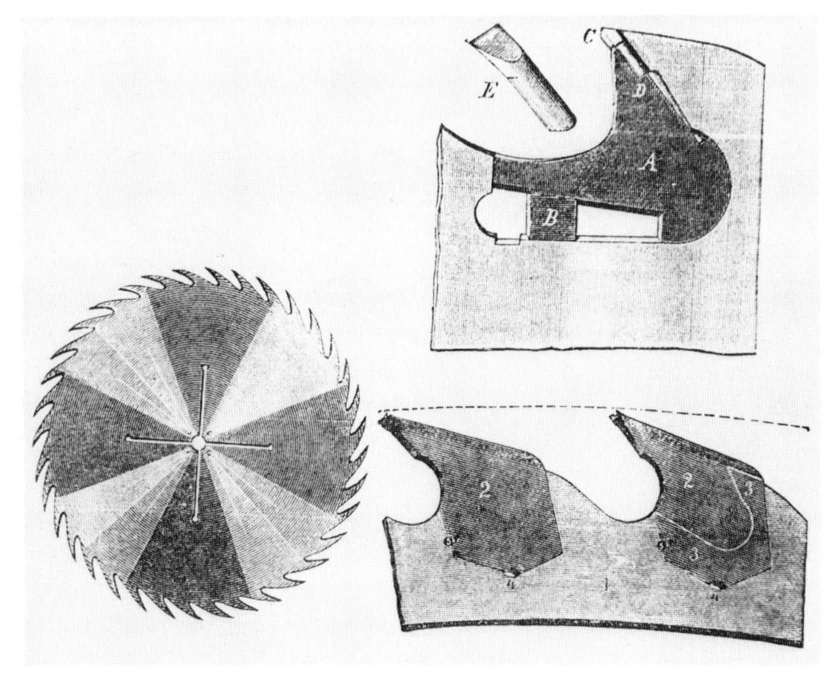

1.7 A portable circular-saw mill, as manufactured in Cincinnati, used two blades to accommodate large logs. At the right is the carriage on which logs were moved through the blades. In the foreground, workers turn a log with peaveys, indispensable tools at lumber camps and sawmills. (*Appleton's Cyclopedia*, 1880, 2:703.)

1.8 At an exhibition in 1918, the West Coast Lumbermen's Association displayed a section of a fir log 4 feet in diameter, cut into lumber and reassembled. Note the similarity of this example to the methods of obtaining rectangular panes from a disc of crown glass (fig. 5.2). (*Scientific American Supplement*, 29 June 1918.)

1.9 The Woodworth planer was capable of smoothing the upper and lower surfaces of boards with rotating cylinders, each having three blades. Edges of the boards could at the same time be smoothed or tooled with tongues and grooves. (*Asher and Adams Pictorial Album of American Industry*, 1876.)

permitting extensions of patents when the inventor "without neglect on his part [has] failed to obtain from the use and sale of his invention a reasonable remuneration for the time, ingenuity, and expense bestowed upon the same."[17] In legal action that followed Woodworth's application for an extension, his opponents questioned the legality of the extension, the validity of the initial patent, and ownership of the rights after an extension. A succession of courts, and eventually the U.S. Supreme Court, determined that upon issuance of an extension all rights reverted to the original patentee, excluding those who may have later purchased interests in the original patent. The Woodworth monopoly was based on restricted distribution of the machines and royalties charged on the amount of lumber planed, and it finally ended in 1865 after opponents presented to Congress a petition 50 feet long.[18]

John Burroughs, the eminent American naturalist, in 1883 contrasted the "piny, woodsy flavor" of American poetry with the pastoral character of English poetry, which dwelt on fields and pastures and seldom spoke of woods.[19] But one should not conclude from this observation that the settlers of New England felt great affection for the forests that surrounded them and extended so far westward. As Theodore Roosevelt wrote, they viewed the land as "a region of sunless, tangled forests . . . with underbrush . . . dense and rank, between the boles of tall trees making a cover so thick that it nowhere gave a chance for the human eye to see even as far as a bow could carry."[20] To the pioneers, forests seemed threatening depths filled with insects, swamps, disease, and marauding redmen. Only by felling trees and clearing the land could space be created for agriculture, houses built, and income produced. Much of the colonial culture was grounded on the use of the region's wood. Barrels for Spanish wines and Barbadan sugar and molasses proved profitable, and New England pines, taller than those of northern Europe, provided masts that need not be weakened by being spliced. A small enterprise of shipbuilding began in New England and prospered. Yet no matter what commercial value could be found in the trees, they occupied land that was wanted for agriculture. For the most part, settlers moved west by wantonly clearing land, planting their crops among the gaunt, leafless trunks of trees that they had girdled and left to rot.

If the forests of North America proved to be a valuable resource, they also proved to be an exhaustible resource. By 1773 Rhode Island had cut all its available firewood and wood was bought from other colonies. A century later the woods of New York and New England had been so depleted that much of their logging activity moved westward to the forests surrounding the Great Lakes.[21] By the end of the shameful Black Hawk War, native peoples of the Lake area were confined to small reservations, rich farmland was made available to settlers, and dense timberland became available to loggers. In the 1850s hard times hit the eastern lumbermen and hastened their migration. Around 1860 the Lake area became the leading source of lumber in the United States. Chicago became the center of the U.S. timber trade, with lumberyards in 1856 already occupying 6 miles along the Chicago River and twice as much 15 years later. The South had a small antebellum timber industry,

largely coastal and marketing yellow pine and naval stores (turpentine, pitch, and rosin). During Reconstruction lumbering expanded, and after 1870 yellow pine from southern woods began to be seen in the North, even on the Chicago market. But when the timber available in the Lake region began to diminish in size, quality, and ease of access, lumbermen and loggers looked to the South and to the West, where vast stands of trees remained untouched.

The forests of Canada and the United States provided a rich source of timber for the westward expansion of those countries in the nineteenth century and a source of exports to Europe. By 1810 timber had replaced furs as the principal export of Canada, largely due to Napoleon's blockades that kept English ships away from Baltic ports. By 1890 the production of sawmills in the United States attained a value one-third greater than the output of all metallic production for that year.[22] In 1906 it was estimated that the lumber industry in the United States cut "approximately 40 million feet, board measure, of lumber, 11 billion shingles, 100 million railroad ties, 4 million poles, 20 million fence posts, 170 million cubic feet of round mining timbers, 3 million cords of pulpwood, 1½ million cords of tanbark, and about 100 million cords of firewood."[23] In the mid-nineteenth century, a succession of federal

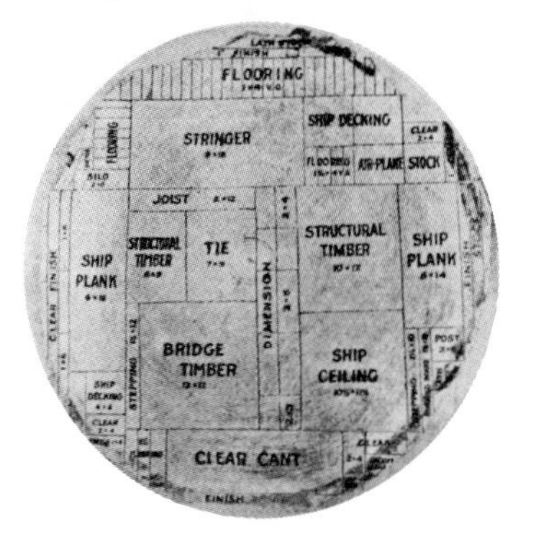

laws—the principal ones including the Swamplands Act of 1850, the Morrill Act of 1862, the Timber Cutting Act and Timber and Stone Act of 1878—included provisions that were used by lumbermen seeking possession of forest lands.[24] The luxuriant stands of pine around the Great Lakes were dwindling and lumbermen of that area looked westward to fresh forests beyond the Rockies. After the California gold rush, some timber had been cut for local use on the West Coast and a few shipments were made to the Orient. When the Northern Pacific Railway Company in 1883 opened a route from Puget Sound to the Great Lakes, the timber of the northwest region suddenly became available. In the fashion of the business world of that time, there came to be timber barons who amassed vast stretches of woodland and operated both logging and milling phases of the business. Reformers, trust busters, and muckrakers loudly protested deforestation and predicted the depletion of the forests. Proponents described reforestation programs and the economic advantage to be gained from expansion of the timber industry. Since lumbermen made most of their profits from land and trees that were government property, this debate became repeatedly the subject of campaign promises and very seldom the topic of effective legislation.

A similar dichotomy is found in the descriptions of logging camps. By the latter part of the nineteenth century, logging camps had moved far from their owners' sawmills, for the machinery of a large sawmill handling large logs was no longer adapted to changing its location as loggers exhausted one mountainside of trees and moved on to another. Some journalists wrote heartwarming descriptions of robust lumberjacks filling the bunkhouses with jolly songs. Other descriptions pictured immigrant workers ill-fed, underpaid, and constantly endangered by the perils of their work.

Large trees had been felled for centuries with axes, because saw cuts became hopelessly clogged by rosin and sawdust. In the 1870s, West Coast choppers abandoned single-edged axes for a double-edged type. This new kind of axe allowed one edge to be used in trimming hard knots and other work that quickly dulled the axe's edge, and it saved the other edge for the work of undercutting, lessening the time required for sharpening the axe (fig. 1.10). It was the 1880s before saws with "raker teeth" were introduced. Typically, these saws bore groupings of about four cutting teeth with pairs of cleaning teeth between them, cut much shorter to pull sawdust from the cut. In addition, sawyers carried with them bottles of kerosene with which resin could be periodically cleaned from the saw blade.[25] The use of saws was said to double the number of trees that could be felled by a pair of lumberjacks, and an additional economy resulted from the fact that sawyers, needing much less skill than choppers, could be paid, according to Minnesota wage practices, little more than half the wage of choppers. In every logging camp, saw filers were charged with the responsibility of sharpening the saws, with due consideration of the length of a sawyer's stroke, the kind of trees being cut, and weather conditions. Power saws for this purpose were not introduced until the 1930s.

It is difficult to describe the way in which trees were cut and brought to a mill. Climate, terrain, the species being felled, the size of the lumbering enterprise, and even local traditions might determine the choice of procedures. Southern locations often were hampered by having to move heavy logs through muddy swampland; northern locations were more governed by seasons, cutting trees in the winter months and floating them to the sawmill when melting snow and spring rains made the rivers run high. A California redwood might offer particular problems, because its brittle wood could be shattered as the tree fell and a log 26 feet in diameter was invariably cumbersome and dangerous to move down even gentle slopes. Such factors could cause adjacent timber camps to use different methods, but at the same time similar methods might be employed by camps far apart in place and time.

For bringing logs from the forest to the sawmill or a waterway leading to the sawmill, the procedures varied according to the terrain. Logging began along the edges of rivers, where logs had to be moved only a short distance to the water. In flat, swampy woodland, such as the forests of the southeast United States, horses or oxen pulled logs along "corduroy" logging roads, surfaced by laying tree trunks across the route. On steeper slopes the "skidway" was often lined with logs to ease the handling, sprinkled in winter with water to make an icy coating and smeared with grease in warmer weather. In some mountainous locales flumes were erected, narrow triangular troughs holding a flow of spring water in which logs or sawn timbers would hurtle downhill for miles, one after another, and splash at the end into a river or lake. When engines were in common use, steam tractors pulled logs and portable "donkey engines" turned reels on which cables were wound to run derricks, hoists, and drag lines to move the logs. If logs that were destined for several sawmills floated in the same stream, they were marked, so that downstream each mill could identify and rescue the logs dispatched from its own logging camps.

On waterways used for other purposes, a more controlled method of moving logs was accomplished by assembling rafts. Units of logs were assembled in crisscross layers with 250 to 500 square feet of surface area. These cribs were linked together as rafts with overall dimensions as much as 135 feet by 160 feet, or even greater when they were assembled for travel down the Mississippi.[26] Workmen overseeing the drift downstream might assemble small huts on the rafts' top sides. On the St. Lawrence,

1.10 Standing on "springboards," fellers make a saw cut about one-third into the trunk and well above the hard grain and heavy sap of the base. Axes were used for undercutting before the workers turned their attention to the other side of the trunk, where the saw and iron wedges were used to fell the tree. By carefully sighting the direction in which the tree would fall, damage to it and other logs could be avoided. (*World's Work*, February 1904.)

1.11 On an Oregon river early in this century, logs float toward a cradle into which they will be loaded by a derrick. The cradle (foreground) is anchored along its back side to a row of pilings. After the cradle is filled and its logs are chained together, the frames on the near side of the cradle will be pulled away and the completed raft freed (background) for its trip to a mill in San Francisco. (*World's Work*, February 1904.)

rafts were numerous and substantial, because the action of tides in the river slowed the journey to the sawmills. In the 1840s, Charles Dickens observed the many rafts of logs on the Saint Lawrence and wrote that their huts with flagstaffs erected beside them seemed "like a nautical street" floating downriver.[27] For transportation on the seas, workers assembled a "cradle" of wood, pointed and shaped like the hull of a broad-bottomed ship. In the fresh water at a river's mouth, where marine borers would not attack it, the cradle was half-filled with logs or trimmed timbers before a network of chains was fastened to hold its shape (fig. 1.11). When chaining was completed, the upper part of the raft curved to almost match its curvature below the water line. The cradle was then removed and the raft, which could be as long as 800 feet, as wide as 55 feet, and as deep as 35 feet, was ready to be towed across the Baltic Sea or southward along the coast of California.[28] Eighty to 100 tons of chains might be required to make a raft of average size sufficiently sound to be towed from the Columbia River of Oregon to a sawmill in San Diego.

The supply of timber has always been closely associated with the development of transportation. Although early demand for timber to be used in shipbuilding was undoubtedly less than that for fuel, both navies and merchant ships required the largest trees of the choice species, and their requirements pressed forcefully for the extension of timbering. During the seventeenth and eighteenth centuries the finest trees of England and her American colonies were hunted out and marked to indicate that they were reserved for the royal navy. It was nearly the end of the nineteenth century before shipbuilding had changed from wood to steel, and by that time the railroads had proved to be a hungrier market for timber. At first locomotives and railroad cars rolled on oak timbers with iron strips spiked on top. After iron rails came into use, they were secured to wooden crossties 6 by 8 inches in crosssection and spaced with 16 for each rail 30 feet long. This meant that the railroads' demand for wooden ties (their major, but not only, use of wood) was roughly the equivalent of providing a paving two inches thick and almost 9 feet wide along every one of the thousands of miles of railroad tracks, sidings, and spurs spread across the broad continent of North America. (European rail networks have been more inclined to use concrete crossties.) Since the average life of a railroad tie might be between eight and fifteen years, depending on the bed on which they rested, the amount of travel they supported, the wood from which the tie was cut, and the climate of the installation, there was a huge annual expenditure for the renewal of ties.

As the price of crossties increased, experiments were made in the application of preservative treatments that might extend their serviceable life, thereby reducing the annual cost of replacement. The situation worsened until in 1907 a railroad engineer grumbled, "It is no longer possible in the United States to purchase 80,000,000 first class cross ties per year." Some of the first railroad companies to investigate the use of preservatives were those that crossed the broad prairies of the western United States, far from forests and therefore burdened with higher prices for ties. Salt had been used in ancient times as a preservative for ship timbers, and wood for buildings was sometimes charred to prevent rot. In the middle of the seventeenth century the German chemist Johann Glauber developed a system by which timbers were charred, coated with tar, and then soaked in an acid resulting from the destructive distillation of wood. In the 1830s many more preservative treatments were developed involving such substances as a "decoction of tobacco leaves" and a "solution of India rubber."[29] Those preservative methods that continued in application included methods of immersion and pressure treatment using copper sulphate, mercuric chloride, and zinc chlorides, processes commonly marketed under their inventors' names (such as Boucherizing, Kyanizing, and Burnettizing).[30]

In the same period a method was patented in Britain for treating wood by pressure with the "dead oil of tar," which was also known as creosote. The carbonizing of coal produced a tar that could compete with Stockholm tar, long imported for caulking ships, and a variety of oils, those heavier than water being designated as creosote. Although some plants were established for the sole purpose of producing coal tar and coal tar oils, the introduction of gas lighting resulted in firms that were interested primarily in utilizing the gases that came from carbonizing coal and anxious to be rid of the odorous by-products.[31] Creosote was one of the first of an extraordinary array of products developed from coal tar as the chemical industries grew in scope and scientific knowledge. In the United States, railroad companies tested both creosote and metallic salts during the 1870s and found creosote most effective for use in coastal areas where marine borers were a problem and metallic salts suited for application in inland areas.[32] Any loss to the timber

industry from the decrease of the demand for railroad ties was certainly offset by the market for pulpwood in the late nineteenth century, when it became a principal raw material for the manufacture of paper.

Traditionally lumber had been seasoned at the sawmill by stacking it in open stacks or under cover, but seasoning wood became more important with the advent of steam-heated buildings and their desert-dry interiors. Kiln drying was attempted around the middle of the nineteenth century, but it was a few decades before a workable system was found. In localities with high humidity and much rainfall, the traditional drying process meant that the sawmills' owners had to maintain a large investment in stored lumber and a rush of orders often could not be filled quickly. Even more influential was the reduction of shipping costs, for drying in a kiln reduced the weight of a carload of lumber by at least one-third. The typical early kiln in Minnesota was described as having steam pipes in a steel chamber, from which hot air was blown into any of several brick structures, each holding about 40,000 board feet of lumber. Air-drying lumber had been inexpensive, but sawdust and trimmings could provide the fuel for drying. The cost of building a kiln was soon paid for by savings in shipping costs, and lumber could be prepared quickly in response to orders and price fluctuations.

In 1832 George Washington Snow introduced to Chicago a method of framing buildings that used light pieces of lumber, usually not exceeding two inches in the smaller dimension, and relying more on nails than the traditional mortise-and-tenon connections. With the introduction of Jacob Perkins's nail machine in 1795, the price of a pound of nails had dropped from 25 cents to 8 cents by 1828 and would be 3 cents by 1842.[33] Balloon framing, so named because of its lightness, tolerated inexperienced carpenters and could be assembled quickly (fig. 1.12). It was, all in all, well suited for the buildings that settlers needed as they moved west. Balloon framing had been developed sufficiently by the 1880s to rely only on nails, without mortises. At the end of the century it was the predominant system of wood construction in the United States, and so it remained until it was succeeded by the western or platform system of framing, in which shorter members were employed.

As lumbermen found forests farther west in the United States, the distance from the sawmill to the consumer of lumber grew, and it became increasingly necessary to have recognized standards by which the quality of lumber could be described to the buyer and the needs of consumers could be communicated to the supplier. Sweden in 1764 had instituted a grading system that recognized four grades of commercial lumber, and seventy years later four similar grades were adopted in Maine. In 1873 the Lumbermen's Exchange in Chicago established grading standards (again using four levels: clear, select, common, and culls), but many regions or individual mills followed their own systems. It was difficult to enforce

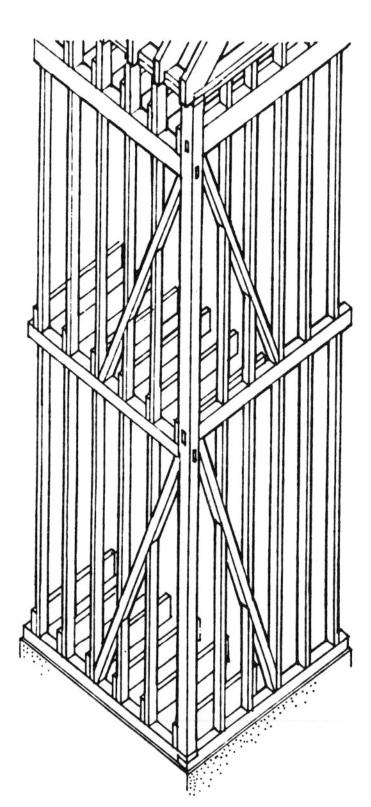

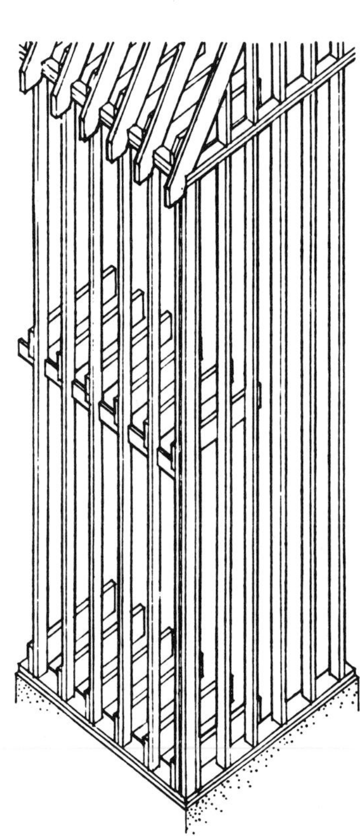

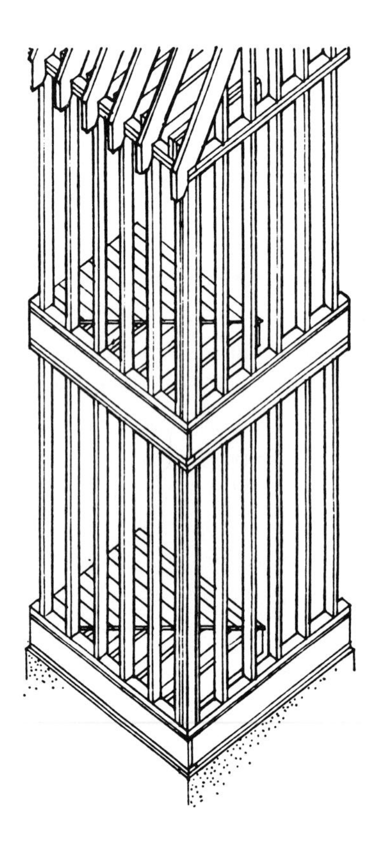

1.12 From left to right in the order of their development and use: braced or eastern framing, balloon framing, and platform or western framing. The simplification of carpentry was a principal factor in this progression. (Drawing by author.)

uniformity among the various lumbermen's associations, especially when the Panic of 1873, which for several years lowered the demand for lumber, was followed by clear signs of depletion in the pine forests of Michigan.

Grading standards were a major subject of debate when the first meeting of the Mississippi Valley Lumbermen's Association was held in 1891. In spite of the organization's name, it represented the northern white pine interests; few of its officers were not from Minnesota or Wisconsin and the most southerly director was from Hannibal, Missouri. Agreement on grading, of course, made it simpler to compare prices and, like many manufacturer's organizations of that period, the Association circulated recommended price lists to its members. As a member of the Northwestern Stave and Heading Association explained, "We are not getting up a trust or anything of the sort. We simply wish to establish uniformity of prices."[34] A year before the founding of the Mississippi Valley Lumbermen's Association, Congress had passed the Sherman Antitrust Act, destined to be more used against labor than against trusts, and one year after it was founded the Association was tried for violation of that law. In one of the first blows to the effectiveness of the Sherman Act, the judge ruled: "An agreement between a number of dealers and manufacturers to raise prices, unless they practically controlled the entire commodity, cannot operate as a restraint to trade, nor does it tend to injuriously affect the public."[35]

Timber companies had long been criticized for the manner in which they had taken advantage of legislation regarding federal lands. The wealth of open land had been openly used as a currency with which the U.S. government could forward certain policies. Grants of land rewarded war veterans, encouraged a westward movement of settlers, attracted immigrants, promoted the drainage of swampland, subsidized the extension of railroads, and financed the foundation of state colleges. Each of these laudable purposes seemed in some way to involve loopholes by which lumber barons were able to gain possession of vast areas of wooded land at extremely low cost. Men were paid to sign up for their 160 acres of wooded land under homestead laws, and then turned their claims over to lumber companies. A law of 1897 permitted landholders to donate their acreage to a protected area of wooded land and receive an equal acreage in another location. Under this program lumbermen traded the land they had already denuded for fresh timberland, and the Northern Pacific Railroad swapped over half a million acres of the least desirable land it had been granted when building the railroad across the northwest (only a fraction of the total amount of land they had received) in return for prime stands of timber in Oregon, Idaho, and Washington.[36]

Allegations and occasional convictions on charges of land grabbing and price fixing were not all that plagued the lumber interests. The IWW (Industrial Workers of the World), formed in 1905, a few years later concentrated its efforts on unionizing lumber workers, one of the few groups in which the IWW had a degree of success. Lumber companies resisted organization of their workers, a conflict that produced bloodshed and years of bitterness that are recorded in labor annals.

Veneers were used, as long ago as ancient Egypt, to apply the color and

1.13 This diagrammatic drawing indicates the relative position of the veneer knife (at left) and pressure bar. One of the spur knives (*A*), which cut the veneer to an exact length, is shown in position on the top of the log. A similar device, replacing the knife with a crayon, could mark the tight-cut side of the veneer, an important consideration when the veneer was assembled into plywood sheets. (A. D. Wood and T. G. Linn, *Plywoods*, 1943.)

pattern of rare woods to furnishings of simpler stuffs. The simplest method of preparing veneer was sawing a thin layer, which could be glued over a wood that was more practical because of its strength, price, or workability. The wide kerfs of early saws meant that much was wasted, but a larger area of precious wood could be shown as veneer than if it were used for the full thickness of the work. By 1805 a circular saw was being used in England for cutting veneer, and later developments produced the segment saw, fine-toothed segments of thin steel being held around the rim of a large center casting. The whole saw was usually at least ten feet in diameter, but it cut only near the rim, where the thin blades were exposed.[37] While they produced veneer about one millimeter thick, segment saws could not provide flitches of veneer that were significantly larger than usual. The veneer slicer, introduced in France around 1830, did little to increase the size of flitches, but by shaving off layers of wood it saved the amount that saws wasted as sawdust. Although a rotary veneer lathe, a machine that rotated a log so that a blade might cut off a continuous band of wood (as paper comes off a roll), was patented in the United States as early as 1840, it was about 1870 before a practical version was in operation (fig. 1.13).

Veneer panels, glued with the grain of layers crisscrossed, were in use long before the term "plywood" originated around World War I. The English furniture maker Thomas Sheraton used such panels in his designs, and maple plywood was used for the wrest blocks of pianos (the piece holding metal pins by which piano strings are tuned). By the middle of the nineteenth century Steinway and Sons had begun using laminated sawn veneers for the curved sides of its pianos, and several manufacturers provided their sewing machines with cases having curved plywood surfaces.[38] Development of the rotary lathe permitted larger sheets of veneer, and in 1884 a factory in Reval (Tallinn), Estonia, started producing three-ply birch seats for bentwood chairs. At the turn of the century, plywood was a frequent material in the mass production of furniture. Estonian and Latvian manufacturers provided Great Britain with the plywood for furniture and for the chests in which tea and rubber were shipped from the Orient.[39] In the United States, factories around the Great Lakes produced plywood for furniture and sliced splints for making fruit baskets.

At the 1905 Lewis and Clark Expedition Centennial in Portland, Oregon, one firm displayed softwood plywood, less costly than the hardwood veneers used for furniture and suitable for structural purposes. The first uses for their product were door panels and the bottoms of cabinet drawers, and most plywood mills in the Northwest were linked to companies manufacturing doors and window sashes. By the 1920s an important market had been found in providing the material for automobile running boards and floor boards. However, because the glues used at that time were not sufficiently water-resistant, plywood was replaced by sheet metal in automobile production, and the exterior application of the material in building construction was limited.

Hide glue, an ancient substance made from the skins and bones of animals, was the first adhesive used to glue veneers, but 1912 saw the introduction of blood albumin glue, which, when applied with steam-heated presses, afforded a more water-resistant bonding material. By World War I, casein glue, made from milk curd, was the dominant type, and it was succeeded in the 1930s by soya bean glue, which was lower in price. In efforts to control costs, simplify manufacturing processes, and approach waterproof qualities, mixtures of such glues were devised to balance their individual characteristics. German companies in 1933 began manufacturing synthetic resin glues, a type that had been proposed in 1912 by Leo Hendrik Baekland, inventor of the first completely synthetic material, Bakelite. These phenolformaldehyde resins could be applied to veneer as a spray or a film. Their maximum waterproof qualities were obtainable with heat, and this discovery quickly revived the hot-pressing techniques that had been employed with blood albumin glues.

A drawback in the structural utilization of wood had always been the linear limitations of its strength. In the direction of its fibers, wood is a relatively powerful material, both supple and malleable. In other ways wood is weak, tending to split and shear easily. As linear structural members, timbers performed well, but at connections the more complex forces were limited by the lateral weakness of the material. By gluing crossed layers of veneer, plywood balanced these capabilities of wood, so long as it was used as a sheet. Laminated wood, a similar gluing of boards parallel, overcame the increasing restriction of available lengths and sizes, but the directional imbalance of strength remained.

The introduction of plywood suitable for use in building construction heralded a new era in the use of lumber. With advances in the chemistry of adhesives it was possible to reconstitute wood fibers, utilizing waste and eliminating many of the structural disadvantages of the linear anatomy of wood as it came from the tree. Chemical treatment of lumber under pressure could alter some of the natural characteristics of wood, enhancing its durability. Most important, steps toward reforestation were initiated. With varying degrees of urgency and success, the lumber industry began a transformation from being an exploitation of natural resources into what may become in essence an agricultural enterprise.

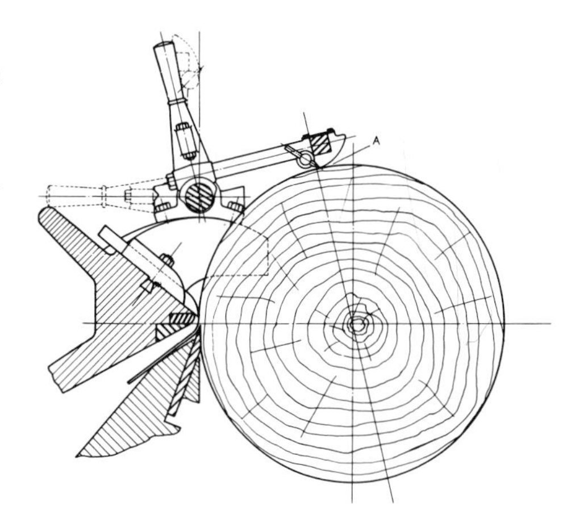

2 Masonry

1689 Blasting employed in English mines

1786 Clay pots used in roof construction of the Théâtre du Palais Royal, Paris

c. 1800 Plugs and feathers used in quarrying

1835 Introduction of brick machines that extrude shafts of clay to be cut by wires

1845 Steam-heated tunnel brick dryers developed

c. 1850 Initial uses of the tunnel brick kiln

1850–80 Popularity of dry-press brick

1851 Model cottage, using hollow brick, exhibited at the Great Exhibition, London

1854 Invention of the wire saw for stone

1858 Introduction of the Hoffman continuous brick kiln

1860s Power drills used in quarrying

Channeling machine introduced in the U.S.

1871 Terra-cotta floor units patented (U.S.) by Balthasar Kreischer and George H. Johnson

1880s Development of the Knox system of blasting in quarries

2.1 In an eighteenth-century stoneworker's establishment, a boy saws slabs from a block of marble. The bucket and ladle were used to pour a mixture of sand and water into the cut. (Diderot, *Encyclopédie*, 1762–1777, "Marbrerie," plate 1.)

The work of masons ranges from lowly sun-dried brick to the most richly veined marble, and the choice among masonry materials has always been largely a matter of cost. Proximity of a certain stone or the absence of stone in a region may, through cost differences, determine the dominant masonry material for each class of building there, may indeed establish a traditional scale by which the relative cost of a building—even the worth of its owner—can be estimated with uncanny accuracy. The presence of clay deposits and a thriving pottery industry may encourage construction in brick. Because the mason's materials are heavy and expensive to transport, in the past there was strong identification of a region's architecture with the local forms of masonry. Even the wall materials associated with periodic and local architectural fashions are most often attuned to financial considerations. When the emperor Augustus boasted that he had found Rome a city of sun-dried brick and left it a city of marble (which was a gross exaggeration), he was remarking only about one aspect of a program of governmental expenditure that set aside budgets for both public spectacles and the restoration of dilapidated temples. On the other hand, when a contemporary rhyme spoke of John Nash's finding early nineteenth-century London "all brick" and leaving it "all plaster," the verse merely recorded the fact that for the pretensions of speculative real estate one could cover brick with stucco and score it to resemble expensive and fashionable stones that had to be shipped there from other parts of England. Transportation costs were always an important factor in the price of masonry. The walls of a crofter's cottage might be made by its owner from rubble picked up in nearby fields, but an emperor's tomb would be carved by skilled artisans from fine stones brought from far away. In thirteenth-century England the charges for carting stone from a quarry near Bath to a building site about fifty miles away were seven times the cost of the stone itself.[1] At the same time, some French stones, which could be brought to England by water, were less expensive than English stone that had to be hauled across land. Often the use of brick and stone as ballast for wooden ships reduced their prices significantly.

Although both of the major categories of masonry materials, stone and brick, were primarily produced by relatively small local enterprises, we must remember an essential difference. Brickmakers seldom made anything but brick. Whether their brick was meant for construction or paving, they were not often also engaged in firing pottery. In contrast, quarries had vast amounts of waste and the stone shipped out for use in buildings and engineering constructions might be as little as 10 percent of the material quarried. Mammoth piles of waste stone could be left, cut into paving blocks, crushed for gravel, or ground to be sand for glassmaking—the choice among such options being determined by the characteristics of the stone and the market for it.

All quarrying methods must vary according to the hardness and internal structure of the stone being extracted, the terrain from which it is to be taken, and the materials available for tools. In ancient times the procedure in almost all cases began with hammering out trenches along the sides of the block that was to be removed, iso-

lating the block from the stone around it and exposing the sides of the blocks that were to be removed next. Trenching was accomplished by crushing the stone with rocks that were harder, chipping it out with iron picks, or making a series of holes with a drill point and breaking out the material between them.[2] After the desired depth had been reached by trenching, the exposed edge of the bottom face of the block was undercut or split by drilling a series of shallow grooves in which wooden wedges were placed or metal wedges were hammered. When wooden wedges were used, they were soaked with water until expansion of the wood split the block from its position. For soft stones, saws might be employed to speed parts of this work. Where large quantities of stone occurred with few flaws, these procedures could be used to form a series of steps up a vertical face of the quarry. Each block, once it had been detached from the surrounding stone, could then be moved forward on the top of the next step and lowered step by step to the bottom of the quarry.[3] The methods described here continued to be used with few significant changes until the eighteenth century. If trenches were to be deep enough to isolate large blocks of stone, they had to be made wide enough for a quarryman to work in them. Toothless metal saws, their cut fed with water and fine gritty sand, could be employed on some stones, but sawing was so laborious that it was commonly used only for slicing blocks after they had been removed from the quarry (fig. 2.1).

The mainstay of later quarrying methods was the jumper, which remained in use in the United States through the end of the nineteenth century for drilling deep holes in certain limestones.[4] The jumper was an iron rod usually 5 to 6 feet tall with steel points at both ends. At mid-height the rod was enlarged in diameter to increase the jumper's weight, which in most cases totaled around 20 or 25 pounds.[5] Placing one hand high on the rod and the other low, a quarry-

2.2 In a Connecticut quarry, workmen split out a block of sandstone. Large iron wedges are placed in a groove cut 4 to 8 inches deep, and workers walk along to hammer each wedge in turn, evening the pressure on the stone. (G. P. Merrill, *Stones for Building and Decorating*, 1903.)

2.3 Even after the use of blasting powder had become customary for splitting out large blocks of stone, dividing the block into smaller pieces was done with plugs and feathers. The steel feathers were half-rounds at the ends that filled the drilled hole and thus evenly distributed pressure as the iron plug was driven in deeper. (G. A. Thiel and C. E. Dutton, *Architectural, Structural and Monumental Stones of Minnesota*, 1935.)

man lifted the jumper and drove it down on the stone. It is recorded that at an English granite quarry a demonstration showed 112 blows of the jumper formed a hole 2¼ inches deep. Echoing off the hard walls of quarries, the clang of jumpers was long a symbol of quarrying.

The alternative to the jumper was a hand-borer team of three quarrymen. One of them knelt and held upright a sharp-edged rod, while the other two alternately struck blows with heavy sledge hammers. Between blows the rod was turned in the hole. There was debate about the relative merits of the jumper and the hand borer. In the 1890s the jumper was strongly favored by many European and Australian quarrymen, but in the United States hand-drilling was generally preferred. Where deep holes were required, it was recommended by many that the first several feet be drilled with a hand-borer and the work completed with a jumper. It was contended that as the hand drill became longer it less effectively transmitted the force of the hammers' blows to the stone at the bottom of the hole, and that when the hole became deeper a jumper of greater length and weight could better be used.[6]

If possible, quarries were established alongside water, so that they might take advantage of water transportation. An escarpment beside water offered the further advantage of presenting its layers of stone for inspection so that the quality and consistency of the stone could be ascertained. While a distant and elevated locale might be a source of unusually fine stone, such a quarry was often abandoned in favor of a river valley cliff of somewhat less desirable stone.[7] Earth and debris above the stone was relatively unimportant if building stone were to be quarried, for the blocks would be trimmed and cleaned before they left the quarry; however, stripping the top was critical for later quarries that produced stone for industrial purposes (as sand for glassmaking or flux for smelting) or for use as crushed stone for concrete. The manner in which building stone was to be extracted was largely determined by the seams that were present. Horizontal seams, arising from the manner in which the material had been originally deposited, may be spaced apart only a few inches or a hundred feet, as with some limestone deposits in Indiana. Where convenient, these seams were used as limits of block size, simplifying extraction. Otherwise, horizontal drilling was necessary to free the bottom face of a block (fig. 2.2). Vertical seams, caused by compressive forces acting within the

stone, usually occur in two systems of parallels in directions that intersect at right angles. In some quarries these vertical seams may be spaced as much as 30 or 40 feet apart.

Grooves and wedges to split off blocks of stone were replaced around the beginning of the nineteenth century by the use of "plugs and feathers." Holes made with jumpers or hand-borers (later pneumatic drills) were spaced about 6 inches apart, and in them were inserted plugs, which were tapered wedges of iron, driven between two iron feathers (fig. 2.3). Since the feathers were flat on one side to hold the plugs and curved on the other to fit against the sides of holes, they assured that pressure would be evenly applied against the stone around the hole. An English granite quarry drilled holes 3 feet deep, but in the 1930s a Minnesota granite quarry found that a pattern combining holes that were 6, 18, and 36 inches deep gave a straight and even break through a ledge about 13 feet thick.[8] Another account tells of splitting rock to a depth of 25 feet if the plugs were hammered in at the end of the day and left overnight.[9] Whether the system employed was grooves and wedges or plugs and feathers, such methods eliminated the immense waste of material and expenditure of labor that was involved in trenching around blocks in order to remove them from the quarry face.

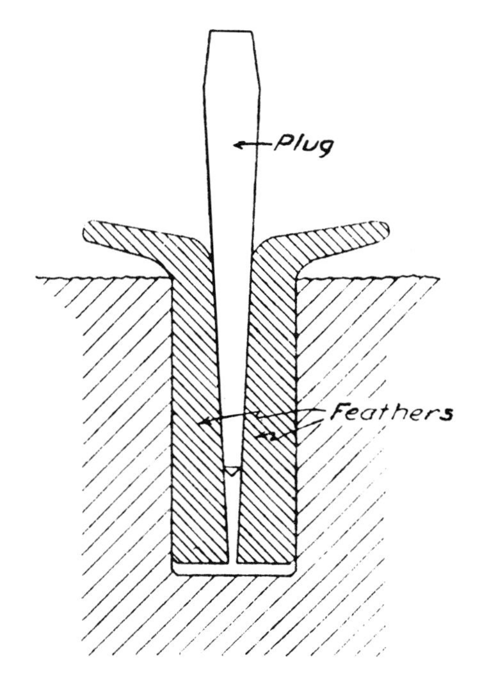

By the 1860s, power drills much like those used in mining were available, both steam-powered and pneumatic (fig. 2.4). Pneumatic drills had the advantage of providing jets of air that emptied the hole of accumulated waste that would reduce the efficiency of a drill, and they were also air-cooled, which eliminated the need of periodically pausing to avoid overheating the drill. One estimate reported that in working granite a power drill and a single operator could in one day accomplish ten times as much work as a team of three hand-borers.[10] Although many small quarries delayed acquiring power-drilling equipment because of its cost, the conversion was inevitable because boring holes was one of the most costly aspects of quarrying. Most power drills were mounted on heavily weighted tripods, but bar drills were mounted on a horizontal steel rod so that they could be quickly repositioned when a line of holes was to be bored (fig. 2.5). The methods used to extract blocks necessarily varied according to the type of stone being quarried, the layout of the quarry, and the purpose for which the blocks were intended. However, a typical quarry would employ wedges or plugs, hand borers, and power drills at the same time.

A steam boiler, mounted on a wheeled frame that rolled along tracks, provided power for the first type of channeling machine, which was introduced in the United States in the 1860s (fig. 2.6). The boiler was connected to a piston that drove a set of sharp-pointed vertically mounted steel blades. As the channeler moved along its tracks, the blades chipped away a channel slightly more than 2 inches wide. As the cut deepened, the blades were lowered, and at certain depths they would be replaced by longer sets of blades. In limestone the channel might be 10 to 12 feet deep and the cut could extend as far as 100 feet across the floor of the quarry. When the cut was to be deep and long, two or more channeling machines could work on a track at the same time, and duplex channelers cut on both sides of the track, usually spacing the cuts about 8 feet apart.

Channeling machines were principally used with stones such as limestone, sandstone, and marble. In granite quarrying the term "channeling" meant, instead, boring closely spaced holes the full depth of the desired block and cutting out the material between. It was a laborious and expensive procedure, but it gave a straighter face than the usual methods of splitting and it avoided shattering the stone, which was always a risk of blasting.

Handguns and artillery were present in Europe by the end of the fifteenth century, the explosive power of gunpowder having been enhanced early in the century by the development of granulated powder. At some time in the seventeenth century gunpowder began to be used for blasting in European mines, and by 1689 it was regularly used in some Cornish mines.[11] Miners drilled deep holes, 3 to 4 feet deep, and poured black powder into them, filling the holes the rest

2.4 Steam-driven reciprocating rock drills of the 1890s were supported on heavily weighted tripods. A crank at the top was used to lower the piston and drill bit. Steam equipment was largely replaced by that employing compressed air, particularly in northern climates where steam cooled too much before reaching the drill. (*Appleton's Cyclopedia*, 1897, supplemental vol.)

2.5 A workman of the 1890s employs a pneumatic bar drill to split a block of granite in a Delaware quarry. Supported by weighted legs at the ends, the bar has a notched track along which the drill is moved. (*Cassier's Magazine*, November 1891.)

2.6 Channeling machines could operate on their own boiler, as shown here, or on steam supplied through a hose. Channelers were particularly useful in extracting a key block, the first to be taken from a fresh layer at the floor of a quarry. After channeling around a rectangle only about 2 feet wide, this small block could be broken off and lifted out, giving room to drive wedges at the bottom of larger blocks after they were channeled at their sides. (*Cassier's Magazine*, November 1891.)

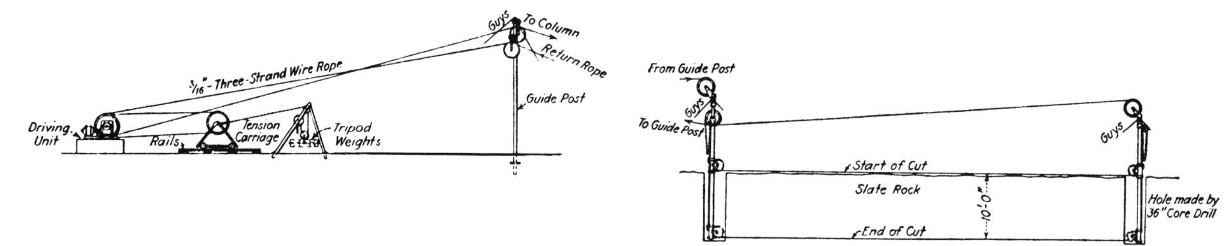

2.7 At the driving end of a wire saw (left), a tripod weight of 800 to 2000 pounds maintained tension in the wire. At the cutting end (right), the portion of the wire in contact with the stone could be mechanically lowered. In slate quarries the cut could be as long as 100 feet and not more than one-quarter inch wide, the diameter of the cable. (O. Bowles, *Stone Industries*, 1939. Courtesy of McGraw-Hill.)

of the way with clay and leading fuses to the powder. The same methods were used for quarrying, but, unlike mining for ore in which a crumbled residue was desired, quarrymen extracting building stone hoped for a controlled split, avoiding damage to the block or adjacent portions of the ledge that was being worked. An account from the middle of the eighteenth century describes the procedure in quarrying:

> When a little hole is made [by the jumper], they put in water as they go on boring, they from time to time clean out the hole in the manner as people clean a gun barrel. . . .
>
> When the hole is thus made of the depth they chuse, they clean it very well out, and make it dry. They then put in a quantity of gunpowder, . . . sometimes not more than half a pound and sometimes two or three pounds. When the powder is in, they thrust down a small wire reaching down to it and up to the surface. They then make up some very stiff clay, or other such matter, and ram it into the hole extremely firm, filling it entirely, and in the strongest manner possible. When the clay or other matter is well in, they draw out the wire; this leaves a little hole quite down to the powder at the bottom. They then fill up this hole also with powder, and lay a train of gunpowder from this. At the end of the train they lay a piece of lighted touch-wood [punk], and leave it to take effect.
>
> It is their business to get away, for the effect is very violent. Several tons of stone are generally loosened by the blast, and frequently pieces of two or three pounds weight are thrown like cannon-shot to a considerable distance.[12]

A method for controlling the force of blasts originated at the sandstone quarries around Portland, Connecticut, principal source of the material used for the brownstones of New York. The beds of this stone were usually 10 to 20 feet deep and blocks were extracted by drilling holes from 10 to 12 inches in diameter. Canisters made of two curved pieces of sheet metal, the cross-section pointed like an American football, were placed in the holes. After the canisters were filled with powder, sand or earth was tamped around them. In granite, a cluster of two or three closely-spaced holes could be drilled, the walls between the holes broken out, and powder poured into that cavity. In both of these procedures the major force of the explosion would act on two opposite principal faces of the shape containing the powder, instead of acting equally in all directions as in a round hole. This caused cracks to form at the ends of the shapes.[13] The Knox system of blasting, introduced in the United States in the 1880s and still in use through the 1940s, drove a steel reamer down a completed drill hole, adding points on opposite sides of the hollow. Thus the hole itself assumed the form of the Portland canister.

The amount of powder to be used in blasting was a matter for judicious consideration, for a misjudgment might result in the wasteful shattering of stone or an irregular split that would greatly increase the labor required to shape the block. Skilled blasting not only separated the block from the surrounding stone, it moved it forward ready to be transferred for scabbling, the process of trimming it to a rectangular shape. At a Portland, Connecticut, sandstone quarry, a block 150 feet long, 20 feet deep, and 11 feet wide, weighing 3,300 tons, was split off and moved out 4 inches by setting off 2 pounds of powder in each of 17 holes; in an English granite quarry by firing 30 pounds of powder in a hole almost 16 feet deep a block weighing about 2,000 tons was moved out a foot.[14]

A later system of dividing stone, developed in Europe through the last half of the nineteenth century but only used in the United States after the 1920s, was the ingenious wire saw, which was useful both in quarrying and trimming limestone, marble, or slate (fig. 2.7). A twisted three-strand wire rope pressed down on the stone as a slurry of water and sand was fed into its cut. The loop of wire ran between two vertical standards, each holding two pulleys, the lower ones descending as the cut was made. Either the saw wire or a driving wire could extend to a steam engine or an electric motor. The standards might be guyed in place on the floor of the quarry or sunk into pits 10 to 14 feet deep and about a yard in diameter. It is reported that in quarrying slate a cut 80 feet long progressed at a rate of about 4 inches per hour. According to comparisons made in the 1930s, in addition to the advantage of its producing the least waste of any method, wire sawing cost about one-third as much as channeling.[15]

Once a large block of stone was detached from its place, it was necessary to subdivide it into workable pieces having dimensions nearer those at which it was to be used. For instance, in a limestone quarry, channels 4 feet apart and 8 to 10 feet deep might be 50 or 60 feet long, cutting out a long thin block. The entire block could be turned on its side by steam power, piles of broken stone being placed beneath to cushion the block's fall and prevent its being damaged.[16] Any of the methods available for extracting blocks (except channeling) might be utilized in subdividing them. Jumpers or pneumatic drills would make a line of holes in which plugs and feathers were driven to break the block into pieces of more manageable size. The extent to which stone was finished at the quarry depended largely on transportation, and in the case of limestone and sandstone, it was far easier and therefore less expensive to work the stone soon after it came from the quarry, before it had dried and hardened.

In Greek and Roman times it was the custom to trim stones at the quarry into a rough approximation of their eventual size, leaving an extra dimension to withstand the hazards of transportation. After the stone had been moved to the construction site, all the faces adjoining other stones were chiseled to their final surfaces, and the stone was set in place. Once the stones has been placed, exposed surfaces were leveled and finished. This procedure involved a trade-off between leaving the stone rough to protect against damage as it was brought to the site, usually a matter of yoked oxen pulling a sledge loaded with the stone, and reducing the weight of the stones that had to be hauled. (If 1½ inches were left on each face of a cube of limestone 4 feet on each edge, the weight would be increased by about one-sixth, almost a ton, and this would be one-third greater if the stone were newly quarried and had not dried.) The same methods were applied in medieval times, although a large part of the more decorative work might be completed at the quarry to conform to the dimensions, drawings, or templates provided by the architect.[17]

2.8 As in this New England granite quarry, stone was extracted in a setting of derricks, cables, and pulleys. The heavy block and the vertical distances within the quarry or leading to a riverbank barge led to the development of intricate systems, most of which were elaborations of the simple boom derrick shown here. (*Asher and Adams Pictorial Album of American Industry,* 1876.)

The least costly stone, of course, came from ruins of the past. Saxons built with stones from the remains of Roman buildings; early Christian and Byzantine builders used the material from classical ruins and even incorporated into their designs the ornament and architectural detail that had been previously carved. Should there be no ruins to quarry, a choice among the available stones was required. In most cases old buildings stood as evidence of the weathering qualities of the stones that were to be found in any region. Certain stones became standards of comparison, respected for the qualities that time had tested. Portland stone, which played a major role in rebuilding London after the Great Fire of 1666, had been used in the city since the fourteenth century, and was long acknowledged as a standard against which other stones might be judged. So long as wood was the fuel most commonly used for heating houses and manufacturing goods, the air of cities offered little threat to stonework. When forests became depleted and wood became costly, coal came to be the principal fuel, cities became crowded and sooty, and the foul air of urban life threatened even the soundest of stones. At the same time, rail transportation made it easier to select stone from quarries farther away without shipping costs interfering with the choice. But many of these stones had been insufficiently tested in the locale for which they were selected.

Caution was not always a sufficient safeguard when selecting stone. Four years after the burning of the Houses of Parliament, a commission was appointed in 1838 to choose the stone from which the new structure would be built. After excluding igneous stones from consideration because of the expense of working them, the commission reportedly visited more than a hundred quarries, inspected many public buildings, and ordered laboratory tests made. Reaching the conclusion that "in proportion as the stone employed in magnesian-limestone buildings is crystalline, so does it appear to have resisted the decomposing effects of the atmosphere," they chose a dolomite limestone from the area of Bolsover Moor in Derbyshire.[18] It was later found that this quarry could not provide stones large enough, and a second investigation led to the use of stone from another locale. Only 14 years after the project was completed, a committee was appointed to investigate "the decay of the stone of the New Palace of Westminster," and one of the problems to be considered was the effect of "the acids diffused in the London atmosphere."[19] Stone from the same quarry had been used for the Geological Museum, where it had not suffered significant damage.[20] Therefore, it has been assumed that the fault lay in the care with which stone was selected from the quarry—a constant problem because of the degree to which the quality of a quarry's stone might vary—and the fact that much of the work had been laid without due regard for the natural bed of the stone.[21] In the nineteenth century, methods for protecting stone were widely discussed. Painting required maintenance and blurred the carvings and ornament; oils and waxes also needed to be applied repeatedly; and neither of these methods fully protected against the damage that could be caused by moisture within a wall and the actions of frost.

After the introduction of steam power in quarrying, there was an exploration of devices by which pieces of stone could be moved about. Many quarries simplified the task of moving stone to barges by building small railways, and, as in mines, light tracks were laid to make it easier for workmen to push heavy loads. Hoists and derricks assumed many tasks for which levers and windlasses had in the past been the only equipment. Because of their origins, granite quarries usually ran deeper than those from which limestone or marble were taken, and lifting equipment was particularly important for them (fig. 2.8). The overhead cableway was introduced in the 1870s and one type was called "Blondin" after a French tightrope artist who had several times crossed Niagara Falls.[22] With a Blondin and a steam-driven hoist, heavy blocks could be picked up and moved to another part of the quarry without the exertion and delay that had previously been required.

Roughly shaped blocks were taken to the mill, where they were worked to attain the exact dimensions shown in construction drawings and the textural finish specified for their visible surfaces. According to the types of stone, there was considerable variation in the tools and procedures used, but in general the methods followed those that can still be observed today in many parts of the world. An experienced workman trimmed the edges with a chisel, squaring the stone and setting its dimensions; the task of leveling the faces and applying the desired textural treatment was usually left to apprentices who chipped away with chisels, points, and hammers faced with sharp steel teeth. With pneumatic tools the procedure was much the same, only speed being gained by mechanization.

Machine finishing was usually done by circular saws or gang saws, parallel blades adjusted in their spacing according to the desired thickness of the slabs that were to be cut. The blades of gang saws for cutting stone had no teeth, and at the end of each stroke they lifted to allow a mixture of sand and water to flow into the cut. Circular saws were usually 5 feet or larger in diameter. The edges of the blade were notched to receive steel teeth, and, for granite, steel shot was used as an abrasive. For granite or limestone a gang saw would cut at a rate of about 6 inches per hour, cutting several slabs as it went; a circular saw cut at a rate from 3 to 16 inches per minute, but made only one cut.[23] In the late nineteenth century, other power tools for finishing, polishing, and grooving stone were introduced, most of them originating in the United States and transferred from there to European quarries. Pneumatically driven chisels and hammers, steam-powered polishers, planing machines, and lathes increased the speed with which the desired finishes could be achieved, although certain attractive traditional textures could not be reproduced by machine.

At the same time that the mechanical equipment of quarries and stone milling plants became plentiful, practical, and varied in function, the adoption of steel frame construction began to decrease the demand for building stone in the United States (fig. 2.9). Tall urban buildings, which formerly required ground-floor walls of extraordinary thickness, were built with thin walls supported on the steel beams of each floor. Quarrymen sadly recognized that their stone was no longer the very substance of buildings, but had instead become a surfacing material. This change occurred much later in European countries, where steel frame buildings came later into use. In many cases the declining use of building stone was offset by the growing use of crushed stone for reinforced concrete in engineering and building construction projects.

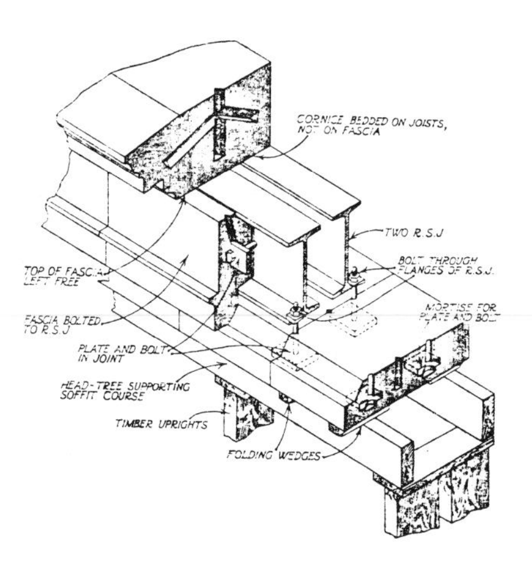

2.9 Using stone as a covering for steel frameworks meant that much less stone was employed than in bearing-wall construction. However, notching the back of the stones to within a few inches of the front surface was often more expensive than carving the details that would be visible. (*Building*, December 1927.)

2.10 Between 1815 and 1850 English production of brick more than doubled, although firms remained small and no significant technological advances were made during that period. Because suitable clay was not to be found everywhere, the expansion of brickmaking was largely due to the construction of canals that could economically bring coal to the kilns and carry off the fired brick. (Bettman/Hulton Archive.)

Fired bricks were seldom used in Greek or Roman buildings, mud bricks or sun-dried bricks being sufficient for the climate. Sun-dried bricks, laid with mud joints, were sturdy and avoided both the tedious work of quarrying stone and the cost of fuel to feed brick kilns. The exterior surface could be plastered to improve its appearance and offer protection from rain. Fired bricks were sometimes incorporated for additional strength at corners and at the jambs of doors. It was the final century of the Republic before wealthy Romans began to include marble in their houses, and only in the age of Augustus did the use of fired brick become significant.[24] Brick walls in Roman construction were combinations of masonry and concrete, with bricks laid at the outside surfaces and the space between filled with concrete poured over layers of rubble or brickbats.

Medieval architecture was dominated by brick in the areas of northern Europe that had little stone, a zone starting in Holland and crossing Germany to the lands around the Baltic Sea. In England, Flemish brick was imported, and when English brickmaking revived in the fourteenth and fifteenth centuries, many of the workmen bore Flemish names.[25] During Elizabethan times, when the price of timber in England rose so high that masonry was no more expensive than wood construction, Londoners built less with timber and turned to brick, just as Paris became a city of stone, which was readily available there.[26] London's enthusiasm for brick buildings and the city's growth caused a multitude of brickworks to appear in the outskirts of the city. In view of a greatly increased demand for brick and high costs of transporting it from other parts, the number of firms participating in the industry increased, and fashion fluctuated among the available colors, which included a variety of reds, a light yellow brown, and a gray that later caught the Georgian fancy.[27]

The traditions of brickmaking changed slowly. From Roman times the work had been conducted in a manner that resembled the rural method by which a farmer in the early nineteenth century made his own brick with the help of a local brickmolder, who might follow some other trade during winter months (fig. 2.10).

When harvest was over, the farmer and his children and hired man would take to the shovel and wheelbarrow and pile the clay loose up to two or three feet high, so that the frost in wintertime might do the disintegrating. Then in the spring the molder would leave his tailor-table and go with his tools to the spot. . . . The raw material would be spread in thin layers over the ground, some water added, and a span of oxen would be driven around and around in the wet clay, doing the pugging, till it was considered ready for molding. The brick molder worked the clay from a table, in single molds without bottom, and his boys or girls carried the filled molds on high edge to the yard, dumping the brick in single rows and returning for another brick. Thus the molder made 2,500 to 3,000 brick in a day, each brick being three inches by six inches by twelve inches in size, sometimes even larger. . . . The kiln consisted of a square hole in the ground, or better, in the side of the hill where one side could be open . . . so as to have ample space for firing.[28]

A more sophisticated and permanent organization was used in 1846, when the Hudson's Bay Company established a brick plant in Vancouver, Canada (fig. 2.11).[29] There the clay was prepared in soak pits, four feet deep and each able to hold enough clay for a molder's daily output. A horse at the end of a sweep drove the pug mill, which was a bin containing a vertical shaft that bore knives to cut any lumps in the clay and work it to a smooth consistency (fig. 2.12). While the horse circled, the molder stood in a pit and took masses of clay as they came from the bottom of the mill. Every cell of a six-brick mold was filled and the excess clay struck off, ready for offbearers to carry the mold to the drying yard, where the blocks of clay were removed from the mold and left to harden in the sun before being stacked in the kiln. After a few hours in the drying yard each brick was bobbed—struck with a flat piece of wood to smooth its surfaces. After a day or two in the sun the bricks were turned for further drying, unless the weather was so sunny and dry that they were ready to make the kiln. A stack of these green bricks was surrounded by previously fired brick over which clay was spread, and wood was brought to commence the firing, which would take 10 to 14 days. A typical simple kiln might hold 20,000 to 50,000 bricks, four to ten days of a molder's work and enough (at 20,000 bricks) to build a small two-story house.[30] Within the kiln, green bricks were stacked in intricate patterns that allowed enough space between them for the hot air to circulate as well as possible, given the need to fully load the kiln. For instance, a rectangular

kiln measuring 27 feet by 11 feet by 12½ feet high inside held about 40,000 bricks. Less than 80 percent of the interior volume was occupied by the bricks; the remainder was taken up by spaces allowed at the walls, between stacks of brick, and between bricks in the stacks, an effort to distribute heat within the kiln as evenly as possible.[31] Nevertheless within a kiln the heat was uneven and, even when the firing was most successful, the hardness of its products varied greatly. Those bricks that were not hard enough to be exposed to the weather on outside walls were, because of their color, usually referred to as "salmon" bricks, and they were set aside to be used in the interior partitions of buildings. Those sufficiently fired to withstand weathering were called "common" bricks. The bricks that had received the most intense heat of the kiln and were discolored or misshapen were called "arch" bricks because of their location in the heat-distributing passages that had been built at the bottom of the stack of brick as it was placed in the kiln. The uses of arch brick were those that required impermeability and are indicated by other names applied to them: foundation, cistern, and cellar brick.[32]

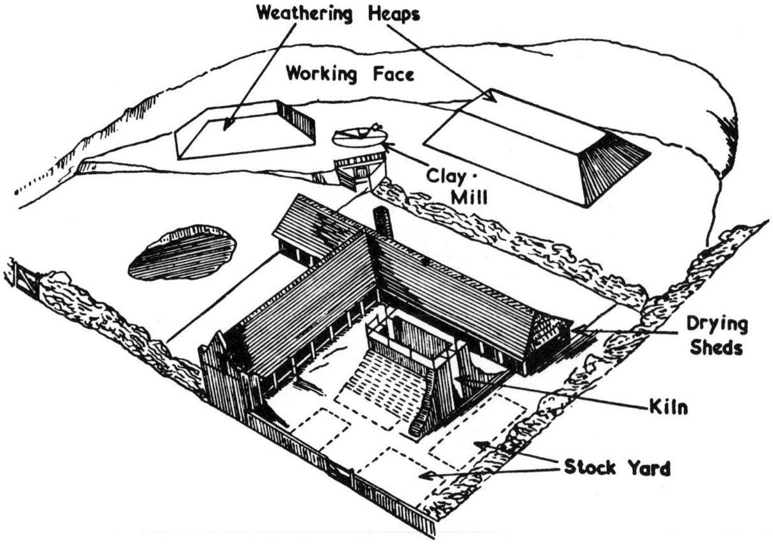

Brick molds, first made of hardwood and later of iron and steel, required treatment by the molder to make certain that the shaped brick would slip easily from the mold. For "sand mold" brick, which had a distinctive surface texture after firing, the mold was dusted with dry sand before being filled with clay, and for "slop mold" brick it was dipped in water. Even after mechanization replaced hand-molding, these processes were duplicated by machines.

By the middle of the nineteenth century, brickmakers had developed mechanical methods of processing clay and forming the bricks that were to go into the kiln. Weathering was often eliminated, and machines prepared the clay as it was dug. The pugging mill was adapted to waterpower and steam power, replacing the horses that had once turned it. For the same purpose, there was the wash mill in which a large trough of clay and water was worked with turning disks. As this process continued, the softened clay ran into a settling basin from which water was later drained. A third machine, the rolling mill, passed clay between two horizontal rollers, and it was sometimes used in combination with other clay-working machines.[33] In many cases, clay was passed through mills more than once in order to insure a fine-textured and plastic material. For some clays these methods did not replace the process of weathering, but at the least they reduced the cost of labor that was required to repeatedly turn over the clay as it lay exposed to frost and rain.

With the seasonal manufacture of hand-molded bricks, molders and their assistants had worked only six or seven months in the year. An English engineer in 1856 wrote of brickmakers,

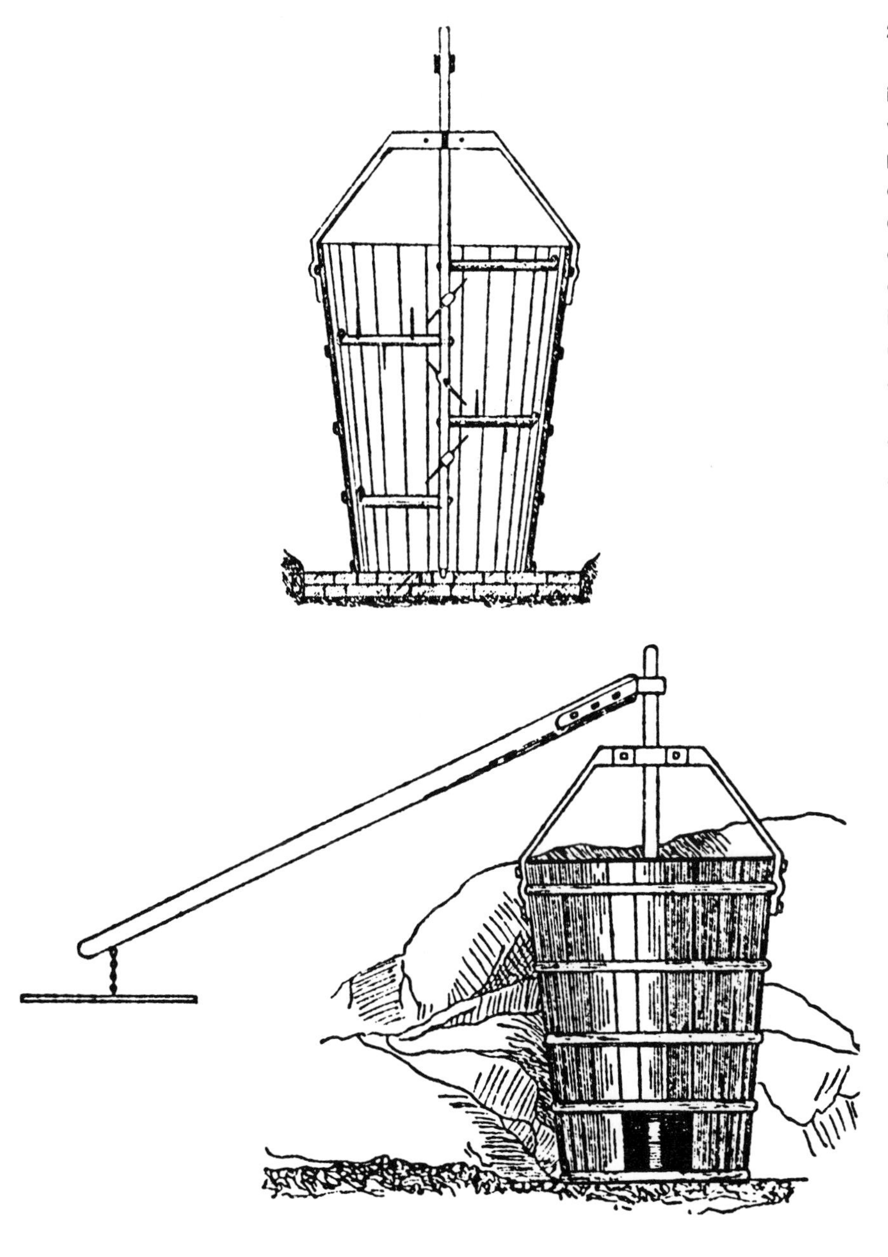

2.11 From top to bottom, this bird's-eye view illustrates the sequence of work at the typical brick plant of the early nineteenth century. The location was determined by the presence of clay at a point reasonably close to a city or an economical mode of transportation. (A. T. Green and G. H. Stewart, *Ceramics: A Symposium*, 1953. Courtesy of the Institute of Ceramics, London.)

2.12 The vertical shaft in early pug mills bore blades or spikes that sliced and mixed the clay before it reached an opening at the bottom. The shaft was turned by a horse that circled at the end of the rod that extends at the left of this drawing. (A. T. Green and G. H. Stewart, *Ceramics: A Symposium*, 1953. Courtesy of the Institute of Ceramics, London.)

They therefore work early and late; and from their having no winter occupation, and on account of their dirty and laborious work, they have such a price per thousand as to compensate them. The result is generally, among this class of men, that, although they are so well off in the summer time, they spend all they earn, leaving the winter to provide for itself. At that season, when they require most to sustain themselves and their families, as well as fuel, they have nothing to fall back upon, and in too many instances no employment; for, as they earn so good wages in the summer, they do not willingly come down to the price of ordinary labor. These are the men whom the introduction of machinery will effect.[34]

2.13 Leahy's brickmaking machine, available in the 1880s, fed clay through a vertical pug mill. At the bottom, the clay was forced into iron molds around the rim of a wheel. As the wheel turned, the bricks were automatically ejected onto a moving band that carried them away to drying sheds. (*Appleton's Cyclopedia*, 1880, 1:248.)

2.14 In the Imperial brick machine, clay fed into the machine (*A*) was mixed and pressed into molds on the face of large wheels (C). As a brick rose to the top position, it was forced out of the mold. This "double machine" was reported to produce up to 100 bricks per minute. (*Appleton's Cyclopedia*, 1880, 1:247.)

By diminishing the importance of weathering, brickmaking became much less a seasonal activity, no longer depending on cold weather for preparation of the material and warm weather for sun-drying the bricks before they were placed in kilns.

For hand-molded brick, much effort was required for preparing clay that was wet and pliable. This "soft mud" flowed from the machines, ejected by pressure from the blades of pug mills or by the auger mechanisms within other equipment. The hand molder placed pieces of this clay in molds and compacted them manually, but in the first half of the nineteenth century mechanical means were also found for this phase of the work. Many different machines were developed to apply waterpower or steam engines to the task. In the "soft mud" process, clay was extruded into molds and the machine pressed the clay down firmly and struck off the top of the molds (fig. 2.13). One or more molds were brought against the orifice of the mill for filling and then moved to make way for other molds. Farther along, the molded brick was ejected by the action of a piston set in the bottom of the mold (fig. 2.14).

A machine, introduced in 1835 and later highly developed, extruded stiff clay in a continuous rod having two dimensions of the finished brick (fig. 2.15). Experimentation was required to achieve an even flow of material through all parts of the rectangular die and so prevent the clay being less dense at the corners of the extruded bar, critical points if the brick were to be durable (fig. 2.16). As the shaped band of clay came from the machine, it moved forward on a tabletop to be sliced into bricks either by blades or by frames across which piano wires were stretched (fig. 2.17). Wire-cut brick, although permitting an easily controlled and efficient process of manufacture, had its shortcomings. When the city of Minneapolis in 1898 sought bids for a million bricks to be used in sewer construction, sand-mold bricks were specified because authorities considered wire-cut brick, the only kind manufactured locally, too variable in size, shape, and quality. Despite the specifications, bids for wire-cut brick were submitted, and (probably because of shipping costs) their price proved to be 15 percent below that of sand-molded bricks.[35]

The manufacture of dry-press brick may have been started in the United States almost as early as the start of the stiff-mud process, but dry-press production became increasingly common from 1850 to 1880. (Nonetheless horse-powered soft-clay methods were the most commonly employed in the United States during that period.) The ability to use finely pulverized shale, marl, or hard clay instead of surface deposits of soft clay and the advances made in steam-powered machinery encouraged development of the dry-press system of manufacturing brick (fig. 2.18). Powdered clay or shale, almost dry, was forced into molds under extreme pressure and, as

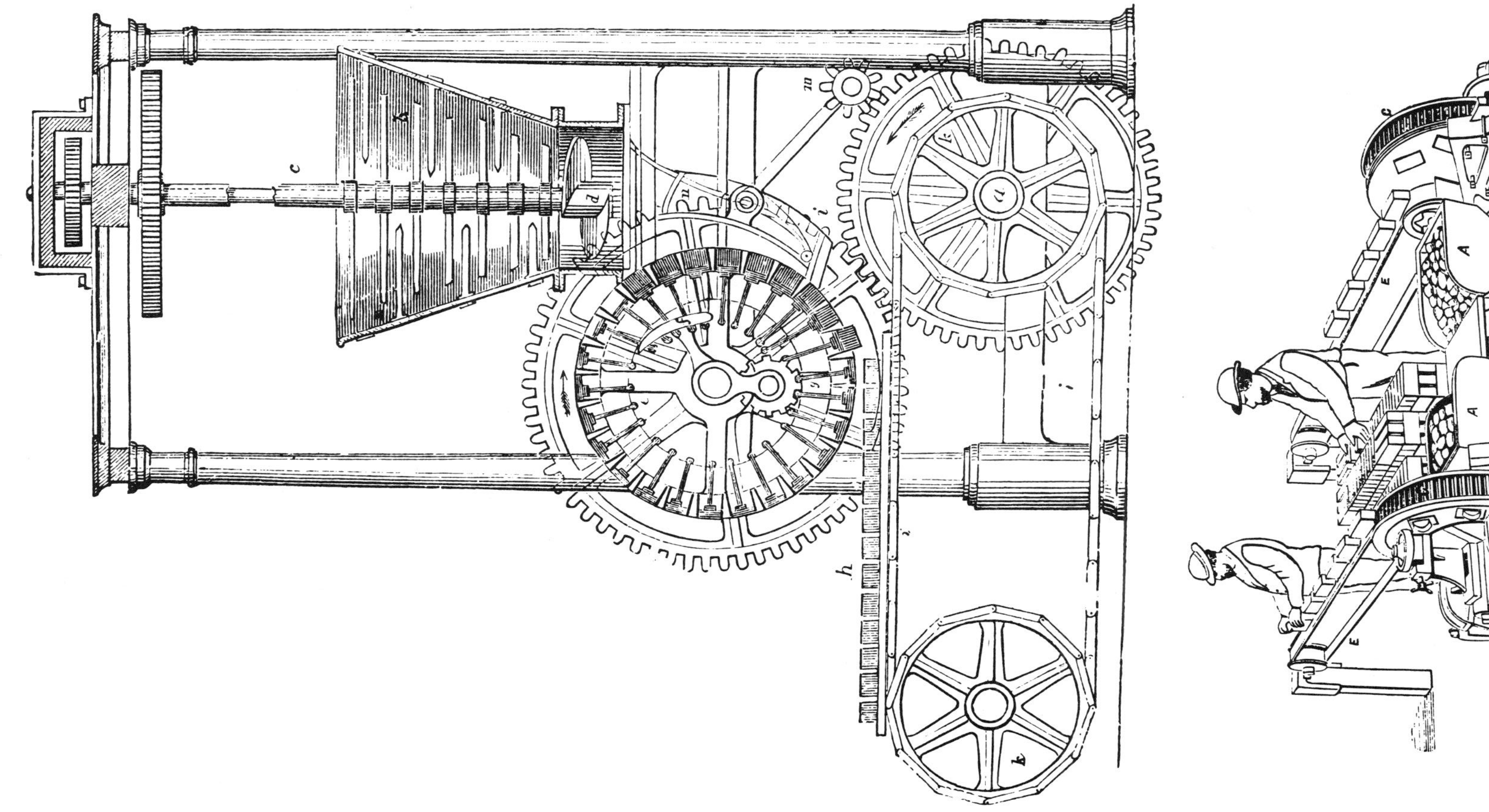
c
b
d
n
h
i
m
k
a

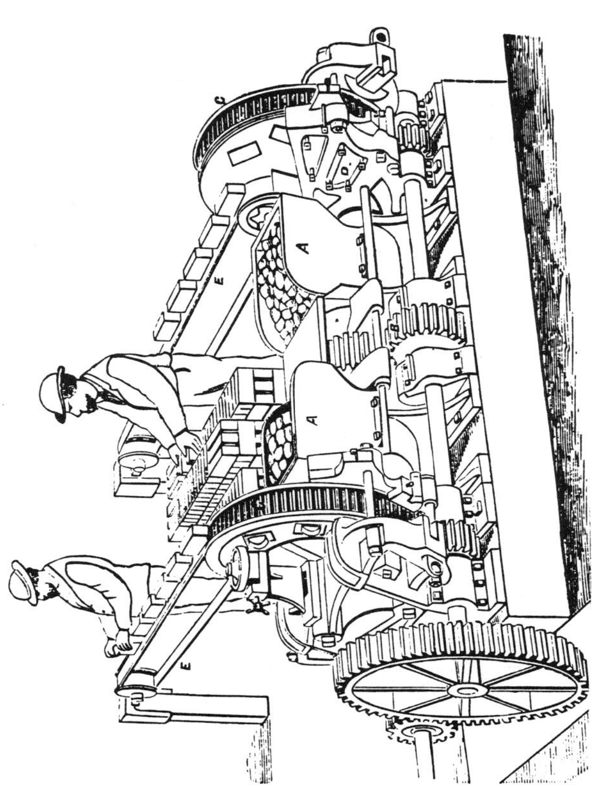
E
E
C
A
A

the molds moved from the heart of the machine, bricks were unmolded onto a table, from which they could be taken directly to the kiln without a period in the drying house.[36] Being soundly pressed into cast-iron molds gave the brick a smooth face and precise edges, and sometimes they were pressed a second time to further compact the material.

> The greatest precaution is always observed in handling the pressed bricks before and after burning, much more attention being paid to these bricks than to common bricks. In carrying the bricks to the kiln they are taken up one at a time, placed lightly on the wheelbarrows, and between each course of bricks on edge there is placed a strip of soft wood or a good thickness of some kind of woolen stuff, such as an old blanket. When the bricks arrive at the kiln they are lightly removed from the wheelbarrows, one at a time, and are very carefully handled.[37]

After the Chicago and Boston fires, many architects and builders abandoned the cast-iron, marble, and granite facades that had fared so poorly in reports of fire damage. Hard brick proved in most cases to be better protection against flames, but the usual red brick gave architects of the period too limited a range of expression. In much of the United States the cream-colored brick of Milwaukee was admired, but it was costly even in the vicinity of that city and much more expensive when shipped elsewhere. Red brick from Philadelphia and Baltimore had been the standard for buildings in New York, but in the 1880s production of light-colored brick increased, much of it coming from producers in Perth Amboy, New Jersey.[38] Construction of the exotic Tiffany mansion in New York in 1882 popularized a hard-fired brick of "mottled shades of brown."

To the search for a variety of color must be added the architects' exploration of "ornamental brick," which could be easily made by shaping the mold in which bricks were pressed. Classical moldings and intricately shaped voussoirs could all be formed of clay, providing ornamentation and detail of the same material as the walls of a building. When pressed brick was introduced into Chicago from Philadelphia, masons were brought west because of their experience in the demanding work of laying it. The popularity of pressed brick in Chicago grew until 1887, when much of it was damaged by a particularly severe winter. Reaction slowed the use of

2.15 Clay was shoveled into the horizontal cylinder of this Chambers brick machine, and water could be added to make a mixture of the desired consistency. As a steam engine outside the workshed turned blades within the machine, the clay was forced along to emerge as a bar of rectangular cross section. A spiral blade cut that bar into bricks. (*Appleton's Cyclopedia*, 1880, supplemental vol.)

2.16 Clay moved unevenly within the dies through which bricks were extruded. In this patented design, the clay entered a channel of curvilinear cross section (upper left) and exited through a section generally rectangular, but slightly narrowed at the center. (*Journal of the Franklin Institute*, January 1883.)

2.17 This "cutting-off" table received an extruded bar of clay from the left. A single wire at left rotated to cut off a length of clay, which was moved on by hand to be cut into ten bricks by a frame of wires. (*Appleton's Cyclopedia*, 1880, 1:246.)

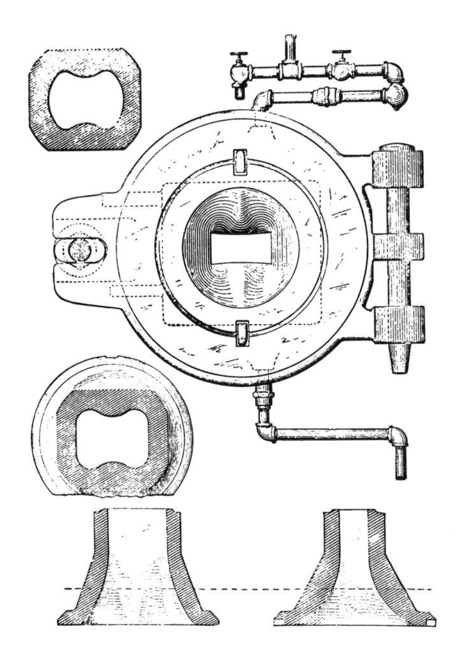

pressed brick in the Chicago area, and in the design of a major civic building brick was abandoned in favor of granite.[39]

With the development of advanced methods for processing clay and shaping bricks came improvements in methods of firing them. The cost of fuel was a major factor in setting the price for brick, and as in other furnace industries the problems were those of achieving economy by mastering the element of time and gaining the greatest possible advantage from the fuel that was expended. The most ancient and elementary form of brick kiln, the scove kiln or clamp, was formed by stacking green brick in the manner described above. After the outside of the kiln was sealed with clay, openings at the sides let fuel be added to the fires, and openings at the top allowed smoke to escape. In order to remove the finished brick, a scove kiln was often simply torn apart, but more advanced types of up-draft kilns had permanent walls and provided grates under the green brick so that heat would rise from a firing chamber below. Although each clamp was used to fire a relatively small quantity of brick, at one time it was estimated that the scove kilns in the United States had as many as a million bricks in them.

Among such intermittent kilns (those that were fired to produce a quantity of brick and later cooled for removal of the brick) were many that varied the locations of chimneys and fires to achieve horizontal or downward movement of heat, but such arrangements did not overcome the fundamental disadvantages of the intermittent kiln. First, much of the fuel was consumed in heating the mass of the kiln itself. As the cost of labor and fuel rose, it was natural that brickmakers should jealously eye the efficiencies that had been developed long before in the furnaces used for making iron and glass. Second, the preparation of clay and the forming of bricks as a seasonal activity was inherently inefficient.

The "season" closed in the early fall; October was the usual month. The entire summer's product was set in a single kiln. Then came "burning time." This was a

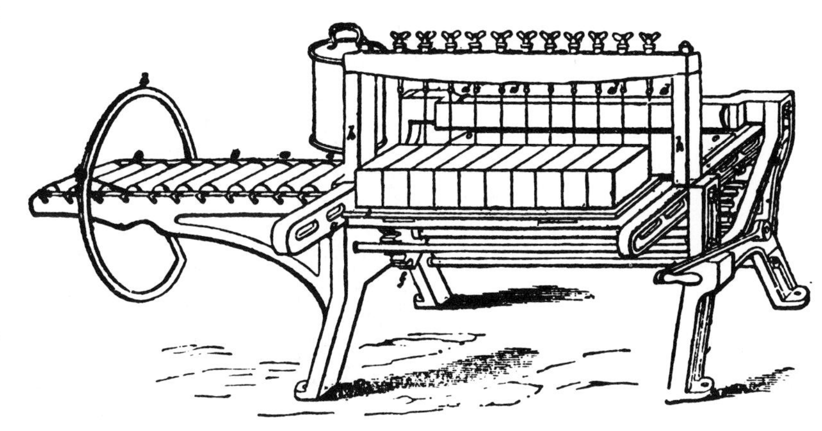

2.18 Mixtures for making brick would often be better controlled when clay was dried and ground to a fine powder before being remoistened. This early form of grinding pan was a roller mill, a basic type that had been used by ancient Romans to grind grain and crush olives. (A. T. Green and G. H. Stewart, *Ceramics: A Symposium*, 1953. Courtesy of the Institute of Ceramics, London.)

2.19 Advertisements for this kiln stated that "radiating and waste heat are used for drying and heating air for combustion." By surrounding the fire with a series of firing chambers, kilns built according to the Hoffman principle were able to operate continuously, directing the most intense heat to each chamber in turn. (*Clay Record*, 29 July 1897.)

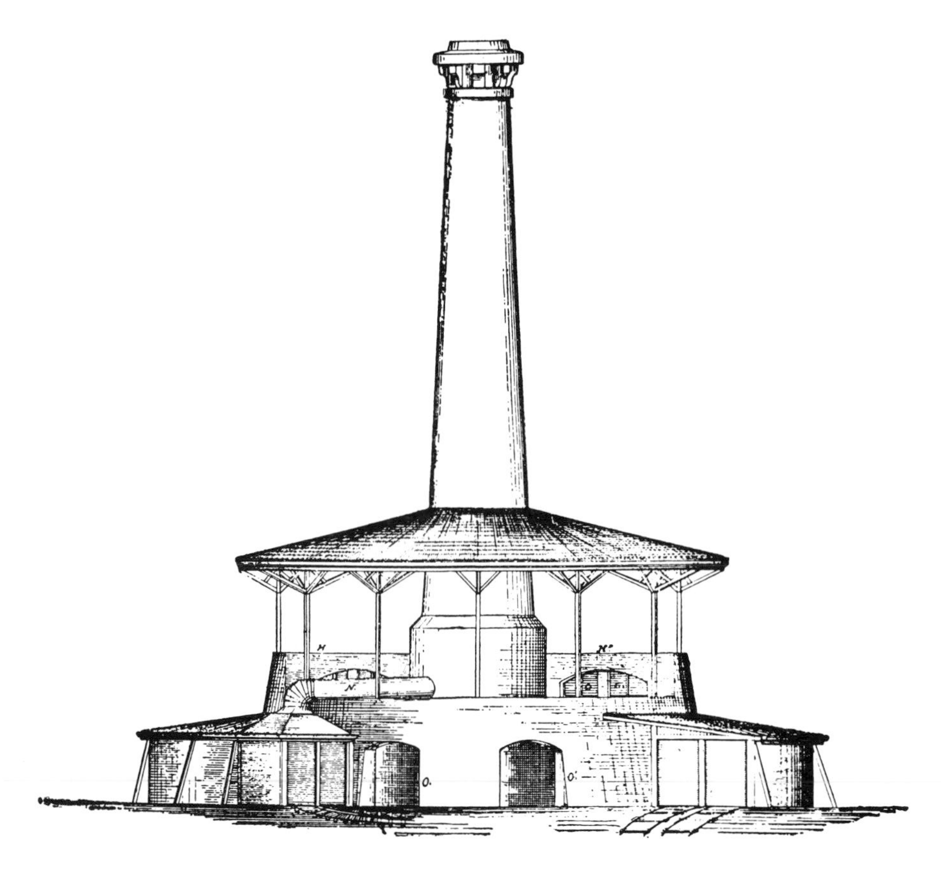

hilarious event. It [was] looked forward to with great interest by certain of the community as an occasion when rum, if ever, was needed to successfully do the job. Everybody then connected with the kiln was happy but the owner of the property; he was commonly overwhelmed with anxiety lest his volunteer help became so utterly drunk as to endanger his kiln by neglect. Such instances were not infrequent. "Boss burners" were in great demand by proprietors of yards, and as the supply of the former was limited and the needs of the latter urgent, there was much rivalry and backbiting over their possession. When one of these "professionals" was secured there was no positive assurance success would attend his efforts. A "good burn" fifty years ago [c. 1840 in Chicago], and even later, was a rarity. It seems more to have been the result of luck than calculation.[40]

Firing in a continuous manner allowed a greater number of small loads in the kilns and thus reduced the risk involved with each firing. With continuous operation of a brickyard, it was possible to assemble a group of workers intimately familiar with their kilns and with the firing characteristics of the clays they used.

The first influential design of a continuous kiln was that of a German, Hoffman, in 1858. It was introduced into Britain about four years later, but as late as 1890 an English visitor reported that this form of kiln appeared not yet to exist in the United States.[41] The Hoffman kiln consisted of a circular ring of kiln chambers with a tall central chimney that was connected to each of the chambers by ducts (fig. 2.19). It was on the level above the kilns that firings were controlled. Fuel could be dropped into any of the chambers, dampers opened or closed to control the intensity of heat in the firing, and openings between the chambers were manipulated to move hot air from one chamber to the next. When a chamber on one side of the ring of kilns was in the midst of a firing, the chamber directly opposite in the circle was cool and being emptied of the fired brick. As workers carried out brick, air to feed the fire entered that chamber and, going through other chambers around the ring, the air was increasingly heated by brick cooling from the previous days' firings. In that way heated air fed the flames, and less fuel was needed to attain the required temperature. Farther around the circle, heat from the fires gradually raised the temperature of green bricks as they approached their turn for firing. At the end of this series of chambers—next to the one being emptied—workmen would be stacking green brick.

Complaints about the Hoffman kiln concerned the limited number of chambers possible around the circle and the inconvenience of stacking bricks in a wedge-shaped chamber. Ovals, oblongs, Ys, Xs, and linear plans were later variations of the scheme, but all were based on the economies possible by sequentially firing a series of chambers. Their popularity was understandable, since the fuel needed to fire a given amount of brick in a kiln of this type was usually less than half that which had been required in intermittent kilns. As late as 1950, it was estimated that about 90 percent of British production of common and face brick was fired in Hoffman kilns or similar continuous kilns.[42]

It was necessary to remove a large part of the moisture in green brick before it could be placed in a kiln, and drying brick was always a process almost as involved as firing it. Drying in the sun was the original and simplest method, but, even in summer, weather made this method unreliable.

About 15 percent of sun-dried brick was damaged and ruined.[43] Roofed drying sheds afforded protection from showers, but they did nothing to relieve the elevated humidity that accompanied rains. Once steam power was introduced in brick plants, the exhaust steam from boilers could be used at night to sustain heat in drying-shed floors, but the construction of iron-floored dryers did little to increase efficiency.

Studies have indicated that the dryers developed later were at least twice as efficient as the "hot floor" method.[44] First there was the tunnel dryer. Its initial form was a coal-fired horizontal flue, extending from the fire to the chimney, in which both bricks and workmen were blackened by the soot. Introduced in 1845, the steam-heated tunnel at first contained wagons of brick moving on rollers, but subsequent designs provided rails on which the wagons moved. It was early recognized that air movement was as important as the heat provided, and fans were installed in the tunnels to make certain that humidity remained uniformly low and the heated air was evenly distributed. A tunnel dryer developed around 1870 exhausted furnace gases through an iron-covered trench in the floor.[45] This caused the heat to be greater where the trench originated and cooler at the opposite end. The rolling wagons of brick and the air that removed evaporated moisture moved in a direction opposite that of the hot gases.

Chamber dryers, a second type, were much like continuous kilns. Hot gases from a burner were channeled so that any heat escaping from the final stages of drying would be employed in the first stages of drying, and the heat released as fired bricks cooled was also used in the work. This system offered several advantages, but the tunnel dryer, in its many forms, remained the simplest to operate.

At about the same time that the tunnel dryer was developed, the tunnel kiln was introduced. In two of the earliest designs, patented around 1850, wagons loaded with brick were moved slowly through a long rectangular chamber. Heat was provided by a furnace at the center and draft by a chimney near the entrance. The top surface of the wagons was covered with slabs of firebrick, but high temperatures within the kilns quickly damaged the metal of wheels, axles, and rails. In an ingenious design of the late 1870s, iron sheets faced with firebrick were mounted on the sides of the wagons and extended into beds of sand alongside the rails. This arrangement sealed off a channel beneath the cars through which cool air was circulated around the wheels.[46] Tunnel kilns were divided into three sections. In the first brick was preheated; in the second it was fired; and in the last the fired brick slowly cooled. This is essentially the action of today's continuous kilns.

All accounts of brickmaking practices spell out the extraordinary range of sophistication to be found among the manufacturing methods that were in use at any given time. There were, of course, contrasts among the procedures used in different climatic zones and variations due to fluctuations of economics and building activity. As a scattered and self-sufficient enterprise, the brickmaking industry long encompassed small establishments with relatively primitive equipment and processes, as well as large companies having considerable technological insight. Since supplies of suitable clay and shale are widely distributed, there was no pronounced geographical concentration of brickmaking activity, and the costs of competitors shipping their brick for long distances often offset the manifest inefficiency of some small local manufacturers.

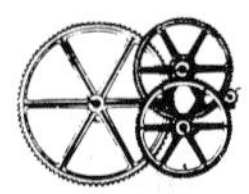

In the Byzantine church of San Vitale in Ravenna, the weight of the dome was considerably reduced by using terra-cotta jars, laid in a continuous spiral embedded in the concrete of the dome. Much later, a Paris tile works around 1785 supplied an architect with clay pots that were closed at top and bottom and hexagonal in shape. The pots (about 7 inches high and 4 inches in width) were set in plaster of paris to build floors spanning as far as 12 feet. The results were reported to the Académie d'Architecture, and Victor Louis, architect to the flamboyant Duc de Chartres, adopted this method for rebuilding the Théâtre du Palais Royal, which had burned a few years before.[47] Materials for the new theater were almost entirely noncombustible, and clay pots were used in partitions, ceilings, and a vaulted roof. Louis employed pots in other buildings, and information about this new building system soon reached England.

Sir John Soane used hollow ceramic pots in 1792 in the domes of the Stock Office at the Bank of England, but a greater contribution to the advancement of building methods was made at that time by William Strutt, eldest son of Jedediah Strutt who operated cotton mills in partnership with Richard Arkwright, inventor of the spinning frame. Two years after Victor Louis's construction of the Théâtre du Palais Royal, an architect friend of William Strutt returned from Paris and wrote him:

> [I] ordered one of each sort of the hollow bricks, of which the Arches are composed, to be sent to me, and I expect soon to hear of their being arrived in London. . . . Unluckily I only saw the building the evening before I left Paris, at a time when I was unwell, so that I have not so perfect a recollection of the Plan as I should have had, had I reviewed it at my leisure. . . .
>
> The roof of the Palais Royale is of framed Iron, with the larger sort of hollow brick to fill up the panes.[48]

The following year Matthew Boulton, partner of James Watt, wrote to Strutt about the use of ceramic pots in floor construction:

> I understand you have some thoughts of adopting the invention of forming Arches by means of hollow pots and thereby saveing the use of Timber in making floors, and guarding against Fire. Allow me to say that I have seen at Paris floors so constructed, and likewise at Mr. George Saunders' in Oxford St., London, who is an eminent Architect . . . I have therefore no doubt but it might be applyd also with great success and security to a Cotton mill.[49]

At the time of Boulton's letter Strutt was beginning work on a six-story mill for his father at Belper. In it, vaults made of hollow pots formed the ceiling of the top story, spanning 7 and 8 feet between the timbers that were the bottom chords of roof trusses.

The eighteenth-century use of ceramic pots in floor and roof slabs was intended to lighten those structural elements, reducing the weight that rested on beams and columns yet maintaining the thickness of the slab. It is said that in the 1850s a foreman working on the construction of the Liverpool Exchange proposed that, instead of wooden laths spanning between I-beams, triangular tubes of terra-cotta be used.[50] This combination provided an incombustible floor,

2.20 In construction of the Liverpool Exchange, tubular hollow tiles with a triangular cross section spanned between iron I-beams. This system provided a formwork over which a concrete floor could be poured. Though incombustible themselves, neither the tile nor the concrete protected the bottom of the beams from fire. (*Proceedings, Institution of Civil Engineers*, 1890–1891.)

2.21 More than six months before the Chicago fire, George H. Johnson and Balthasar Kreischer received a patent on a fireproof floor system of terra-cotta. The design was much the same as that used years earlier in the construction of the first story of the Cooper Institute in New York, but terra-cotta "filling strips" were added to protect the bottoms of iron beams. (U.S. Patent no. 112,926.)

but the smooth bottom surfaces of the terra-cotta members proved to be a poor key for the application of plaster ceilings (fig. 2.20).

In the same period, flattened tubes of clay were used in the first story of the Cooper Institute, New York, one of the earliest instances of an iron-framed floor. The system was patented by Frederick A. Peterson, architect of the building, and the terra-cotta was shaped by hand. These tiles spanned about two feet between the iron beams. Little use seems to have been made of Peterson's system. For larger spans, the patent of an English manufacturer of clay products, Joseph Bunnett, provided a shallow arch of tiles with a tie rod across the top to prevent the thrusts of the arches displacing beams that supported them. By this method a room as wide as 12 feet could be spanned with a single vault. After the Chicago fire, an experimental arch of this type was built by architect W. W. Boyington, but it is not known that the system ever found application in the frantic rebuilding of the city.[51]

Around 1860 hollow terra-cotta—or structural tile, as it came to be called—was little used in floor construction. For spanning between iron beams, the more common solutions of that period in the United States were brick arches or curved sheets of corrugated iron filled over with concrete; in England, concrete arches; and in France, light iron bars with a slab of plaster cast between them.

The system patented in England by Maurice Abord, a Frenchman, was much like Peterson's design, although the open cells in Abord's tiles were relatively small. In the English patent, dated 1866, the tiles were shown in use with wooden beams, but a U.S. patent obtained in the same year indicated its application to a span between iron I-beams. The Abord system seems the first to cover the bottoms of the supporting beams and thus fireproof them. This precaution was neglected in a patent of the following year, which produced a flat arch by assembling light tiles made in the shape of traditional voussoirs, the wedge-shaped units that make up an arch. This form is strikingly like the designs that later dominated the field.

About six months before the 1871 Chicago fire, Balthasar Kreischer, a firebrick manufacturer in Staten Island, New York, and George H. Johnson, a Chicago engineer specializing in the construction of grain elevators, patented a floor construction that was essentially the same as that patented by Peterson 16 years earlier (fig. 2.21). The only significant addition was the provision of grooves on the tops of the tiles to hold wooden runners on which flooring could be nailed. Kreischer tile was used in the Kendall Building and other projects in Chicago and elsewhere, but legal action resulted in a court decision that, in comparison with other designs, the Kreischer patent lacked the requisite originality.[52]

Patent offices were deluged by slight variations of floor arch systems, and lawsuits for patent infringement were frequent. In 1888 an English patent was granted the "Fawcett Ventilated Fire Proof Floor," a system that closely resembled the triangular tiles used almost thirty years before in the Liverpool Exchange (fig. 2.22). The following year a patent of Julius Homan did little more than convert triangular tubes to trapezoidal shapes. When Fawcett sued Homan for infringement of his patent, the defendant charged that at least eleven earlier patents covered all important characteristics of the Fawcett system, and the court dismissed the charges.[53] However, on appeal the decision was reversed on the basis that a new assemblage of old elements qualified for the protection of a patent.

The use of concrete at the turn of the century caused a radical alteration in the function of structural tile floor systems. Earlier methods had employed structural tile to bridge between elements of a metal framework; now concrete spanned the distance, as well as serving as the material with which the floor surface was prepared. The Kleine system, first patented in Germany in 1892, allowed generous spaces between tiles. Metal straps lay in the bottom of these spaces and, once concrete had been poured over the assembly, each joint acted as a slender reinforced concrete beam. In such systems the primary function of tiles was to form concrete into an effective structural shape.

The use of voids in masonry for walls began when an Englishman, Benford Deacon, in 1813 patented a perforated brick that was intended for "conveying air up chimneys from the kitchen fire boiler to the attics, &c., in walls."[54] In his drawings the perforated bricks were shown around a flue, with their perforations continuous so that air warmed within them rose to rooms overhead. The openings were to be cut by hand out of green brick. Little use seems to have been made of Deacon's invention in England, but similar brick later appeared in France, where Prince Metternich saw them and had samples sent for study at the Imperial Polytechnic Institute in Vienna. After a variety of tests to determine the strength of the bricks, a Viennese architect tested them in construction.

Most hollow brick devised in the first half of the nineteenth century had the great disadvantage of being produced by tedious handcraft methods, and because a frequent reason for using light masonry units was the reduction of weight in vaulting and domes, many shapes departed from the traditional rectangular form of

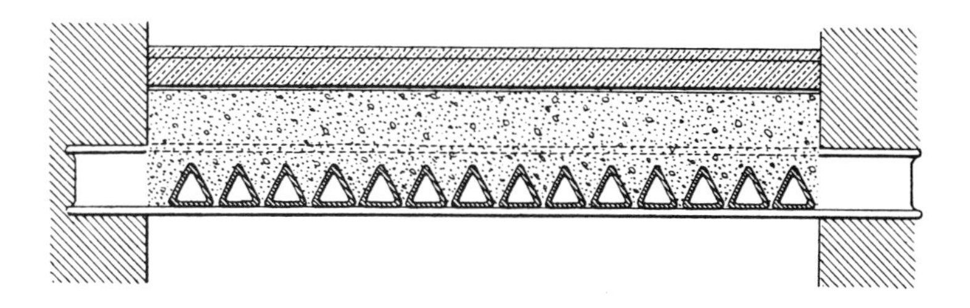

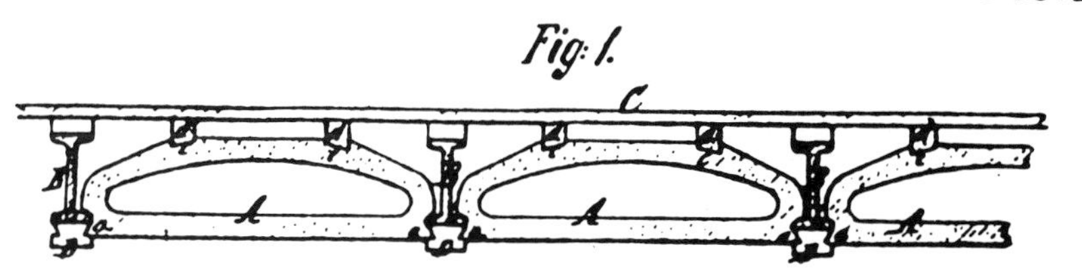

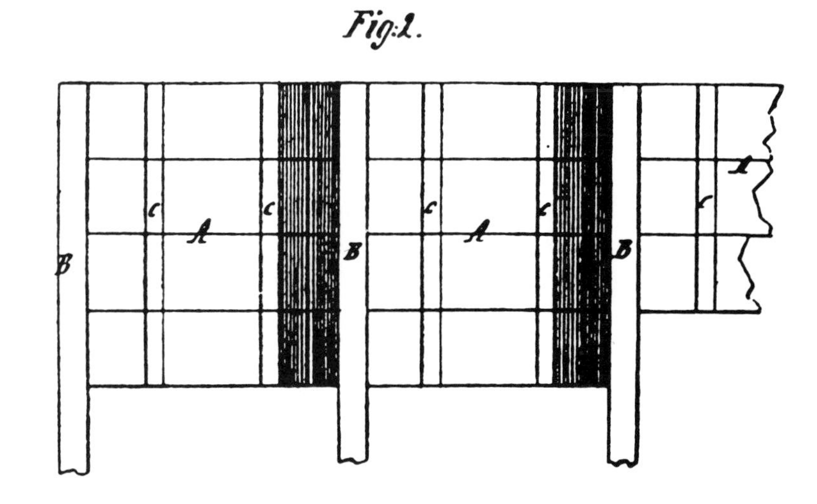

THE "FAWCETT" VENTILATED FIRE PROOF FLOOR.

PLAN OF STEEL JOISTS AND TUBULAR LINTELS
FIXED READY FOR CONCRETING.

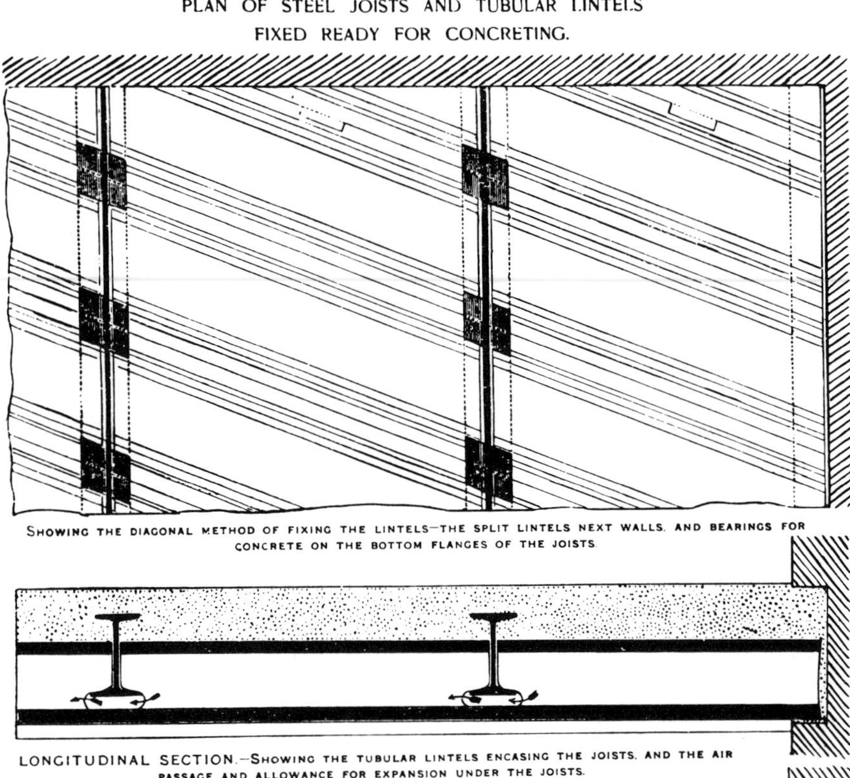

SHOWING THE DIAGONAL METHOD OF FIXING THE LINTELS—THE SPLIT LINTELS NEXT WALLS, AND BEARINGS FOR CONCRETE ON THE BOTTOM FLANGES OF THE JOISTS.

LONGITUDINAL SECTION.—SHOWING THE TUBULAR LINTELS ENCASING THE JOISTS, AND THE AIR PASSAGE AND ALLOWANCE FOR EXPANSION UNDER THE JOISTS.

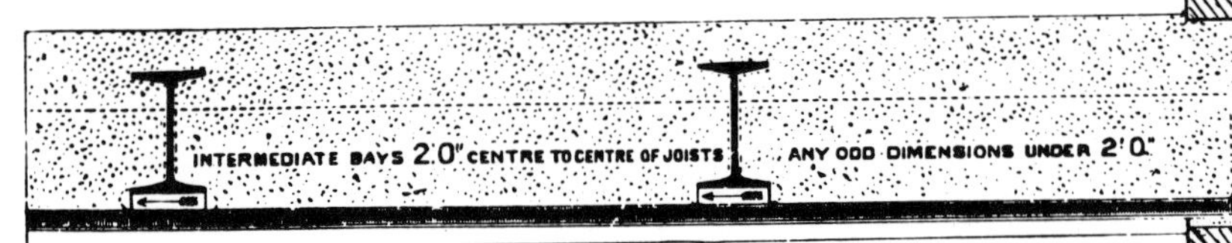

LONGITUDINAL SECTION.—SHOWING THE CONCRETE BEARING ON THE BOTTOM FLANGES OF THE JOISTS, AND THE COLD AIR PASSAGE AND ALLOWANCE FOR EXPANSION UNDER THE JOISTS.

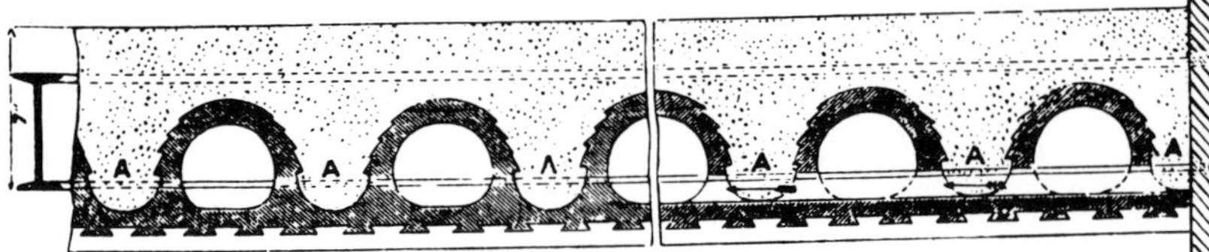

TRANSVERSE SECTION.—SHOWING THE CONCRETE BEARING ON THE BOTTOM FLANGE OF THE JOISTS AT A. NOTE—THE CONCRETE DOES NOT GO UNDER THE JOISTS AT A. SEE LONGITUDINAL SECTION.

TRANSVERSE SECTION.—SHOWING THE AIR PASSAGE UNDER THE JOISTS.

SCALE 1½ IN. TO THE FOOT

2.22 This illustration from a catalog shows the Fawcett floor system, in which the form of tiles covered the iron beams and provided a passage for air beneath the beams. Placing the tiles diagonally was presumed to simplify their installation, but this method was dispensed with in later years. (*Transactions, British Ceramic Society*, February 1959. Courtesy of the Institute of Ceramics, London.)

2.23 The first patent of Beart employed a die within which hexagon-headed bolts were suspended from crossbars. Later development of the design improved the flow of clay through the die. (*Transactions, British Ceramic Society*, February 1959. Courtesy of the Institute of Ceramics, London.)

brick. Grooved ends and interlocking shapes afforded an opportunity to make relatively rigid connections in curved roof shapes.

A mechanical means of shaping hollow bricks was patented in 1843 by a Frenchman named Collas. His machine consisted of a chamber that held clay as a piston pressed it through a die. Within the opening of the die, cores were suspended by crossbars, and these produced perforations that ran through the extruded block. Other holes could be punched by hand in the sides of block while the clay was still soft. These additional holes were intended to assure a firm bond between masonry units and the mortar in which they were set.[55]

Two years after the Collas patent, Robert Beart perfected a system making it possible to produce hollow brick economically and efficiently (fig. 2.23). Collas's patent had lapsed, apparently without commercial application, but the Beart patent was soon put to use. Bricks with round holes through their least dimension began to be used by architects. In 1848 Henri Jules Borie obtained a French patent for a machine very similar to that of Beart. After taking out a British patent, Borie exhibited the machine and its products at the Great Exhibition of 1851, where his display attracted attention and won several medals. In order to protect the market for his product, Borie brought suits for patent infringement against almost every manufacturer of perforated brick in England and France and, despite the similarity of his machine and others, Borie's suits were surprisingly successful.

In 1844 Prince Albert fostered the establishment of the Society for Improving the Condition of the Laboring Classes and consented to be its first president. The Honorary Architect of the Society, Henry Roberts, patented a hollow masonry unit suitable for the construction of small houses. Roberts's block corrected these three faults he found in ordinary brick wall construction: bonding between masonry and mortar was often weak, joints running through the wall conducted moisture to the interior, and header courses (bricks with their long dimension extending through the wall in order to tie the two wall surfaces together) were costly of labor.[56] All of these problems were eliminated by Roberts's design, and the size of the blocks (11½ inches long and 3½ inches high) meant that a mason was required to lay fewer bricks, 40 percent fewer in most cases. A model house for workers using these blocks was erected in the vicinity of the Great Exhibition in 1851, and other projects were built for occupancy, including an entire street of houses in the East End of London.

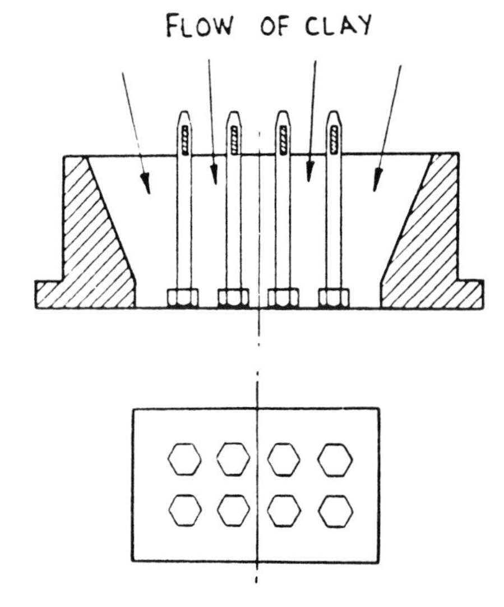

Roberts's block proved to be more expensive than the brick cavity wall, which had originated sometime before 1821. It is reported that by the 1850s about 80 percent of workers' houses in Southampton were being built with cavity walls.[57] The two faces of such walls were held together by either headers or metal ties that extended across the cavity.

McGRAW-HILL

3

Terra-Cotta

1769 Coade factory commences producing terra-cotta in London

1869 Chicago Terra Cotta Company organized

1870 Terra-cotta imported from England for the Boston Museum of Fine Arts

1871–81 Construction of the Natural History Museum, London, revives the use of terra-cotta

1878 Boston English High and Latin School employs terra-cotta provided by the Chicago Terra Cotta Company

1886 New York Architectural Terra-Cotta Company founded

1920 Twenty-four terra-cotta factories operating in the U.S.

In classical architecture, terra-cotta was principally employed for roofing tiles and floor coverings. Early Greek temples sometimes had ornamental details of terra-cotta and faced brick and wood with tiles, but in later temples stone largely replaced terra-cotta for these purposes. From the earliest times, it was the ease of shaping it that made terra-cotta attractive in architecture. The Etruscans not only used roof tiles but encased beams and covered brick walls with panels of terra-cotta. Since they displayed a liking for terra-cotta sculpture, it is not surprising that their use of the material in buildings was often highly decorative in both pattern and color. Again, the use of terra-cotta—except for roofing and floors—gave way to the brick, concrete, and stone construction of the Romans, although wall surfacing of tile remained popular. Some of these appear to have been fashioned by pressing clay into molds.

Medieval architecture made little use of terra-cotta except for roofing and flooring. The latter, as well as occasional wall tiles, were often stamped with patterns, taking advantage of the clay's potential for decoration. Even roof tiles were less in use at this time, for in much of the medieval world thatch and wood shingles made less expensive roofs. So long as clay pots and jugs were the indispensable containers in everyday life, however, a knowledge of firing and glazing was maintained.

"Artificial stone" was the most common name under which terra-cotta was manufactured in eighteenth-century England. In 1722 Richard Holt and his partner, architect Thomas Ripley, patented "a compound liquid metall, by which artificiall stone and marble is made by casting the same into moulds of any form, as statues, columns, capitalls."[1] Carved stone ornamentation was expensive, and lead casters had long produced work that, when painted, could serve as a less costly substitute for stone. Holt's artificial stone was even more economical, and he appears to have made a good business of supplying architectural details, which could be made according to a client's design or selected from stock patterns. Business was sufficiently brisk to invite competition and to require that he conceal the formula for his mixture of clays and other ingredients. Among Holt's unsuccessful rivals was Batty Langley, a pushy carpenter-builder-architect whom Holt accused of endeavoring to get the secret from his workmen, who were "tamper'd with, decoy'd into Publick-Houses, that being Drunk, they might be the more easily [questioned]."[2]

Holt's company declined and was taken over in 1769 by George and Eleanor Coade, who had come to London from Lyme Regis. The following year George Coade died and management of the factory was continued by either his wife or their unmarried daughter, who was also named Eleanor.[3] Almost as soon as the Coades began operation of the factory, their work attracted the attention of influential buyers. Horace Walpole ordered a Gothic gate for Strawberry Hill, the Twickenham residence he had begun almost twenty years before. Through Walpole, Coade stone became known to his architect, James Essex, and to Sir William Chambers, whom Walpole asked to intervene when Mrs. Coade's bill for the work was higher than expected.[4] Chambers seems to have been favorably impressed with the factory when he visited it in 1772 and, as a Commissioner of the Board of Works and architect to the King, he found occasion to use Coade stone in many designs, such as the twenty-nine urns placed atop the parapet at Somerset

House. Artistic direction of the Coade factory was placed in the hands of John Bacon, who had apprenticed with a London potter and had studied at the Royal Academy. Some of the work was completed by him and the remainder seems to have been done under his supervision, although he continued to be active as a sculptor in marble. Two popular ornaments were the Borghese and Medici vases modeled at two-thirds of actual size from the originals that Robert Adam had brought back from Italy.[5] Many architects employed stock designs selected from Coade catalogs, and often molds for designs executed from architects' drawings were used to add stock designs to the catalogs. The Coade works also supplied ornaments for buildings designed by Sir John Soane almost throughout his career. His orders included bas-relief plaques, urns, figures, many balustrades, and the bases and capitals of columns. After the firm in 1818 began manufacturing *scagliola*, an imitation of the veining and color of rich marbles, Soane and other architects ordered columns shafts of that material.[6] After Mrs. Coade's death in 1821, the plant was continued until it closed in 1839 by William Croggon, a distant relative who had assisted her. A visitor from the *Somerset House Gazette* in 1824 described the conduct of the work:

> Some articles are first formed roughly to give them the external shape in a mould, they are then polished by the chisel while in a soft state, which they endeavor to preserve by wrapping the block carefully in wet cloths. In some cases particular enrichments prepared in matrices are added, and in others the whole is nearly the work of the hand. . . . After the figure is completed in all its parts, it is cut into several pieces for the conveniency of introducing it into the oven, and is afterwards put together, firmly cemented, and iron rods introduced into the arms or other parts that may require to be strengthened.[7]

The result was a work of stoneware, pale cream in color and sharp in detail as a result of carving done as the casting dried. Coade stone and similar terra-cotta products resisted weather excellently. In 1868, almost fifty years after terra-cotta columns and ornamentation were installed on St. Pancras church in London, where four terra-cotta caryatids were mounted around cast-iron columns, a writer contrasted "the sharpness and freshness of the terra-cotta with the worn, bleached, and disintegrated stone."[8] In addition, glazed terra-cotta was washed clean by rain, an important consideration in cities that were growing increasingly smoky.

The process of making terra-cotta was described in 1896 in the *Yale Scientific Monthly*.[9] The extent to which methods varied through the years or from place to place was minor. Clay was weathered under sheds and then dried, ground, and screened to remove any grit or foreign matter. It was mixed with water, passed through a pug mill, and forced through a sieve; if it were to be used for a slip (a coating to provide a smooth finish to the fired work), it was mixed with large amounts of water so that larger particles would settle to the bottom of the vat. To the clay was added pulverized terra-cotta or firebrick, which limited shrinkage during firing. The proportions used for this mixture varied with the quality of the clay, and sometimes the amount of powder almost equaled that of clay. After being set aside to "ripen," the mixture was again put through a pug mill in which it was kneaded by rotating blades. Bubbles in the clay were removed by spreading it on a slab and beating it thoroughly with iron rods, and then the clay was ready to be molded.

Molded pieces, those to be repeated often enough to warrant preparation of a matrix, were shaped by pressing pieces of clay into the mold and forming the bracing webs of clay necessary to maintain the shape of the piece. Sculptural work was modeled and then hollowed out to leave a shell, which was cut with a wire into pieces of a size that would be manageable in the kiln (figs. 3.1, 3.2). After being carefully and slowly dried in a warm room, the pieces were placed in a kiln, deeply carved pieces being packed in sand for protection. About 48 hours were required to fill the kiln, an equal time to eliminate moisture in the clay, 60 to 80 hours to fire the clay, and an equal period to cool the kiln slowly—nine or ten days altogether (fig. 3.3). It took only four to six weeks to fill an order for terra-cotta for a small building project after a design was approved, though any delays might interfere with scheduling the construction. In the United States it became the practice of many architectural offices to obtain bids for the provision of terra-cotta before seeking general bids for construction.[10] Natural clay produces white, buff, or red terra-cotta, but minerals may be added to achieve a broad range of colors. In Chicago much of the work was a "grayish buff" produced by a clay dug in Brazil, Indiana, and used to duplicate the Joliet limestone that was popular in the area.[11] Another location provided a clay that baked to a color matching red brickwork, desired for the orders of many eastern architects. The firm of McKim, Mead and White—particu-

3.1 In the modeling room of the Federal Terra Cotta Company, New Jersey, a master craftsman displays a decorative urn. From such clay models were made the plaster molds in which terra-cotta was molded. (W. A. Starret, *Skyscrapers and the Men Who Build Them*, 1928.)

3.2 This photograph of Louis Sullivan's ornamentation for the Bayard Building, New York, has lines added to show the manner in which the design was divided in order to facilitate firing. (*Sites*, no. 13. Courtesy of Dennis L. Dollens/SITES.)

3.3 A view from the 1920s shows the outdoor storage of stock patterns at the Atlantic Terra Cotta Company's works. Each of the kilns seen in the background held 30 to 50 tons of terra-cotta in a single firing. (W. A. Starret, *Skyscrapers and the Men Who Build Them*, 1928.)

larly Stanford White—did much to broaden the range of colors available in terra-cotta and brick:

The late Stanford White, upon one of his visits to the Perth Amboy Plant, to talk over with Mr. Hall a special size Roman brick he wanted made, noticed the bricks used in the Company stable, and in the old Hall residence. . . . He said that was just about what he wanted for color, only with more spots.

. . . They proved a great success, and many large orders followed their use in the Tiffany residence. These bricks were known in the market at that time as "Tiffany Brick."

Some years later, Mr. White gave Mr. Hall a fragment of a Roman brick which he had brought with him from Italy, and asked him to try and match it. . . . They turned out satisfactory to Mr. White, who ordered them used in the Boston Library. This was the origin of the so-called "Old Gold" brick and terra-cotta. . . .

In 1889, McKim Mead and White asked for white terra-cotta. This was made by spraying a buff body with white burning clay.[12]

Ridged and scored surfaces, imitating the marks of stonecutters' tools, became popular at one time, particularly among those architects working in the Romanesque fashion of H. H. Richardson.

Through most of the nineteenth century, terra-cotta was employed principally as "artificial stone," an economical substitute for stone carving. Its lightness had proved to be advantageous for hidden uses such as floor systems and fire protection, but when exposed to view it was most often required that the ceramic material resemble stone. There was, in fact, some argument in England about whether terra-cotta should even attempt the fineness of detail that was found in stonework.[13] One group of architects and builders favored the improvement of clay with a variety of admixtures and the use of carving to sharpen details. The other group, mostly associated with projects in the South Kensington district of London, opposed tampering with natural clays and preferred the loose modeling they believed appropriate for clay—leaving more precise forms to the stonemason.

A designer's color preference could easily be satisfied so long as it corresponded to the colors that could result from local clays. In the New York region it was assumed until the 1880s that only red terra-cotta could be produced. Early attempts to produce buff shades were flawed by a portion of each firing taking on a strong pink tone. This was found to be caused by excessively high temperatures. (The science of pyrometry being in its infancy, it was common practice to hang a copper wire in kilns and assume that when the wire melted the fire had reached the required temperature.) Gray terra-cotta was soon introduced, and those three colors dominated the New York market throughout that decade. More exotic colors of terra-cotta and brick were occasionally produced (fig 3.4). However, such variations were usually short-lived, more expensive, and difficult to manufacture. When white terra-cotta was first requested, a slip was sprayed on buff clay, but this coating held soot and soiled easily. A white matte glaze could not be produced, so for a time glossy white finishes were dulled by sandblasting. It was the turn of the century before fresh discoveries in pyrometry and the chemistry of glazes made it possible to produce polychrome terra-cotta at the scale required for extensive use in architecture.

After World War I, a period of prosperity, architectural experimentation, and the wide-ranging exploration

3.4 **Although its architectural style is difficult to classify, the Fulton Theater, New York, achieved overall patterns, architectural detail, and even a pictorial frieze through the use of polychrome terra-cotta. (*Architecture and Building*, July 1913.)**

3.5 **Most of the terra-cotta blocks used in the McGraw-Hill Building (1931) measured about 51 inches by 16 inches by 4 inches. Usually, finished terra-cotta was shipped loosely packed in hay, but the glazed blocks for this project arrived at the construction site in individual boxes of corrugated cardboard and were not removed from their containers until ready to be set in place. (*Architectural Record*, April 1931.)**

of eclectic options of style encouraged further development of polychrome terra-cotta. With their need for novelty, movie houses—Classical, Italianate, Moorish, Mayan, and Oriental—proved to be a fertile area for the use of colorful terra-cotta. The Philadelphia Museum of Art (1927) required that companies bidding to supply the terra-cotta needed for its building demonstrate their ability to match the colors that had been determined by intensive study of ancient Greek architectural polychromy.[14]

A building material of this sort, durable yet capable of the full range of coloration, invited painterly experimentation. Peter Behrens, the most influential of German Expressionist architects, designed the lobby of the I. G. Farben Dyeworks as a lavish and fitting display of the artistic use of color. The lobby, entirely of brick, changed gradually from blue-green at the lower levels to yellow-orange at the top—as a painter would create gradations of hue on a canvas. Other buildings employed exterior shading of terra-cotta glazes, usually from dark below to light above.[15] In some designs, bands of brilliant color delineated the forms, framed openings, and accented skylines.

In 1931 the McGraw-Hill Building was nearing completion in New York (fig. 3.5). Thirty-three stories high, this structure was the first major project to be sheathed in machine-made terra-cotta, with blocks forming blue-green spandrels that were about half of the building's wall surface.[16] At the top of the McGraw-Hill Building the company's name was spelled out in orange-and-white terra-cotta letters 11 feet high.

3.6 Construction of the Natural History Museum, London, involved an unprecedented amount of terra-cotta and an unusual variety of pieces. The company that produced the terra-cotta was forced to expand its plant rapidly. For this project's picturesque qualities, blocks were included that had been fired darker than was usually judged acceptable by other architects. (*British Architect and Northern Engineer*, 14 June 1878.)

The Natural History Museum in London, part of Prince Albert's plans for a cultural center in South Kensington, pioneered in the use of terra-cotta as the major surfacing material of a building. In 1864 a competition for the design of the museum was won by Captain Francis Fowke, a military engineer who had become something of a specialist in museum design after his work on the Museum of Science and Art, Edinburgh, and the expansion of the National Gallery in Dublin. Execution of the winning design for the Natural History Museum, a massive mixture of various elements from the Italian Renaissance topped with a dome and a dozen Baroque towers, was taken over by other hands after Fowke's death in 1865. The task of developing the design was assigned to Alfred Waterhouse, a young Manchester architect who had only recently set up his practice in London. One of the first changes he made was the conversion of Fowke's design from Renaissance forms to those of the Romanesque. Waterhouse was convinced that terra-cotta should be the major material for the project because of its decorative advantages, lower cost, and resistance to the sooty air of Victorian London (fig. 3.6.). Once the decision on material had been made, a change of architectural style was easily justified. At that time the techniques of making terra-cotta could not assure a uniform product, and small pieces with the variation of color that was natural to the material lent themselves far more readily to the medieval spirit than that of the Italian Renaissance. An additional advantage of terra-cotta was the ease with which it was possible to incorporate a wealth of decoration using the naturalistic forms that had been popularized by John Ruskin and Owen Jones. The opportunity to ornament the Natural History Museum with designs derived

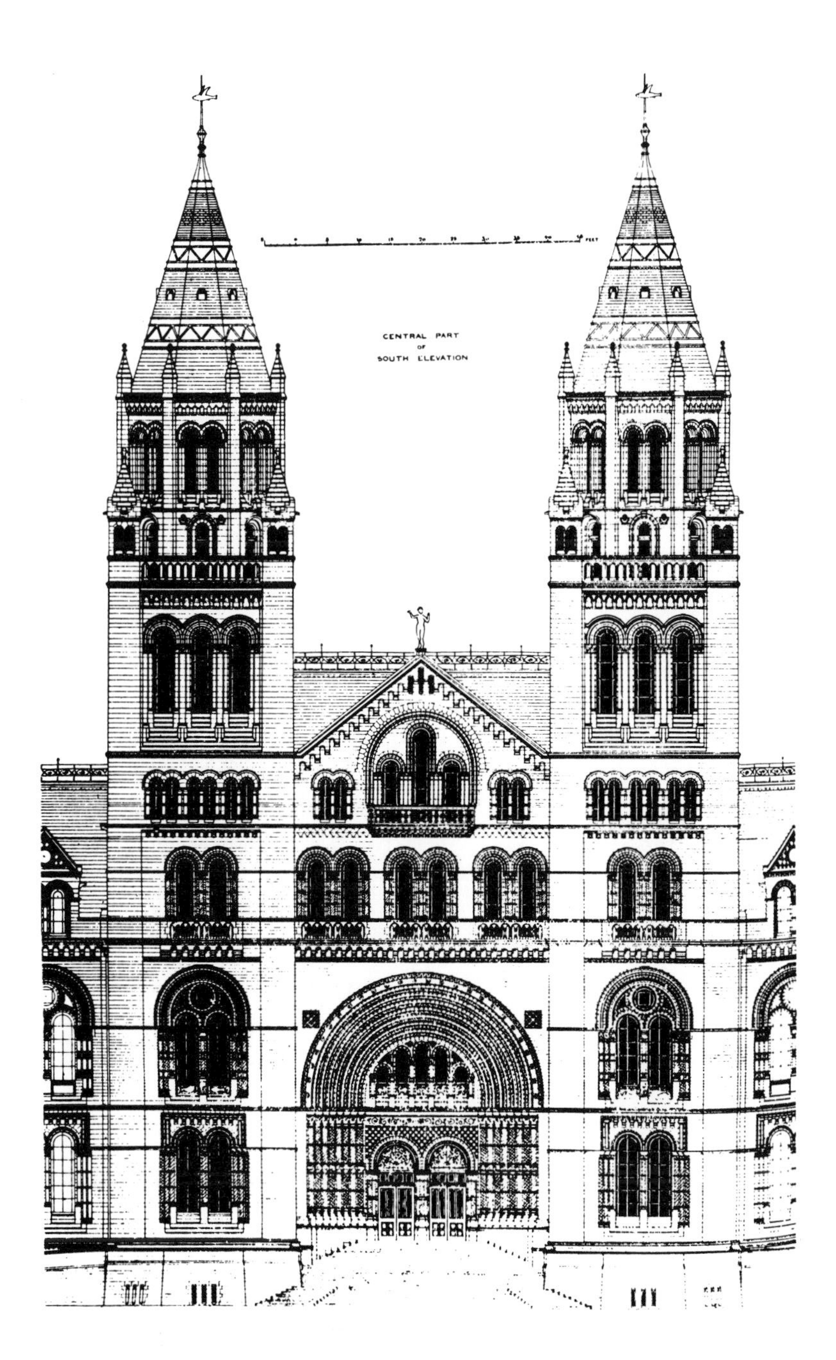

from plants and animals was irresistible. The museum director provided specimens and Waterhouse developed designs for the terra-cotta ornaments, using living material as models for the embellishment of the walls surrounding the zoological exhibits on the west side of the building and extinct models for the geological section on the east side. Much of the decorative work for the Museum was modeled by art students, which reduced costs and may have misled people about the economy of the material.[17]

When James Renwick, a leader of the Gothic Revival in the United States, attempted to revive the use of terra-cotta in 1853, he was assured by masons that the material would not survive the winters of New York. In fact, his terra-cotta lasted very well, but after Renwick had used it on three houses, the manufacturer he had persuaded to execute the terra-cotta work returned to the manufacture of sewer pipes (fig. 3.7). About ten years later Horatio Greenough, the sculptor, returned from Italy with intentions of introducing the use of terra-cotta. After hearing Renwick's account of his experiences, Greenough abandoned the project.[18]

At about the same time, Richard Upjohn, another leader of the Gothic Revival in the United States, employed terra-cotta on two New York buildings. Window trim and cornice details on the Trinity Building, made by another manufacturer of ceramic pipes, lasted well for many years, but the terra-cotta cornice of a bank building, produced by still another plant, was ruined by frost in its first winter. Interest in the use of terra-cotta faltered in New York, and 25 years passed before it was revived.

In Chicago the firm of Hovey and Nichols, dealers in seeds and flowers, decided to enter the business of manufacturing garden urns and statuary as a result of one partner's travels in Europe. They bought a company located in Indianapolis, where coal and clay were available, but soon they discovered that shipping finished ceramics was much more costly than shipping the raw materials required to make them. Consequently, the pottery works was moved to Chicago in 1868. The following year the company was reorganized as the Chicago Terra Cotta Company with a number of new investors that included Sanford E. Loring, an architect who had briefly been a partner of William LeBaron Jenney.[19] New kilns were constructed and the company readied itself for increased production with some minor examples of architectural terra-cotta added to its line of garden furnishings. Loring began to participate more actively in the company; when the superintendent of the works quit, he wrote John Marriott Blashfield, a leading English manufacturer of architectural terra-cotta, asking for his help in discovering a replacement. The letter was passed to James Taylor, one of Blashfield's employees who had superintended the preparation of terra-cotta for Sir Charles Barry's buildings at New Alleyn's College, Dulwich. Taylor, who had already made up his mind to emigrate, thus became superintendent of the Chicago Terra Cotta Company.

The company's plant escaped damage from the Chicago fire in 1871, and it was prepared to take an active part in the rebuilding that followed. Taylor had quickly brought the equipment and procedures to the level of the latest English practices. Sanford Loring withdrew from his architectural practice to become president of the company, and its range of architectural products was rapidly enlarged.

In Boston, competition drawings for the Museum of Fine Arts were submitted in 1870, and a few months later the firm of Sturgis and Brigham was chosen as architects for the project. John Sturgis, son of a wealthy Boston merchant, had studied in England, and five years after establishing his practice he had taken the practical-minded Charles Brigham as a partner.[20] Sturgis spent four years in England during the period when the Museum of Natural History was built and Sir Charles Barry completed New Alleyn's College, both outstanding examples of the use of terra-cotta. It

3.7 James Renwick designed this building in 1853, using terra-cotta for window trim and the frieze of its cornice. Early terra-cotta was unglazed and consequently soiled like other building materials, but 75 years after the building was completed the terra-cotta was described as having "lines as clean-cut as on the day [it] came from the kiln." (*American Architect*, 20 November 1925.)

was not surprising that Sturgis recommended terra-cotta for the Museum of Fine Arts, although the material had rarely been used in the United States. When he spoke on the subject of terra-cotta at an annual convention of the American Institute of Architects, Sturgis presented his audience with the equivalent of an elementary textbook on the material, using much of the information that Barry and Blashfield had presented to the Royal Institute of British Architects three years before.[21] When it came time to contract for the terra-cotta to be used for the Museum of Fine Arts, it was Blashfield's English factory that supplied the material, there being no qualified competitor in the United States.

When the Boston English High and Latin School was planned in 1878, the city architect was instructed to use terra-cotta for the building's ornamentation. A short time before, the Chicago Terra Cotta Company had provided trim for two Boston houses, the city's first use of terra-cotta manufactured in the United States. The contract to supply terra-cotta for the school was awarded to the Chicago firm, and Sanford E. Loring, president of the company, hurriedly made arrangements to establish a branch in Boston. Leasing space and facilities from the Boston Fire Brick Company, Loring arranged that all clay required for the project should be shipped there from Chicago, ready for the modeling, molding, and firing to be done in Boston. It was necessary to find someone who would serve as Loring's deputy in Boston and supervise the work. James Taylor had left Chicago two years earlier to settle in New Jersey, where his brother Robert directed the work of the Perth Amboy Terra Cotta Company. James Taylor was persuaded to leave his position at Eagleswood Art Pottery and assume direction of the work in Boston, and after completing the contract for the English High and Latin School he continued his work in that city under the sponsorship of two local investors. Moving from their leased location, the fledgling company built a new plant, but within a short time Taylor left them to assume direction of the Boston Terra Cotta Company, newly formed by merging the Boston Fire Brick Company and another firm.[22]

After a building in New York burned in 1882 with the loss of five lives, its owner, Orlando B. Potter, set about building a new structure that was to be totally fireproof in the terms of that period. The Boston Terra Cotta Company supplied over 500 tons of materials for the project, and Taylor frequently encountered Potter at the construction site, for the owner was in the habit of keeping a close eye on the progress of his building projects. A New York guidebook of 1892 provided a description of the new Potter Building:

> It has eleven stories, and was the first building in the midst of the great newspaper section to be erected of such a height. The Potter Building possesses two unusual features: first, it was the first office building erected in this city which was ornamented elaborately with terra-cotta; second, it was the first to have the iron and stone work covered with hollow bricks as a protection against fire. It is one of the most substantially constructed and fireproof office buildings in the city. The owner so ordered its constructions that it would endure practically forever.[23]

The Potter Building was also the last tall building constructed in New York without using a steel skeleton. During their encounters at the building site, Taylor convinced Potter that there was an opportunity for a new company to manufacture terra-cotta in the

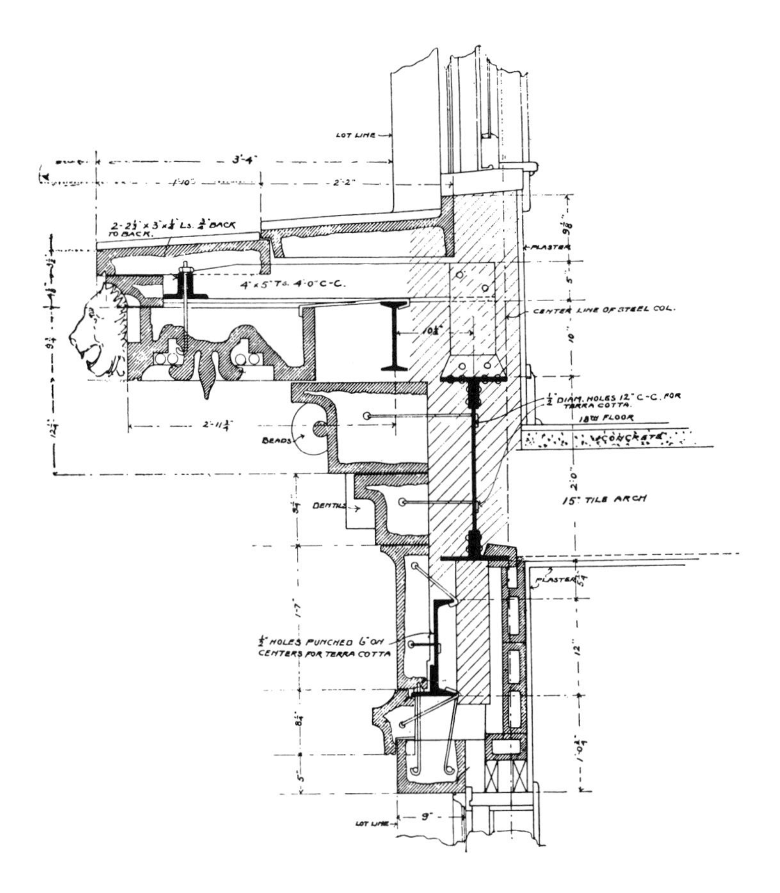

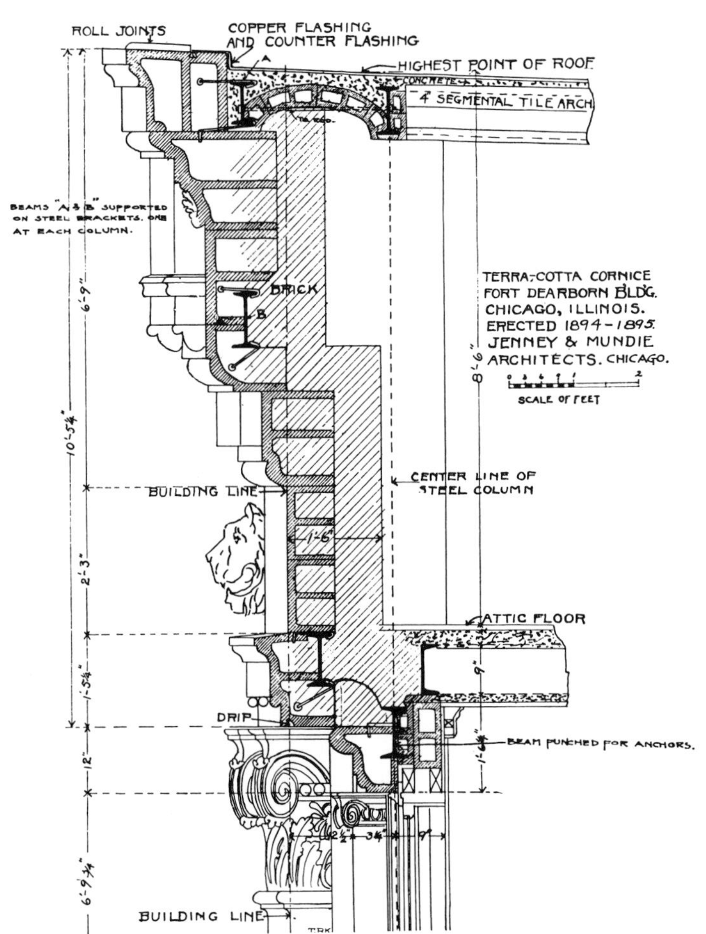

3.8 In publishing details of terra-cotta cornices, William LeBaron Jenney stressed the importance of filling the backs of terra-cotta pieces wherever possible. Cornices with strong projection, popular at that time, were usually supported by cantilevered I-beams, and soffit pieces of terra-cotta were held in place with metal hangers. (*Brickbuilder*, June 1897.)

decoration, few tested its potential for elaborate ornamentation as thoroughly and consistently as F. H. Kimball, who used intricate Moorish detail for the Casino Theater on Broadway and developed friezes depicting American Indian life for the Montauk Club, Brooklyn, in 1890.[25]

Shortly before the Chicago Terra Cotta Company established its Boston branch to fulfill the English High and Latin School contract, a group of its employees left to establish True, Brunkhorst and Company in Chicago. Sanford Loring's management of the Chicago Terra Cotta Company soon collapsed, leading to the company's failure in 1879 and Loring's return to architectural practice. Its competitors, retitled as the Northwestern Terra Cotta Works, fell heir to the clients that Loring had developed over the preceding years. Other manufacturers of terra-cotta were soon launched, and by 1920 there were 24 companies operating from Atlanta, Georgia, to Seattle, Washington, and 28 other companies had failed, merged, or withdrawn from the brisk competition.

The depression of the 1930s took its toll. The orders received by major U.S. terra-cotta manufacturers in 1933 were but a tenth of those received a decade before.[26] When building activity resumed after World War II, terra-cotta had become a curiosity. Other materials, including cast stone, were available for architectural elements, and fashion had abandoned the ornamental work for which terra-cotta had been so well suited. Its high labor costs hampered its competition with other materials more adapted to mechanized production.

3.9 Standard architectural ornamentation could be offered by manufacturers of terra-cotta. As with cast-iron building fronts, repeated use of molds was necessary in order to recover their cost, and the custom designs made for architects often found their way into catalogs of stock designs. (New York Architectural Terra-Cotta Company, 1887.)

3.10 Terra-cotta was well-suited for sheathing steel skeleton construction, as in three adjacent buildings for wholesale millinery companies in Chicago. The firm of Gage Brothers and Company agreed to pay additional rent to compensate for the employment of Louis Sullivan, Chicago's master of terra-cotta ornament, to design the facade of their portion of the structure. (*Brickbuilder*, December 1899.)

New York area, and in 1886 the New York Architectural Terra-Cotta Company was formed with Potter as one of two partners and James Taylor as superintendent (fig. 3.9).

The revival of terra-cotta in New York had begun in 1877 when George B. Post designed a residence on 36th Street using red terra-cotta shipped from Chicago.[24] The first public building to use terra-cotta and the first significant contract of the newly organized Perth Amboy Terra Cotta Company was the Brooklyn Historical Society, another work of Post, who led New York architects in their experimentation with terra-cotta ornament. His order of terra-cotta for the Product Exchange Building (1882) totaled over 2,000 tons and required the Perth Amboy company to construct additional kilns and workrooms and hire more craftsmen, some attracted from England. Of the other architects who employed terra-cotta

PLATFORM LEVEL
240 0
60
3.0
4.0
2 9
4 0
11.3
SECTIONAL PLAN THRO' b b

4

Iron and Steel

1709 Coke employed in smelting iron

1781 The relationship of carbon content and the strength of steel is discovered

1783 Use of rollers to bind lumps of wrought iron introduced by Henry Cort

1784 Henry Cort produces wrought iron by "puddling"

1796 Cast-iron beams and columns used in a Shrewsbury mill

1809 Cast-iron dome covers the courtyard of the Halle aux Blés, Paris

1835 Construction of the iron fishmarket behind Hungerford Market, London

1849 Installation of the first cast-iron facades produced by James Bogardus

1851 Crystal Palace, London, and Lime Street Station, Liverpool

1855–56 Henry Bessemer and William Kelly patent processes for the manufacture of steel

1859 Three-high rolling mill awarded a patent (Britain)

1879 Standard shapes for rolled steel established by German manufacturers

1884–85 Construction of Home Insurance Building, Chicago, often credited as the first example of skyscraper construction

1889 Galerie des Machines, Paris, employs arches of wrought iron

1893 Chicago regulations limit building heights to ten stories

4.1 Diderot's *Encyclopédie* shows a blast furnace of the late eighteenth century. Iron ore and fuel were placed in the tall brick furnace. Water-powered bellows (below left) insured a hot flame; molten iron flowed out at the bottom (above center). The heart of the furnace would usually last about 30 weeks before it required rebuilding, but if the outlet for molten iron became plugged for a short time a new furnace could quickly be ruined. (Diderot, *Encyclopédie*, 1762–1777, "Forge," plate 2.)

Although iron had been as precious as gold during the Mycenaean period of Greek history, in later centuries it became more available.[1] The structural use of iron in Greek buildings is limited to a very few examples in connection with beams and many applications in masonry walls and columns. In beams, rectangular iron bars, as large as 36 square inches in cross section, were set in grooves at either the top or bottom of the stones of architraves, and they apparently served as additional beams, supplementing or safeguarding the load-bearing capacity of the stone beams. Such uses are relatively rare, but much iron was to be found within Greek walls. In both walls and columns, iron dowels guarded against horizontal movement between stones, just as iron cramps hooked a stone to those that were placed beside it. Molten lead was poured around such dowels and cramps to insure their fitting snugly in the holes and grooves carved for them. I-shaped cramps in the walls of the Parthenon measure about a foot in length and 4 inches in width, and appear to have been heated when they were placed in grooves partially filled with lead.[2] Although concrete and the arch were the mainstay of Roman builders, iron continued to be used as dowels and cramps in their masonry.

In medieval buildings, iron played an important role within the fabric, usually hidden but sometimes visible. Rods between piers and columns stabilized the work, and at the Sainte-Chapelle (Paris, 1244–1247) curved straps were fastened on each side of voussoirs, "holed to take the ends of round iron bars like the rungs of a ladder."[3] The most intricate of Gothic work required considerable augmentation with iron. For a large window at Westminster 126 cramps were furnished, weighing a total of 143 pounds.[4] To deter rust, which endangered masonry as it expanded, it is said that iron was boiled in tallow, just as linseed oil was applied to stone that did not weather well and resin was sometimes used in mortar. In addition, records indicate that iron was painted with pitch, varnished, or dipped in tin to protect it from rust.[5]

Rods within masonry often took the form of complex frameworks knitting together the stones. In the spire of Salisbury Cathedral, iron bars were linked together to reinforce the structure. Christopher Wren in 1669 wrote of the iron bars in the spire: "These [are] so essential to the standing of the work that, if they were dissolved, the spire would spread open the walls of the tower, nor could it stand one minute."[6] Outward thrusts imposed on walls by vaults, domes, and other roof shapes remained a problem through the Renaissance. The elegant high melon-shaped domes favored by Baroque architects followed the precedent established by Filippo Brunelleschi's dome for the Florence Cathedral, where three iron chains were set in the masonry to counteract outward thrusts of the dome. In supports and buttressing for such ambitious constructions, carefully designed systems of tie-rods and cramps guarded against the strains produced by settlement of the structure and the pressure of winds. Even in wood construction the supplementary use of iron assumed a critical role, with connections replaced or supplemented by iron straps. In all of these cases, however, iron served as an assistance to a major structural material. Only later did the production of iron advance sufficiently for it to become a major material of building.

Cannons were the major product of the English iron industry in the sixteenth century, but toward the end of the century governmental regulations limited the production and export of cannons and severely restricted cutting wood for fuel, in order to make certain that timbers would be available for shipbuilding.[7] Ironmakers, accustomed to making charcoal by cutting away the forests surrounding their furnaces, were soon forced to purchase fuel from charcoal makers, who planted acorns and other nuts to produce crops of small trees which they harvested to produce charcoal. The process required covering a hemispherical mound of logs and branches with earth or sod, igniting the wood, and controlling the heat by adjusting an opening at the top of the mound. Charcoal was the usual fuel for smelting until the eighteenth century, for it provided a high temperature with no smoke and little ash. Although experiments were often made in the use of coal and peat, neither of these fuels proved successful at that time. Every fuel contained oils and minerals, which were less troublesome for other trades in which the fuel and the raw material were kept apart. In smelting, on the other hand, fuel and ore were stacked together and impurities in a fuel strongly influenced the quality of iron produced. Therefore, charcoal became the principal fuel of iron smelters in ancient times, and when charcoal became scarce and expensive, attempts were made to find another fuel for the furnaces. Brewers, who had found that using coal to dry malt produced a foul-tasting beer, began in the 1640s to employ coke made from coal by methods quite similar to those by which charcoal was made. It was more than a half-century before coke would be used to smelt iron.

Abraham Darby, operator of a Bristol brass and iron works, in 1707 leased iron furnaces at Coalbrookdale in Shropshire. There he began to manufacture iron pots and kettles according to his patented method of casting them in molds of sand. In 1709 Darby started to use coke in firing his furnaces. The iron furnace of that period consisted of a vertical cavity in a large mass of masonry with water-powered bellows to inject air at the bottom (fig. 4.1). Alternate layers of charcoal and ore were laid in the cavity. Smoke went out an opening at the top and melted ore ran down through the bright coals and flowed out at the bottom of the furnace. Once lighted, a furnace was charged at the top with additional fuel and ore, the firing continuing day and night until the masonry deteriorated and it was necessary to build a new furnace. Besides being much less costly than charcoal, coke was less likely to crumble, and this made it easier to maintain the flow of air through larger and taller furnaces. Because the molten metal ran down a greater distance through the fuel held in a larger furnace, it became hotter, more liquid, and more able to fill fine details of any mold into which it might be poured.

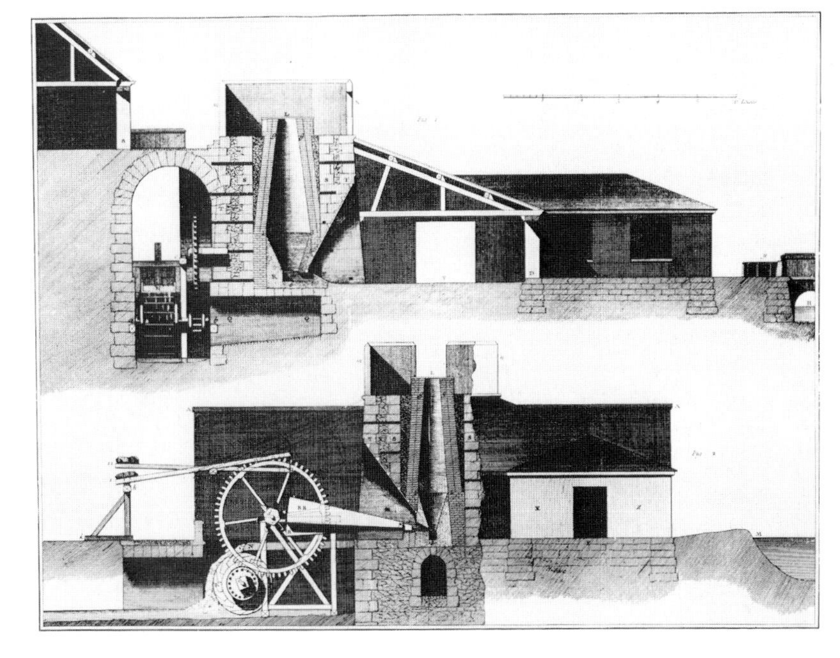

The development of metal-framed buildings can best begin with the history of English textile mill construction at the end of the eighteenth century.[8] The typical textile mill of the period was five or six stories high with heavy walls of masonry. Timber beams spanned between the walls, supporting wooden joists and floor boards. The width of mills was usually about 28 feet, sometimes a single span but most often having one or two rows of intermediate columns, and the length was usually around 120 feet. With wooden floors, lanterns, lint, and oily machines, the risk of fire was great, and high insurance rates encouraged mill owners to investigate metal construction.

William Strutt in 1792 planned a cotton mill for his father's hosiery factory at Derby. In plan the building was a traditional mill, six stories high with masonry walls. On the top floor, shallow arches made of hollow clay pots were set between the bottom members of the wood trusses. This eliminated the need for columns on that floor. On other floors cast-iron columns divided the width of the mill into three spans of 9 feet. In nearby Milford, a four-story warehouse built by Strutt was constructed in a similar manner with plaster covering the bottoms of the wooden beams and sand and tiles laid as the floor (fig. 4.2). The heavy timber beams were bored to receive metal pieces that connected the top of one column to the base of the column above. In this way shrinkage of the wood beams did not alter the structure. Tie rods between the column heads provided lateral bracing. The structure of the Milford warehouse was protected more on the basis of flammability than fire resistance. While timber beams were shielded from flames, the cast-iron columns were left exposed.

The wooden beams employed at Milford had triangular strips of wood fastened along their bottom edges to hold brick arches. The use of iron beams with similar projections characterized a second phase of English mill construction. In a flax-spinning mill built in 1796 at Shrewsbury, iron columns and beams were combined. The

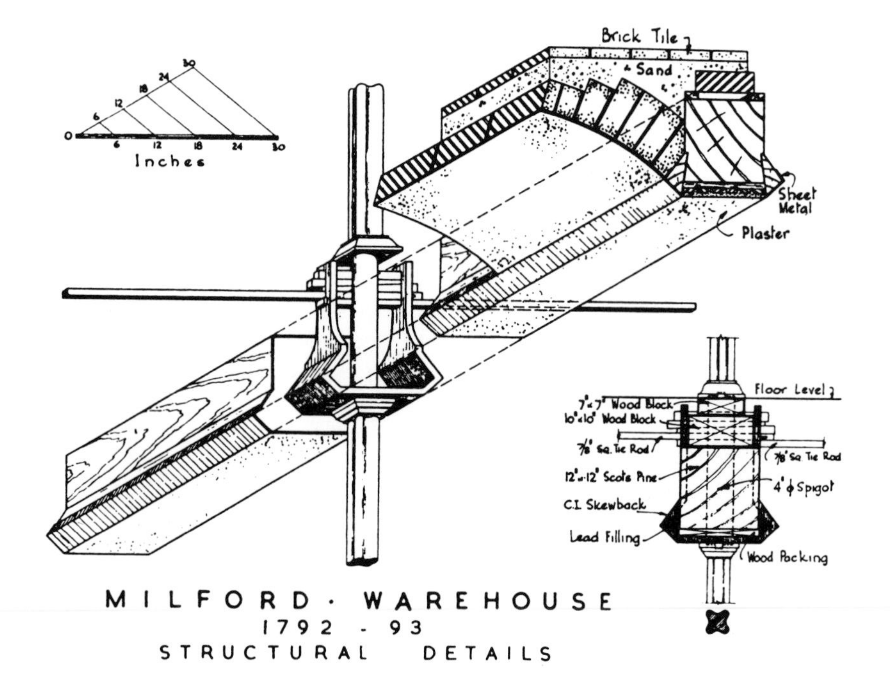

4.2 In Strutt's Milford warehouse, cast-iron columns were connected vertically by short lengths of metal and laterally by tie rods. Sheet metal and plaster protected the lower side of the wooden beams. (*Transactions, Newcomen Society*, 1955–1956.)

beams were vertical plates, 11 inches high, that thickened on the bottom to provide a triangular portion, 5 inches wide at the base. The shape suggests that it was designed as a seat for brick floor arches much more than it implies any structural understanding of iron beams.[9] These beams extended over four spans of 9½ feet and were cast in two pieces, which were bolted together a short distance to one side of the center column.[10]

Twenty years later, with the use of iron beams and the improvement of their shapes, the floor area that could be supported by a single beam's span and a single column was doubled. This advancement effected distinct savings in the cost of the iron needed for mill construction, the cost of the material having been a deterrent to broader adoption of iron for that purpose, and it allowed greater freedom for the mill operator's arrangement of machinery. Once the use of iron beams had been tested and accepted, the next phase in the development of metal construction was the theoretical understanding of the beams themselves.

Early in his career, before he became deeply involved in the design of ships and bridges, William Fairbairn designed two English textile mills. In 1824, when work on these mills began, Thomas Tredgold published the second edition of his *Practical Essay on the Strength of Cast Iron*, reporting on the experiments he had made. His proposal as the most efficient shape for iron beams was a slender I-beam (fig. 4.3). Tredgold erroneously assumed that cast iron was equally strong in tension and compression, whereas it is about five times stronger in compression. His I-beam had only slightly more of its metal at the bottom, where tension stresses are present, than the narrow inverted T-beams used over twenty years before. In one of his mills Fairbairn employed a beam 18 inches deep at the center of a span measuring 20 feet.[11] During the same period Eaton Hodgkinson, a British mathematician, began an intensive theoretical study of beam action, and in conducting tests to corroborate his theories he was aided by Fairbairn. Hodgkinson's analysis of the action of tension and compression within a beam proved to be accurate. According to his calculations the ideal beam shape would have had a small upper portion to take compression stresses and a heavy lower portion to take tensile stresses, the two connected by a thin vertical plane. Such differences in the dimensions of a casting would have cooled unevenly, causing stresses within the material, and therefore the theoretical shape had to be adapted to the requirements of its manufacture. This adaptation was often the case during the period before cast-iron structural units were supplanted by those that combined cast iron for compressive stresses with wrought iron for tension.

When work began in 1887 on the alteration of the old grain market of Paris for use as the Bourse de Commerce, an inscription was discovered on one of the columns:

> Philibert Delorme, architect, in 1540 conceived the idea of a wood framework; this system, long neglected in Paris, was used for the first time in the construction of this dome in 1782.
>
> That structure burned in 1802 and was replaced in 1811 by the present dome.[12]

The original Halle aux Blés, built in the 1760s, was a circular building arranged around a circular courtyard about 130 feet in diameter. Its simple facade, with tall arched openings at

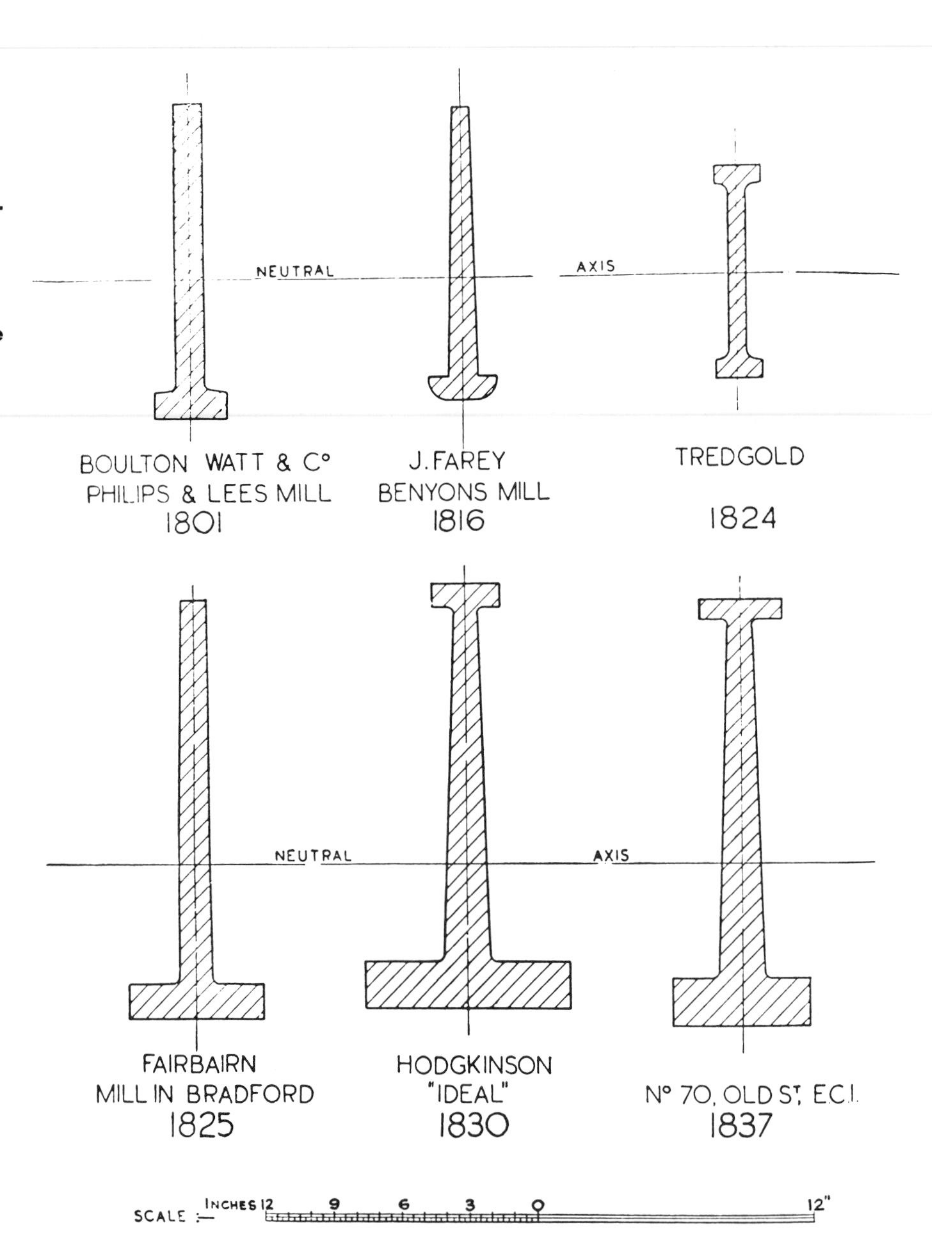

4.3 These sections of cast-iron beams demonstrate by their shapes the increased understanding of beam action and the low tensile strength of cast iron. (*Transactions, Newcomen Society*, 1940–1941.)

4.4 The metal dome of the Halle aux Blés rested on the inner wall of the ring-shaped building. The iron dome built in 1811 copied the form of a wooden dome built almost 30 years earlier. (*Construction Moderne*, 21 December 1889.)

ground level, was interrupted only by the incorporation of a Doric column 90 feet high, which had served as an astrological observatory for Catherine de Medici's palace on the site. Over 15 years after the market was built, it was decided to roof the courtyard so that it could better serve for storing and displaying grain and flour. In 1782 proposals were sought for roofing the courtyard. François Joseph Bélanger submitted a design for a dome of iron; two other architects, Molinos and Legrand, together presented a scheme for a wooden dome. The wood design, which followed a drawing of Philibert Delorme, was executed. This was the wooden dome that burned in 1802.

Five designs were submitted for replacing the wood dome of the Halle aux Blés, but all were rejected as impractical.[13] Later Bélanger submitted plans for an iron dome, and it too was based on the scheme published in 1540 by Delorme.[14] Cast iron was chosen as the material by the Minister of Interior because it was considered much more economical than wrought iron.[15] This decision seems to have been strongly influenced by Napoleon's enthusiasm for iron construction and the high cost of the wrought iron that had been used over a decade before in the construction of the Théâtre Français. The dome for the Halle aux Blés was practically as large as that of St. Peter's in Rome, and it was long a major monument of Paris (fig. 4.4).

Even while its new dome was being planned by Bélanger as architect and J. Brunet as engineer, the development of new market facilities for the grain trade began to diminish the use of the Halle aux Blés.[16] In the 1880s it was decided to convert the building to a stock exchange, and during the process of reroofing the dome its construction was carefully studied, since none of the original drawings remained. The dome was made of 51 ribs, each divided into 6 castings of iron (fig. 4.5). Fifteen horizontal rings of cross bracing tied the ribs, and a sixteenth formed the base of a lantern about 37 feet in diameter. Thick plates of copper were placed in the connections of the castings so that the soft metal would compensate for any unevenness of the surfaces being joined.[17] Careful examination after 75 years showed no deterioration other than a crack in the bottom ring of cross braces and rust in the dome's lantern.

The dome of the Halle aux Blés was no doubt a model for the cast-iron dome of the London Coal Exchange, built in 1849. While the Coal Exchange dome had a diameter of 59 feet, less than half that of the Halle aux Blés, its ribs were integrated with a variety of ornamentation in the cast iron that continued on supports and the railings of galleries. The lower half of the dome in the Halle aux Blés had been obscured by an allegorical mural; a much smaller band was plastered at the bottom of the Coal Exchange dome, and this band was divided into panels, allowing the lines of the ribs to continue to the supports.

Although the use of iron and glass to roof narrow passages between

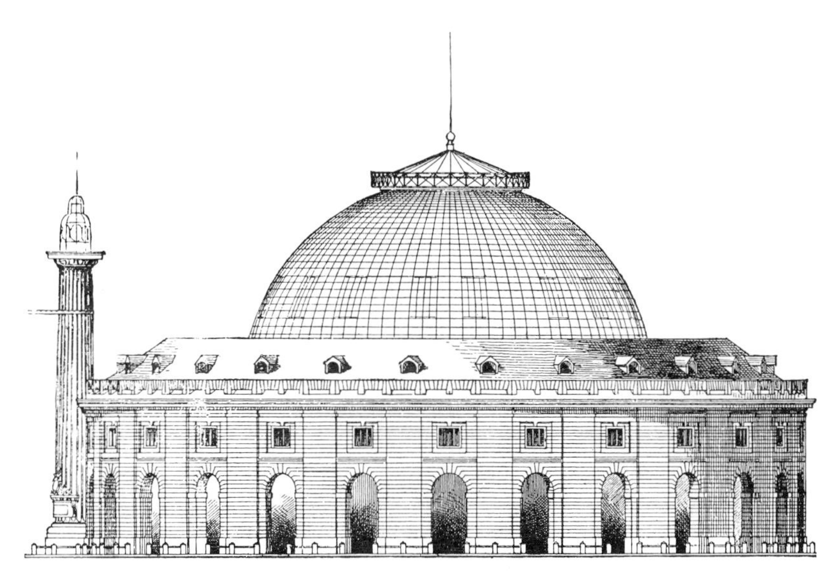

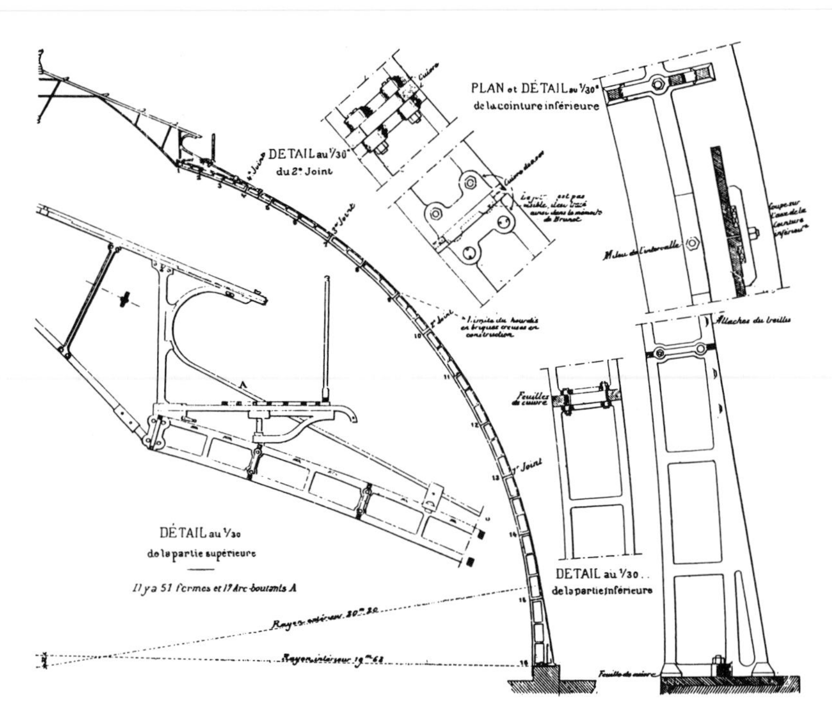

shops had been developed in France during the latter half of the eighteenth century, it was principally the English architect Charles Fowler who exploited metal in the design of markets. In 1826 Fowler built the Covent Garden market, combining a stiff Greek Revival facade with a fruit market roofed by glass held in ribs of iron. A little later he built Hungerford Market, which was demolished in 1860 to make way for the Charing Cross railway station. While this original building of Hungerford Market was stone, roofed in timber, in 1835 Fowler used iron to cover a fish market located behind in a courtyard facing the Thames (fig. 4.6). The span of 32 feet and cantilevers of 6 feet were roofed with sheets of zinc, mounted on pieces of tarred felt to avoid electrolytic action between the two metals.[18] Lateral stability was provided by diagonal rods in the clerestory and iron brackets at the heads of the columns, so that the structure of the market roof required no masonry walls or piers to maintain its rigidity. Fowler designed other iron market buildings, but in the church he built at Honiton, Devonshire, he confronted the difficulty of using metal in a more finished space. The church's nave was spanned by iron members with tile and cement filling between. When parishioners gathered in the church during winter months, the heat and moisture that rose from them caused condensation to form on the iron ribs and drops of water fell on the heads of the congregation. It was necessary to add a roof of more customary construction.[19]

Metal construction of market halls was to be found almost everywhere by the end of the nineteenth century. This was principally the result of Napoleon III's construction of the Halles Centrales in Paris. In 1853 the emperor halted construction of a masonry market building so formidable that it was referred to as "The Fortress of the Marketplace."[20] In its place the same architects, Victor Baltard and Félix Callet, about a year later began construction of a market building made of iron. Baltard's design provided the light and air needed; as the first market to serve so large a population, it became a model.

The ability of cast iron to assume decorative detail was most often exploited by direct imitation of stylistically traditional castings that assumed the appearance of stone. Sometimes the shapes took on thinner dimensions when formed of cast iron, dimensions far from those typical for stone, but even then paint might be mixed with grit to give a stonelike texture to the surface of the iron. This was common in the cast-iron facades manufactured in the United States by James Bogardus, Daniel Badger, and others. In 1836 Bogardus had gone from New York to London, where he promoted his engraving machine, one of his many inventions (others included a sugar-grinding mill, several clock designs, and a gas meter). It seems reasonable to assume that during the four years he spent in England, he would have become familiar with Fowler's use of metal in market halls as well as with the other experiments in iron construction. Bogardus continued inventing when he returned to New York, focusing much of his attention on grinding equipment. By 1847 his manufacture of milling machines had grown sufficiently to require a larger building. Foundations were laid, but work was halted while Bogardus prepared cast iron for the facade of a five-story building, the other parts being constructed of brick and timber. A row of three storefronts followed. The castings for these projects were made by four different foundries, according to patterns of Classical derivation that were provided by Bogardus. He stressed the ability of cast iron to reproduce decorative detail: "Were only a single ornament required, it might perhaps be executed as cheaply in marble or freestone: but where a multiplicity of the same is needed, they can be cast in iron at an expense not to be named in comparison, even with that of wood; and with this advantage, that they will retain their original fullness and sharpness of outline long after those in stone have decayed and disappeared."[21] When these projects were completed, work was resumed on Bogardus's workshop, a five-story structure occupying the standard New York lot of 25 by 100 feet. Its cast-iron facade was made in the same design as had been used on the storefronts that came before it. After completing his own building, in 1851 Bogardus prepared the arches and columns on the facade of the Baltimore Sun Building, according to a new design by the New York architect Robert G. Hatfield.

Preparation of cast-iron building panels was relatively simple compared with the accuracy required to cast and prepare machine parts, but the work of planing and grinding the surfaces that were to be joined was a major cost factor. (Earlier the English had solved this problem, though not well, by filling joints with "rust cement," a paste made with iron filings, sal ammoniac, sulphur, and a little water.) After the small-scale drawings of the architect were completed, the foundry's work began:

> Large scale drawings are made, followed by full-size drawings of the principal parts. Then the patterns are prepared. In the foundry the pieces are moulded in sand and castings are made. Cleaning, chipping and filing next follow. The ends of the cast columns are cut off true and smooth in a double-ended rotary facing machine. In the fitting shop, the columns are laid on their backs, spaced the right distance

4.5 Details for the 51 ribs of the dome for the Halle aux Blés, Paris. Note the thick pads of copper that were inserted at the points where sections of a rib were bolted together. (*Construction Moderne*, 3 November 1888.)

4.6 Charles Fowler chose to use cast iron in the Hungerford wholesale fish market because he believed it would be more sanitary and less odorous than wood or stone. Structural members also served as gutters, and alternate columns served as roof drains. (*Transactions, Royal Institute of British Architects*, 1835–1836.)

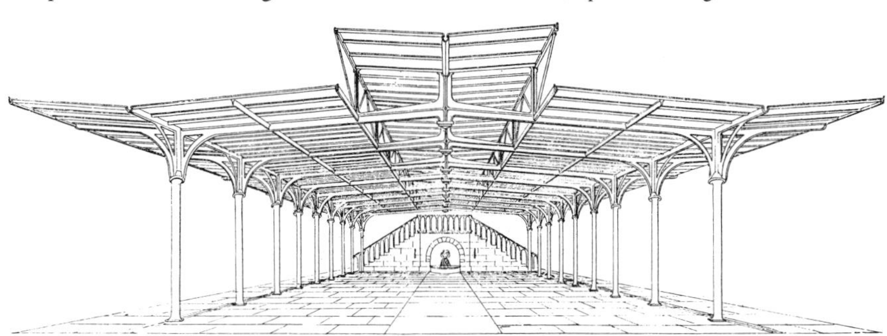

apart, bolted together story upon story. The light castings, the arches, the soffits, the sill, the ornaments are all fitted in their place and bolted or secured fast. Lying on the floor the iron front is thus put together in all its parts. A surface of oxide of iron paint is given to the work. The parts are then separated, care being taken to mark each piece so that it can be put back in its proper place.[22]

By 1858, eleven years after his first building project, Bogardus's catalogs listed more than 30 projects, including five in Chicago and one each in San Francisco and Havana. The best known was the printing plant built for the publishing house of Harper Brothers after its previous building burned in 1853. The new building was designed by John B. Corlies, an architect and contractor (fig. 4.7). The upper four floors of its facade were cast in the molds used for the Baltimore Sun Building, but new designs, rather more ornate, were made for the ground-floor pieces. The interior framing was laid out by Corlies. Between columns ran decorated composite girders made of cast iron and wrought iron, and they in turn supported rolled wrought-iron beams in inverted T-shapes with an enlargement along the top edge. These beams were placed less than 3 feet apart, and their flanges carried shallow brick arches over which a concrete floor was poured.[23]

Bogardus was not alone in the development of cast-iron construction, and at the time that the Harper Brothers Building was under construction a San Francisco foundry advertised its own line of cast-iron storefronts. In fact, Daniel D. Badger had erected a cast-iron storefront in Boston six years before Bogardus's first effort. In 1846 Badger moved his foundry to New York, where he combined the fabrication of building elements with the manufacture of rolling metal shutters, a necessary protection for show windows before electric lighting permitted merchants' displays to be safely lighted all night. One of Badger's largest projects was E. V.

4.7 The use of cast iron for the facade of the Harper Brothers Building provided a simple, repetitive design that was also inexpensive. However, the low initial cost of cast-iron fronts was offset by the need for masonry backup and periodic repainting. (*History of Real Estate, Buildings and Architecture*, 1898.)

4.8 Front and back views of a portion of a Bogardus cast-iron storefront, with a section that indicates the manner in which the parts were bolted together. If spaces behind were not filled with masonry, they became dangerous ducts through which fire might spread. (J. Bogardus, *Cast Iron Buildings: Construction and Advantage*, 1858.)

Haughwout and Company's store on Broadway, in which Elisha Graves Otis installed his first passenger elevator. In spite of the banking panic in 1857, the sale of cast-iron fronts boomed, perhaps because of the surge of westward migration. The Cornell Iron Works in 1859 supplied the facades for A. T. Stewart's store on Broadway, occupying an entire block, 200 feet by 328 feet and 85 feet high. John Kellum, architect for the store, employed cast-iron columns and wrought-iron girders on the interior, but the floors were framed with wooden beams. The building was completed in stages, and the rhythm of bays and arches varied in horizontal dimensions in the different phases of construction, a significant problem since repetition was the only identifiable basis for the design. The owner was said to have compared his structure to "puffs of white clouds," but others found little to favor in iron painted white and window shades of bright blue. One critic sneered: "When these shades are pulled down, the architect, if, indeed, it ever had one, must be congratulated upon having manufactured the most purely ugly and conspicuously offensive structure in New York City if not in the whole continent."[24] Nevertheless, Stewart's store was commercially successful, and a fourteen-story addition was built after it was bought in 1896 by the Philadelphia merchant John Wanamaker.

After the Boston fire of 1872, it was reported that fire had spread through cavities behind cast-iron facades, but since many such buildings had floors and roofs constructed with wooden beams, joists, and decking, this seemed only a secondary problem. Later fires further proved the need for blocking off the vacant space behind cast-iron panels, but it was 1885 before New York building regulations addressed the problem.[25] By the end of the century, many other U.S. cities had adopted legal requirements for filling behind cast-iron panels. In New York, brickwork 8 inches thick was demanded; wherever similarly stringent requirements were made, the iron castings might become merely facings for full masonry walls, losing their structural function and much of their economic advantage.

The cast-iron storefront was on the whole peculiar to the United States and was spurned by most European architects. When an English architect reported to colleagues in 1882 on his tour of cities in the northeastern United States, he observed that several cast-iron fronts had been built in London. His listeners were assured that "the passer-by certainly does not admire them, and only looks at them, if at all, as eccentricities of a peculiarly vulgar kind."[26] Cast-iron storefronts afforded large window areas, an important factor for retail sales establishments, but the inexpensive imitation of stonework was as strong a

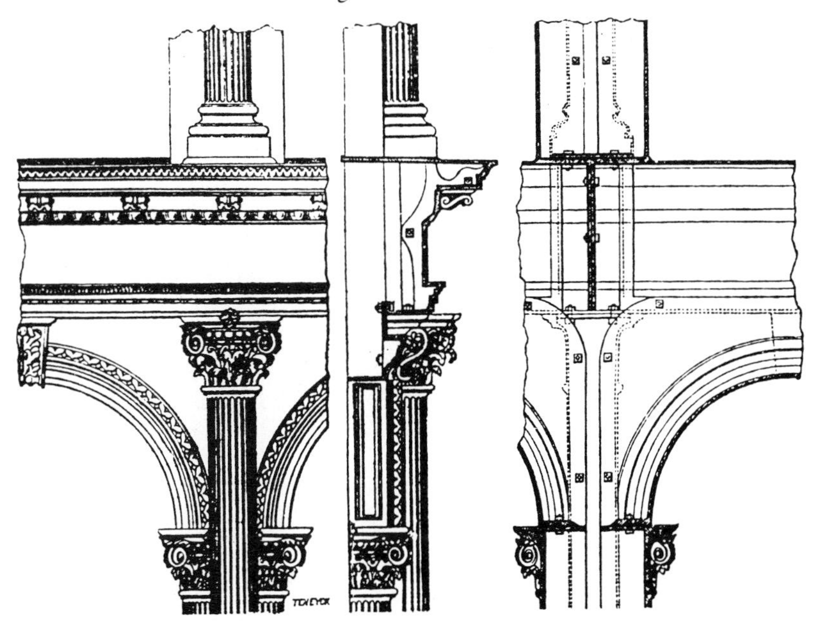

factor in their use. The need to paint the iron was often viewed as an opportunity for a new tenant to

4.9 ***Gleason's Pictorial*** **in 1853 published this drawing of a "new iron building" erected in Philadelphia. Cast-iron fronts were popular for commercial structures of this sort, where masonry party walls could be combined with a front that provided large areas of windows and inexpensive architectural details. Glass-filled grillwork beneath the ground-floor display windows could provide light to the basement. (*Gleason's Pictorial,* 16 July 1853.)**

employ a distinctive color, an advertising advantage that might well outweigh the cost of maintaining paint on the facade. Ten years after the English report, a writer in the United States put forward the notion that the decline of cast-iron fronts resulted from the fact that architects "let the cheapness of the material—cast iron—run away with their good sense," leading them to engage in frivolous ornamentation.

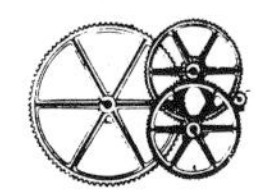

For the nineteenth-century builder, cast iron and wrought iron had complementary advantages and shortcomings. Cast iron was produced inexpensively and there were many local ironworks from which castings could be obtained. By casting in sand and other materials, it was possible to obtain structural members of a distinctive shape without incurring great expense, and architects could easily indulge their own decorative tendencies or duplicate traditional forms. As in firing ceramics, care had to be taken that the design did not require significant variations of thickness, which could cause a casting to warp as it cooled.

Cast-iron members were usually bolted together; a precise joint between the pieces could be obtained by planing and grinding the surfaces to be joined, an expensive procedure, or by using some plastic material as a grout between the surfaces. Nevertheless, the bolted connections were not always absolutely rigid. In fires, columns and beams of cast iron often cracked when chilled by water from firemen's hoses.

Although the usual nineteenth-century wrought iron had a compressive strength much less than that of cast iron, its tensile strength was about three times that of cast iron and almost equal to its compressive strength. Cast iron was a logical material for use as columns because its strength in compression permitted small columns to replace the large masonry piers that had occupied so much floor space. At the same time, its low strength in tension made cast iron less advantageous for beams. It is not surprising that the use of cast-iron columns continued, particularly in the United States, long after that material had been replaced by wrought iron for other portions of the structural framework.

Cast iron contains around 3 percent carbon, wrought iron usually less than 0.1 percent. The advantage of removing carbon from pig iron lies in the ease with which wrought iron can be shaped and its strength in tension. There was an ancient method by which wrought iron was made directly by melting ore and blowing air over the ore, thus providing oxygen for burning out the carbon. The process of converting pig iron to wrought iron involved several metallurgical problems. Iron smelted with coal or coke was much less costly than that smelted with charcoal, but both coal and coke iron contained deleterious substances. Coal-smelted iron contained sulphur, which caused it to be brittle; coke-smelted iron contained silicon. Melting the pig iron with coal eliminated most of the silicon, but added sulphur from the coal. At first the iron was placed in crucibles or "pots" with lime as flux to draw out the sulphur. In England the potting process, as it was called, produced about as many bars of malleable iron as were made with charcoal. A better method was found, but it did not begin to replace potting until around 1795.

4.10 The puddling process resulted in a ball of iron interlarded with molten slag, the latter removed by hammering or compacting the iron. The "crocodile squeezer," shown here, pressed the iron as it was gradually moved nearer to the fulcrum of the squeezer's upper jaw. (*Appleton's Cyclopedia*, 1880, 2:195.)

In 1784 Henry Cort, who had entered the iron business only nine years before, patented a process that allowed more efficient preparation of wrought iron, the method known as "puddling." A shallow tank was filled by melting pig iron or pouring molten iron from another furnace. Through the use of a reverberatory furnace, in which fuel was kept separate from the iron and heat was provided by flames drawn over the tank, the iron received no carbon from the fire. Once the iron was liquid, a "puddler" stirred it, allowing air to reach the carbon it contained. The purified metal had a higher melting point than the impurities in it, and therefore as they were stirred the two substances separated. The process has been compared to churning cream to make butter. One of the best descriptions was provided by James J. Davis, Secretary of Labor under Presidents Harding and Coolidge, who had been a puddler in his youth:

> My spoon weighs twenty-five pounds, my porridge is pasty iron, and the heat of my kitchen is so great that if my body was not hardened to it, the ordeal would drop me in my tracks.
>
> Little spikes of pure iron like frost spars glow white-hot and stick out of the churning slag. These must be stirred under at once; the long stream of flame from the grate plays over the puddle and the pure iron if lapped by these gases would be oxidized—burned up.
>
> Pasty masses of iron form at the bottom of the puddle. There they would stick and become chilled if they were not constantly stirred. The whole charge must be mixed and mixed as it steadily thickens so that it will be uniform throughout. I am like some frantic baker in the inferno kneading a batch of iron bread for the devil's breakfast.[27]

Taken from the furnace in balls usually weighing about 80 pounds, the iron was compacted by rollers or hammers that were driven by water or steam, forcing out the molten slag and shaping the wrought iron into bars (fig. 4.10).

The processes of puddling and rolling iron had appeared in other patents, but Cort's contribution was the combination of those procedures into an economical method of producing wrought iron. Unfortunately Cort gained nothing from his work. Needing financial assistance, he took a partner to whom he assigned his patent; when it was discovered that his partner's funds had been obtained by misappropriating government finances, the patent was useless.

Wrought iron could be rolled into plates or rods, and it was in these forms that it found its early uses. Plates were riveted to form boilers, and rods served as ties for masonry and for timber trusses. During the period from 1830 to 1850, many efforts were made to combine cast iron and wrought iron in a manner that would overcome the limitations of cast iron and exploit the characteristics of wrought iron. An 1847 study of the use of cast iron in railway bridges, surveying prominent engineers, found that on an average they considered the maximum feasible length of a cast-iron beam to be about 45 feet, with 60 feet the highest figure given by any of these experts.[28] To overcome this limitation, bridge engineers built up beams of cast-iron sections and added wrought-iron rods, bending them down to the bottom of the beam at the center of its span. This form, which attempted to combine the catenary of a suspension bridge with a simple beam, was generally unsuccessful when used for bridges and was seldom employed in building construction. In trusses the distinction between compressive functions and tensile elements was more clearly identifiable, and composite

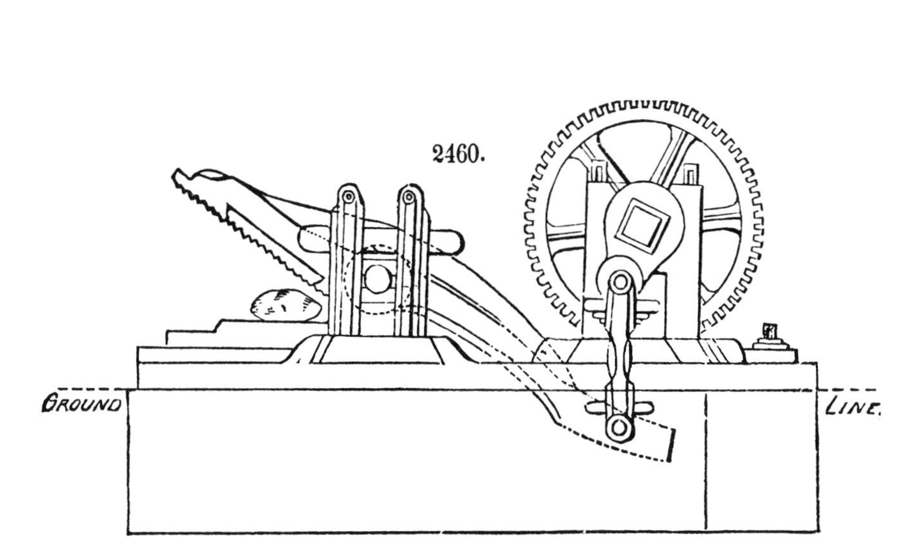

assemblies were more successful in such applications.

With the development of rolling mills, relatively efficient structural shapes could be fabricated, but they were limited in size. To attain a girder longer than the 60 feet that was a maximum in cast iron, it was necessary to use riveting techniques that had been developed by boilermakers and shipbuilders. Girders could be made up of plates, angles, and Z-bars, but the cost of such work was high. By the time that wrought-iron beams deeper than 7 or 8 inches were available, steel could be had.

An English advertisement in 1832 pictured a warehouse built in the vicinity of the London docks and extolled the merits of the material of which it was made, "Patent Corrugated Iron" as marketed by the firm of Richard Walker. The warehouse, over 200 feet in length, was covered by a vault of fluted metal that spanned 40 feet between two U-shaped cast-iron beams, which also served as gutters.[29] Previous structures of a strictly utilitarian purpose had been vaulted with iron sheets cast with rectangular ridges to stiffen them. Walker's material was wrought iron, produced in larger quantities with a new method by which the material was rolled into sheets instead of being formed by hammering. The first corrugated sheets were formed by a laborious process that made only one furrow at a time, requiring much labor and extreme precision for a workman to complete a sheet of corrugations satisfactorily. In 1844 a British patent introduced a machine that would form corrugations by passing flat sheets between rollers that were fluted longitudinally. If an arched sheet of corrugated metal were desired, the sheet could then be passed through a machine with three rollers that had ridges around them. Later patents further developed the method of rolling corrugations and added machinery for shaping corrugations by means of

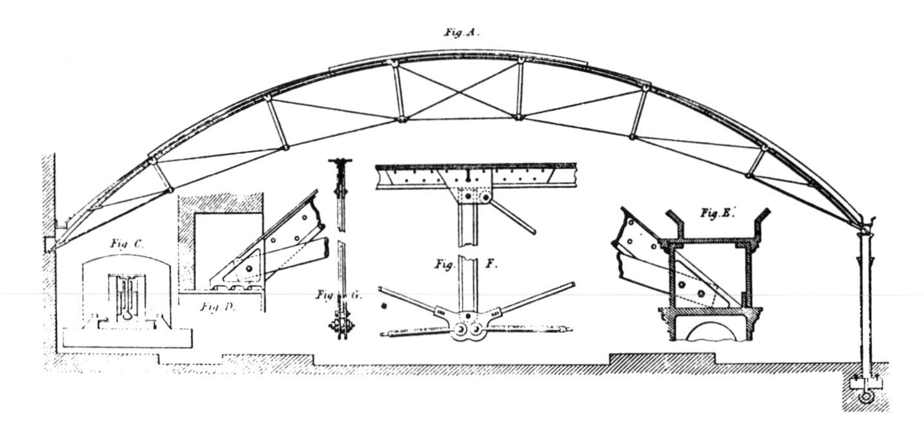

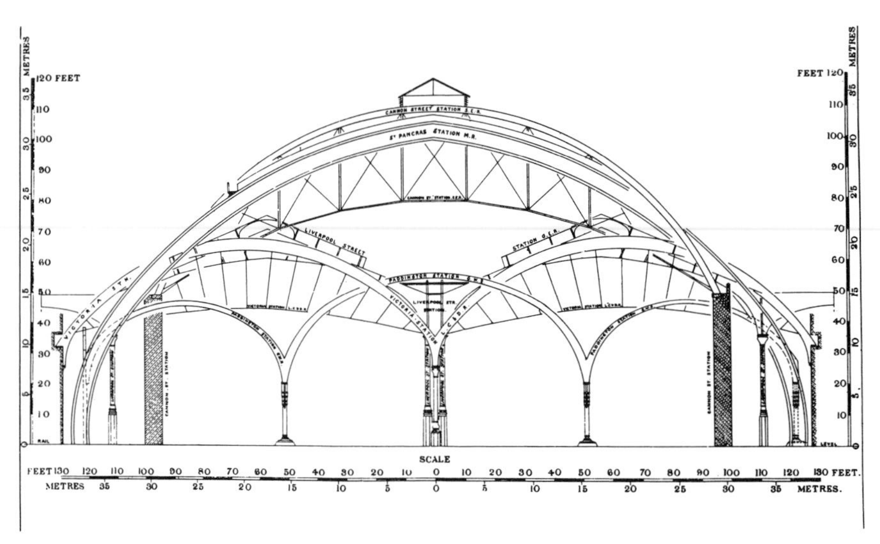

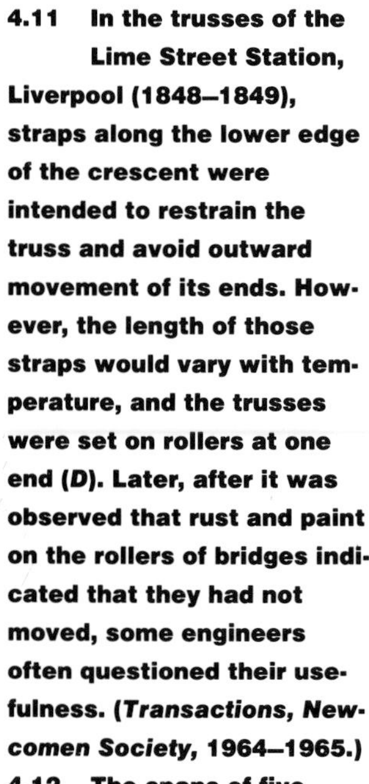

4.11 In the trusses of the Lime Street Station, Liverpool (1848–1849), straps along the lower edge of the crescent were intended to restrain the truss and avoid outward movement of its ends. However, the length of those straps would vary with temperature, and the trusses were set on rollers at one end (*D*). Later, after it was observed that rust and paint on the rollers of bridges indicated that they had not moved, some engineers often questioned their usefulness. (*Transactions, Newcomen Society*, 1964–1965.)

4.12 The spans of five major English train sheds are here superimposed for comparison. This composite diagram can be compared with the diagrams of figure 4.13. (A. T. Walmisley, *Iron Roofs*, 1888.)

dies, usually stamping only a few ridges and furrows at a time in order to avoid stretching the sheet.[30]

Iron sheets were subject to rust and weathering, especially damaging because of the metal's thinness, and tin was the first material used to provide a protective coating. In the sixteenth century the Germans beat iron into sheets with mechanized hammers and, after cleaning the surface, dipped the sheets in melted tin. By the middle of the eighteenth century a finer quality of sheet iron for tinning was provided by the rolling process. In early French experiments it was discovered that coatings of zinc produced a hard protective covering that bonded well to the iron beneath. Little was done with this knowledge until the French chemist Ernest Sorel rediscovered that process in his experimentation to find ways of protecting sheet metal from rust, an important investigation if metal were to be used for sheathing the hulls of ships. Sorel's patent of 1837 outlined a process of dipping cleaned sheets of iron into molten zinc, which was protected from vaporization by having a layer of sal ammoniac or resin floating on top.[31] It was Sorel who introduced the term "galvanize," apparently convinced that the bond between the iron and zinc was the result of some form of electrolytic or galvanic action.

In the construction of prefabricated structures, an appropriate concern because of colonial expansion around the world, the simplicity of iron frameworks had been contradicted by the necessity of applying traditional roofing materials. Iron sheet roofing, corrugated to lesson the weight required and galvanized to protect it from weather, answered both the problems of shipping the structures economically and of maintaining them in the colonies. Its use increased as droves of adventurers responded to the lure of gold, discovered in California in 1849 and in Australia two years later. Other ores and further colonization caused the development of a vast market for corrugated metal, for in the vicinity of riches few cared to engage in carpentry. Warehouses and gold rush construction found the use of corrugated iron to be eminently sensible. For the most part, the material remained restricted to use in coarse utilitarian buildings, but in New York and other cities corrugated iron often covered the roofs of major buildings—though hidden by the parapets and cornices.

For the Great Exhibition of 1851, the Crystal Palace was erected in London's Hyde Park, and two years before Lime Street railway station had been completed in Liverpool. There were other kinds of buildings with roofs of metal trusses, but few have matched the prideful display of structure to be seen at expositions and railway stations in the late nineteenth century. Lime Street station spanned a train shed 154 feet wide with crescent-shaped trusses, at that time one of the most popular types, particularly in England and Germany (fig. 4.11). The top lines of the crescents and the trusses' vertical members were shaped for compression, the bottom lines were made of flat iron bars, and diagonals were usually round rods.[32] The roof of Lime Street Station was covered with rough glass and sheets of metal, corrugated and galvanized. A greater span, 211 feet, was achieved a few years later in New Street Station, Birmingham. Lime Street Station had only one diagonal rod in each panel of the truss, but New Street Station had diagonals in both directions, a design that better provided for the horizontal forces of wind.[33]

Another early system of roofing train sheds was an arched truss in which the curve of the roof was followed by triangulated lattice work. Because this form of truss did not have a tension rod along its bottom line, it was necessary to compensate for outward pressures at the bases of the truss. In the Gare de l'Est, Paris, built in 1852 with the modest span of 100 feet, a network of rods filled the curve, serving as a bowstring between the ends of the trusses. In Victoria Station in London, rods from each vertical member of the trusses supported the shallow upward curves of the bowstrings that connected the heads of columns on which trusses rested. Such provisions against outward thrust were eliminated in the design of the train shed for St. Pancras station, London, built in 1866 (figs. 4.14, 4.15). There the trusses, spanning 240 feet, took the form of a slightly pointed arch. Instead of resting the trusses on columns or walls, their curvature was extended to the floor, thereby providing resistance to the outward thrusts of the arch. The arches of St. Pancras were much like those that had been used a few years before to provide a roof spanning 108 feet over the furnaces of a Berlin gasworks, except that the German structure used pivoting pin connections at the base of its arches and St. Pancras had a rigid connection at that point.[34] (The Berlin gasworks may have been the first to use pin connections to provide for the flexing of arches under wind pressure and temperature changes.) St. Pancras was expensive in initial cost, and the chairman of the Midland Railway Company said that, "although the proprietors were proud of it as a triumph of engineering skill, it was an utter abomination as a railway station from an economical point of view."[35] Nevertheless, similar train sheds continued to be built. The 1870 Grand Central Station in New York was similar in structure, although less elegantly formed and shorter in span.

The Hauptbahnhof at Frankfurt-am-Main, built from 1879 to 1888, replaced three separated railway stations with a single building combining three arched train sheds, each having a span of 183 feet. This was the introduction of the three-hinged arch, which employs a pivoting connection at each base and at the crown of the roof. This improvement isolated each of the pieces that made up the truss from transfer of the complex forces that resulted from wind or temperature changes, and thus it greatly simplified structural design of the roofing system. The largest train shed of this type was the Broad Street Station in Philadelphia, which had a span slightly greater than 300 feet.

Early in the twentieth century, interest in long-span train sheds

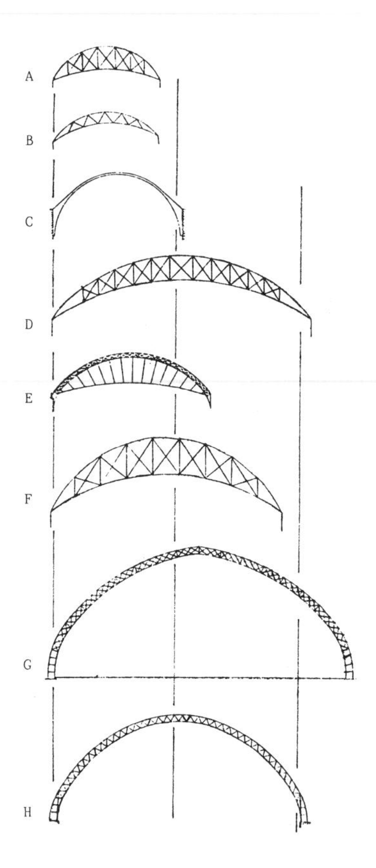

4.13 Eight nineteenth-century British train sheds in chronological order. Spacing of the vertical lines measures 100 feet. (*A*) London Bridge Station, London; (*B*) Lime Street Station, Liverpool; (C) King's Cross Station, London; (*D*) New Street Station, Birmingham; (*E*) Victoria Station, London; (*F*) Cannon Street Station, London; (G) St. Pancras Station, London; (*H*) St. Enoch's Station, Glasgow. (Drawing adapted from *Builder*, 22 September 1906.)

4.14 In St. Pancras Station (London, 1866), iron arches spanned 240 feet, a distance that was long unequaled. The pointed arch was believed to offer advantages in the structure's resisting wind pressure. (*Iconographic Encyclopedia*, 1889, vol. 5.)

4.15 The arches of St. Pancras Station extended to ground level where a rigid connection was made, instead of the rollers, pivots, or rockers that were often used for long-span construction. The vertical member that crosses the arch establishes the location of walls for service facilities along the side of the train shed. (*Engineer*, 7 June 1867.)

MIDLAND RAILWAY EXTENSION—ST. PANCRAS STATION—DETAILS OF MAIN GIRDERS.
MR. W. H. BARLOW, C.E., F.R.S., ENGINEER.
PLATFORM LEVEL
FRONT ELEVATION

4.16 **After the Great Exhibition closed, the Crystal Palace was dismantled and moved from London's Hyde Park to a suburban site, where it was reassembled. Pools, plants, and fountains were installed for the pleasure of paying visitors. (*Iconographic Encyclopedia*, 1889, vol. 5.)**

4.17 **The Crystal Palace and the Great Exhibition it housed were a success by several prime criteria of the Victorian age. The building was inexpensive, costing two-thirds the estimated price of a brick building; it was completed quickly; and the exhibition made a profit. Prefabricated and standardized construction of iron and glass made these successes possible, but the large arches of the roof (background of this drawing) had to be made of wood. (*London Illustrated News*, 16 November 1850.)**

waned. A British engineer explained in 1906,

> In justice to the designers of railway station roofs, it must be pointed out that the fashion for ostentatious spans did not originate with them, but with traffic managers, who thought it desirable that the whole interior area should be absolutely unimpeded by columns or interior supports. Fifty years of experience have served to dispel this notion, and we are now on the point of returning to the least ambitious ideal which possessed engineers in the early days of the railway system.[36]

After half a century invested in the competitive construction of broad spans, railroad stations began to be made of repeated small spans, usually around 50 feet and seldom over 150 feet.

The rapid acceptance of iron construction demanded that there be speedy advances in theoretical knowledge and practical experience. During the 1840s, the railway mania in England had required that bridges be built quickly, and the knowledge that was gained from those successes and failures contributed to the understanding of the structural requirements of long-span construction in buildings. The most famous bridge engineers occasionally designed arches and trusses for buildings, but most of the latter were the work of young men of practicality who received their training during the flurry of construction activity. For instance, Rowland Mason Ordish began as a draftsman in the London office of an engineer who was engaged principally in designing bridges. After Joseph Paxton's first sketches were accepted for the building of the Great Exhibition, the Birmingham engineers Fox, Henderson and Company undertook calculations and the preparation of detailed drawings that were necessary to translate Paxton's ideas into a building. Ordish participated in this work, and he played a larger role in the reassembly of the Crystal Palace when it was moved to Sydenham. While with the firm of Fox and Henderson, Ordish prepared construction drawings for the New Street railway station in Birmingham, the longest span of the time.

In his own engineering practice Ordish later prepared designs for a series of bridges, exhibition structures, and train sheds that include a surprising number of the period's significant metal constructions. Either in developing architects' designs or as a consultant engineer, Ordish contributed to a series of projects that included winter gardens (including that of the Dublin Exhibition in 1865), many bridges, the dome of Royal Albert Hall, and railway stations in Amsterdam, Glasgow, and London.[37] The roof for St. Pancras station was designed by Ordish in collaboration with the railway company's engineer, and around 1870 Ordish acted as engineer for a cast-iron kiosk to be erected in India, the design by Owen Jones. When there was concern about the stability of the vaulting over the octagonal chapter house at Westminster Abbey, Ordish added an iron roof framework from which the medieval vaulting was suspended.

In almost every country there were practical men of this sort—builders, engineers, ironmakers—who developed great skill through experience in metal construction and who acquired a degree of theoretical knowledge. When the rapid growth of iron construction began, there were no experts. Theoretical explanations of trusses and arches were subject to many divergent opinions, and often the practical men, learning from experience, were the source of the most reliable information. On the whole, they were not unaware of mathematical theories as they were developed,

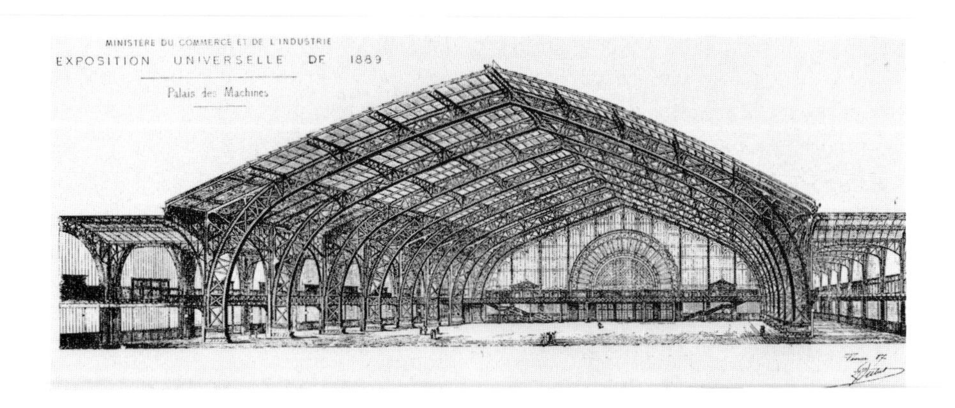

4.18 The Galerie des Machines of the 1889 Exposition in Paris was the largest span built since St. Pancras Station over two decades before. This drawing, which was published in 1887, does not indicate the pin connections of a three-hinged arch, but its publication was accompanied by details showing pin connections (fig. 4.19). (*Construction Moderne*, 6 August 1887.)

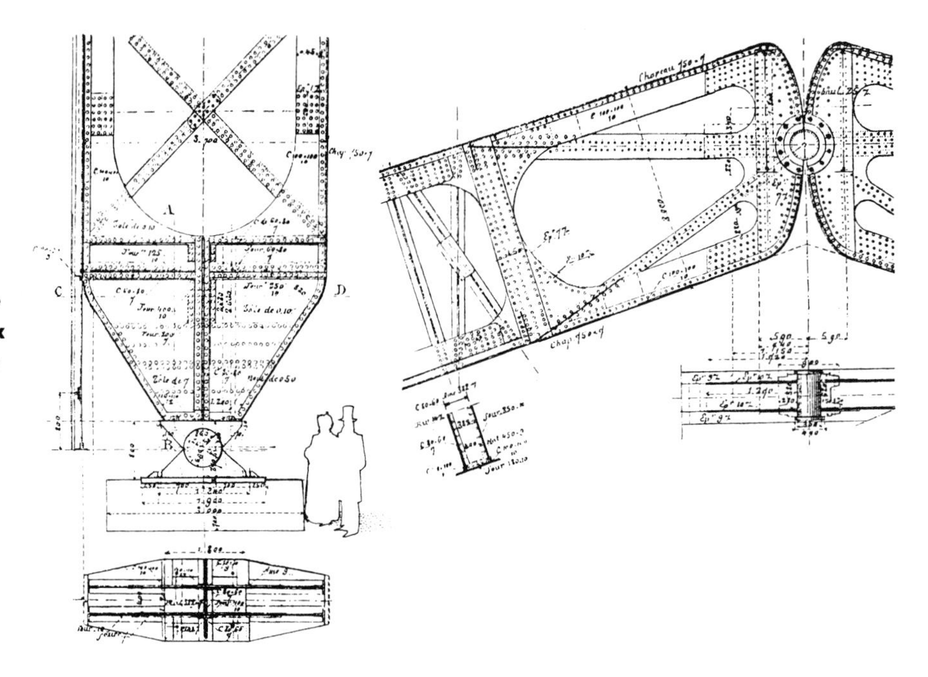

4.19 Details of the three-hinged arches used in the Galerie des Machines demonstrate the vast number of rivets used to form them of wrought-iron sheets and angles. The arches were actually constructed like box girders with about 16 inches inside between the two faces. (Human figures have been added to the drawing to suggest the scale of the structure.) (*Construction Moderne*, 6 August 1887.)

but the practical man's decisions were firmly guided by the experience gained in constructing railways, bridges, railway stations, and the spectacular iron-and-glass exhibition buildings that were monuments to the technological advances of the last half of the nineteenth century.

The Crystal Palace suffered from an excessive amount of light, requiring shades in some areas and tentlike shelters for certain exhibits (figs. 4.16, 4.17). The two exhibitions that followed (Dublin, 1853; New York, 1853–1854) avoided this problem by limiting glass areas to clerestories and sidewalls. Both were built much like the Crystal Palace, iron lattice girders and arches providing most of the structural framework. At Dublin, as in the Crystal Palace, laminated wood was used for the arches having the longest spans. The Palais de l'Industrie, principal building of the Paris Exposition Universelle in 1855, used a similar central vault with flanking galleries. About one-third of its central roof was glass, and inadequate ventilation often caused sweltering heat in the exhibition areas. All of this was wrapped in a stone wall, three stories high and detailed in the accepted academic style. Seven years later the London International Exhibition had a similar surround of masonry, resulting in a building disliked by both traditionalist architects and progressive engineers.

The plan of the 1867 Exposition Building in Paris was based on a system for classifying exhibits, but the circular principle of organization was converted to an ellipse in order to fit within the limits of the site. Only the next to last ring, the Galerie des Machines, required a significant span, over 111 feet, and it was much higher than the rest of the building. Instead of using the customary methods of resisting outward thrusts of the arches, the columns were extended as high as the crown of the arches and connected with tie rods. One engineer later evaluated the system as "an expedient and on the whole an expedient of little rationality since the tie rods placed above do not really relieve the foundations, they must resist the same horizontal pressure, just as though the tie rods did not exist."[38]

Between 1867 and 1900 there were thirty-three international fairs held, averaging one per year. The sites varied from 3 acres (Kimberley, South Africa, 1893) to 685 acres (Chicago, 1893); attendance ranged from less than a quarter-million (Brisbane, Australia, 1897) to over 32 million (Paris, 1889). Of these, the Expositions Universelles held at Paris in 1878 and 1889 mark major advances in the development of long-span construction in France. The earlier of the two exhibitions had a Galerie des Machines that was over 2,000 feet long with structural elements of 107-foot span, spaced about 50 feet apart. The latticework of those ribs ran between the slopes of the gable roof and a bottom curve in the shape of a pointed arch. The deep haunches provided by this shape stiffened the arch against the lateral pressure of winds, and such arches did not require tie rods or buttresses. Around a decade before, this form had been used in England for St. Pancras Station and in Germany for the Palmen-Garten in Frankfort. In France it was unusual: "From the point of view of construction, it is a novelty; from the point of view of stability, a tour de force."[39]

162The system was further developed in the Galerie des Machines at the Paris Exposition of 1889, a building so large and powerful that it could compete against the Eiffel Tower for the public's attention (figs. 4.18, 4.19). Its length was more than a quarter-mile and the three-hinged arches spanned

375 feet. At the top and both feet of each arch, steel rollers were installed, 14 and 20 inches in diameter. Steel had been considered for construction of the arches, but for economy wrought iron was used instead.[40] The space was so vast that above the displays of engines and machines an overhead platform rolled the length of the hall carrying visitors past the exhibits.

Visible ironwork might be acceptable for train sheds and temporary exhibitions buildings—particularly when the displays were machinery—but there was great reluctance to reveal iron structure in other kinds of buildings. The Bibliothèque Ste.-Geneviève in Paris, built in the 1840s, provided one of the first instances in which an iron structure was made visible in a building of dignity and permanence. In the library's reading room the architect, Henri Labrouste, placed a row of slender iron columns down the center of the long rectangular space. The semicylindrical vaults that spanned between the row of columns and the outside walls rose from iron arches of ornamental openwork. These vaults were simply plaster filled between the outer curves of the

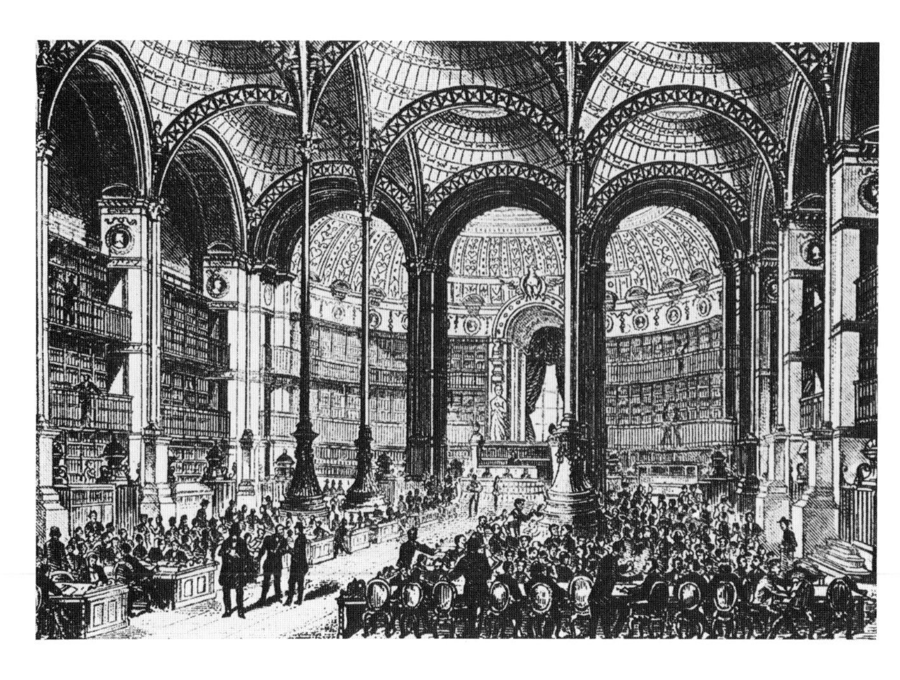

4.20 The reading room that was added to the Bibliothèque Nationale (Labrouste, Paris, 1862–1868) had no facade, for it was fitted among other buildings of the library. Because of this restriction, toplighting was necessary, and the solution was nine domical forms, each with an oculus. Above the domes, a gabled roof protected the reading room. (*Iconographic Encyclopedia*, 1889, vol. 5.)

arches, and the gabled roof over both vaults was supported by vertical iron members placed at the quarter points of the iron arches. This was not a solution offering structural innovation, but it did much to introduce the decorative potential of structural ironwork.

The reading room of the Bibliothèque Nationale in Paris, built over a decade later, offered a clearer statement of Labrouste's use of metal construction (fig. 4.20). A square space, the reading room was covered with nine domical shapes, equal in size. A gabled roof was again supported by verticals from the iron arches that held the domes. Around the outside of the room, iron columns stood free of the masonry walls, allowing the metal structure to expand independently. At the end of the century, Auguste Choisy ended his classic *Histoire de l'architecture* with a paragraph on the Bibliothèque Nationale:

In a masonry shell, which shelters the interior from changes of temperature, there is placed an iron framework in which the connections to the shell allow the metal's free expansion; and on this framework is supported an iron skeleton that carries three rows of three domes with enamel decoration. Not only in the forms but in the effects of color it is apparent that a new system of proportions is dawning, in which the laws of harmony will be none other than the laws of stability.[41]

Steel is an alloy of iron and carbon, but this fact was long unrecognized. In the making of metals, carbon was inescapably present in the fuel by which ores were melted, but only in iron and steel did its presence play a role so vital in determining the properties of the material that was produced. At the smelter's workplace various forms of carbon were everywhere, and it is understandable that this ubiquitous substance would not be quickly identified as the critical ingredient of the process. Through much of the eighteenth century, metallurgical thought was based on the phlogiston theory, which hypothesized the existence of a substance that embodied the qualities of fire. Since steel was made with great quantities of heat, it was long believed that phlogiston was a major factor in determining its character. The part oxygen played in combustion was discovered in the 1770s, and in 1781 Tobern Bergman, a Swedish analytical chemist, published the results of 273 tests on iron and steel.[42] Although Bergman clung to the phlogiston theory, he identified the quantities of plumbago (carbon) in cast iron, wrought iron, and steel, and so realized the importance of these proportions. Several years later a French chemist was able to write that "the conversion of iron to steel operates principally because either there is formed in it or it receives [from outside] an appreciable amount of plumbago."[43]

Chemical analysis did much to govern smelting practice and permit the preparation of metals from different ores. The fervor of the Industrial Revolution and colonial expansion assured rapid communication between scientists and ironmakers, but it was almost a century before microscopic analysis opened the next phase in understanding the nature of iron and steel.

A common early process for making steel involved three steps in which the chemical constituents of iron, especially carbon, were altered. First, cast iron was produced, having a relatively high percentage of carbon, between 2 and 4.5. This was melted

4.21 Bessemer's converter, as shown in U.S. Patent no. 16,082 (11 November 1856), corresponds to that in British Patent no. 356 (12 February 1856). At the lower left the converter is positioned for molten pig iron to be poured in; the upper left shows the position at which air was blown through valves at the bottom of the converter; and at the lower right the steel can be poured out.

4.22 Kelly's U.S. Patent no. 17,628 (23 June 1857) shows a converter drawn more crudely than that in the Bessemer patent. However, there remain the essentials: a place to pour in iron (*B*), openings through which air can be injected (C), and another out of which the finished steel may drain (*D*).

and stirred to burn out the carbon, producing wrought iron, which usually contained less than 0.1 percent carbon. The wrought iron was then melted and carbon added to produce steel, which had less than 1.7 percent carbon. Large quantities of fuel were required to complete this process and therefore steel demanded a high price and was used sparingly, usually only where a hard cutting edge was required.

During the Crimean War, Henry Bessemer invented a projectile that would rotate when fired from smoothbore ordnance. Trials conducted by the French government showed that the cast-iron barrels of guns were unable to fire the heavy projectile safely, and Bessemer thereupon launched his study of metal production. In 1856 he presented his findings to the British Association for the Advancement of Science in a paper deceptively titled "On the Manufacture of Iron and Steel without the Use of Fuel." In the patent he had taken out the previous year, Bessemer described his "converter," which he proposed as a fast and economical method of making steel (fig. 4.21). In an 1860 patent, the Bessemer converter consisted of a large container mounted on pivots. It was tilted to be filled with molten pig iron. Then it was moved to an upright position, and a blast of air was blown upward through the iron. Bessemer described this stage of the process: "A succession of mild explosions, throwing molten slags and splashes of metal high up into the air, the apparatus became a veritable volcano in a state of active eruption."[44] When sufficient carbon had been burned out of the iron, the converter was tilted once more to pour out its contents.

Major English ironmakers hurried to obtain licenses permitting them to use the Bessemer process, but time after time their trials were unsatisfactory. Perhaps the explosions that Bessemer had witnessed might have been stopped at a point when a correct amount of carbon remained in the metal, but in practical terms such precise control of a rather violent event was out of the question. The inventor had contemplated adding a measured amount of carbon at the end of the process, but that did not solve the problem. It was found that the iron melted in Bessemer's own experiments had been unusual in containing very little phosphorus; but licensees used their own iron which carried more phosphorus. Also, in the process the iron gained an excessive amount of oxygen.

These two chemical characteristics of the Bessemer process caused its output to be useless, until Robert Forester Mushet introduced into the process a compound that was patented in 1856. Mushet's *spiegeleisen* was a combination of iron, carbon, and manganese. When this compound was added molten to the iron, the manganese drew out oxygen and the carbon content was increased. Mushet, a reclusive and rather quarrelsome metallurgist, is said to have written, "Bessemer metal without Mushet = Iron; Bessemer metal with Mushet = Steel," and his statement was accurate.[45] The problem of phosphorus remained. Some English ironmakers obtained iron having a sufficiently small amount of phosphorus, and Swedish iron, recognized then as the purest, caused no difficulty. But most iron contained phosphorus, including that of France, Germany, Belgium, and the United States.

Although it was not realized at the time, phosphorus was retained in the iron because the Bessemer converters were usually lined with materials that

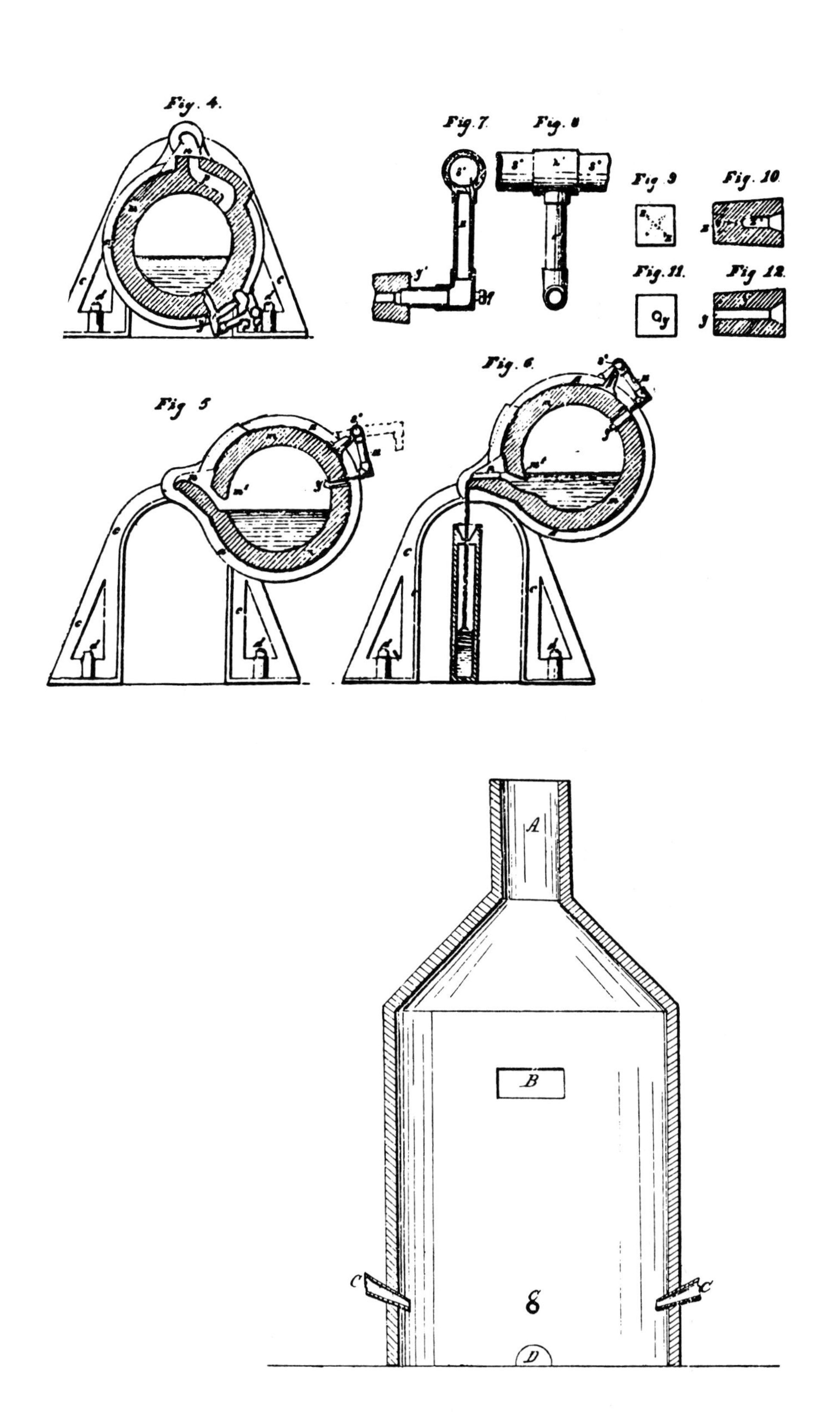
Fig. 4.
Fig. 7
Fig. 9
Fig. 10
Fig. 11.
Fig 12.
Fig 5
Fig. 6.
A
B
C
C
8
D

4.23 Much of the mass of an open-hearth furnace consisted of tunnels in which fresh air passed through red-hot grilles of brick, which had been previously heated by escaping gases. A damper (C) switched the tunnel functions. The metal was melted on a sand bed (*T*) laid in an iron box, which had cooling air passages beneath to prevent its being burned out by the heat of the process. Compare this furnace with that used in glassmaking (fig. 5.15). (*Appleton's Cyclopedia*, 1880, 2:807.)

were chemically acid. In 1879, at the age of 29, Sidney Gilchrist Thomas, a clerk in a police court, and his cousin Percy Gilchrist, a chemist, discovered that base (alkaline) materials when used as a lining for converters could remove phosphorus from the iron. Only then, over 20 years after Mushet's introduction of *spiegeleisen*, did the Bessemer process—or more truly, the Bessemer-Mushet-Thomas-Gilchrist process—produce reliable steel. The Bessemer process was still restricted to using low-phosphorus (0.1 percent or less) iron if the converter were acid-lined; and to high-phosphorus (1.5 percent or more) if base-lined. This eliminated the use of the many irons with an amount of phosphorus that fell between those figures. By this time the open-hearth process, which did not rely on the material as a source of heat, could utilize iron of any phosphorus content.[46]

Bessemer's first British patent for making steel by forcing air through molten iron was dated 17 October 1855; his tilting converter received a British patent on 12 February 1856; and a more developed form of the converter received its British patent on 1 March 1860. A U.S. patent matching the second British patent was dated 11 November 1856.[47] These dates are significant because at the end of 1856 William Kelly, an American, instituted interference proceedings and claimed his own invention of the process had been prior to that of Bessemer. Bessemer's address to the British Association had taken place in August 1856 and was published in *Scientific American* on 13 September 1856. On 30 September 1856 Kelly had written *Scientific American* describing his own experiments. Hearings were held, and in April 1857 the Acting Commissioner of Patents decided that Kelly's invention had preceded Bessemer's and instructed that the U.S. patent be issued to Kelly if Bessemer did not appeal the decision within 60 days. No appeal was made.

Kelly had briefly studied metallurgy at what is now the University of Pittsburgh, and in 1846, when he was 35 years old, he and his brother bought the Eddyville Furnace in Kentucky. In the following year William Kelly began experiments on his "Air-Boiling Furnace" (fig. 4.22). As Kelly described it: "I conceived the idea that, after the metal was melted, the use of fuel would be unnecessary—that the heat generated by the union of the oxygen of the air with the carbon of the metal would be sufficient to accomplish the refining and decarbonizing of the iron."[48] The test of his first furnace was delayed for several years by the company's construction of a new blast furnace. In 1851 experiments were resumed:

On looking for the cause of failure in my experimental furnace, [I] found that my chief trouble lay in the melting department, not in the important matter of blowing into the iron, so that the question presented itself to my mind, Why complicate my experiments by trying to make pig metal in a furnace not at all suited to the business? Why not abandon altogether the melting department and try my experiments at our new blast furnace, where I could have the metal already melted and in good condition for blowing into? I fully believed that I could make malleable iron by this process. In my first efforts with this object in view I built a furnace consisting of a square brick abutment, having a circular chamber inside, the bottom of which was concave like a moulders' ladle. In the bottom was fixed a circular tile of fire-clay, perforated for tuyeres [air inlets]. Under this tile was an air-chamber, connected by pipes with the blowing engine. This is substantially the plan now used in the Bessemer converter.

The first trial of this furnace was very satisfactory. The iron was well refined and decarbonized—at least as well as by the finery fire. This fact was admitted by all the forgemen who examined it. The blowing was usually continued from five to ten minutes, whereas the finery fire required over an hour. Here was the great saving of time and fuel, as well as the great encouragement to work the process out to perfection.[49]

Shortly after he received his patent, Kelly's company was bankrupt, a victim of the depression of 1857.

It was 1864 before the first Bessemer steel was produced in the United States. Around 1870 the principal early U.S. patents expired, but until that time organizations of manufacturers, such as the Pneumatic Steel Association, held and made available the U.S. patents of Kelly (for the process), Bessemer (for the machinery used), and those of other inventors who had made major contributions to the development of the Bessemer process, as it was known. Between 1880 and 1890 the production of Bessemer steel in the United States increased fourfold and its share of the growing market for iron and steel doubled.[50] By 1890 it accounted for over half of the total iron and steel output.

In 1861 William Siemens, one of an international family of inventors, received a patent on the gas producer, which was the final step in his development of the open-hearth furnace (fig. 4.23). By making gas from solid fuel, the producer allowed the use of cheap forms of coal, and the gas flames could be extended over a shallow tank of metal. Escaping gases were drawn through tunnels containing masonry grilles that held the heat. Entering air was admitted through another tunnel of similar form. A damper permitted switching the intake and exhaust functions between the two tunnels, preheating the air that fed the gas flames and extracting heat from gases as they were expelled from the furnace. In this way the temperature of the gas flames could be raised to around 3000° F. The effect of melting iron at such high temperatures was much the same as that of puddling; carbon was burned out of the molten metal. The ability to use scrap iron in the open-hearth furnace was a significant economy. Open-hearth production of steel was not nearly so "interesting and exciting" as the Bessemer process, and the United States was slow to replace the methods and

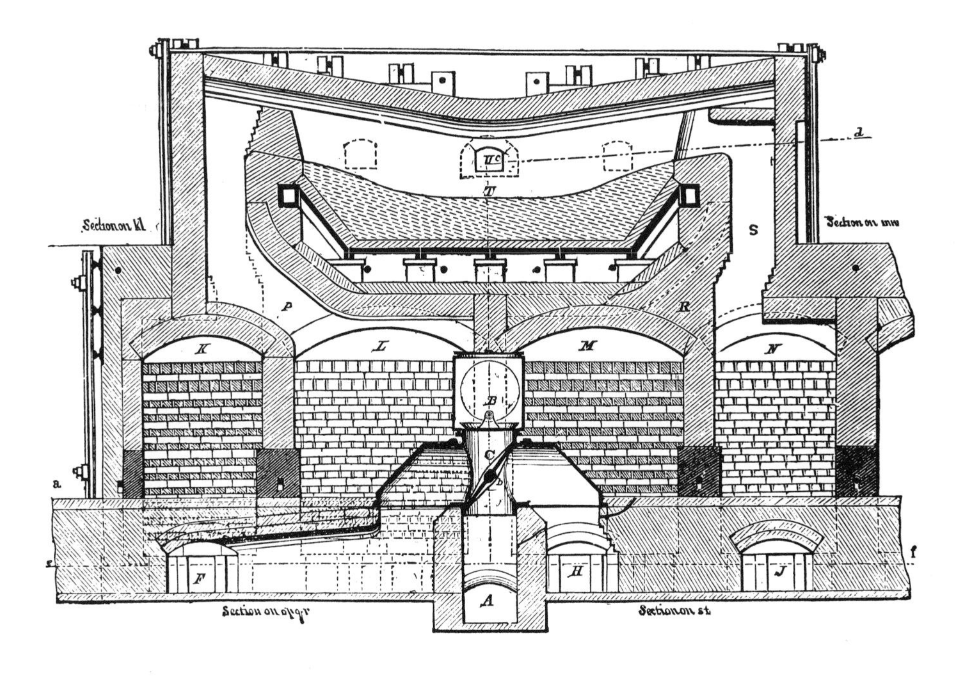

equipment it had adopted only a few decades before. Because of tariffs and price controls within the industry, it is hard to determine the relative costs of producing Bessemer and open-hearth steel in the United States. A governmental investigation around 1905 indicated that the difference in the total production costs for the two methods was probably less than 3 percent.[51]

Casting was a poor method for making thin and intricately shaped pieces of metal. By the seventeenth century there were machines in which flat sheets of soft metals, such as lead or tin, were passed between rollers to produce sheets that were larger, thinner, and more uniform than could be made by hammering. In the same period an Italian invention drew lead rods between steel rollers to form the H-shaped rods needed for fabricating windows.[52] Slitting mills used ridged rollers to cut straps and rods from sheets after they had been flattened between smooth rollers (fig. 4.24). Henry Cort in 1783 introduced grooved rollers through which wrought iron could be passed to amalgamate the lumps as they came from the puddling process. Cort's rollers produced bars of wrought iron 15 times faster than had been possible by hammering. With steam engines used as power, slitting and rolling became mechanized early in the nineteenth century, but for forcing out slag and compacting the ball of puddled iron, steam-powered lift hammers were not used until the middle of the century.

Bars, rods, and straps were rolled for those applications in which slender members of wrought iron were to be used in combination with pieces of cast iron, and plates and angles were rolled for the construction of boilers. Larger structural shapes were beyond the limits of practical production; 200 pounds was the largest mass that could be rolled, producing an I-beam 12 inches high and 6 wide but only about 5 feet long.[53]

Much of the early rolling of wrought iron was done in an effort to produce rails for the period's frantic construction of railroads. In 1820, wrought-iron rails were rolled in the form of a bulbous-headed T, the vertical leg being held by cast iron brackets. The familiar American T-rail, with a rounded top supported by an inverted T-shape, was introduced in 1830. Robert Livingston Stevens, president and chief engineer of the Camden and Amboy Railroad in New Jersey, designed the shape while sailing to England, where he went to buy rails and a locomotive. Upon reaching his destination, Stevens tried unsuccessfully to have the shape manufactured by major English ironworks, and months were spent before a Welsh ironworks succeeded in producing the rails.[54] For several decades most T-rails were imported from Great Britain, although some were rolled in the United States. The Trenton (New Jersey) Iron Works, one of the many investments of Peter Cooper, worked for at least five years before it successfully produced a T-rail in 1854. This rail, made 7 inches high, was not a successful addition to the types rolled at the time, but its use in building construction introduced the I-beam to the United States. It was planned to use this structural shape in framing the floors for Cooper Union, Peter Cooper's institution for training workingmen, but that work was delayed by the rush to replace the burned printing plant of Harper Brothers. A second delay came when the U.S. government ordered beams for an extension to the Assay Office in New York.[55] After these postponements, the wrought-iron beams were rolled for the completion of Cooper Union.[56]

Structural angles, channels, and I's were first shaped on two-high rolling mills, which had two grooved rollers,

4.24 In this early water-powered rolling and slitting mill, flat strips of metal were made progressively thinner by the rollers at the right. After they were heated in the furnace at the rear, they were cut into rods (left). Power for this work was supplied by a water wheel turning the shaft at the right of the drawing (*E*). (Diderot, *Encyclopédie*, 1762–1777, "Forges," plate 3.)

4.25 The sensuous curves of Art Nouveau designs were particularly suited to execution in wrought iron. In this railing, designed by Victor Horta, the ease of bending wrought-iron straps is exploited, and rivets are emphasized as a punctuation of the pattern. (*Art et Décoration*, January 1897.)

one above the other. A first set of grooves at one end of the rollers roughly shaped the iron as it passed through, and each of the succeeding sets of grooves along the rollers refined the shape until it reached the desired dimensions (fig. 4.26). Successive passes through the mill required that the length of iron be lifted and returned to its starting place after each pass. The metal cooled during this time-consuming process, even more quickly when small sections were being rolled, and as a result flaws could appear in the flanges. The British long preferred "reversing" mills, in which time and labor were saved by changing the direction of the rollers' rotation each time the iron passed through the mill. In 1859 a British patent was granted for a three-high rolling mill for plates and sheets in which three rollers were mounted vertically. The middle roller was geared to rotate in a direction opposite to those at the top and bottom. By raising and lowering the plates, they could be quickly passed back and forth through the mill without the wrenching mechanical strain of reversing the gears of the machine or the costly process of reheating the metal.

During the 1850s, British ironworks manufactured over three-fourths of the rails used by U.S. railroads. The rails shipped to the United States were usually the lower quality of British production, and British-made rails sold at a price well below those made in the United States. A worker in Pennsylvania rolling mills said that "had it not been for the use of putty, oxide of iron and the absence of inspectors there would have been but few rails shipped."[57] This ironworker, John Fritz, was destined to establish U.S. leadership in the development of rolling mills. The Cambria Iron Works, where Fritz was employed, produced only rails, and through the years the area's higher grade ores had been exhausted, leaving an ore that proved to be "red-short," brittle when red-hot. The losses due to flanges cracked in rolling were great, and in 1857 Fritz was instructed to build a new two-high mill. His dogged insistence that a three-high mill should be built at last persuaded the officers of the company. When completed, the mill was successfully tested on a Saturday morning, but that afternoon the building burned. The mill was rebuilt and proved so successful that U.S. companies built others within a short period. As Fritz commented at the end of the century:

With the introduction of the three-high mill in 1857, the commencement of the great improvement in rolling mills and

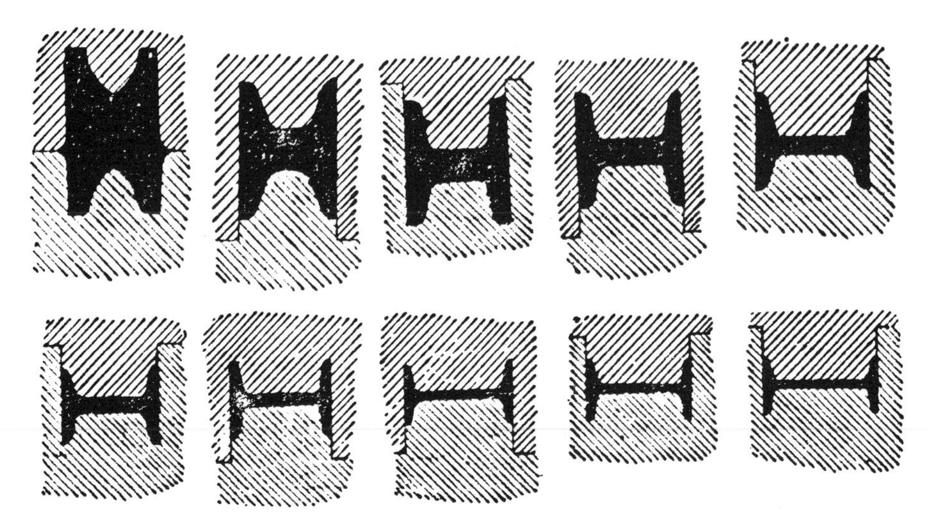

4.26 To shape beams or girders, a series of rolls was required. The rolling mill was equipped with rollers having several grooves, each shaped to execute a single step of the process. Since the metal started at white heat and was finished at a "low" red heat, it was necessary that rolling be done quickly. The upper row of roller grooves, as shown here, was used to "rough" the shape of a beam, and the lower row finished the member. (*Encyclopedia Britannica*, 1911, "Rolling Mills.")

4.27 The truss that collapsed in 1905 at Charing Cross Station, London, failed near the bottom of the third vertical from the left. Investigation showed that the cause was a hidden flaw within the tie rod, a flaw that reduced the rod's effective area to about one-third of its total cross section. (A. T. Walmisley, *Iron Roofs*, 1888.)

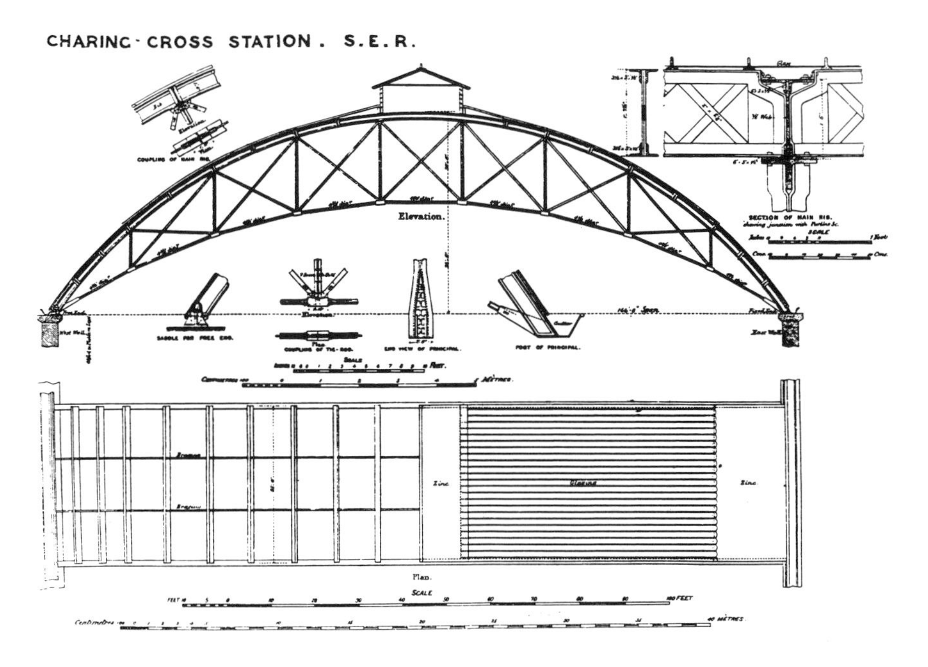

> machines connected with them took place. The rolls were made larger in diameter, better fitted up, and a more powerful and a much better class of engines was introduced, larger and better heating furnaces were built, and many labor-saving devices were introduced. But with the marvelously increased production of Bessemer steel it was evident that a larger ingot must be used in order to prevent congestion in the pit furnaces and rolls. This, of course, involved the building of larger, heavier and more rapid working machinery.[58]

The rolling of structural sections from Bessemer steel became mechanized through the introduction of electric motors and the use of many labor-saving inventions of John Fritz and his brother George. By 1900 there would be around ten workers on the floor of a rolling mill that handled 3,000 tons of steel per day, an amount equal to a year's production before introduction of the three-high mill.[59]

As methods of rolling wrought iron and steel shapes improved, manufacturers began publishing sales literature to describe the shapes they produced and indicate their prices. These publications were not always helpful. One writer characterized continental sales catalogs as "misleading" and having "little information on much space."[60] German organizations of architects and engineers, who had become impatient with the inconsistency of the data they received, insisted on standardization, and in 1879 the manufacturers who served them set standard dimensions and soon afterward published their *Normal Profile Book*, showing the rolled shapes available for structures and shipbuilding. The shapes designed for this purpose were widely adopted in other countries that used the metric system.

In the 1880s the Carnegie Steel Company published its own standard shapes and discouraged orders for other shapes by establishing minimum quantities and imposing high charges for such orders.[61] The other steel makers that had joined the Carnegie pricing combine soon adopted those standard shapes. The formal adoption of standard sections by the Association of American Iron and Steel Manufacturers took place in 1895. Nine years later the British Engineering Standards Association established similar standards. Standardization would have been impossible without the full development of rolling practices and the increased ability to control the qualities of the steel and iron produced.

In the early 1850s, at the time when architectural use of iron became significant, at least half of the world's output of pig iron was produced in Great Britain. France had not yet undergone extensive industrial growth and its railway system lagged in development. Only at the middle of the nineteenth century did French industrialization begin to demand appreciable quantities of iron, and it was then that large-scale producers of metals began to replace a scattering of small French ironworks. German industrialization was more rapid, and there the governmental role was far greater than in other major European countries or the United States. With the exploitation of rich coal deposits in the Ruhr basin, beginning in 1840, Germany moved toward increased production of iron, assisted by much scientific knowledge of metallurgy and the frequent importation of English experts. The peak of English iron production came in 1880 and thereafter it remained relatively constant for many years. During the period from 1880 to 1908, worldwide production doubled, largely due to the fact that Germany's output increased four times and that of the United States more than five times.[62]

From 1865, when the Bessemer process and improved rolling mills were established in the United States, to 1907 production of pig iron in the United States steadily grew, reaching about 26 million tons, more than 30 times the production at the start of that period.[63] A major problem in U.S. production of iron was the distance between the mills and new supplies of Lake Superior ore; the mills were located where there was an ample supply of coal but the deposits of ore had been exhausted. By 1907 there were reports that the quality of Lake Superior ore had deteriorated, but the supply was immense, and only 12 percent more ore was needed to produce a ton of pig iron than had been needed ten years before.

Despite the enthusiasm with which metal construction was accepted, not all experiments in iron and steel construction were successful. The mathematics of theorists had not been reduced to accepted formulas for the use of iron, and there was considerable disagreement about the action of structural shapes and the characteristics of the metal used. It has been said about the 1840s in England:

> It must also be remembered that bridge engineering made rapid progress, because after the inauguration of the railways, this industry was of sufficient importance to attract the best engineering brains in the country. Professional engineers, however, found nothing interesting in the simple columns and beams supporting the floors of buildings of solid construction and limited height; while the architect, with his multifarious duties, could not find the time, even were he so disposed, to master the intricacies of the new materials.[64]

The majority of engineers at that time came from a background of practical work as millwrights, instrument makers, or other crafts, and although England led in the production and utilization of iron, France and Germany were much farther advanced in the education of engineers.

A cotton mill at Oldham, Lancashire, collapsed in 1844, killing 20 workers. The structure failed at one end where, instead of the iron beam being evenly loaded by brick arches,

4.28 When there was concern about the Crystal Palace's rigidity under the footsteps of visitors to the Great Exhibition, soldiers were summoned to test the girders that were to be used. A few testing laboratories could verify the strength of materials and test the key parts of an assembly, and often bars of iron or bags of sand were stacked on portions of a completed structure to determine its strength. For an additional system used to test girders of the Crystal Palace, see figure 15.7. (*Illustrated London News*, 1 March 1851.)

transverse secondary beams rested on it in the center of the span. This produced a stress greater than that in typical spans of the building. The report prepared by a governmental investigation makes it apparent that the builder of the mill had been unaware that the beams in question had been stressed more than twice as severely as others in the mill.[65] Other mill buildings failed, usually because of similar misunderstanding of the loads that bore on the members, as well as naive interpretations of the physical characteristics of iron beams and columns.

Another kind of problem was exemplified by the failure of the train shed at Charing Cross railway station, London, in 1906. Crescent-shaped trusses, built in the early 1860s, spanned 164 feet and were made of wrought iron, tension being taken by tie rods between 4 and 5 inches in diameter (fig. 4.27). One of the tie rods failed because of "concealed flaws, which may grow in size under continued tensional stress, although in itself that stress may not be exceptional." The fault had not been detected when the tie rod was tested before assembly of the truss. The investigator's report recommended the structural safeguard of redundancy by "duplication or strengthening of the main tension members."[66] Certainly, if there were any question of the material's soundness, two or more members used together reduced the risks that might result from an undiscovered flaw.

Direct testing of structure was the usual manner of verifying the stability of a design and the quality of materials. Before the Great Exhibition opened in 1851, soldiers were marched about the galleries of the Crystal Palace and cannon balls were rolled along them to prove for the public the reliability of the structure (fig. 4.28). Two years later, when wrought-iron lattice girders arrived in Ireland for erection of the Dublin Exhibition building, officials noted that all the girders' diagonals were made of flat bars. Test girders were loaded with heaps of stone, and the bars bent under the compressive forces present. Systems of testing promised an increasing ability to avert failures.

Conservatism was a frequent, though often misguided, guard against failures. In 1859, when the Bessemer process for making steel was in its early stages of development, the Board of Trade prohibited the use of steel in English bridges, a restriction that remained in force until 1877. Unlike the early use of reinforced concrete, the introduction of steel building construction did not have the advantage of competitive enthusiasm on the part of manufacturers or engineers, being overshadowed by activities in construction of railroads and bridges.

The sole distributors of rolled joists to the building trades were merchants who neither possessed nor professed any technical knowledge. Compound girders and the majority of joists with workmanship were imported from the Continent, or at times such work as holing or cleating was performed locally by blacksmiths.[67]

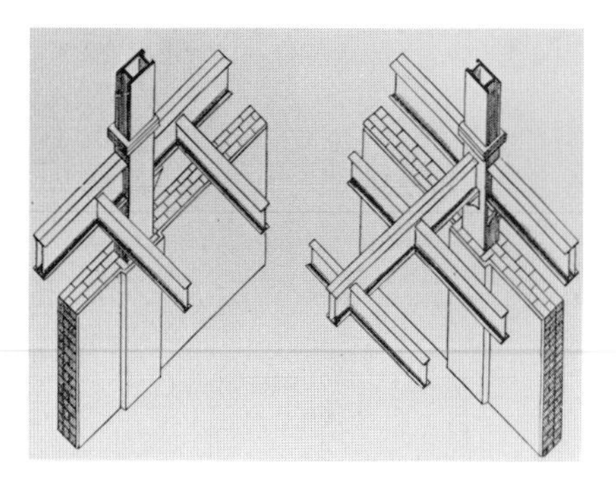

4.29 There were basically two ways of laying out skeleton construction. In the drawing at left, girders run along the perimeter of the building, supporting both the wall construction and floor beams. In the drawing at right, girders on the perimeter carry only the wall, and other girders perpendicular to them support the floors. (*History of Real Estate, Buildings and Architecture*, 1898.)

4.30 The Reliance Building (Chicago, Burnham and Root, 1895) employed terra-cotta facing over masonry walls. The framework (left) and its finished state (right) are shown for a spandrel wall beneath a window. (J. K. Freitag, *Architectural Engineering*, 1895.)

Steelmakers in the United States were more aggressive as salesmen. In 1895 an English speaker pointed out that "when anyone required an hotel or building even fourteen storeys high, he was informed by Mr. Carnegie's engineers not only how to do it, but that they were ready to undertake to do it for him."[68]

Early in the nineteenth century, wrought iron and steel could not assume forms as intricate as those that had been cast. Even elementary shapes, such as angles and T's, were beyond the capabilities of the earliest rolling mills, which could produce little more than straps and rods. In the 1830s there were attempts to fashion inverted T's by using U-shaped bolts or straps to fasten a vertical plate to a horizontal plate, but such expediencies did not utilize the full strength of the metal.[69] It was more effective to shape a T by riveting together two angles, formed by hammering plates of wrought iron. Such combinations depended largely on the use of hot rivets. Cold rivets had long been known, but hot rivets, introduced in the shipbuilding industry, provided stronger connections for the construction of bridges and buildings. As they cooled, the rivets shrank, firmly holding metal plates together and insuring frictional forces that prevented movement in the connection, a much firmer connection than had been provided by bolting cast iron. Cherry-red rivets were not so hot that they would alter the quality of the wrought iron around them, and girders and columns could be assembled from small and simple shapes. A riveted girder, about 19 inches high and 7 inches wide, was displayed at the French Exposition of 1849. Over 20 years later the situation was described by an engineering official of the government: "Since at that time there was not yet in France any application of this sort of construction and since no one could foresee the extent of advantages arising from it, nobody paid attention to this example and it was not mentioned in the report of the [exhibition] jury."[70]

England, more advanced in iron production and more actively expanding its railway system, was quick to adopt riveting. Construction of railway bridges and buildings had long been dominated by the use of trusses, arches, and girders that were made by bolting cast-iron members with tension rods and straps of wrought iron, but this combination was much less popular after the failure of the Dee Bridge in 1847. The following year the structural use of riveting was introduced in construction of the tubular wrought-iron bridge over the Menai Straits. Box girders, with plates and angles riveted together in a hollow rectangle, afforded great resistance to lateral bending in the bridge. There were shapes devised that provided tubular forms at the top of inverted T's, but they were seldom used. Instead, the I-shaped girder made of four angles and a single vertical plate was the one most often used in building, because it was easily assembled and could be painted to prevent rust.

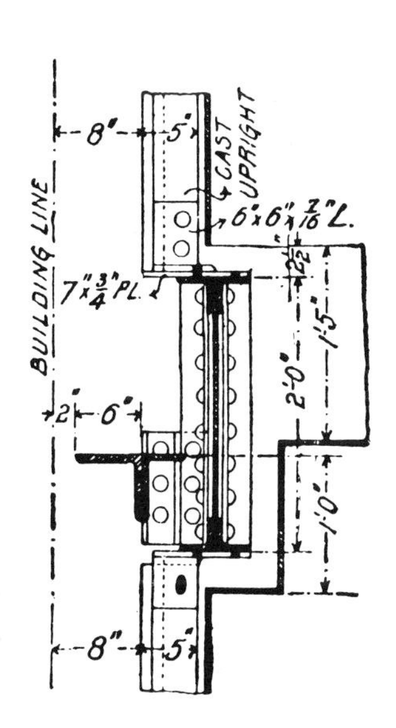

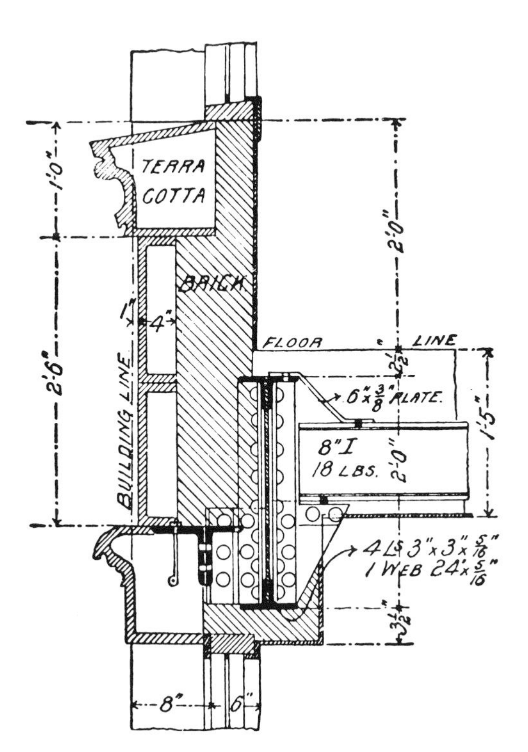

At first, riveters worked in pairs, each wielding a hammer. By alternating their blows, they could quickly spread the point of the red-hot rivet. Steam-driven or hydraulic riveting hammers were stationary and had little use in building construction. A compressed air hammer, introduced in 1865, was applicable, but it did not become common until the following century when tools operated by compressed air were more generally used.[71]

Among the heroes of skyscraper construction were the riveter and his crew. As the uppermost beam of a skyscraper was swung into place, passersby on the street could spy the riveting crew setting it in place as their forge was moved to a roost made of a few planks across beams below. A 1930 newspaper story described the work of the riveting crew:

Four men make up a riveting gang: the heater, the catcher, the driver and bucker-up. A youth in training for a riveter's career—he is called the rivet-jack—does the miscellaneous errands. He goes for new supplies and fills the water bucket. He stands beside the heater, who presides over the forge, a soft-coke burning chalice with a rotating blower that rouses the fire to efficient malevolence.

The heater puts eight or ten large rivet spikes into his forge at once, burying them in the flaming coke. When the rivets are red hot and the rest of the gang is ready for them he takes a pair of slender three-foot tongs and picks out a likely looking specimen. With an easy underhand throw he tosses it through the air, for all the world as though he were merely discarding it. The flame-colored lump of steel sails straight for the catcher's head. He observes its flight with attentive nonchalance, his arms at his sides. When the red

4.31 **Steel framework and terra-cotta surfacing meant that buildings could be erected quickly. The photograph at left shows Louis Sullivan's store for Schlesinger and Mayer (Chicago) as it appeared on 23 March 1903; center photograph on 31 March; and right photograph on 11 April. (*Brickbuilder*, May 1903.)**

hot missile is almost within six feet of his eyes . . . he lifts his right hand with an easy unhurried motion and catches the rivet in a battered tin can. Then he fishes it out with a pair of tongs he holds in his right hand, taps it on the beam he is standing on to remove any cinders and sticks the rivet into the steel plate. . . .

A moment later the roar of a riveting hammer goes up. The bucker-up holds the rivet in place with the dolly bar, a heavy cylinder of steel capable of withstanding the shattering drive on the other end. Then the driver takes the yard-long riveting hammer with its trailing air hose and smashes the end of the rivet into a cap. That takes only a few seconds. The metal cools quickly, and for a minute or two New York's most popularly denounced noise is stilled.[72]

The work combined those factors of danger and skill that excite popular admiration. In the 1920s a championship riveting gang was acclaimed for setting a record by driving 308 rivets in 37 minutes.

There were several structural advantages to the use of arc welding, which had developed from the electric arc lamp, and there were savings in the manpower required, but noise was the factor "behind most of the inquiries which are continually coming from business men's associations regarding the possibility of doing away with riveting."[73] In the narrow canyons of streets in a downtown area, the nerve-wracking clang of riveters could quickly become an issue. Welding with an electric arc had been introduced around 1885, and metal wires later replaced the use of carbon rods. It was the end of the century before acetylene was available and the welding of structural steel became practical. Some techniques of welding were mostly developed in the United States, where there was need for welded pipelines to bring natural gas to northern metropolises, and oxyacetylene welding techniques became highly developed in France. Welded boilers could withstand higher pressures than those assembled with rivets, and after 1920 welded ship construction was introduced and exploited by Germany, where "pocket battleships" could be constructed within the limits imposed by treaty at the end of World War I.[74] It was 1940 before electric arc welding of steel was well developed, spurred by wartime needs.

A first step toward skyscraper construction was the use of cast-iron interior columns rather than thick masonry walls and piers. Next, wooden girders were replaced by iron beams and wooden floor framing filled between. Then the search for fire-resisting construction led to spanning between iron beams with brick arches, hollow terra-cotta units, or reinforced concrete. A system of metal framework was developed that was fundamentally orthogonal yet could be adapted to the irregular forms that might be required by irregularly shaped building sites.

First this system took the form of "cage" construction, in which the iron and steel framework carried all floor and roof loads of a building while the surrounding masonry walls were self-supporting. This form of construction was strongly associated with New York and that city's architects, and that association may have resulted from the fact that New York building regulations stipulating wall thicknesses made little distinction between load-bearing walls and those that were nonbearing.[75] Some architects strongly favored cage construction, and in 1896

George B. Post insisted that if an exterior wall were sufficient to protect the metal frame of a building from water penetration and fire, the wall would be thick enough to support itself.[76] This, of course, could only be true within a very limited range of building heights.

The second system, surviving to the present, was called "skeleton" construction, and it supported masonry walls on the metal framework (fig. 4.29). Traditionally the thickness of masonry walls had varied with their height; a handbook from the middle of the nineteenth century recommended a thickness of at least one-twelfth of the wall's height.[77] With buildings of many stories this rule of thumb was not followed precisely, because linkage to other parts of the structure, intersecting walls, and the shape of the wall itself usually added stability. When the masonry of exterior walls was supported at each floor level, their thickness needed only to be sufficient to stiffen a height of around 12 feet against wind pressures and to protect the building's frame from water and fire.

Skeleton construction permitted flexibility in scheduling construction of a tall building (fig. 4.31). Once the steel framework was in place, work could proceed without the necessity of slowly moving upward until the structure's full height was reached. In the 1890s at least one Chicago building was enclosed in masonry from the top down, and in New York, when a contractor encountered delays in delivery of cut stone for the lower stories of his project, the masons completed the upper stories first.[78]

Columns of cast iron, the first sort to be used, were cheap and their many advocates recommended them as being more resistant to fire and rust than either wrought iron or steel. However, a single unseen flaw within a casting could cause failure of an entire column, and thicknesses were so unreliable that small holes were sometimes drilled to check them.[79] Manufacturers of cast-iron columns recommended increasing the factor of safety used, but that would have required heavier columns, making them fully as expensive as wrought iron or steel. Once buildings were so high that wind-bracing was an important consideration, it was generally accepted that the rigid riveted connections of wrought-iron columns were to be preferred to the bolted connections required for cast-iron columns. By 1890 cast-iron columns were no longer used in "important buildings."[80] Even before wind-bracing was strongly considered, cast-iron columns competed with other types. In 1862 the Phoenix column was introduced by the Phoenix Iron Company of Phoenixville, Pennsylvania. The Phoenix column was assembled by riveting together four, six, or eight pieces of wrought iron, each a segment of a circular circumference with outward flanges at the ends of the curved surface (fig. 4.32). In a later "improved" Phoenix column, vertical plates of metal were added in the joints and ran the full height of the columns. These plates were sometimes extended as brackets to which beams could be secured.[81] In 1862 the Z-bar column was patented in Germany, and it was introduced into the United States by 1888.[82] Four Z-bars were attached to a plate, two at each end of the plate, so that the resulting cross section resembled the letter H. An additional type, the lattice column, consisted of two channels (right-angle C-shapes) spaced as a rectangle with diagonal flat bars bridging the gap between them. In 1891 a Chicago engineer reported that the Z-bar column was the one most used there, but in 1924 another Chicago engineer stated that H-columns (rolled steel sections and a recent addition to the alternatives for columns) were prevalent.[83]

Calculation of the stresses in metal columns was complicated by the fact that they were often complex shapes built up from a variety of rolled shapes (fig. 4.33). Even if iron or steel members were part of the assembly, other pieces were added to make the section more nearly square and obtain a larger proportion of the metal far from its center. It was well into the twentieth century before rolling methods had advanced sufficiently to make it possible to produce wide-flange sections for columns, square H-shaped members with metal concentrated at the outer edges.

Before the advent of the passenger elevator in the later nineteenth century, commercial buildings were constructed to a maximum height of about six stories (the limit imposed by the effort required in stair-climbing), although rental area above the fourth floor was seldom profitable for the buildings' owners. Early elevators made it reasonable to extend construction to a height around 10 to 12 stores, but, according to Dankmar Adler, at that time the high price of iron discouraged construction above 100 feet tall.[84] There were other difficulties. At the height of 12 stories, exterior walls that were needed to carry floor loads occupied an excessive portion of a building's ground-level floor area, the level of the building that leased for the highest figure per square foot.

By the end of the nineteenth century, skyscraper construction methods had eliminated the necessity of massive ground-level piers of masonry, but it was widely assumed that economically sound construction of office buildings would never exceed a height between 16 and 20 stories, because beyond that height the rentable floor area consumed by elevators became excessive. Only after the development of the electric traction elevator and the introduction of zoning systems for elevator service did heights become virtually unlimited by technology.

When the height of Chicago office buildings with elevator service had reached nine stories, the upper levels were still not popular with tenants, and buildings with their foundations engineered to support nine stories were often initially limited to seven (fig. 4.35). In time, however, tenants came to prefer the light, quiet, and airy offices available at the upper levels of tall buildings. In fact, noise was a factor of sufficient importance that some owners themselves paid for the application of asphalt to the cobblestone streets alongside their buildings in order to secure quiet for offices located in the lower stories.[85] Considerable prestige was attached to businesses' having offices in new skyscrapers, and there was extraordinary profit to be made from building even 12-story structures on land that had been bought at prices that reflected the income expected from six-story buildings. More and more skyscrapers were built.

As early as 1889 there was resistance in Chicago to the construction of taller buildings.

> It came first from the owners of property on the edge of the main business district and on secondary business streets who wanted business to expand laterally rather than vertically. Second, it came from the owners of some skyscrapers already erected who wanted to enjoy a monopoly advantage. Third, it came from owners of old buildings who objected to the assessment of their land for taxation on the basis of its use for a tall building.[86]

In 1893 these groups secured passage of an ordinance limiting the heights of buildings, but by then an oversupply

4.32 The World Building (New York, 1891) used an iron and steel interior frame in which richly decorated stair railings contrasted with the rows of rivets on patented columns. The 13 stories of outside wall were load-bearing and at street level the wall was 9 feet thick. (*Engineering Magazine*, May 1891.)

4.33 The two top rows of column sections are those made by riveting together plates, angles, and Z-bars. The bottom two rows show special and patented shapes, the last two being tied together by lattices, indicated by broken lines. (W. H. Birkmire, *Skeleton Construction in Buildings*, 1894.)

of office space and a nationwide financial panic had already brought a halt to the construction of skyscrapers in Chicago. In some other cities of the United States there was similar opposition to building tall.

Until recently, the passion to build taller was on the whole limited to the United States. A British engineer could state that in 1904 "steel skeleton construction had barely got beyond the stage of rumours from America."[87] In the United States, however, activity responded to economic changes, a natural reaction since economics was the impetus driving the desire for tall buildings. Chicago in 1893 was the first major U.S. city to impose height limitations. Through the years those restrictions fluctuated from 130 to 260 feet, and in 1927 they allowed a height of 264 feet on the building line and greater height with setbacks. At the other extreme was Detroit, where it was spelled out that "Fireproof construction buildings shall not be limited in height."[88] Debate continued about the proper and economic heights of skyscrapers. Each decline in construction—World War I, the depression of the 1930s, and World War II—led to a demand for additional office space, and each boom saw buildings constructed higher than before. In 1902 the *Atlantic Monthly* forecast that "the mania for mere bigness is bound to give place to a better conception of corporate eminence," but that has not yet proved true.[89]

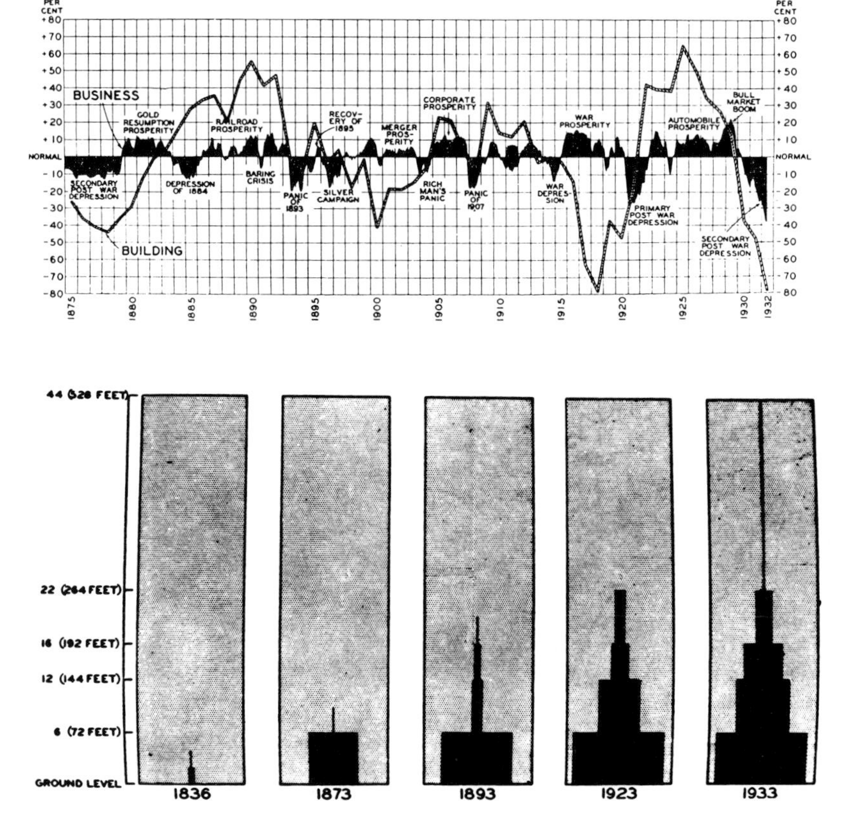

4.34 Building activity in the United States, when charted with general business conditions, indicates the periods of intense office building construction and the economic context in which they occurred. Conditions in an individual city may have varied significantly from the overall national conditions. (*Journal, American Statistical Association,* June 1933.)

4.35 A graph of the space occupied by buildings in the central business district of Chicago, 1836 to 1933. The black areas indicate the space filled by building masses at different heights of construction. (H. Hoyt, *One Hundred Years of Land Values in Chicago,* 1933. Copyright University of Chicago Press.)

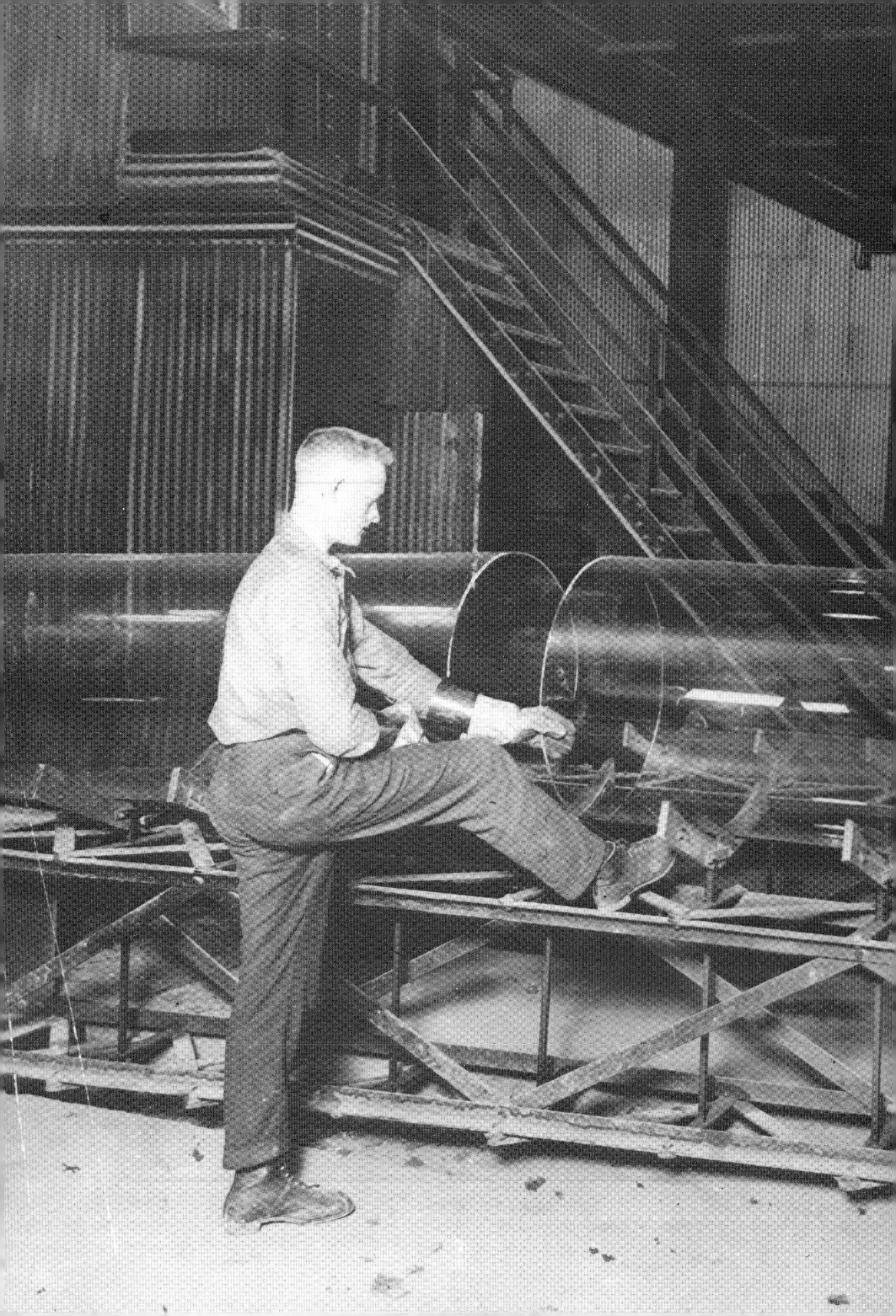

5

Glass

16th c. Both crown glass and cylinder glass manufactured in France

1589 Fifteen glasshouses operating in England

17th c. Increased use of cone furnaces in England

1615 Proclamation of James I forbids use of English timber in glassmaking

1665 Colbert attracts Venetian glassmakers to France

1690 English taxes to fund war levied on glass and windows

1848 Construction of glass-roofed Jardin d'Hiver, Paris

1850 Contract for glazing the Crystal Palace, London, awarded to Chance Brothers and Company

1856 Frederick Siemens's regenerative furnace, in which waste gases heat entering air

1860s Introduction of continuous tank furnaces

1890s Majority of U.S. manufacturers affiliate as the American Window Glass Company

1902 Machine for drawing glass cylinders patented (U.S.) by John H. Lubbers

1914 Production of drawn sheet glass commences in Belgium

1915 Production of drawn window glass begins in the United States using a development of the Colburn machine

1924 Production line for making plate glass at Ford Motor Company's River Rouge plant

1952 Pilkington Brothers Ltd. begins experiments leading to the float process

Wealthy Romans filled their windows with shutters, glass, parchment, or thin translucent slabs of alabaster. When used for this purpose, glass was cast by pouring the melted material into shallow molds or onto smoothed pieces of stone, and the sheets produced by this method ranged in thickness from ⅛ to ½ inch. Or perhaps, according to another opinion, most Roman window glass was made from blown cylinders, slit and flattened in a way that will be described later.[1] In the medieval period, linen or paper were more common materials for closing window openings. For the few who could afford glass, the window sash became a valued property. As late as the fifteenth century the casements of an English house usually belonged to the tenant rather than his landlord, and when a tenant moved he took his sash with him.[2] A sixteenth-century report on the condition of an English castle recommended that its windows be kept in place only during the presence of the lord of the castle and removed to storage upon his departure.[3] In Scotland, country houses of that period had no glass in their windows, and a royal palace had shutters in the lower half of its windows and glass only for the upper half.[4]

By the end of the eleventh century, Venice had assumed leadership in European glassmaking. Two hundred years later the craft had achieved a magnitude that attracted the attention of city officials. The Council of Ten commanded that all glass furnaces within the city be razed and rebuilt elsewhere, a ruling that caused glassmakers to relocate on the five small islands of Murano, just north of Venice. A severe fine was set for any glassworker who left Venice to work elsewhere, and in 1474 this punishment was changed to the penalty of death. The glassmaking industry of Murano prospered and became so large that more than 8,000 workers were employed, and the shops in which they worked filled a full mile along one street.[5]

In spite of the death penalty that had been imposed, Venetian glassworkers made their ways to other countries. Colbert, the finance minister who established the foundations of French industry, dispatched agents to Venice with instructions to smuggle out glassworkers who could bring Venetian glassmaking techniques to France. About a century earlier, Elizabeth I had granted a license for a Venetian craftsman to manufacture drinking vessels in a London workshop. Such licenses included the prohibition of competing imports, and English dealers in imported glassware loudly objected to the queen's action, warning of the immense quantities of wood that might well be consumed by the glassmaker's furnace. Thus began the migration to England of artisans from Italy, Holland, and France, many of the last group being Huguenots who fled religious persecution. Most countries enacted statutes that were meant to stem the flow of technical information and skilled craftsmen to other countries, but such measures invariably failed.

Always glassmakers were among the most prestigious of skilled workers. Those who made vases, goblets, plate glass, or mirrors were highly respected; the makers of bottles for wine and beer ranked low among glassmakers, although still higher than most other craftsmen. Venetian glassmakers had been raised to nobility for their craft, and some of those in France were made nobles either because of or in spite of their occupation. It is written that in Normandy the nobles who made window glass proudly disdained to blow bottles, a common use of poorer materials remaining from the production of window glass.[6] Bottles they left to the commoners employed in their glasshouses.

The social status of glassmakers may have resulted in part from the active market for their services. Window glass for buildings in England and northern Europe was purchased from other countries, but only if the building were sufficiently important to warrant the extreme cost. French glassworkers were brought to England as early as 675 to produce glass for a church, and "they not only did the work required but taught the English how to do it for themselves."[7] Despite this instruction, less than a century later German glassmakers were brought in to make windows for an English monastery.

In France, two kinds of window glass were manufactured during the seventeenth century. The glassworkers of Lorraine, many of whom came to northeast France from Bohemia, blew cylinders, which were split and flattened. The Norman glassworkers of northwest France blew a glass sphere and altered it into a disk, a method that originated in the Near East in the fourth century.[8]

Normandy or crown glass, as it was called through its long period of popularity in England, was thin, and it glistened with a brilliant finish that had been glazed by the fire. Glass cools to a rigid state without crystallization, hence it can be gradually softened by heat without reaching a melting point at which it is suddenly altered from solid to liquid. This permits slight variations in its temperature to result in degrees of change in the plasticity of the material. To produce crown glass, the watery molten glass was cooled to a thick syrupy consistency. A mass of this viscous glass was gathered on the end of a heavy blowpipe. The gatherer dipped the pipe again and again, always taking glass from the center of a ring of fireclay, floating in the molten glass, from which floating impurities had been cleared. By rolling the mass against a slab of stone, iron, or wet wood, it was formed into a pear shape or a cone as it was blown. Repeated brief returns to the heat of the furnace softened the mass so that glassblowing could continue (fig. 5.1). Care was taken to shape a pointed projection on the bottom of the vessel being blown and to maintain it as the blower spun and enlarged his work. When blowing was completed, the shape of the glass resembled "an enormous decanter with a very flat bottom and a very short neck," having the pointed projection in the exact center of the bottom. An iron rod was attached to the bottom point with a small lump of melted glass, and the blowpipe was broken off by touching the glass around it with a cold piece of iron.

Then began the remarkable transformation of a glass vessel into window panes. The glass was spun in front of flames, until centrifugal forces widened the hole where the blowpipe had been removed. Slowly the vessel flattened into a circular disk with its rim doubled over, and then it "flashed" into a flat disk, making a sound that was compared to "that produced by quickly expanding a wet umbrella."[9] This circle of shining glass had remarkably little variation in thickness, except at its very edge and its center.

The completed disks of crown glass were placed on edge in an annealing oven with the center projection (bull's-eye, as it was known) spacing each disk so that the heated air of the oven could freely circulate around it. To

prevent brittleness, it was necessary to lower the temperature of glass slowly, a process requiring one or two days, depending on the amount of glass placed in the oven.

Disks of crown glass were sometimes as large as 6 feet in diameter. A disadvantage of crown glass was the unavoidable waste that resulted from cutting rectangular panes out of a circular shape. Before commencing his task, the cutter carefully inspected each disk for imperfections within the glass and scratches on its surfaces. The first cut was made a few inches to one side of the bull's-eye, and each cut thereafter had to extend through the piece remaining at the time (fig. 5.2). Even when cutters followed the charts provided by handbooks, waste would be about 10 percent, but all scraps were saved for use in filling future pots for the furnace. Waste might be higher when it was necessary for the cutter to make certain that only the smallest panes would contain the bubbles, flecks of dust from the glasshouse floor, or streaks caused by an uneven melting of the glass mixture. A perplexing problem was "hum," a white film that appeared as the disks of glass came from the annealing oven. As Henry Chance remembered in an address before the Society of Arts in 1856:

> The history of this "hum" is curious. It arose, probably in the first instance, from the deposition of sulphur from the fuel upon the surface of the glass. It became associated with the process of annealing, and buyers fancied that the more "hum" there was upon the glass, the better the glass was annealed. The manufacturers of crown glass, ever ready to accommodate themselves to the fancies of their customers, have taken the trouble to produce an additional "hum," by the introduction of sulphur in the kiln. The members, however, of the Glass Jury of the Paris Exhibi-

tion, not being in on the secret of this hum, stoutly maintained that glass thus clouded must be bad glass.[10]

Although the film of hum could be brushed away, it was usually removed by immersing the glass in an extremely mild acid bath.

Crown glass declined in popularity on the European continent during the eighteenth century, being supplanted by cylinder glass, but in England its luster caused crown glass to dominate the market through much of the nineteenth century. Although crown glass could not attain dimensions as large as those of cylinder glass, its surfaces were clear and bright, unmarred by damaging contact with other surfaces. The English tax system in some ways favored crown glass over cylinder glass, but as late as 1856 it could be stated that it was the extreme "briliancy of surface which has enable crown glass to maintain in England its position."[11] In 1880 crown glass was still produced in England, although it was no longer manufactured in the United States.

The other kind of glass, cylinder glass (or broad glass as it was also known), was made by blowing a round-ended cylinder and trimming off the ends to form a simple tube (figs. 5.3, 5.4). The tube was then slit once lengthwise and manipulated under heat until it lay flat. Because of its contact with the surface on which it was flattened, cylinder glass could not achieve the gloss of crown glass, but the rectangular shape from which panes of cylinder glass were cut permitted larger sheets to be cut than had been possible with crown glass. Cylinder glass was considered by many to be a reasonable compromise between crown glass and costly plate glass.

As with crown glass, the gatherer dipped the flared end of a blowpipe into the molten glass within a floating ring of fire clay. For making single-strength glass, which was 3⁄32 inch thick, the pipe would be dipped about three times until a mass of 15 to 20 pounds had been gathered. For double-strength glass, which was ⅛ inch thick, the pipe was dipped into the viscous glass about five times, resulting in a mass that weighed around 35 pounds. After each gathering of glass the mass was shaped, usually by rolling it in the concave hollow of a moistened wood block, until it had reached a pear-shaped form, the dimensions of which would determine the diameter of the cylinder to be blown.

When the gatherer had completed his task, the pipe and its accumulated glass were passed on to the blower, who stood on one of the platforms that extended outward from the furnace. Alongside these platforms ran trenches about 10 feet deep, and the mass of gathered glass was suspended into the trench. As the blower swung it back and forth the weight of the

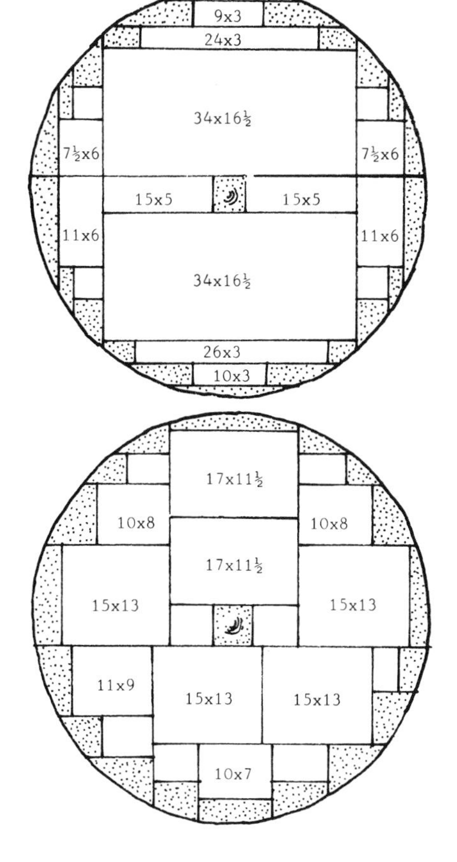

5.1 By spinning, manipulating, and reheating the vessel that had been blown, glassworkers converted it to a flat circle of glistening crown glass. When the glass had hardened, it was detached from the rod and laid in a hollow shaped in sand. (Diderot, *Encyclopédie*, 1762–1777, "Verrerie," plates 13, 14.)

5.2 *The Crown Glass Cutter and Glazier's Manual* (1835) contained a host of diagrams showing ways to divide a disk of glass. All of the diagrams have approximately the same proportion of glass wasted, but the location of blemishes and a demand for certain sizes of panes could recommend the selection of one cutting pattern over others. (Drawn from W. Cooper, *Crown Glass Cutter and Glazier's Manual*, 1835.)

5.3 The making of broad (cylinder) glass is described in this eighteenth-century illustration. In the center is shown the slitting of the cylinder, preparatory to its entry into the flattening oven, from which it will be taken as a flat rectangular sheet. (Diderot, *Encyclopédie*, 1762–1777, "Glaces souflées," plate 36.)

5.4 At the factory of Chance Brothers, blowers made cylinder glass for installation in the Crystal Palace. At the far left, workers gather and prepare the mass of molten glass, which will be swung, reheated, and blown larger by the master craftsmen. (*Illustrated London News*, 21 December 1850.)

glass pulled the blown vessel into a long tubular shape. Periodically the work was returned to the fire for reheating. If the glass were overheated, threatening to lengthen so quickly that the walls of the cylinder would become too thin, the mass was swung overhead, cooling as the weight of the glass halted overextension of the cylinder. This continued—blowing and swinging, reheating, blowing and swinging again—until the walls of the cylinder reached the desired thickness. A blower's skill was judged largely on the accuracy with which he produced glass of a prescribed thickness.

The size of cylinders increased through the years. For 1844 an average size of 40 inches long and 9 or 10 inches in diameter was reported, but about 40 years later the typical cylinder was 60 inches long.[12] The largest cylinders were blown only of double-strength glass, and their weight could be handled by few workmen. It required extraordinary strength for the blower to swing the heavy cylinder like a pendulum, to hold it before the fire, and occasionally to whirl it upward into a vertical position. Once a cylinder of single-strength glass was completed, it was allowed to cool slightly. Then the mouthpiece of the blowpipe was sealed as the end of the cylinder was heated until expanding air within the cylinder blew out the softened end. For double-strength glass a lump of molten glass was placed on the end of the cylinder, its heat having the same effect.

It was the next phase of the work that most influenced the quality of cylinder glass. The finished cylinder was trimmed at both ends and slit down its length by a diamond or a red-hot iron running along the surface. With its slit upward, the cylinder was heated until it lay flat on the table. As the cylinder began to soften in the flattening oven, it was opened until it lay on the table "like a sheet of rumpled paper."[13] The flattener used a tool to rub the sheet of glass until it lay flat, and it was then moved into the annealing oven, where its temperature was slowly lowered. It was essential that the surface on which the glass was flattened should withstand high heat and maintain a finish as smooth as possible. The most common solution of this problem was the use of large tiles of fireclay, carefully leveled and polished. This "flattening stone" lay on a bed of sand and was 4 to 6 feet in each dimension, larger than the sheet of glass it would produce. The French often placed a sheet of glass over the stone, providing a smoother surface, although the glass had to be replaced periodically. To prevent adhesion of the two glass surfaces, the flattener would throw lime into the fire, producing a fine film.

Because of stresses within the flattened cylinder, its inside circumference being shorter than its outside circumference, sheets of cylinder glass were always very slightly bowed when they cooled. To avoid breakage in shipping, it was necessary to pack the sheets with those subtle curvatures parallel, and panes of cylinder glass were set in window sash with the curve bowing outward.[14] The flattening process was subject to myriad flaws that influenced the quality of the product. In 1856 the problems of quality control were described:

> Standing before the table of the "assorter," your eye lights upon a piece which, blown under an evil star, has imbibed in the glass-house every possible defect. The founder, skimmer, gatherer, and blower, have all stamped their brand upon it. It is seedy,—the vesicles, which were in the crown tables rounded by the rotary motion of the piece, [are] here elongated by the extension of the cylinder; it is

5.5 At the far left, a copper roller is ready to spread the glass that is being poured onto the casting table. After the glass hardens, bars are removed from the edges of the sheet, and the table is rolled to an annealing oven into which workers push the sheet of cast glass. (Diderot, *Encyclopédie*, "Glaces," plates 24, 25.)

stony, disfigured with stony droppings from the furnace; stringy, thin threads of glass meandering over its surface; "ambitty," covered with stony speckles, symptoms of incipient devitrification; conspicuous with gatherers' blisters and blisters from the pipe—badly gathered; badly blown—thin here, thick there, and grooved with a row of scratches; and on this abortion the flattener chances to have exerted his most exquisite skill; it has passed through his hands unscathed, flat as a polished mirror, yet, from its previous defects, entirely worthless. Next comes before you a piece whose beginning was miraculous,—no seed, no blisters; it prospered under the hands of the gatherer and blower, and left the glass-house a perfect cylinder. But the crappie [tool] of the flattener marked it; the fire scalded it; dust fell on the lagre [flattening stone], and dirtied it; scraps from the edges of the preceding cylinder stayed upon the lagre, and stuck to it; the stone scratched it; and the heat of the annealing chamber bent it. Such are the difficulties to which every cylinder is subject—those of the glass-house, and those of the flattening-kiln.[15]

As methods were refined, the quality of cylinder glass improved. There was fortunately a considerable market for low-quality glass to be used in greenhouses, skylights, and other applications for which perfect clarity would have been a foolish luxury. English manufacturers had the opportunity of sending their worst glass to colonies, where tariffs penalized foreign competition. In fact, in grading English glass at one time the level just above "coarse" was listed as "Irish."

A market for the finest of glass rose toward the end of the seventeenth century. Courtiers, nobles, prosperous merchants, and ladies of society in France delighted in rooms with large polished mirrors and coaches with broad windows. Thick polished plates of glass were in high demand. Crown glass could not provide the desired dimensions from its half-circles, and for cylinder glass of the desired thickness, sizes were necessarily limited because the weight of a greater quantity of glass would have been unmanageable in the blowing process.

In 1676 the French government gave official support to Norman glassmakers' efforts to produce plate glass by casting. They had previously polished cylinder glass, but casting permitted them to manufacture sheets of glass that were both larger and thicker. As compared with blown glass, the casting process had the advantages of taking half the time, replacing blowers with workers at less than half the blowers' wages, and requiring fewer workers overall.[16]

For casting, a pot was heated in the furnace before molten glass was ladled into it. By levers or a system of pulleys the heavy pot was then moved to the casting table. The glass was poured between iron bars that marked out the size of the sheet and established its thickness. A heavy iron cylinder was quickly rolled over the viscous mass, achieving a uniformity of thickness seldom obtained in blown glass (fig. 5.5). Once sufficiently cooled, the sheet of glass was transferred to an annealing oven, where its temperature was lowered over the course of a week or more. Casting tables were first made with copper tops supported on masonry pedestals, a French construction that was soon adopted by the English. These surfaces sometimes cracked and in the 1840s it was possible to substitute large cast-iron plates, which were supported on rollers so that they could be wheeled from the furnace to the annealing ovens.

Once plates of cast glass had cooled in the annealing oven, they were laid on flat slabs of stone and fastened in place with a bed of plaster of paris.

Over the sheet was mounted a wheel, with one or more smaller pieces of plate glass cemented to its surface (fig. 5.6). Between the two glass surfaces were fed water and sand, increasingly fine sand being used as the grinding progressed. After grinding was finished, the glass was polished with a paste of rouge, a ferric oxide (fig. 5.7). When those surfaces were smooth and fully polished, the sheets of glass were turned to grind the other sides. Grinding was a long, tedious, and costly task until steam power was first employed toward the end of the eighteenth century, and the expense of bringing the glass to a fine polish far outweighed the economies of the casting process.

A series of improvements to this method of making glass elaborated the simple process. At the start of the twentieth century, there was introduced a continuous annealing oven in which a series of preliminary ovens were followed by a tunnel about 300 feet long. The time required to cool the glass was thereby reduced from over 200 hours to little more than 3. In addition, extensive mechanization of the grinding and polishing process reduced the time required to obtain a high gloss and greatly simplified the delicate task of turning sheets of glass so that their second surfaces could be finished.

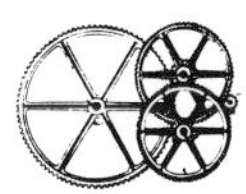

In Daphne du Maurier's novel *The Glassblowers*, a father warns his daughter, "If you marry into glass . . . you will say good-bye to everything familiar, and enter into a closed world."[17] Glasshouses were usually isolated deep in the forests that provided their fuel, and the typical work force numbered between 25 and 50. Before the French Revolution the production of crown window glass in Normandy was restricted to four families having the rank of nobility.[18] Usually many of the workers in the glasshouse were related, and the reluctance of noble glassmakers to marry their daughters to plebeian artisans meant that there were frequently blood ties among the owners of nearby glasshouses. Whatever the reason for glassworkers having been granted nobility in some countries, their economic status was below that of other nobles, who were forbidden to engage in any occupation other than agriculture.[19] In some instances owners worked alongside the craftsmen in their glasshouses, but others limited their activity to supervision of the work.

Three workers customarily made up the team charged with preparing the disk or cylinder from which panes of glass would be cut. The required mass of viscous glass was taken from the furnace by the gatherer, youngest member of the team. The material was prepared by the assistant blower, and when readied it was passed to the master craftsman. Often this team was hired as a unit and the blower paid the other members of his group. The manager of a glasshouse provided the other workers that were required. A stoker tended the furnace and was commonly second in command at the glasshouse. Potmakers prepared the crucibles in which the glass was melted. For cylinder glass, flatteners were essential, and all types of glass were cooled by those who tended the annealing ovens. Basketweavers and packers prepared cut panes to be hauled away by carters, who also brought in the fuel gathered by a force of woodcutters. All of these workers were provided lodging and food by the owner of the glasshouse. Wives and children of the workers were hired, being paid one-half and one-third, respectively, of the wages paid men for the same work.

Between the owner or manager of a forest glasshouse and his workers there was a relationship somewhat similar to that of a feudal lord and the workers of his manor. The widows of workers and their children were pensioned or provided with jobs. When business was good, workers deeply resented the importation of foreign employees, but the manufacturers found it difficult to prevent their employees' leaving for jobs elsewhere.

When the London Company established its settlement at Jamestown, Virginia, the second ship sent there brought a group of Poles and Germans who had been selected to develop the manufacture of "pitch, tar, glass, mills, and soap-ashes."[20] The glasshouse they constructed soon fell into disuse, and in 1621 (only one year before the colony's first phase was ended by massacre) Italians were brought to Virginia to make bottles and provide beads for trade with the Indians. There were other attempts to make glass before the colonies revolted, but none lasted long, and the need for window glass was satisfied by imports or the substitution of oiled paper or cloth.

The first successful window glass plant in the United States was established in Boston shortly after the Revolutionary War with generous support of the Massachusetts legislature. About a decade later a German workman was employed, and the quality of the crown glass improved until "Boston window glass" was sometimes claimed to be superior to imported glass.[21] When a company was begun in Utica, New York, the owners dispatched agents to Boston with instructions to lure away indentured workers, but these recruits were arrested before crossing the state boundary. A new plant in Virginia attracted a large number of Boston artisans but soon closed, and several small glasshouses were begun by former employees of the Boston Crown Glass Company.

Pennsylvania was to become the center of early glass production in the United States; glassmaking there was begun by Albert Gallatin a few years before he was named Secretary of the Treasury by Thomas Jefferson. This glasshouse was isolated about 90 miles south of Pittsburgh, and it did not influence development of the industry in that area so strongly as a factory later established in Pittsburgh. Operating eight pots for the production of window glass, the second company may have been the first in the United States to employ coal as fuel, a great risk for a new factory in a young industry. Pittsburgh glassmaking flourished until its glasshouses employed 169 workers in 1815, but four years later, at the lowest point of the depression that followed the War of 1812, there were only a fourth as many workers.[22] Coal and sand were available in the vicinity of Pittsburgh, but some other materials had to be brought from far away. Saltpeter came from caves in Kentucky and also from Calcutta, and pot clay was brought from New Jersey and Holland.

The major investment in establishing a glasshouse was the careful construction of the furnace. The German metallurgist Agricola described glass furnaces in his treatise on mining, *De re metallica*, which was published in 1556. The furnace Agricola described had originated in southern Europe and was shaped like a large beehive. In the lowest of three tiers, pots of raw materials were heated until the ingredients fused; the pots were then lifted to the middle level, where their contents melted and were available to the glassblowers; and the finished work was placed in the top level for annealing (fig. 5.8). Sometimes the functions of the top and bottom levels were fulfilled by separate furnaces. The beehive furnace and rectangular versions of it persisted for several centuries. Usually they were sheltered by barnlike structures from which the smoke

5.6 Cylinder glass could also be polished to eliminate traces of the flattening process. At left, small pieces of glass are fastened to stone slabs for polishing the large pieces on the tables. Several small pieces of glass could be fastened to the spokes of a wheel (right), which was spun to polish them as well as the sheet beneath. (Diderot, *Encyclopédie*, 1762–1777, "Glaces," plate 39.)

5.7 For the finest polishing of plate glass, always required for mirrors, rouge and felt pads were employed. Arcs of bent wood applied a constant strong pressure for the polishing. (Diderot, *Encyclopédie*, 1762–1777, "Glaces," plate 46.)

5.8 European furnaces of this sort were used from the sixteenth to the nineteenth century. The fire was built at the bottom level, glass was melted in the middle level, and the work was finished at the top. Beside each blower was a wet block of wood or stone against which the work was turned to achieve the desired shape. ("German" furnace, 1752. Courtesy of School of Materials, University of Sheffield.)

5.9 The masonry mass of a furnace usually occupied the central bay of the wooden structure that sheltered early glassmaking operations. Vents permitted smoke to be dissipated through the roof. (Diderot, *Encyclopédie,* 1762–1777, "Verrerie en bois," plate 4.)

5.10 This French furnace for making plate glass had space for four pots (*M*). Heat passed into kilns at the corners, which were used to fire pots and the box ladles in which glass was taken to the casting table or to preheat materials that would be melted later. (Diderot, *Encyclopédie,* 1762–1777, "Glaces," plate 6.)

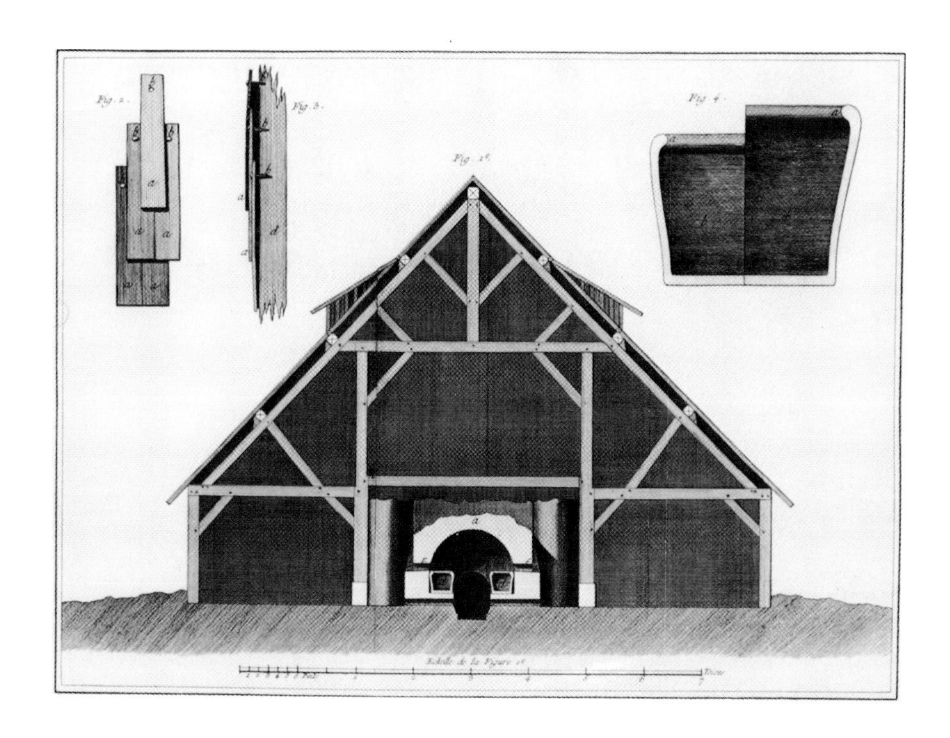

escaped through louvered openings in the roof (fig. 5.9).[23]

The use of coal as fuel led to development of the English cone furnace. There were several conditions that governed the design of efficient coal-burning furnaces. To generate sufficient heat for melting and refining the materials, coal required a stronger draft than wood. Profits rose as more pots and glassblowers could be fitted around a furnace. Waste heat could be utilized for related purposes, such as drying newly made pots, preliminary heating of charged pots, or annealing the finished glass (fig. 5.10).

The cone furnace was a tapered masonry structure, open at the top and measuring as much as 100 feet in height (figs. 5.11, 5.12). The entire cone acted as a chimney, creating a strong draft that pulled air through the furnace. Underground tunnels brought air to the coal grate, permitting large quantities of air to reach all parts of the fire, while the floor around the furnace remained unobstructed. Because soot from the conical top surface often fouled the pots of molten glass, a shallow inner dome was added with flues built around its edges. These peripheral flues drew flames from the central fire to the

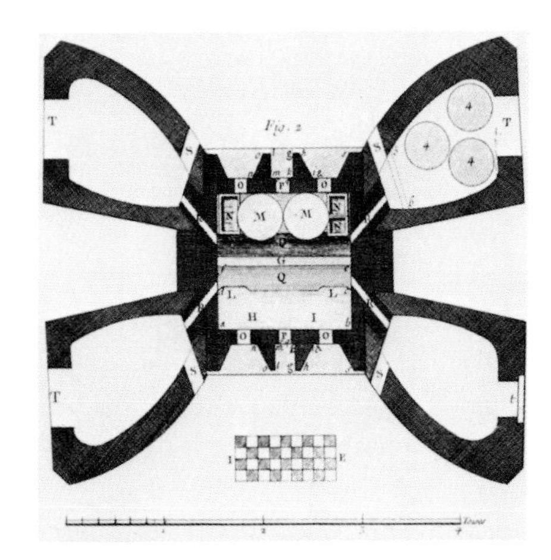

edges of the furnace, where the pots were placed. Around the base of the cone annealing ovens were built, and pots were dried on the top of the furnace. Ovens around the furnace heated pots and melted their contents before they were placed within the fire itself. These conical glasshouses sometimes fell without warning, but they continued to be used in England through the eighteenth century and well into the nineteenth. The French made little use of the conical shape, probably because their forests lasted longer than those of England and the conversion to coal came much later.

All furnaces consumed great quantities of fuel, and early glasshouses were located in forests, because it was cheaper to bring sand and the other materials to the furnaces than to transport fuel. The hungry fires were fed by felling trees around a glasshouse, until the distance to fresh fuel became so great that it was time to rebuild the furnaces in another part of the forest.

The earliest precautions against deforestation were the result of glassmakers' and ironmakers' denuding vast areas of ancient woodland. By the last years of Elizabeth I there was a remarkable increase in the use of wood to build ships and extend England's trade and to house a growing population. Consequently, the prices of firewood rose dramatically. The "Proclamation Touching Glasses," issued by James I in 1615, forbade the use of English timber by glassmakers. Experiments were made to substitute coal as a fuel in glassmaking and iron smelting, the latter a much larger industry. Smoke from coal colored the glass that was melted in open pots, and in covered pots it took longer and required more fuel to melt the batch. For some types of glass, such as that used in bottles, coal was soon adopted in England, but many manufacturers of window and plate glass continued to use wood until the middle of the eighteenth century.

France, with its broader areas of wooded land, waited well into the nineteenth century before a shortage of wood forced the adoption of coal as a fuel for glassmaking. As late as 1829, the Manufacture Royal des Glaces at St. Gobain melted glass with coal but used a wood-fired furnace for fining, the process in which bubbles were eliminated by stirring glass at high temperature and low viscosity.[24]

The use of coal put an end to small glassworks situated in forests, and the

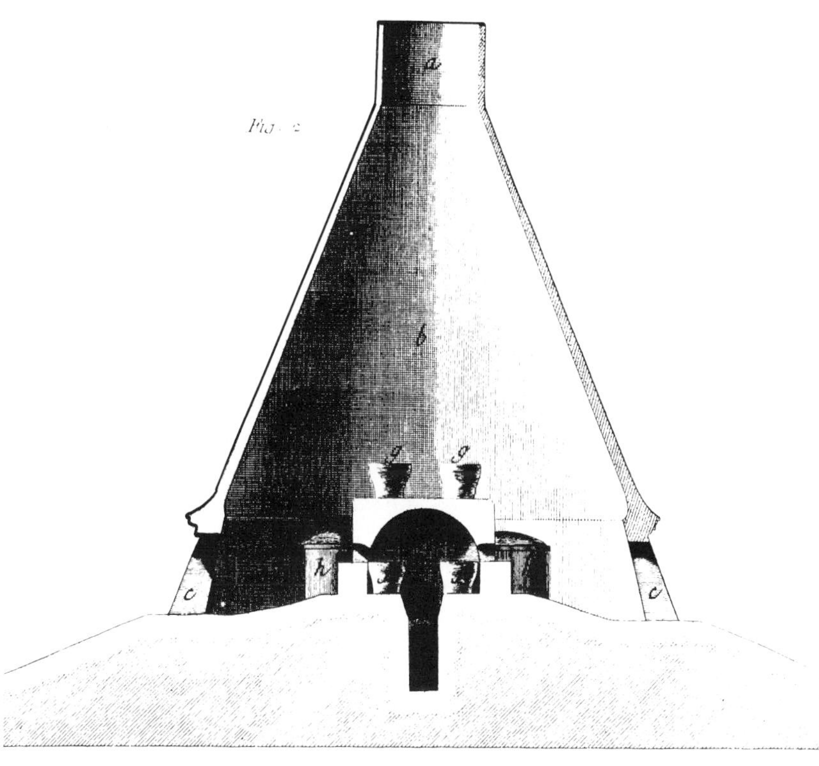

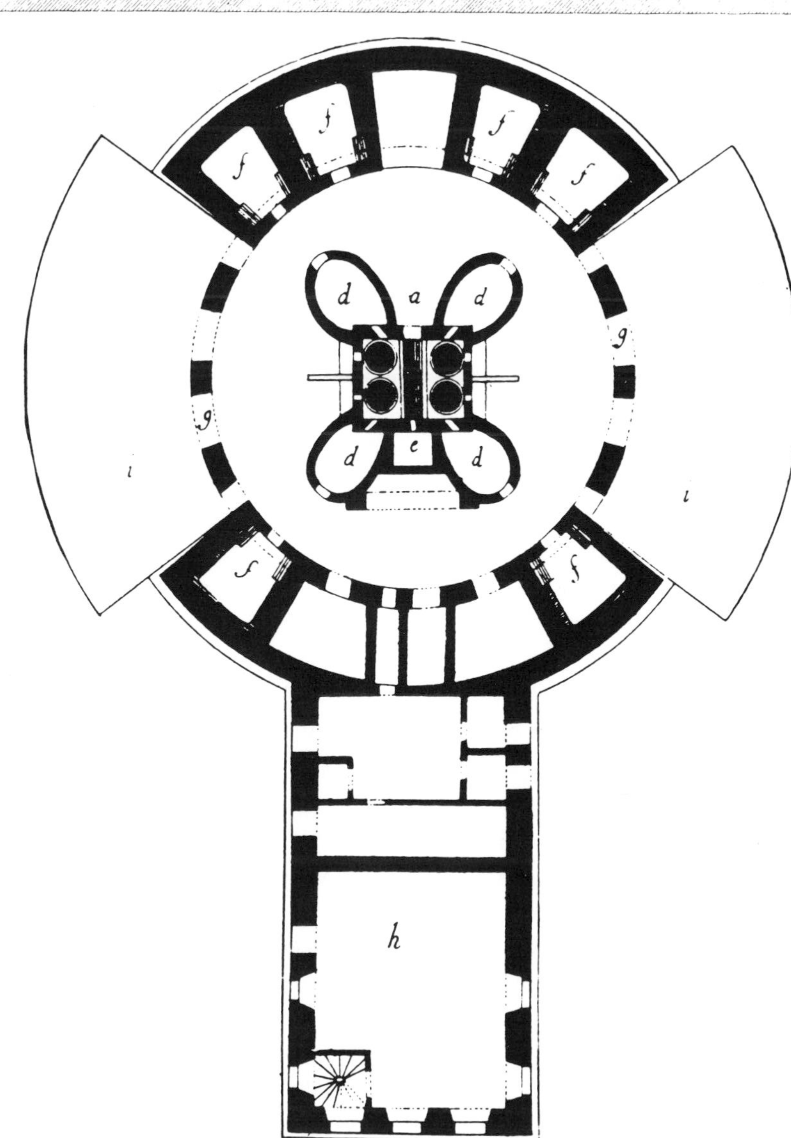

5.11 The dramatic silhouettes of English cone furnaces were frequently seen along the canals and railroads of the eighteenth and nineteenth centuries. A chimney in itself, this form of glasshouse afforded the strong draft required by the use of coal as fuel. (Nineteenth-century engraving. Courtesy of School of Materials, University of Sheffield.)

5.12 A working furnace stood in the center of the English cone furnace, and air reached the fire through an underground tunnel. Ancillary spaces were often attached to the simple form of the furnace, providing storage for materials and rooms in which the business of the glassmaker might be transacted. (Diderot, *Encyclopédie*, 1762–1777, "Verrerie anglaise," plates 2, 3.)

industry's locations came to be more commonly determined by the availability of coal and access to shipping. In the United States the area of Pittsburgh and the Monongahela River became a center of glass manufacturing, with window glass almost half of the area's glass production during the nineteenth century.[25] Even when timber or bituminous coal was close at hand, in the middle of the nineteenth century a typical window glass plant paid more for fuel than for any other material used, about one-sixth of the total cost of ingredients.

Table 5.1 Various Compositions of Glass

	Sand	Lime	Soda
1880 (U.S.)	**73%**	**13%**	**13%**
1892 (Austria)	**68.5**	**10.5**	**21.5**
1925 (U.K.)	**72–74**	**12–14**	**12–14**

Window glass is made principally of sand, lime, soda, and other materials that usually amount to less than 1 percent of the mixture. Chemical analyses vary, but not greatly (table 5.1).[26] Sea and river sand were early employed in glassmaking, and where good sand was not available, ground flint and quartz might be used. Much of the sand used for glass, including France's famous Fontainebleau sand, was obtained by grinding quarried sandstone. Where local material was not satisfactory, sand was imported for making glass of the best quality. For plate glass some English manufacturers obtained Belgian sand, and German glassmakers long employed French sand.

A major early source of the soda needed for glassmaking was the ashes of kelp or plants growing at the shore. The French government offered a prize for the invention of a method for making soda from salt, and Nicolas LeBlanc won the prize in 1792. His process simplified the manufacture of soap, a major step in the improvement of public health, and provided soda for glassmaking. Toward the end of the nineteenth century a Belgian chemist, Ernest Solvay, introduced a method of producing soda from the ammonia by-products of manufacturing illuminating gas. The Solvay process produced a less expensive soda that soon replaced LeBlanc soda, which had often contained impurities that colored glass.[27]

Lime, a necessary ingredient to harden window glass and protect it from weathering, was to be found in most regions. In some locales, lime was prepared from mollusk shells, a method that had been used since Roman times. Every batch of glass also included cullet, waste glass resulting from breakage in the glassworks and from cutting sheets to size. Cullet took advantage of waste and sped the melting of the materials.

A greenish color resulted from the presence of iron in sand, but when manganese was used to counteract it the glass tended to acquire a purplish tone. Arsenic was often used to bleach the glass, eliminating the greenish cast of the cheapest kinds. The addition of magnesia assisted in controlling the viscosity of molten glass. A slight excess of manganese lent a pink cast to the glass, a shade that was briefly popular in the United States because the tinted light that fell through those windows was thought to be flattering to women.

A major expense of the glassmaker was preparing the pots in which glass was melted, costly when the pots proved to be sound, even more costly when they failed. Henry Chance described the problem in 1856:

> It was truly remarked to a manufacturer, at a period when such calamities were frequent, "Your pots break, because they break."

The breakage of a pot often disturbs the furnace to such an extent, that the breakage of others follows, and many weeks will sometimes elapse before the disorganization thus produced can be rectified. The loss of the pot and the "metal" contained, is nothing as compared with the injury which the glass in surviving pots, and these pots themselves, are apt to sustain.[28]

At that time the average useful life of a pot was considered to be seven weeks. Later in the United States a pot, when fired with coal, was expected on the average to last four months, and a major advantage of the control available when firing with natural gas was the fact that pots might even survive six months of use. Since it required several months to prepare a pot for the furnace, the manufacture of pots was a significant part of the manufacture of glass.

Pot clay was carefully chosen. Each European nation had its own most respected clay, and in the United States before World War I more than half the clay was imported from Germany, until the cessation of imports led to the development of American clays.[29] As late as 1887, clay would be soaked for months before a barefoot workman trampled it into a smooth mixture without air bubbles. Over a period of almost six weeks the pot was built up until it was almost 3 feet high and 3½ feet wide at the top, with the walls of the vessel between 3 and 4 inches thick. The pot was fired after drying for four to eight months, and the inside surface was glazed with molten glass. The period between mining the clay and finishing the pot might easily total three or four years.

During the seventeenth and eighteenth centuries, glass and windows were still viewed as luxuries, and as such governments found them attractive sources of revenue. The English government, searching for means to finance the battles of William III, in 1690 imposed taxes on both glass and windows. The glass tax was discontinued nine years later, but the window tax remained until 1851, when it was replaced by a tax on inhabited house.[30] Factories, warehouses, commercial buildings, and empty houses were exempted from the window tax, and the number of taxable windows in a residence (the number above ten in a dwelling after 1747, and above seven after 1766) determined the tax, which was paid by tenants rather than their landlords. Between 1776 and 1808 the tax on a house with ten windows grew sevenfold, and often some window openings in large houses were filled with brick in order to reduce taxes. It was estimated then that, as a result of the window tax, English houses had half as many windows as were present in comparable European houses.[31]

The English tax on glass was reinstated in 1746, and it applied to imported glass as well as the output of English manufacturers. Because it was levied by the weight of glass produced, this tax encouraged continuation of the manufacture of crown glass, which was thinner than cylinder glass. The tax was calculated on the weight of entire disks of crown glass, but even after the inherent waste of cutting panes of crown glass there remained an economic advantage. Later a rebate of one-third of tax charges, based on the amount of waste in cutting crown glass, was also granted for cylinder glass, although there was little waste in cutting it. Thus the rebate was, in effect, a reduction of tax for cylinder glass, and this financial incentive persuaded many manufacturers to shift from crown glass to cylinder glass.[32]

The English tax on glass was high, often equal to half the cost of production and once going as high as three times the cost. In 1777, when funds were needed to fight the American Revolution, the glass tax was doubled, and another major increase took place in 1812, during the costly Napoleonic Wars. The English building boom of the 1820s, which stirred a large increase in the production of window glass, was fostered by the window tax being lowered by half and the permitted number of untaxed windows being increased.

English glassmakers opposed the window tax much more strongly than they opposed the tax on glass itself. Repeal of the tax on glass would have exposed English glassmakers to competition with foreign sources. But it was the tax on glass that was repealed in 1845, probably because it produced less than a fourth as much revenue as the window tax. Prices of window glass fell by more than half, and at the same time sales were greatly increased by the orders of London builders who were hurriedly completing buildings in the months before the Building Act of 1845 imposed new regulations on construction. For this greatly increased level of glass production there were too few qualified workmen, a shortage that forced Belgian manufacturers to raise the wages of their workers in order to prevent their leaving for jobs in England.[33]

In the United States cheap fuel was available, but that advantage was outweighed by high wages. A box of window glass made in England often cost two-thirds the price of an equal quantity produced in the United States.[34] Protective tariffs began in 1820, and imported window glass became an even smaller portion of consumption in the United States, dropping to about 2 percent around 1830. Under strong tariff protection until 1846, the window glass industry improved in the number and qualifications of workers available, but procedures were little improved. Then tariffs on glass were lowered to 20 percent, and duties continued low until the Civil War, when high tariffs on imported window glass were a means of increasing government revenues. Until the tariff act of 1913 brought about a drastic reduction of duties on window glass, U.S. industry worked under distinct advantages. Imports still came from Belgium, particularly in the small sizes of panes that were least profitable for American manufacturers.

At the end of the nineteenth century, prices for window glass in the United States were largely determined by prices of competing imports. Along the seacoasts, prices for domestic window glass were low in order to compete with the imports that arrived there. Railroad charges for transporting imports increased the price of imports inland, and prices for domestic glass increased accordingly, although the glass factories might be close at hand. For instance, the price of window glass in Pittsburgh, then a center of glass production, was appreciably higher than in Boston or Philadelphia. Protection was so thorough that in a two-year period of the 1890s manufacturers' profits tripled, although at the time the wages of glassworkers in the United States were two to four times those of foreign workers.[35]

After World War I, the reduction of tariffs produced sharp competition with European manufacturers, particularly those in Belgium. Foreign producers mechanized their production, matching the improvements made in the United States. Transatlantic shipping costs were so low, in comparison with rail charges, that in major American ports the shipping costs for Belgian glass could be lower than the cost of bringing glass from certain American centers of domestic production.

By the 1880s the unions that had been established by blowers, gatherers, cutters, and flatteners in the American glass industry joined the national organization of the Knights of Labor. The aggressive stance of glassworkers is apparent in their affiliation with this militant group rather than with the craft union of the American Federation of Labor.[36] At the same time, the American Window Glass Manufacturers Association was formed. The Association established price schedules, determined production quotas for all member plants, and negotiated with the union for all member companies. For over 20 years, the bilateral monopoly of these two organizations, management and labor, controlled most of the glass industry in the United States. In time, this comfortable relationship was threatened, principally by the manufacturing organization's being unable to control its members. During the 1890s about 20 of the largest window glass producers merged into the American Window Glass Company, which developed the Lubbers system of mechanically blowing cylinder glass and thereby eliminated many of its competitors.[37] Prices of window glass were elevated, but because window glass was so small a part of building costs, price increases as high as 50 percent did not cause a significant reduction of purchases.

The remaining independent companies in 1909 formed their own selling organization, but when it merged with the American Window Glass Company the combination was dissolved under anti-trust laws. Introduction of the Fourcault and Libbey-Owens methods of drawing sheet glass broke the tyranny of the American Window Glass Company, which had been largely based on its control of the Lubbers patents. Charges of violating anti-trust laws and conspiring to restrain trade were leveled at both the unions and the manufacturers. Most of the problems arose as a result of the industry's extremely rapid conversion from handcraft methods to mechanization, which made it possible to produce window glass at half the cost and eliminated the jobs of many highly trained craftsmen.[38] With self-imposed limits placed on the amount of glass that plants could produce, their operation had become so unprofitable that in 1922 one-seventh of American window glass factories did not plan to operate the next year.

Monopolistic maneuvers were not limited to the United States. American investors in 1902 explored the possibility of purchasing all of the Belgian glass factories, which were then supplying around 80 percent of American imports of window glass. Shortly thereafter Belgian manufacturers organized a glass trust, and all but one agreed to "the determinations of a committee with reference to selling prices, purchasing raw materials, and fixing wages."[39]

There was an old saying that three generations were necessary to make a glassblower, the elite of the glassworkers. The wages paid blowers were usually "higher than those for almost any other class of labor."[40] In fact, the owner of one American glass factory expressed the opinion that many glassblowers lived at a level of comfort rivaling that of their employers. An 1880 report on Belgian workers describes them as "well housed and well fed. . . . Among the dwellings of the well-to-do, the finest buildings belong to them. The blower is part of that class of workers who, by reason of their high salaries and the comfortable situation in which they find themselves, have acquired good manners and a degree of education."[41] The same report described their physical condition:

They are usually large or at least almost always exceed the average; even the young apprentice blowers, called *gamins*, are distinguished by their height from other workers of that age and they carry the stigmata of their profession on their cheekbones, the tip of the nose, the chin, or the forehead.

The heat of the fire, which they go very near to, has reddened those parts of the face that protrude. . . .

The walk of the blower is remarkable. He walks with arms held away from his body; a position due to the large muscles in his arms and particularly in his shoulders. At the same time he gives a characteristic swing to his body, which reminds one of the swing he gives to the cylinder of glass.[42]

Emphysema, neuralgia, tuberculosis, and cataracts were foremost among the common maladies of the blowers, none of them surprising in consideration of the environment in which they worked. Alcoholism was a common problem and, on the whole, glassblowers were said to age quickly and die young.

The U.S. Census of 1880 showed that in window glass factories, where blowing the cylinders that produced sheet glass required greater ability than any other major branch of the industry, the requirements of experience and skill were reflected in wages. Master blowers received the highest daily wages (averaging $5.47); flatteners, whose work to a great extent determined the quality of the product, received the next highest wage ($3.82); cutters, less important for cylinder glass than they had been for crown glass, received the third level of wages ($3.14); and gatherers, who usually aspired to become blowers, earned an average of $2.72.[43] These wages can be compared to the average wage of $3.90 for managers of glasshouses in which they worked.

Before the introduction of the continuous tank furnace, a blower's team could not start its work until the glass was melted, but the time that this would take could not be accurately predicted. Filling the pots and melting their contents usually took about 24 hours. Once the glass was ready, a boy was sent to inform the blower's team, day or night, to appear at the glasshouse and begin their work. It might take about 8 hours for the pots to be "worked out," and only four or five work periods could be fitted into the week. Belgian glassblowers of the nineteenth century followed a somewhat different schedule, reporting for four 12-hour periods during a week but never working on Sunday. Most of them lived on small farms some distance from the glasshouses that hired them, and their weekends were reserved for planting and harvesting. Scheduling production was even more difficult because glasshouses always closed during the summer months, when the heat of the furnaces was absolutely unbearable. By 1899 many glassworkers' unions in the United States were sufficiently strong to prohibit glassmaking in July or August and to stipulate limitations on the quantity of glass to be produced, the number of apprentices that would be accepted, and the hiring of foreigners in the factories.

"Little Ice Age" is a term often applied to the period from 1550 to 1850, when Europe was subject to harsh winters and brief cool summers. To capture precious warmth during the short growing seasons, the Dutch developed the greenhouse.[44] The extravagance of hothouse fruit was attractive to wealthy merchants, and during long gray winters warm plant-filled rooms were a welcome respite from dank interiors. The most wealthy built orangeries in which their displays of tubbed orange trees were less a horticultural pursuit than a backdrop for entertainment—the luxury of fine gardens being provided indoors for enjoyment in inclement weather.

Manor houses of Elizabethan England usually had included long many-windowed galleries providing a place for strolls that could not be taken in cold and wet gardens. The old slowly walked up and down the gallery, while an attendant read to them; children played their games; and ladies chatted as they promenaded back and forth, passing the idle hours required by their station. By the early part of the nineteenth century, conservatories became a requisite of upper-class housing. Attached to a house, these glass rooms brightened adjoining spaces and provided leafy retreats from formality. House plants came to be a popular decorative element in the last half of the nineteenth century, lending a romantic informality to any room. Among the English middleclass it was said that each £100 of a family's income was indicated by a leaf on the parlor aspidistra, a house plant that had little to recommend it except its ability to survive the fumes of gas lighting.[45]

Large public conservatories were built and attracted streams of visitors who chatted, sipped refreshments, and listened to music among the exotic plants. Probably the most successful of these was the Jardin d'Hiver, built in Paris in 1848. It encapsulated formal gardens and a picturesque *jardin anglais*, which could be viewed from galleries above. The structure was cross-shaped, 300 feet in one direction and 180 in the other, and the roof was well above the top of a Norfolk Island pine that stood 50 feet high.

The ultimate pleasure garden was the Crystal Palace, designed for the 1851 Great Exhibition of the Industry of All Nations.[46] As the first of a succession of iron-and-glass exposition structures, the Crystal Palace pioneered a vast scale of construction and relied heavily on the experience of its designer, Joseph Paxton, in erecting greenhouses for the horticultural enthusiasms of the Duke of Devonshire. Construction of the project demanded speed and technological ingenuity (figs. 5.13, 5.14). Charles Dickens described the commitment that was required:

> Two parties in London, relying on the accuracy and good faith of certain iron-masters, glass-workers in the provinces, and of one master carpenter in London, bound themselves for a certain sum of money, and in the course of four months, to cover eighteen acres of ground with a building upwards of a third of a mile long (1,851 feet—the exact date of the year) and some hundred and fifty feet broad. In order to do this, the glass-maker promised to supply, in the required time, nine hundred thousand square feet of glass (weighing more than four hundred tons) in separate panes, and these the largest that ever were made of sheet glass; each being forty-nine inches long. The iron-master passed his word in like manner, to cast in due time three thousand three hundred iron columns varying from fourteen feet and-a-half to twenty feet in length; thirty-four *miles* of guttering tube, to join every individual tube together, under the ground; two thousand two hundred and twenty-four girders; besides eleven

hundred and twenty-eight bearers for supporting galleries. The carpenter undertook to get ready within the specific period two hundred and five *miles* of sash-bar; flooring for a building of thirty-three millions of cubic feet; besides enormous quantities of wooden walling, louvre work, and partition.[47]

The contract for glass was awarded to Chance Brothers and Company. A rival bid, only slightly higher, offered sheets 2½ times as large and over 1½ times as thick, made by rolling glass on a casting table. Although this glass might have offered greater protection from hail and would have required fewer wooden sash bars for mounting, its greater weight was feared. Chance Brothers supplied sheets 49 inches by 10 inches, produced by the cylinder process, three pieces being cut from each cylinder.

Lucas Chance owned a small glassworks, and in 1830 he and his partner had traveled to France to observe the production of cylinder glass at Georges Bontemp's factory outside Paris. Crown glass was favored in England, but the taxes on glass made it more profitable to manufacture cylinder glass. With blowers and flatteners sent by Bontemp, Chance began making cylinder glass in 1832. Production expanded, but the contract for the Crystal Palace required that in a few months they produce about twice the normal output of the five furnaces that were devoted to making cylinder glass. With the assistance of Bontemp, who had closed his Paris works following the Revolution of 1848 and joined Chance Brothers, additional workmen were recruited from France. Between August 1850 and February 1851 about 300,000 sheets of glass were provided for the Crystal Palace, and at the same time the company continued to fill its other orders.[48]

The efficiency of glass production was drastically advanced by improvements in furnaces. In the first decades of the nineteenth century, much attention had been given to achieving a scientific understanding of heat, a study fostered and financed by the opportunities for immediate application in

5.13 The Crystal Palace covered 17.5 acres of London's Hyde Park and enveloped full-grown trees. Prefabricated members of iron and wood supported the structure, and glass enclosed it. This was the largest order of glass that had ever been filled by a glass manufacturer, and much of it had to be replaced when it was broken by workmen rushing to complete the building within the scheduled time of 17 weeks. (E. Walford, *Old and New London,* 1872–1878, vol. 5.)

5.14 The glazing wagon that sped the roofing of the Crystal Palace rolled on wheels placed in the gutters that drained the roof. With so much glass to set in place, little effort was wasted. In fact, a worker set an average of 108 panes in a single day. The wagon had arched members overhead for canvas to protect the workmen, so that work need not be interrupted during the short construction period in the fall and winter of 1850–1851. (*Illustrated London News,* 7 December 1850.)

steam engines and smelting. At that time the accepted theory of heat pictured it as a sort of invisible substance referred to as "caloric." With the foundation of the science of thermodynamics, a relationship could soon be made between heat and mechanical energy, and the mysteries of steam and furnaces began to be unraveled. Frederick Siemens in 1856 patented a regenerative furnace, one in which entering air was heated by waste gases. Beneath the furnace two tunnels were built with grilles of firebrick across them. Hot gases left the fire through one of the tunnels, and fresh air entered through the other tunnel. When the walls of the tunnel exhausting gases reached a high temperature, a damper switched the direction of air flow. Entering air was then warmed by the masonry surfaces of the heated tunnel, while escaping heat raised the temperature of the cooler tunnel (fig. 5.15). This system required about half as much fuel as previous furnaces.

With the use of natural gas or producer gas, flames could be directed at the top of a tank furnace. In the "day tank" a batch of material was processed much as in a pot furnace: melted, refined, and then cooled to a viscous state. This work was done at night and a tank of molten glass was ready for blowers at the start of their workday. As with any furnace or kiln that is heated intermittently, a large part of the fuel used in a "day tank" was actually expended on warming the mass of the furnace, rather than in melting glass.

A tank furnace for continuous operation was soon developed. The first segment of the tank was used to melt materials with the heat of gas flames overhead. A dividing wall between the first and second chambers allowed melted glass from the bottom of the first chamber to flow into the top of the second chamber, where it was refined by being heated to a higher temperature at which bubbles and impurities would rise to the top. After it passed another dividing wall, glass in the third chamber cooled to the viscosity required for blowing.

The tank furnace used for glassmaking, with gas heat from the top, was essentially the same as the open-hearth furnace used for making steel, and as the glass moved from one end of the tank to the other, it passed through all the traditional stages of preparation. Early tanks kept molten glass only about 10 inches deep. By 1879 that depth was almost doubled,

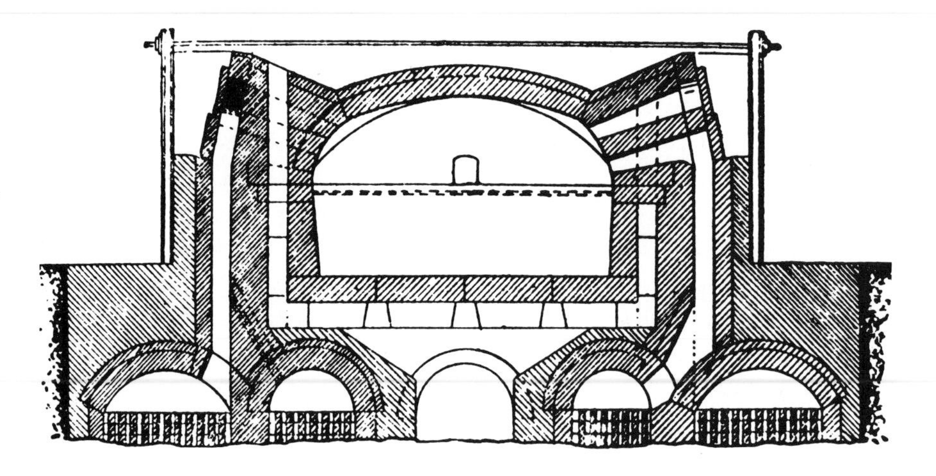

5.15 Air passages around the Siemens Continuous Tank-Furnace prevented temperatures that might damage the tank's walls. Tunnels beneath the furnace preheated entering air and expelled hot fumes from the fires. The tank itself (center of drawing) was divided into three sections. In the first, materials inserted through a door at the rear were melted; in the second, heat was raised and the material refined; and glass was drawn from the third, which was cooler. This continuous tank furnace used in glassmaking closely resembles the open-hearth furnace used in steelmaking. (*Appleton's Cyclopedia*, 1897, supplemental volume.)

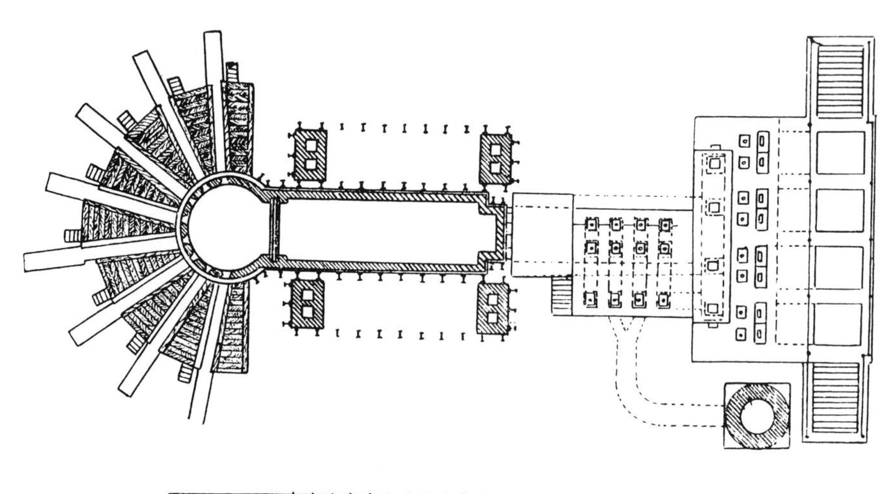

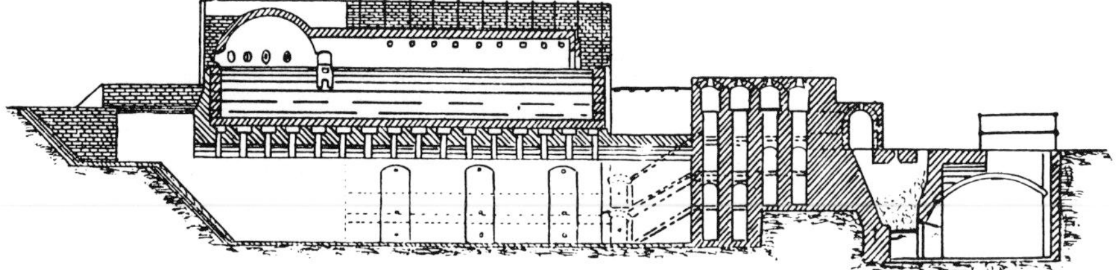

5.16 Before the introduction of mechanized glassblowing, the continuous-tank furnace was highly developed. In this Belgian furnace the tank was fed from the right, and the prepared glass was kept in a circular portion of the tank, from which the platforms and trenches for eight blowers radiated. During a day of continuous operation, 24 blowers manned the furnace. (*Engineering Magazine*, November 1898.)

and in a Belgian furnace about 6 feet deep the glass almost solidified at the bottom of the tank. In practice, usually only about a twelfth of a tank's capacity was used in a 24-hour period, and each of the "working holes" at the end of the tank could be manned by a daily schedule of three shifts of blowers.[49]

The advantages of continuous tank furnaces fired by regenerative systems were many (fig. 5.16). By 1880, less than two decades after its first trials, the advantages of the continuous tank were spelled out in a U. S. Census report on the manufacture of glass:

1. Increased power of production, as the full melting heat may be employed without interruption, while with the old method of melting nearly one-half of the time is lost by cooling and settling the metal, the working out of the glass, and the reheating of the furnace.

2. Economy in working, as only one-half the number of men are required for the melting operations.

3. Durability of the furnaces, owing to the uniform temperature to which they are subjected.

4. Regularity of working and improved quality of the glass made.

5. Convenience to the men and advantage to the manufacturers, as owing to the continuous action the metal is always ready for the blowers, and the gatherers can draw the metal from a practically constant level.

6. For the manufacture of window glass the working-out end of the furnace may be so arranged that the blowers can work without interfering with the gatherers.[50]

Near the end of the nineteenth century, a Belgian continuous tank furnace served as many as 45 blowers, 15 in each shift. In the United States, where introduction of the continuous tank furnace had been closely associated with the shift to natural gas as fuel, this was a common capacity and some furnaces had the space for 20 blowers at a time, arranged around the circular end of the tank.

The glass industry had long needed a fuel that could provide high temperatures with little contaminating residue. Producer gas first offered these advantages. To provide this gas, coal was burned in a sealed container separate from the furnace. When air and steam were passed through the thick layer of glowing coals, the gas obtained usually contained about 80 percent of the energy in the coal, roughly double the energy obtained in making coal gas.[51] Manufactured gas had obvious advantages, especially if the gas producers were located near the glass furnace so that the initial heat of the gas was not diminished. The combination of producer gas and the regenerative furnace was quickly adopted in Europe, particularly in England and France, but few were installed in the United States because their construction was costly, and the low price of fuel in America did not demand stringent economy in its use. Producer gas required shutting down weekly to clean soot from the chimneys, and many European manufacturers later converted their plants to the use of fuel oil.

French and Belgian glass had long excelled in quality, largely because of those countries' artisans' superior skill in filling and firing furnaces, but the discovery of vast deposits of natural gas in the United States gave the latter a great advantage. Manufacturers in the United States were persuaded to move their plants to areas where natural gas had been found. During the 1880s, the first decade in which natural gas was used, the number of glass factories in Ohio and the value of their output more than tripled. In Indiana during the same period, the number of glass factories rose from

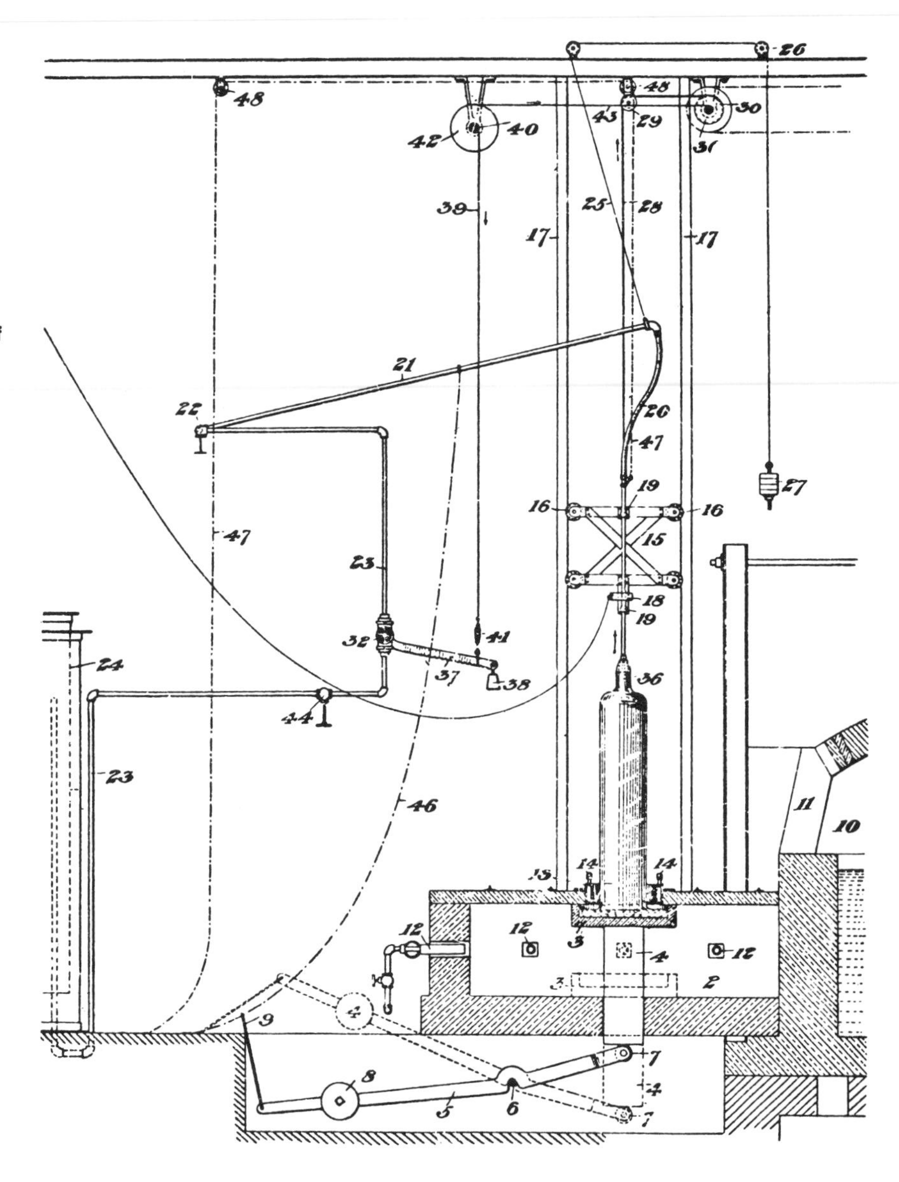

5.17 **In the Lubbers glass-blowing machine, air was fed through a blowpipe that moved vertically (21), drawing with it a measured quantity of glass that had been ladled into a heated vessel (3). Success of the process depended largely on regulating even movement of the blowpipe and stable air pressure. (U.S. Patent no. 702,013.)**

four to 21, and the state's production almost quadrupled.[52] As the gas supply decreased in the Pittsburgh area, towns in Ohio and Indiana lured glassworks by offering cash incentives and free land for the construction of the factories. Edward Libbey, faced with labor problems and the high price of coal in Massachusetts, moved his entire plant to Toledo, Ohio, when civic leaders offered a free factory site. Unfortunately, before the introduction of welded steel pipelines in the 1930s, a local supply of natural gas might soon be exhausted, leaving glasshouses and other furnace industries without the fuel that had been the reason they came there. When natural gas was found in Indiana many Pittsburgh manufacturers moved there, but when these supplies of natural gas had been consumed, some returned to the Pittsburgh area, some transferred to Kansas and other new gas fields, and those remaining in Indiana converted their factories to make producer gas from coal.

European countries, not having extensive deposits of natural gas, relied on coal and oil. In the United Kingdom, from 1950 to 1965, the amount of coal used in glass production dropped drastically and the amount of oil used increased more than fourfold. The use of electricity grew in regions that were rich in hydroelectric power, but usually it served only as a supplement to other fuels.[53]

During the last half of the nineteenth century there were repeated attempts to invent machines that could replace the glassblower in the production of both cylinder glass and bottles. Compressed air and measured quantities of molten glass were employed to control the blowing process for making bottles, but mechanically imitating the deft movements of the glassblower who made window glass was much more difficult. In 1896 John H. Lubbers, a flattener working at a South Pittsburgh window glass factory, began experiments on the use of compressed air to blow cylinder glass, working with the support of his employer and a group of investors.[54] By 1903 the English company of Pilkington Brothers was considering investment in the Lubbers process, but officials of the concern received a report advising against negotiating with Lubbers (characterized as "a clever man but generally drunk"), because the system was not yet developed to the point of practical application.[55] In the same year, Lubbers's patents were purchased by the American Window Glass Company, which controlled some 85 percent of the U.S. industry's output. By 1911 around half of American window glass was produced by machines. The manager of a plant in the United States explained the popularity of the process: "The machines make poor glass . . . but they make so much more glass that they can pick out a great deal of good glass and sort it very carefully."[56]

In the Lubbers process, refined glass was ladled into a large flat dish of fireclay, which was heated from beneath (figs. 5.17, 5.18). The flared head of a vertical blowpipe was lowered and held there until glass had solidified around a flange on the inside rim of the blowing head. As machinery raised the blowpipe, air pressure enlarged a bubble until it attained the circumference desired for the cylinder. Once formed, this shoulder dimension could be maintained if constant air pressure was maintained as the blow-

5.18 **This critical phase of the Lubbers process determined the diameter of the cylinder to be blown by slowly lifting the blowpipe. Once this task had been completed, it was a simpler matter to raise the blowpipe while maintain a constant pressure within the cylinder. (Courtesy of Pilkington Brothers, Ltd.)**

5.19 **Cylinders completed by the Lubbers process lie near the area where cylinders are drawn. As they rose, cables were secured around the cylinders to prepare for the delicate work of lowering them to a horizontal position. (*Compressed Air Magazine*, July 1921.)**

5.20 **Once the Lubbers cylinders had been gently lowered, they were cut into several sections. These were then slit and taken to a flattening oven. From that point the procedures were the same as those used with hand-blown cylinders. (Courtesy of PPG Industries.)**

5.21 After the introduction of the Lubbers method for mechanically blowing cylinders of window glass, there were many patents filed that described means by which the tall cylinders of glass might be safely lowered from their upright positions. Like L. R. Schmertz's patent, most depended on a system of cables and pulleys to swing the cylinder to a horizontal position, resting on a cradle. (U.S. Patent Office, Official Gazette, 9 July 1908, no. 890,306.)

pipe drew glass upward. A motor slowly raised the blowpipe, and so a tall cylinder was produced, cooling as it rose. When its full height had been attained, the cylinder was ended by rapidly lowering the dish of hot glass or speeding the vertical movement of the blowpipe until the walls of the cylinder drew thin and ended.

At first the Lubbers machine produced cylinders roughly 5 feet long and 1 foot in diameter, about the size of hand-blown cylinders. Since a large part of the time required to produce a Lubbers cylinder was absorbed in ladling a fresh supply of glass and starting the blowpipe's upward movement, there were obvious advantages in increasing the size of the cylinders blown. First, length was increased to around 25 feet, but the size was limited by the difficulty of flattening the glass as a single large sheet. Soon the diameter of cylinders was increased, and the long tubes of glass were cut into lengths that would be manageable in the flattening oven (figs. 5.19, 5.20). Eventually cylinders over 40 feet tall and more than 30 inches in diameter were produced. Before flattening, these cylinders were cut into 10 or more segments by wrapping a red-hot electric wire around the cylinder. Blowing a cylinder of single-strength glass required about 35 minutes, and double-strength about 50 minutes.[57]

The mechanical blowing of window glass was constantly endangered by drafts in the glasshouse, for if one side of a cylinder were even slightly cooler it would be drawn thicker throughout its entire height. Residue from a previous charge of molten glass could cause streaks in a cylinder, but this problem was soon solved by the introduction of a reversible container that could be flipped over, so that residue would drain away while a new charge was being ladled into the upper part. In early trials it was discovered that hand control of air pressure and the movement of the blowpipe did not satisfactorily govern the thickness of cylinder walls. An intricate system of automatic controls was developed to insure the variations of pressure and vertical speed necessary to produce a cylinder of uniform thickness.

Not the least of the problems in processing mechanically blown cylinders was that of lowering a tall vertical cylinder of glass to a horizontal position, where it could be divided and slit for the flattening process. U.S. patent records of the period show many inventions that used slings and cables to bring a cylinder as tall as 40 feet to the cradle on which it was to be divided (fig. 5.21).

Mechanically blown cylinders of window glass required that the relatively few workers needed be highly skilled and highly paid. Continuous tank furnaces were usually enlarged because a great amount of glass was required for the mechanical process. With as many as 12 machines operating around the clock from a single furnace, extraordinary amounts of glass could be produced. A French report indicated in 1926 that a single machine could manufacture around 2,000 square feet of window glass in a day.[58] It was the early 1930s before cylinder machines in the United Kingdom and the United States had been completely replaced by sheet drawing machines, which eliminated the flattening process, the flaw inherent in manufacturing cylinder glass, by hand or by machine.

In 1904 a French writer surveyed the whole of his country's glass industry for the *Revue des Deux Mondes:* "He did not find in it, as in so many other industries, that constant application of mechanical improvement in order to economize the human material at work. On the contrary, there seemed to him to have been but little progress in that respect since the days of the old Egyptians."[59] This comment preceded the vast changes that were made by mechanization of the production of window glass. Potmakers had been eliminated by the introduction of the continuous tank furnace. Blowers and gatherers were replaced as the Lubbers system of machine-blown cylinders came into use, and flatteners would no longer be needed with the advent of drawn glass. The workers no longer required were those who had been at the top of the wage scale. After mechanization, a union survey showed that of about 600 glassworkers recently laid off in the United States, roughly 100 remained unemployed and an equal number had obtained employment within the glass industry. The remainder, two-thirds of them, had found employment in other fields or operated their own small businesses, among which saloons and billiard parlors were most frequent.

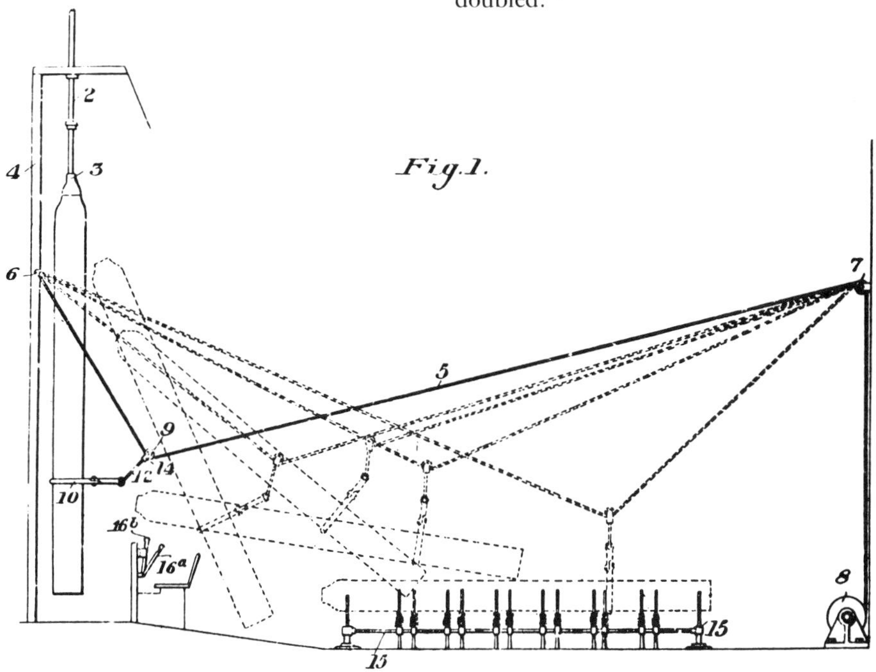

Changes in the window glass industry can be seen in data regarding American production in 1899 and 1925, dates that include many of the changes mentioned above. During this period the 100 window glass factories in the United States were reduced to 42, while the total number of workers employed declined insignificantly and the output of glass more than doubled.[60]

Throughout the last half of the nineteenth century, many schemes were patented to draw flat sheets directly from a mass of molten glass. Cylinders had been successful because internal air pressure maintained the dimensions of the cylinder. If the cylinder was to be replaced by flat sheets, thus avoiding damage from flattening, it was necessary to find a method that would prevent the sheet's narrowing and thickening ("tailing off") as it was drawn. In addition, it was necessary to limit the contact of the soft ribbons of glass with parts of the machine that might mark its surface. A solution was discovered by Emile Gobbe, a Belgian glassmaker. By floating on viscous glass an open box-shaped vessel made of fireclay, a sheet could be drawn upward through a slit in the bottom of the clay *débiteuse*. As water rises in a leaky boat,

5.22 The Colburn patents for drawing sheet glass, like others, used a *bait* to start drawing the glass upward. The more difficult task was preventing the sheet of glass narrowing as it was drawn. This patent included the claim of "imparting movement to the surface portion of the molten mass away from the medial line of the sheet." For that purpose, two revolving elements are placed below the drawing level to pull the glass outward. (U.S. Patent no. 821,780.)

5.23 Early drawn glass was often marred by rough edges of the slit in the *débiteuse* through which it was drawn. In the Colburn system no slit was used. Two rollers moved the molten glass toward the point where the drawing was done. Immediately above these rollers the glass was cooled, to be reheated again (*B*) in order to bend to a horizontal direction. The band of glass moved forward onto rollers (C), and overhead sprocket chains (*F*) drove grips that pushed the sheet of glass along the production line. Obviously the timing of all phases of the machine had to be carefully synchronized to produce clear glass of a consistent thickness. (*Scientific American Supplement*, 16 May 1908.)

the weight of the fireclay forced a band of glass to rise and, so long as that glass was drawn upward at the same rate that it appeared, a continuous band of glass could be obtained. To start the flow of glass, a *bait*, usually a piece of glass or metal mesh, was inserted into the slit at the bottom of the *débiteuse*. Gobbe's invention was improved by Emile Fourcault, an engineer who had succeeded his father as director of one of Belgium's largest glassworks.[61] The process was patented in 1902 after several years of experimentation, but another decade remained before the Fourcault process, as it came to be known, was sufficiently developed to be placed in production. At a glasshouse in Dampremy, in the coal-mining region of southwest Belgium, a small tank furnace was built, and through a two-year period one, two, and three machines were tested. It was estimated that glass from this experimental plant cost about 40 percent less than that produced by the usual methods of the day.[62] The system's performance in tests was sufficiently favorable to lead the Fourcault-Frison company to install a large continuous tank furnace with eight machines to draw glass.

In the Fourcault system, once a bait of metal fabric started the flow, it was necessary to cool the sheet of glass before it would be touched by the rollers that drew it upward. For this purpose, water-cooled tubes were placed on either side of the glass a short distance above the *débiteuse*. These tubes lowered the temperature of the glass sufficiently to achieve a viscosity that prevented the band of glass narrowing and avoided its surfaces being marred by the asbestos-sheathed rollers that pulled it higher within an enclosed and insulated shaft. Fins inside this shaft, more than a dozen spaced about a foot apart, caught any broken glass before it fell into the tank, and the temperature of the drawn glass was gradually reduced as it moved upward. By this arrangement the use of an annealing oven to slowly cool the glass was eliminated. As the band of glass rose from the topmost rollers, some 30 feet above the tank, it was cut into lengths of about 6 feet, and these were taken to the cutting room where they were divided into panes of the desired size.

By incorporating the annealing process and eliminating flattening ovens, the Fourcault system conserved fuel and produced sheet glass with a clean, brilliant surface. More than two-thirds of labor costs present in the manufacture of handmade cylinders were eliminated.[63] Thickness could be accurately controlled by varying the speed at which sheets were drawn from the *débiteuse*. There were often long streaks produced by impurities in the mass of melted glass, but with experience these imperfections were reduced in number and size.

Unfortunately, large-scale production of drawn sheet glass began at Dampremy in 1914, the year that German troops invaded Belgium. It was 1919 before a Fourcault machine operated in England, and five years later there were only two Fourcault plants in the United States. Once the postwar building boom had subsided, hard times restrained the window glass industry's investments in new equipment.

In 1898 Irving Colburn had started experiments in a small Pennsylvania community, trying to develop a machine that would blow lamp chimneys and drinking glasses. After two years Colburn shifted his attention to the development of an improved machine to blow cylinder glass, an unsuccessful project that lasted several years. In 1906, five years after the Fourcault patent, Colburn found investors and formed the Colburn Machine Glass Company, intent on discovering a way of drawing sheet glass (figs. 5.22, 5.23). A Colburn machine was sold but, after a well-publicized start, the machine failed in operation. Colburn continued experimenting until his company closed in 1911, and its patents were bought at auction by the Toledo Glass Company. Four years of development and testing in Toledo prepared the way for the formation of the Libbey-Owens Sheet Glass Company and the construction of a plant in Charleston, West Virginia, to start commercial production of window glass using the Libbey-Owens equipment, as it was then called.

Colburn's machine, as described in 1908, drew a sheet of glass directly from the surface of a shallow tank, the draw being started by a metal bait. To prevent the sheet narrowing as it was drawn, Colburn provided a rotating sphere of fireclay at each edge of the draw, pulling the glass outward as the sheet rose and was cooled. (Libby-Owens replaced this device with a pair of small ridged rollers.) By grasping the edges just before the glass had cooled, the width was maintained without making an imprint. Shields on either side of the rising band of glass protected it from the heat of the furnace. After being drawn about 8 feet upward, the glass was slightly softened by a row of glass flames and then rolled over a cylinder to a horizontal direction. A series of belts carried the band 200 feet to the cutting table, and on the way it was annealed. The bending roller, made of a nickel alloy, was air-cooled to maintain a faint red glow, for if the roller were too hot glass adhered and if too cool the surface of the glass would be marred.[64]

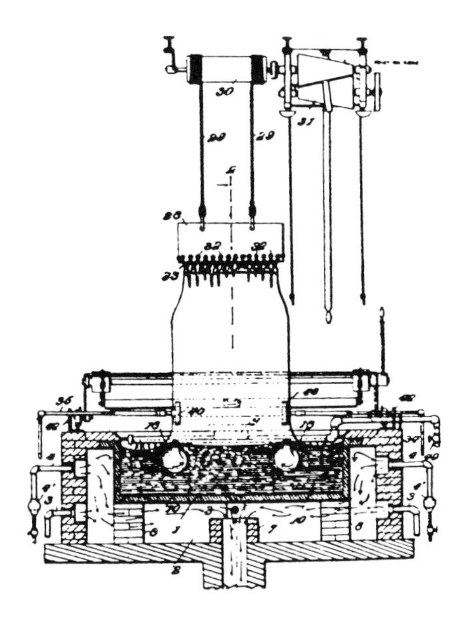

The speed of the Libby-Owens machinery varied according to the thickness of glass to be drawn. In 1925 thicknesses ranged from 1/24 to 5/16 inch, and speeds from 100 to 25 inches per minute. Savings in labor had been discovered in the 1908 tests, for which it was reported: "There are no gatherers, blowers, snappers, or flatteners. The cutters and superintendent are the only skilled men employed, and the cutters will some day give way to automatic devices. With three men and six boys more glass and better glass is made by the machine than can be made by thirty-nine men with the cylinder process."[65]

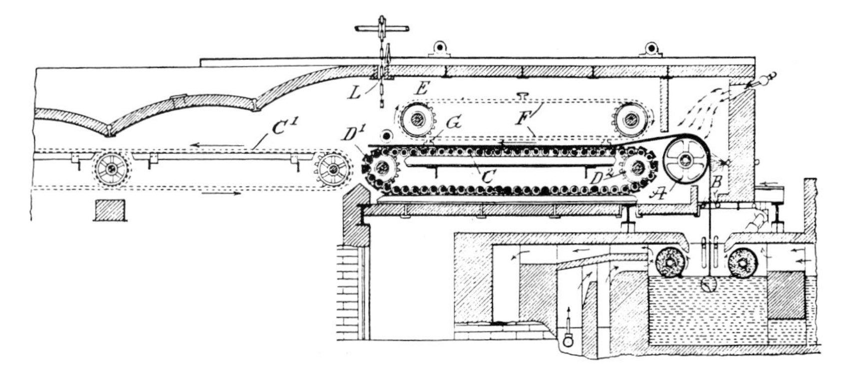

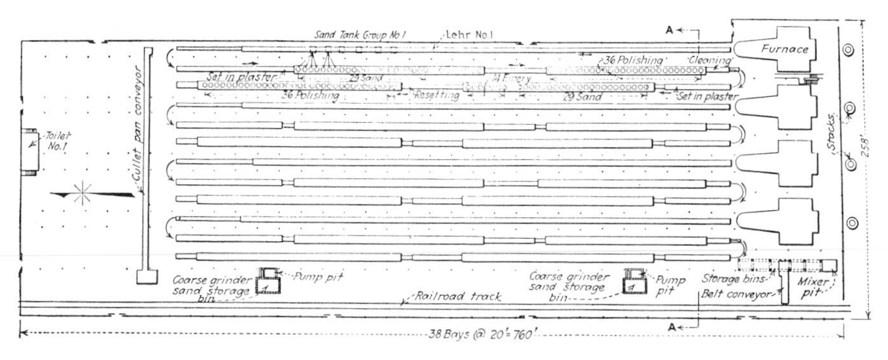

5.24 At Ford's River Rouge plant, each of the four furnaces (right) fed glass to a production line that extended the length of the building (about 1,500 feet) three times before the polished glass was ready for installation as automobile windshields. The first length was the annealing furnace, a journey of about three hours during which the temperature of the glass was controlled by gas jets. The second length ground and polished one side of the sheets of glass; and the third length completed the other side. (*Engineering News-Record*, 4 September 1924.)

"One tank furnace would feed four [drawing] machines, giving a daily output of around 250 tons so that the glass for the Crystal Palace would absorb approximately two days' production."[66] Nevertheless all problems had not been solved. The Fourcault system was troubled by deterioration of the slit through which glass was drawn. Glass made by the Libby-Owens system was often marred by contact with the bending roller, in which it was difficult to maintain the critical temperature. In the 1920s a new system was developed by the Pittsburgh Plate Glass Company, combining features of the two earlier systems. Glass was drawn from the tank much as in the Libby-Owens system, but it continued upward as in the Fourcault system. All three systems were in production in 1950, when Fourcault equipment was used for about 72 percent of the sheet glass produced world-wide, Colburn equipment for about 20 percent, and Pittsburgh Plate Glass systems for the remainder.

After the systems for mechanized production of window glass were developed, a new market for plate glass materialized. About 4 million automobiles were built in the United States during 1923, and roughly half of them—in fact, about half the automobiles driven anywhere in the world at that time—were Model T Fords produced by Henry Ford's assembly line methods. Ford had early encountered difficulty in obtaining an adequate supply of plate glass for the windshields of his automobiles. Since the start of the century the number of plants producing plate glass in the United States had not increased, though their output had almost tripled. The use of plate glass for store windows provided an ever-expanding market, and the increasing enclosure of automobiles promised to elevate the demand far beyond the industry's capacity. In response to these factors, prices for plate glass more than doubled in the United States, and imports increased.

In 1922 Durant Motors, Inc., bought the American Glass Company, and in 1926 the General Motors Corporation would buy the Rishers Brothers plant. Ford bought the Allegheny Plate Glass Company near Pittsburgh, but its output proved inadequate for his needs. At his Highland Park factory in Detroit, trials were made producing plate glass by a continuous process without grinding or polishing, but after two years of experimentation under inexperienced direction the project was abandoned, leaving behind a great mound of cullet that had been produced by the trials. A new start was made, acknowledging the need to grind and polish the material in the traditional manner, but nevertheless operating as a continuous process, the production system for which Henry Ford was famous. Once initial breakdowns had been repaired, the Highland Park plant produced plate glass at the rate of 1,250,000 square feet annually, but that was only 7 percent of the amount the Ford Motor Company required.[67]

A plate glass plant capable of producing 10 million square feet of glass per year, eight times as much as made by the Highland Park prototype, was built in 1924 at Ford's River Rouge plant (fig. 5.24). At one end of the building, four tank furnaces were fired by gas, and the escaping fumes heated the passages through which air was admitted to the flames. At each furnace a spout below the top level of the molten glass allowed a band of glass about 12 inches wide to flow out and continue between a pair of water-cooled rollers, which controlled the thickness of the ribbon and increased its breadth to 20½ inches. From the rollers the glass began its travel along a belt, through a gas-fired annealing tunnel oven 520 feet long, which in three hours gradually lowered the temperature of the glass from 1,150° F to about 80° F. At the end of the tunnel the glass was cut into pieces approximately 9 feet long, ready to start the nine-hour process of grinding and polishing both surfaces.[68]

As in the traditional manner of making plate glass, pieces were set in plaster. Chain-driven tables passed under 29 rotary grinders using successively finer sand and 14 grinders using emery or powdered garnet (fig. 5.25). Then the glass was rouge-polished by 36 felt-surfaced buffers. The sheets were stripped from the tables, turned, and once again set in plaster for the process to be repeated on their other surfaces. Remembering the dramatic moments of blowing window glass and casting sheets of plate glass, the periodical *Glass Worker* wistfully commented, "The spectacular has been eliminated."[69]

All of this process was little more than an ingenious conversion of traditional techniques to modern assembly line procedures in which the pieces of glass could be moved along at the rate of 35 inches per minute. At first as much as half the glass was lost due to breakage in the annealing and grinding processes, but in time losses were somewhat reduced. Broken sheets of glass were cut to make windows smaller than windshields.

As the continuous process was developed, it effected a saving of one-third in labor costs required to produce polished glass. However, by 1929 only half of American plate glass production came from plants using the continuous process. Since in that period three of the four continuous process glass plants were owned by automobile manufacturers, one must conclude that the continuous process dominated the manufacture of automobile glass, while most glass for storefronts was made by older methods.

The Ford process of making plate glass provided a system of continuous grinding and polishing, but it was necessary to set the sheets of glass in plaster and turn them by hand, a time-consuming task. The English firm of Pilkington Brothers initiated experiments on a truly continuous system that would simultaneously grind both sides of a band of glass, and in 1935 that machine was put in commercial operation. From the tank furnace, glass ran between two rollers. The sheet continued through an annealing oven, then between grinding disks above and below. At first the glass advanced at the speed of 45 inches per minute, but improvements soon attained a speed of 200 inches per minute.[70] The twin grinders were the source of most difficulties, and often it was necessary to cut the band of glass into sheets and finish it in the manner of the Ford process.[71]

None of the processes developed for sheet or plate glass provided the "fire-finished" surface that had always been admired in crown glass. There were experiments that attempted to remelt the surface of the glass, but once the glass was again brought to a soft state it required support and any surface underneath was certain to leave some imprint.[72]

5.25 Water and sand from the River Rouge grinding machines ran into a channel in the factory floor and were pumped from there into a storage tank. At the tank, the sand was separated into seven sizes for reuse. In the same way, the emery particles used for finer grinding were graded into five sizes. (*Glass Industry*, January 1923.)

In 1952 Alastair Pilkington of the British firm of Pilkington Brothers Ltd. began experimentation on a process in which a vat of liquid metal was used to support the glass ribbon pouring from a tank furnace. By placing gas flames overhead and floating the glass on melted tin, heat could be applied to both the upper and lower surfaces, simultaneously achieving a "fire-finished" gloss on each side of the glass. Seven years were taken in developing the process to the level of commercial application. The complex chemistry of the interface between melted glass and melted tin required careful analysis and repeated experiments, and to prevent oxidation it was necessary that the tin and glass be brought together in a gas-filled chamber. At the end of the vat, the glass was cool enough to allow it to move off the surface of melted tin into an annealing oven without damage to its surfaces. A major advantage of float glass production was its speed. No longer controlled by the time-consuming operations of grinding and polishing, plate glass could move as fast as 45 feet per minute, about 16 times as fast as Ford's River Rouge plant. Furthermore, the high percentage of breakage, inherent in the grinding process, was eliminated.

Glass in architecture has traditionally been associated with the interruption of a solid wall or roof to provide vision and daylight, ventilation being more a function of the sash than of the glass. The use of large wall panels of glass was considerably advanced with the art of fashioning Gothic church windows, with the leading of the colored glass braced by iron bars, but the Crystal Palace and many of the exhibition buildings that followed had almost all their external surfaces, in some cases even the roof, covered with glass. From this came a new architectural expression, which one observer hailed as the "Ferro-vitreous Art" of the future.[73] With the introduction of skyscraper construction, supporting the walls of each story separately on the steel frame, exterior walls became nonsupporting, merely curtain walls—a term already in use at the time. With feeble gas lighting or even the first electric lights, it was necessary that windows in office buildings be large to take in enough light to fill the depth of offices. In some cases, the expanse of glass was interrupted by little more than the depth of floor beams and the width of columns. This was particularly seen in office buildings constructed on party-wall sites, where the front wall, and perhaps a skylight, had to provide almost all of the daylight within. In the Hallidie Building in San Francisco (1918), a wall of glass four stories high was placed approximately 3 feet in front of the floor beams, braced by concrete window sills less than 3 inches thick. Only eight years later the machine shop of the Bauhaus in Dessau sheathed three sides in glass set away from floors and columns. Transparency had become an architectural medium, challenging the solidity of past architecture; and, in a world that associated fresh air and sunlight with physical and spiritual health,

openness was festive. In a practical sense, tighter-fitting sashes and steam radiators allayed the chills and drafts that had always been associated with windows.

In addition to its transparency, glass afforded a clean, hard, smooth surface to withstand the fouled atmosphere of smoky cities, where soot settled on building surfaces to make pale tones of brick a dull gray and washed down walls to blacken the ground floors of buildings. In the 1930s an opaque colored glass facing was introduced by major manufacturers. Available in rich colors ranging from white to black, sheets of this material could be attached to walls by mastic or mortar, and it became particularly popular for resurfacing old ground-level commercial spaces for chain restaurants, candy stores, and other businesses for which it was desirable to present an up-to-date and hygienic appearance. Glass was enthusiastically accepted as a wall finish beyond its association with the admission of light.

The rush to erect office and commercial buildings after World War II and mounting costs of labor caused the virtual abandonment of masonry facings for such structures in the United States. The government's investment in factories to meet its wartime demand for aluminum provided a postwar source of extrusions that simplified the construction of curtain walls. Metal panels were often incorporated in these surfacing systems, but through the years glass in its various forms became the most accepted material. Air conditioning systems permitted the construction of sealed buildings in which walls no longer served a ventilating function. Given the economy of the glass-filled curtain wall, glass manufacturers assumed the challenge of altering their products in response to the problems that were inherent in such a wall. Glare in the interior was reduced with tinted glass; heat gain was lowered with heat-absorbing glass; and heat loss was lessened with double-glazing. With the introduction in 1966 of reflective glass, basically a one-way mirror said to transmit only about 10 percent of the solar energy striking a wall, buildings were no longer sculptured masses but became instead illusory forms with thin-lined grids marking off extensions of the sky.

1756 John Smeaton experiments to determine hydraulic cement for the Eddystone Lighthouse

1796 Roman cement patented by James Parker

1820 Natural cement discovered in construction of the Erie Canal; patented (U.S.) by Canvass White

1824 Portland cement patented (Britain) by Joseph Aspdin

1858 Systematic testing of Portland cements for use in construction of the London drainage system

1870s German scientists improve the chemical analysis and processing of cements

1871 David O. Saylor obtains patent (U.S.) for production of Portland cement

1890s Introduction of cement made from blast furnace slag

6.1 The rebuilding of the Eddystone Lighthouse in 1756 was the impetus for John Smeaton's investigation of hydraulic cements. Because waves and fire had destroyed three previous lighthouses and the rock was awash at high tide, the use of the best hydraulic cement was essential, even with dovetail connections between the stones. Smeaton's structure lasted more than a century. (Eighteenth-century engraving, reproduced from R. W. Lesley, *History of the Portland Cement Industry*, 1924.)

It was the Romans who made the transition from lime mortar to cement mortar. Of these mortars Vitruvius wrote:

> Limestone when taken out of the kiln . . . is found to have lost about a third of its weight owing to the boiling out of the water. Therefore, its pores being thus opened and its texture rendered loose, it readily mixes with sand, and hence the two materials cohere as they dry, unite with the rubble, and make a solid structure. . . . There is also a kind of powder which from natural causes produces astonishing results. It is found in the neighborhood of Baiae and in the country belonging to the towns round about Mt. Vesuvius. This substance, when mixed with lime and rubble, not only lends strength to buildings of other kinds, but even when piers of it are constructed in the sea, they set hard under water.[1]

Lime mortar became strong through the chemical action resulting from its contact with air; but with the addition of pozzolana—the natural compound of silica alumina and iron oxide to which Vitruvius referred—a hardening reaction was achieved independent of air and deep within large masses of mortar.

For medieval buildings, simple lime mortar appears to have been customary. There is evidence that in the Rhine valley, tarras, a volcanic stone, was added to the mixture. A powder of ground brick or tile, which Vitruvius had recommended as a substitute for pozzolana, was also employed to strengthen mortar. From contemporary records and from ruins it is evident that medieval walls, like Roman walls, were made with a filling of mortar and rough stone. The Romans used brick and most medieval masons used stone to form the two faces of a wall, and between these surfaces mortar and scraps of masonry were placed. In both periods it does not appear that the stones and mortar were mixed together before being put in the wall.

Through their studies of Vitruvius's writings and the remains of Roman structures, later architects had constant reminders of the Roman development of a strong hydraulic cement (one that would set under water). When Sir Christopher Wren was building St. Mary-le-Bow after the Great Fire of London, "he sank about 18 feet deep through made ground, and then fancied he had reached the natural soil and hard gravel; but, upon examination, it appeared to be a Roman causeway of *rough stone*, close and well rammed, with Roman brick and [rubble] at the bottom for a foundation and all *firmly cemented*."[2]

Fourteen miles southwest of Plymouth harbor, long one of England's busiest ports, three ridges of rock called Eddystone lie near the surface of the water. The water around them is rough, even when the sea is calm. Shortly after the Eddystone Light burned on a December night in 1755, John Smeaton was chosen to direct the rebuilding of the lighthouse. Only 31 years old, Smeaton had two years before been made a Fellow of the Royal Society, and he had entered the field of engineering after several years as a maker of marine and astronomical instruments.

At high tide the rock on which Eddystone Light was to be rebuilt was covered by the sea, and the choice of mortar was critical under those conditions of construction (fig. 6.1). Smeaton began a series of experiments to determine the cement that should be used. First, he found that the use of lime made from "imperfectly burnt" stone was wasteful. Second, he found that the hardness of the stone from which lime was made had no relation to the hardness of the mortar produced—an erroneous assumption that went back to the writings of Vitruvius. He tested a variety of stones in turn as sources of lime. Tests were made on mortar also containing pumice stone, coal cinders, brick and tile dust, scale from smiths' forges, and several other substances that were sometimes employed to impart hydraulic properties to mortar. The most effective proved to be tarras and pozzolana. After different mixtures of the mortar ingredients had been tested, Smeaton estimated their cost per cubic foot. Tarras would have been the choice because of cost, except for an unforeseen circumstance:

> In making enquiries how some [pozzolana] might be procured from [Italy], I very fortunately learnt, that there was a quantity of it then in the hands of a merchant in Plymouth; which had been imported as an adventure from Civita Vecchia during the time Westminster Bridge was building; and which he expected to have disposed of for that work to a good advantage; but failed in his speculation: for having found that tarras answered their purpose, neither Commissioners, Engineers, nor Contractors, would trouble themselves to make a trial of it [pozzolana], and therefore refused it.[3]

A low price quoted by the merchant made pozzolana Smeaton's choice, for with lime from Aberthaw's blue stone it would make the most economical mortar, balancing strength and cost.

In his experiments Smeaton had found that hydraulic cement depended on "a considerable quantity of clay" being in the stone, although clay added to pure lime did not result in

hydraulic qualities.[4] Without knowledge of the complex chemistry of cements, Smeaton had determined a fundamental characteristic of natural hydraulic cement by one of the earliest exhaustive studies of a building material.

For some reason Smeaton's methodical investigations in preparation for the construction of Eddystone Light were not known to Bryan Higgins, an Irish physician and chemist who operated a school of chemistry in London. After a series of carefully conducted laboratory studies he patented a form of stucco containing lime, sand, and bone-ash, a mixture he employed in several houses. But Higgins's stucco proved not to withstand weathering well, and it was overwhelmed by the competition of James Parker's Roman cement.[5]

Little is known of Parker before his 1791 patent for using peat as a fuel for burning brick or lime. Nothing came of this venture, but five years later he

6.2 Bottle kilns such as these were filled with fuel and cement stone to produce the clinkers from which Portland cement was ground. Disadvantages of the bottle kilns were the fuel wasted in heating and cooling the mass of brick and the costly hand sorting of material when it was imperfectly fired due to variable winds about the kilns. (A. J. Francis, *Cement Industry 1796–1914*, 1977.)

obtained another patent for "A certain Cement or Terras to be Used in Aquatic and other Buildings, and Stucco Work."[6] This material, first known as "Parker's Cement" and later as "Roman cement," was made from "noddles" found at various places in England, particularly in the low cliffs where London clay reaches the shore. Known locally as cement-stones and technically as septaria, these kidney-shaped stones contain veins and a center of claylike material. To produce Roman cement they were shattered by boys wielding hammers and carried off to a bottle-shaped kiln that held about 30 tons with the coal necessary to burn it. After the kiln had been fired for about three days, the stone that had been sufficiently burned could be taken out at the bottom of the kiln and additional stone and coal added at the top.[7] The burned stone was then ground between millstones and sieved before being put in barrels for shipment.

Shortly after Parker received his patent, the directors of the British Society for Extending the Fisheries and Improving the Sea Coasts of This Kingdom authorized that his product be evaluated by the Society's engineer, Thomas Telford, who had considerable experience in the construction of canals and harbor facilities. Telford's report to the directors was so favorable that Parker quickly published it in a promotional pamphlet. For some reason, almost two years later Parker sold his patent to Samuel Wyatt and his young cousin Charles Wyatt. A brother of the prominent architect James Wyatt, Samuel Wyatt had become an architect after starting as a builder, and it was he who years before had experimented with the use of Higgins's stucco.[8] The Wyatt factory prospered, at least until 1810 when Parker's patent expired. Successive members of the Wyatt family inherited direction of the company, and the development of other products after 1810 suggests that their sales of Roman cement declined in the face of the competition that quickly materialized.

A British patent taken out in 1824 by a Leeds bricklayer, Joseph Aspdin, described a method of making cement and applied to this product the term "Portland cement." The use of such cement was claimed to produce a mass equal in strength and appearance to the limestone taken from quarries near the town of Portland on the southern coast of England. The modern definition of Portland cement may not be applicable to the product Aspdin patented. Portland cement today is "made by burning at a high temperature—to incipient fusion of the material—a definite mixture of limestone with clay or shale, and finely grinding the resulting clinker."[9] It is doubtful that the cement produced under the 1824 patent of Aspdin was burned at a temperature high enough to produce clinkers, and his patent does not state the proportions of the ingredients employed, so we cannot with certainty identify it as a true Portland cement by modern definition.

In 1825 Aspdin established a cement factory in a suburb of Leeds. The kilns used for burning the raw materials were built of masonry in the shape of a bottle, about 36 feet high and 17 feet in diameter near the bottom (fig. 6.2). Each firing required that the entire mass of brick be reheated, and certain velocities and directions of wind could result in the amount of coke consumed being well over half the weight of the clinker made. In addition, a good percentage of the product might be imperfectly burned, which would require tedious and costly inspection and sorting by hand.

In the 1840s, controversy about cements was keen in the London area, and England led in the development of this material. J. B. White and Sons had samples of their cement tested in a 75-ton hydraulic compression

machine in 1847, and the results were published the following year in *The Builder.* Publication of these tests stirred the firm of Robins, Aspdin and Company (one of whose partners was William Aspdin, Joseph Aspdin's son) to commission and publish their own tests. The compressive strength of their cement was much higher than that for White's product. Because these tests are similar to those of tests made in 1860–1862 for the London Drainage System, which are known to represent samples of true Portland cement, it seems reasonable to assume that in 1848 Portland cement was being made by both companies, and perhaps also by the four competing cement companies in England (figs. 6.4, 6.5).

Joseph Aspdin's principal competitor for the title of "inventor" of Portland cement was Isaac Charles Johnson, who began in 1835 as manager of the J. B. White and Sons cement works. Around 1845 Johnson's employers failed in negotiating an arrangement with William Aspdin and his partner to participate in the manufacture of Portland cement. In an interview made when he was approaching seventy, Johnson clearly indicated that he had been willing to engage in a bit of early industrial espionage, and apparently William Aspdin realized the likelihood of such efforts:

PATENT
PORTLAND
CEMENT
WORKS

There was no possibility of finding out what [Aspdin] was doing, because the place was closely built in, with walls some twenty feet high, and with no way into the works, excepting through the office. . . .

[Aspdin] had a kind of tray with several compartments, and in these he had powdered sulphate of copper, powdered limestone, and some other matters. When a layer of washed and dried slurry and the coke had been put in the kiln, he would go in and scatter some handfuls of these powders from time to time as the loading proceeded, so the whole thing was surrounded by mystery.[10]

Johnson had a chemical analysis made of Aspdin's cement but this was of little use, and a series of experimental mixtures produced no clues.

By mere accident, however, some of the burned stuff was clinkered, and, as I thought, useless. . . . However, I pulverized some of the clinker and gauged it. It did not seem as though it would harden at all, and no warmth was produced. I then made mixtures of the powdered clinker, and powdered lightly-burned stuff, which did set, and soon became hard.[11]

By the 1850s there were six companies in England manufacturing Portland cement, and it had begun to overtake the popularity of Roman cement.

After Smeaton's experiments there were many trials and many errors made by practical men, but little was done to provide a theoretical explanation of the properties of cement. A French military engineer, Louis Joseph Vicat, in 1818 published *Recherches expérimentales sur les chaux de construction, le béton et les mortiers*, an account of his studies of cements and the conclusions he had drawn from them. He was at that time directing construction of a bridge across the Dordogne River, the first large project of the period for which concrete foundations were made without the use of pozzolana.[12] Vicat had set about investigating those factors that would result in a mortar capable of setting under water. By mixing lime, chalk, and clay of different types and in different proportions, Vicat prepared small blocks of the test materials. In simplest form, his conclusion was that "no perfectly hydraulic mortar exists without silica, and all lime which can be so denominated [hydraulic] is found by chemical analysis to contain a certain quantity of clay."[13] Where Smeaton had searched for the most advantageous natural materials for a hydraulic cement, Vicat's conclusion implied that the key was in the planning of mixtures. In fact, Vicat declared that "we can give to the factitious [artificial] lime whatever degree of energy we please, and cause it at pleasure to equal or surpass the natural limes."[14]

Vicat's ability to foretell the future is remarkable. He was aware of the changes that improved knowledge of the chemistry of cement would make in the industry that supplied it:

That which is in England very improperly termed Roman cement is nothing more than a natural cement resulting from a slight calcination [heating] of a calcareous mineral [limestone, chalk, or marble], containing about thirty-one percent of ochreous clay and a few hundredths of carbonate of magnesia or manganese. . . . Its use will infallibly become restricted in proportion as the mortars of eminently hydraulic lime shall become better known and in consequence better appreciated.[15]

Vicat devised a method that was long used to determine the setting

6.3 By 1843 William Aspdin, younger son of Joseph Aspdin, appears to have begun production of a true Portland cement. The nine years that followed saw Aspdin involved in a succession of partnerships and bankruptcies. With the organization of Aspdin, Ord and Company, the company was moved from London to these new buildings near Newcastle, where fuel and wages cost less. The plant was later taken over by I. C. Johnson. (Borough Library, Gateshead.)

6.4 In 1855 William Aspdin began the construction of Portland Hall at Gravesend, east of London on the Thames estuary. Coated with Portland cement stucco, the house was to have extensive gardens ornamented with statuary and architectural details made of the firm's product. After its completion, Portland Hall became embroiled in Aspdin's intricate financial crises. (Borough Library, Gravesend.)

6.5 The testing room of an English cement works in 1850, though crude by modern standards, indicates the early concern with the complex chemistry of cement. Only seven years earlier, a faulty chemical analysis of a competitor's cement had led I. C. Johnson to experiment vainly with the inclusion of burned bones in his cement. (*Journal, Portland Cement Association*, 1960. Courtesy of Blue Circle Industries, London.)

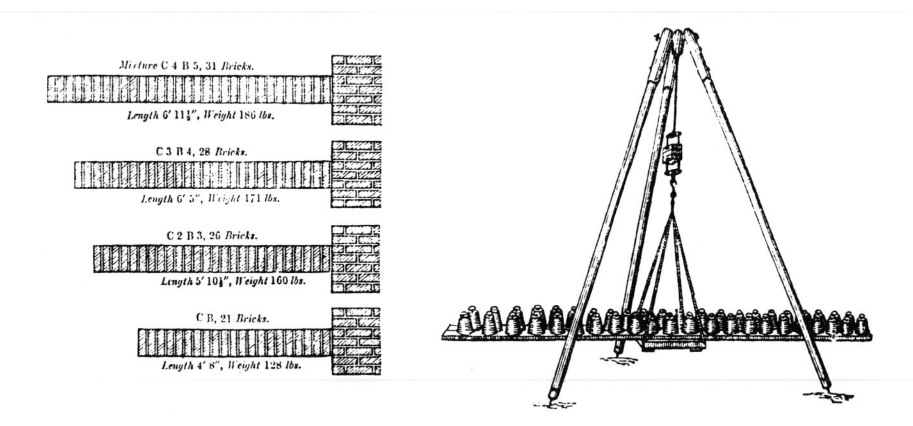

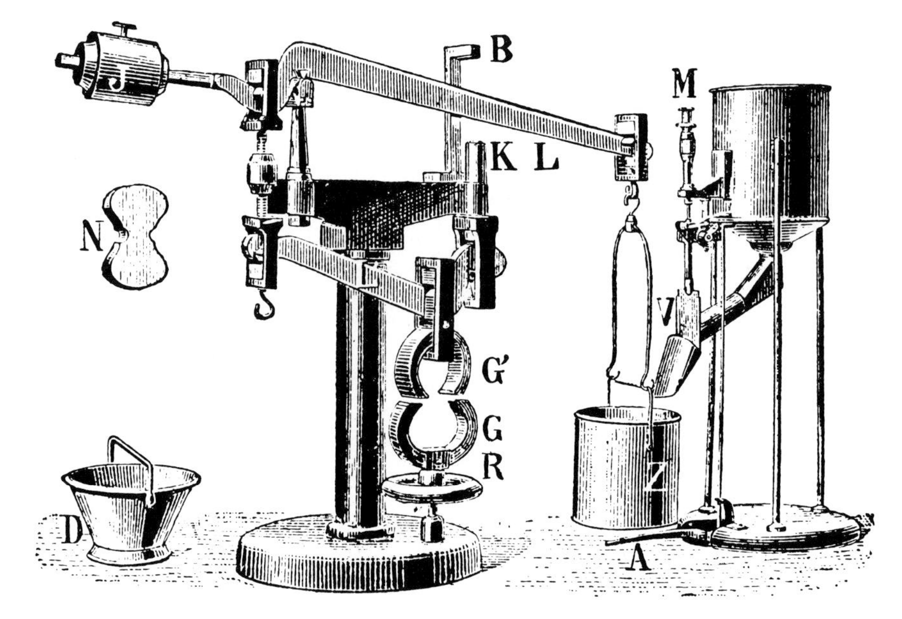

6.6 Two of the testing procedures used by Pasley were (left) adding bricks to a cantilever from a wall, each brick held in place with pressure for 5 to 10 minutes, and (right) determining the weight required to pull apart three bricks about 40 days after they were joined. In both tests, a "C4B5" mixture for the cement (4 parts of chalk to 5 parts of blue clay) was strongest. (C. W. Pasley, *Observations on Limes, Calcareous Cements*, 1847.)

6.7 Cement's resistance to tension was long tested by equipment developed from that originated by a French engineer. A mixture of one part cement and three parts sand was molded (*N*) and placed in the machine (G, G′). Lead shot was poured into a bucket (Z) suspended from a system of levers that multiplied its weight by fifty. (*Science, Progrès Découvert*, 1 October 1932.)

6.8 Over the century from 1850 to 1950, the compressive strength of Portland cement increased by more than five times. A. W. Skempton has charted test results from this period, and with such evidence has identified three major periods of development. (*Transactions, Newcomen Society*, 1962–1963.)

time of a cement. A thick paste of water and the lime to be tested was placed in a container and covered with water, whereupon tests with a weighted, blunt knitting needle indicated when the sample had solidified. A weighted needle was dropped about two inches, and its penetration into the sample showed the relative hardness.[16] With similar concerns, General C. W. Pasley of the British Royal Corps of Engineers tested the adhesion of mortars made of different cements by attaching brick after brick to a wall, each beyond the other to form a cantilever (fig. 6.6). Each brick took only about six minutes to set up, and any cement that allowed 18 to 20 bricks to be added before they fell was considered acceptable for use, although the maximum performance was over 30 bricks.[17] This system of testing mortars is said to have been in common use from 1830 to 1850.

In preparation for construction of the new main drainage system for London, John Grant, chief engineer of the Metropolitan Board of Water Works, in 1858 began methodical testing of Portland cements. To determine tensile strength, small briquets of cement paste were molded in a figure eight shape (fig. 6.7). When inserted between two clips, the briquets would be pulled until they broke. To determine their compressive strength, molded blocks of cement were crushed in hydraulic presses. Grant's discoveries were immediately translated into specifications for the materials to be purchased for work on the drainage system. In 1859 a new requirement appeared in the specifications for the Southern High Level Sewer: "The whole of the cement to be used in these works and referred to in the specifications is to be Portland cement of the very best quality, ground extremely fine; weighing not less than 110 lbs. to the striked [leveled] bushel and capable of maintaining a breaking weight of 400 lbs. on an area of 1½ inches by 1½ inches equal to 2¼ square inches, seven days after being made in an iron mold . . . and immersed six of these days in water."[18] Grant believed strongly that the weight of cement related directly to its strength and for years specific gravity was a common requirement in specifications.

In much of the last half of the nineteenth century, advances in testing methods centered in Germany,

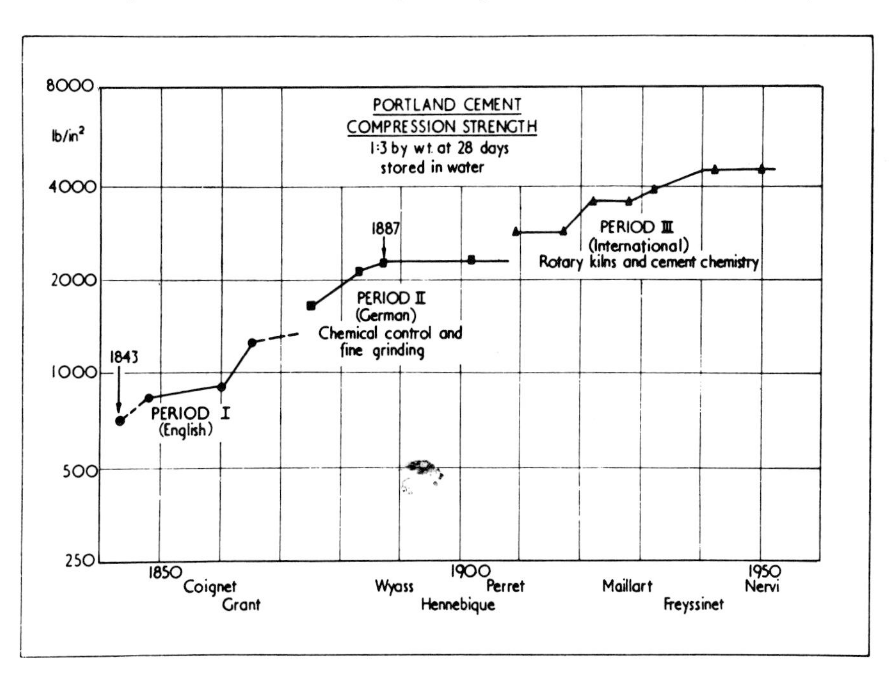

6.9 Isaac Charles Johnson patented his chamber kiln in 1872. While operating the cement works that had been vacated by the bankruptcy of Aspdin, Ord and Company, Johnson found a way to use the heat of a kiln to dry slurry. The upper and lower parts of the bottle kiln were moved apart as much as 100 feet, connected by a horizontal chamber in which slurry was poured about 8 inches deep. (A. J. Francis, *Cement Industry 1796–1914*, 1977.)

6.10 Shaft kilns of this sort were much like bottle kilns but continuous in operation. They were fed at the top with fuel and dry blocks of material, burning took place at mid-height, and the calcined material was taken out at the bottom of the shaft. (C. A. Davis, *A Hundred Years of Portland Cement*, 1924.)

6.11 Frederick Ransome built his rotary kilns at several English cement works. This one, built for the Arlesey Lime and Portland Cement Company, had a cylinder about 5 feet in diameter and 25 feet long. The rotary kiln worked well at lower temperatures, but, once the raw materials were burned to the degree required, the gummy substance adhered to the firebrick lining of the kiln. (*Proceedings, Third International Symposium on the Chemistry of Cement*, 1952.)

where the manufacture of Portland cement had begun in the 1850s (fig. 6.8). Improvements were largely the result of advances in kiln design that increased the uniformity of output and the introduction in 1871 of systematic chemical analysis of all the raw materials used. A greater proportion of lime and higher kiln temperatures resulted in a harder clinker. By 1875 it was recognized that the larger grains of ground clinker did little to strengthen cement, and within two years only 3 or 4 percent of the particles in the typical German cement were too large to pass through a sieve with about 6,000 holes in a square inch. All of these technical improvements were assisted by the German government, its testing laboratory, and the facilities of the country's technical universities.

The chamber kiln for manufacture of cement originated with a patent of Isaac Johnson in 1872, and it was used with the "wet process" in which powdered ingredients were mixed with water, dried, and then burned. To construct a chamber kiln, a rectangular chamber as long as 100 feet was built at the side of a usual bottle kiln (fig. 6.9). The top of the kiln was closed and a chimney was located at the other end of the horizontal chamber. In this chamber sufficient liquid slurry was poured to load the kiln when it was dried, and while one load was fired, its exhaust heat dried the slurry for the next firing. Later improvements stacked two or three chambers, decreasing the required length and increasing the kiln's efficiency. Two factors worked against continued use of the chamber kiln. First, a large amount of labor was required to fill the chambers with slurry, remove the dried powder, and load the kiln, and when the fluctuating price of coke rose the chamber kiln was no longer economically sound. Second, after the development of a wet-grinding process that used less water, a smaller area was needed for drying slurry.[19]

The shaft kiln was generally limited to the Continent, where dry grinding was more common. (The choice between "wet" and "dry" grinding processes was largely based on the materials to be used.) As in early bottle kilns, raw materials were entered at the top and clinkers fell to the bottom (fig. 6.10). Fans were developed to increase the draft, and mechanical grates prevented clogging. The shaft kiln was economical of labor and fuel, but it did not produce a uniform firing at a time when the control of quality was a primary factor in the competition among the many companies manufacturing cement.[20]

Before the rotary kiln, all designs had been based on the bottle kiln, the elementary means of burning the ingredients for cements. The best known rotary kiln was that designed in 1885 by Frederick Ransome, who had earlier obtained a patent for making "artificial stone." The first Ransome kiln consisted of an iron cylinder, 21 feet long and 3½ feet in diameter, mechanically rotated on rollers (fig. 6.11).[21] The cylinder was slightly tilted, powdered material being poured in at the upper end and gas flames situated at the lower end of the kiln. It was hoped that this arrangement would permit continuous operation with more reliable quality. A series of very costly tests were conducted at several English cement plants over a two-year period, and they all failed. As the powder was fired it melted, caked together, stuck to the firebrick lining of the kiln cylinder, and thus impeded the movement of material within the cylinder. Also, any effort to recirculate hot air in order to raise temperatures resulted in

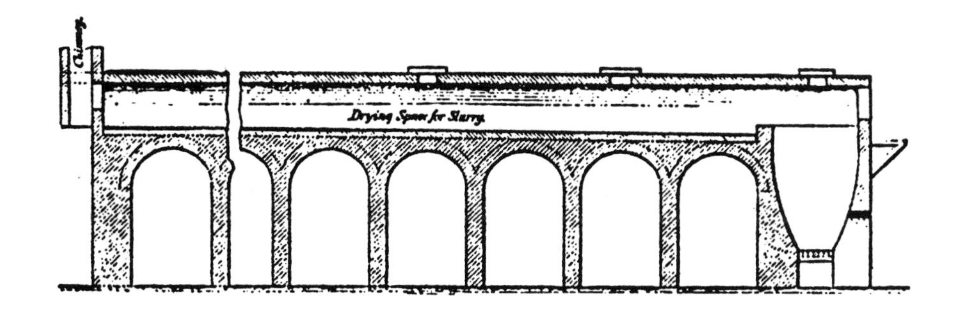
Drying Space for Slurry

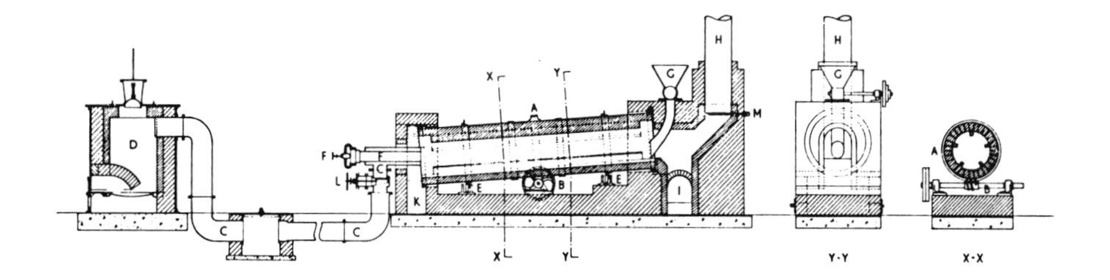
Y-Y
X-X

the fine powder clogging any passages through which the hot air flowed. While Ransome's tests and revisions continued, a young civil engineer, Frederick Stokes, in 1888 patented an improvement, using the regenerative principle of industrial furnaces. The Stokes rotary kiln included a burning cylinder 35 feet long, from which hot gas was exhausted into another cylinder, slightly longer, the outside of which was coated with a slurry of the powdered material.[22] The liquid mixture was dried, the chunks of material it produced then entered the burning cylinder, and from there clinkers traveled into a cooling cylinder.

Variations and improvements followed. The Continent led in finding the advantages of longer rotary kilns, and what had first been a length of 60 feet grew to as much as 250 feet by 1924. In the United States the development of longer rotary kilns was led by the Edison cement works, and by 1922 more than three-quarters of U.S. plants had rotary kilns longer than 100 feet.[23] The knotty problem of cylinder linings was unsolved by experiments with bricks of magnesium or bauxite. After some years, a method was found to line cylinders with ground clinker, to which the burned material would not adhere.

The first great route into the western reaches of the United States was the Cumberland road, and the second was the Erie Canal. Connecting Lake Erie and the Atlantic via the Hudson River, the canal sped the flow of settlers to the West and established New York as the principal port of the east coast. Construction of the canal was supervised by Judge Benjamin Wright, a lawyer previously responsible for engineering two small canal projects, and one of his assistants was a young surveyor, Canvass White. In 1817, after a year working with Judge Wright, White went to England at his own expense to spend several months tramping along more than 2,000 miles of canals. White interviewed engineers, obtained drawings of canal construction, and apparently probed all aspects of canal building.

Commissioners for the Erie Canal made no budget provision for the purchase of imported hydraulic cement, which had been strongly urged by Judge Wright, and stonework laid in common lime mortar in 1818 quickly showed signs of deterioration. When their suppliers of lime burned one local stone and delivered the lime made from it, it was discovered that the mortar did not crumble in water—usual with ordinary lime—and the supervising engineers were informed. Tests confirmed that when this stone was burned, powdered, and mixed with sand and water it would indeed harden, even when submerged in water. After he had investigated the material, Canvass White was granted a patent in 1820 for the first natural cement to be manufactured in the United States, a cement made from stone that had lime, silica, and alumina in proportions producing a hydraulic cement. White's patent was later bought by the state of New York, which resulted in the removal of all manufacturing restrictions related to the patent.[24] In the year that the Erie Canal was opened, White opened his cement works at the village of Chittenango, New York. Leaving his younger brother in charge of manufacturing "White's Water-Proof Cement," Canvass White went on to follow a career as an authority on the construction of canals.[25]

Hydraulic cement proved essential to the construction of other canals, and most such projects began by establishing a quarry that could provide stone for manufacturing natural cement. The Delaware and Hudson Canal led to the opening of cement production in the Rosendale area of New York, where at the end of the nineteenth century 42 percent of the American natural cement was produced.[26] Such cement plants had the considerable economic advantage of a large and long-lasting initial market supplying the canal builders, and this start often provided them enough financial stability to allow expansion of their market areas after the canal had been completed.

In 1903 a critic of natural cement wrote gloomily in *The American Architect*:

> It is well known to cement-men that natural plants turn out a product that varies very perceptibly from day to day, . . . [as] the quality of rock in the same quarry frequently differs in its chemical constituents. . . . However, every pound of the variable "raw" material is subjected to the same treatment. . . . The method of calcination in a natural plant, after half a century of experience, is astonishingly crude; . . . nor is it certain that the ordinary vertical kiln now in use will *ever* produce a high-testing cement.[27]

The criticism was harsh, but its reasoning was irrefutable. Markets for natural cement had increased with the nation's growth. Even when no canals and few buildings were being built because of economic crises, the ceaseless westward expansion of railroads afforded a fair level of business. At the same time, the German cement industry had been rapidly advancing, while American manufacturers remained unwilling to learn more of Portland cement or to invest in a drastic improvement of their methods of production.

The construction of iron and steel ships had begun an upward climb around 1860, but in the decades remaining in the nineteenth century there were still many wooden ships afloat. Liners to carry tourists and valuable cargo might be built of metal, but more mundane goods were usually carried in wooden vessels, which were unable to flood their compartments with water as ballast. They came to America and loaded with agricultural products, particularly grains and cotton, for the trip to Europe. For the westward voyage the ships' masters were willing to take on heavy loads for little payment. The arrival date of such shipments was somewhat unpredictable, because the departure date and the cargo to be taken on were often known only at the last moment. On the other hand, there might be no charge for shipping barrels of cement to the United States. In fact, when highly profitable cargo was scheduled for the return trip, purchasers of cement could be paid as much as ten cents a barrel to hurry the unloading.

At first the dominant share of imports was English. Later the German makes became more popular because of their high quality, and Belgian brands were used because they were extremely low in price. Even with savings in transatlantic shipping, the price of imported cement might be as much as twice the price of good quality Rosendale cement from New York, and railroad rates were high for shipping cement into the Mississippi River valley, even during the periodic price wars among railway companies. In spite of the erratic scheduling of shipments from Europe, importers endeavored to receive much of their cement during the summer so that the barrels could be shipped westward on canals, which charged a lower rate at that time of the year.

American-made Portland cement was first marketed by the Coplay Cement Company in the Lehigh River

6.12 When cement was shipped in wooden barrels, manufacturers in the U.S. and some other countries marked their product by stenciling trademarks on the barrelheads. Stenciling cloth bags was more difficult and, when bags began to replace barrels for shipping, printed paper labels became more popular as a means of marking both bags and barrels. (*Concrete*, April 1923.)

area of Pennsylvania. The company began producing natural cement in 1865, and six years later David O. Saylor, one of the owners, patented a method of making Portland cement. Several years of experimentation had convinced Saylor that his plant could be adapted for the necessary steps of manufacturing Portland cement: grinding the stone into a fine powder, adding lime when his stone required it, obtaining a clinker with high heat, and grinding the clinker into a fine powder. By 1876, "Anchor Brand," as Saylor's Portland cement was known, had won the highest award at the Centennial and had been specified by U.S. government engineers for use in jetties constructed by the engineer James B. Eads in the Mississippi River's delta area. There were also large sales of "Improved Anchor" cement, in which the company's clinkers for Portland cement were mixed with the stone for its natural cement.[28]

An American design for a rotary kiln to make Portland cement was installed in 1886 by José de Navarro, a New York entrepreneur and owner of the Union Cement Company. After two years of trials the kiln proved unsatisfactory and a Ransome kiln was substituted when the company moved to the Lehigh Valley in 1888.

By the 1890s many cement plants had begun shifting to the production of slag cement, which had originated in Germany almost 30 years earlier. In the United States slag cement, which was officially classified as a "puzzolan" cement, resulted from finely grinding the waste of blast furnaces and mixing it with powdered lime. It was commonly sold under the name "Steel Portland Cement," but a government board insisted that it should not be classed as a Portland cement. In response to this, steel companies, which had understandably become enthusiastic investors in cement making, started experiments to improve slag cement. They found that by grinding both slag and limestone, the clinker resulting from the mixture made a true Portland cement. By 1912 cement made of this mixture was about 13 percent of the total U.S. production of Portland cement, but sales began to drop after that time.[29]

By the end of the nineteenth century, production of natural cement had virtually ceased in Europe. In the United States, however, Portland cement was only 28 percent of domestic production in 1896, and the amount of Portland cement made in the United States was half that imported.[30] At that time, about two-thirds of the U.S. production of Portland cement came from the Lehigh Valley of Pennsylvania. By 1923, over 90 times as much Portland cement was manufactured, but only a quarter of it came from the Lehigh Valley.[31] The cement industry had scattered, with a growing production from plants west of the Mississippi River.

Just as ingredients are added to produce a Portland cement of the desired quality, other additives may be used to obtain desired results. It was discovered early that the addition of gypsum retarded the setting of concrete, which is sometimes useful in construction. Other additives can accelerate the rate at which concrete gains strength with time. Since the action of cement as concrete hardens is extremely complex, the chemical determination of cement characteristics has become a focus of attention.

STAR PORTLAND
BONNEVILLE PORTLAND CEMENT CO.
WORKS SIEGFRIED PA
KEYSTONE
PORTLAND
CEMENT
MERAMEC PORTLAND CEMENT & MATERIAL CO.
THE MERAMEC BRAND
ST. LOUIS
UNITED STATES CEMENT COMPANY
BEDFORD PORTLAND
U.S. CEMENT
BEDFORD, INDIANA
THE BRIER HILL IRON & COAL CO.
BRIER HILL
PORTLAND
CEMENT
YOUNGSTOWN, O.
LAWRENCE CEMENT CO.
ESTABLISHED 1832.
IMPROVED SHIELD CEMENT
SALES OFFICE NEW YORK.
STRONGEST OF AMERICAN CEMENTS.
BUFFALO CEMENT CO. LIMITED
TRADE MARK.
WORKS MAIN AND AMHERST STS. BUFFALO, N.Y.
PORTLAND
TOLTEC
CEMENT
A P C
NEW STATE BRAND
B C
BADGER BRAND
PORTLAND CEMENT
PORTLAND CEMENT COMPANY
RED ROSE BRAND
PORTLAND, OREGON
THE U.S. PORTLAND CEMENT CO.
YOCEMENTO, KANSAS.
HUDSON PORTLAND CEMENT CO.
HUDSON, N.Y.
HUDSON BRAND.
ART PORTLAND CEMENT CO.
ROMAN
PORTLAND
PORTLAND
EMPIRE PORTLAND CEMENT CO.
EMPIRE
PORTLAND CEMENT
WARNERS N.Y.
GERMAN-AMERICAN
OWL CEMENT
PORTLAND CEMENT
WORKS
CHICAGO-LA SALLE
NORTHAMPTON PORTLAND CEMENT COMPANY.
WASHINGTON PORTLAND CEMENT COMPANY.
HECLA
PORTLAND
CEMENT
OMEGA PORTLAND CEMENT CO.
JONESVILLE, MICH.
UNION
ROSENDALE
CEMENT
300 lbs
AMERICAN CEMENT Co
EGYPT.
PA
PEOPLES PORTLAND CEMENT CO.
SPOKANE WASH.
UNITED STATES PORTLAND CEMENT CO.
Portland
Concrete
BRAND
Cement
CONCRETE COLO.
SOUTHERN CALIFORNIA CEMENT CO.
SUNSET
PORTLAND CEMENT
BRAND
RIVERSIDE, CALIFORNIA.

7

Reinforced Concrete

1836 Uses of concrete described in George Godwin's paper at the Royal Institute of British Architects

1854 System for reinforced concrete patented (Britain) by W. B. Wilkinson

1867 Joseph Monier patents (France) his system of reinforced concrete, exhibits at the Paris Exposition

1873–76 Construction of the Ward House, Port Chester, New York

1875 Residential construction with concrete panels patented (Britain) by W. H. Lascelles

1877 Thaddeus Hyatt publishes *An Account of Some Experiments with Portland Cement Concrete*

Monier patents (France) a method of reinforcing concrete beams

1884 Ernest L. Ransome patents (U.S.) the twisted square bar as the basis of his system

1892 Patents of François Hennebique give rise to an international enterprise

1900s Flat slab construction developed by C. A. P. Turner and Robert Maillart

Introduction of construction using precast structural elements

German, French, and British regulations published for design of reinforced concrete structures

1903 Ingalls Building, Cincinatti, Ohio, the first skyscraper of reinforced concrete

Auguste Perret's apartment building at 25b Rue Franklin, Paris

1903–06 Housing built in Liverpool using John Brodie's system of precast panels

1910s "Unit systems" of precast concrete construction introduced by Ernest L. Ransome and John E. Conzelman

1918 Duff Abrams publishes results of his tests on the water-cement ratio in concrete

1922 Z-D thin-shell dome constructed in Germany for use as a planetarium

7.1 A drawing published in 1859 shows concrete, a mixture dry enough to be carried in a basket, being tamped in wooden formwork. (*Über Land und Meer*, 19 January 1859.)

Concrete was used in constructing the walls around a fourth-century B.C. Roman city some forty miles east of Rome, and by the second century B.C. the new material began to be used for buildings in Rome.[1] In constructing walls, Roman concrete was in some respects merely mortar, for bricks were laid for the faces of walls and mortar was poured between over chunks of stone and broken brick. But in vaulting, which became the dominant theme of Roman architecture, concrete was clearly used according to its own nature, that of a plastic material to be molded until it developed sufficient strength to stand alone. Once the masonry walls reached the required height, wooden formwork for vaulting was set in place. The scarcity of wood in much of the Roman empire demanded economy and ingenuity in preparing formwork. In some cases it is believed that a portion of a long vault would be built, and the forms lowered and moved on to construct the next part of the vault. Certainly, in all cases it was desirable that formwork be reused as much as possible. Alternate floors of multistory buildings were often supported by concrete vaults while the floors between would be framed with wood beams.[2] Planks that shaped the soffit of a vault sometimes were covered with one or two layers of flat square tile, but it is debated whether these tiles had the function of giving additional strength to the forms or simply provided a surface for plastering the underside of the vault once the wooden formwork had been removed.[3] It was the technique of concrete construction that constituted the basis of the spatial order found in Roman architecture.

Medieval architects relied on stone in most construction. For foundations, Gothic builders usually filled a trench with stone rubble and tamped it hard to serve as a level for the first course of masonry, but for "some important buildings, a better foundation was made with strong concrete of rubble and lime-mortar."[4] Entire walls might be made of rubble, to be plastered over, but often the outside surfaces were of carefully fitted ashlar, with mortar and small stones used to fill the inside of the wall.

Much of the early study of cement was directed at the improvement of mortar, particularly that used in the massive harbor works built for the eighteenth century's active maritime trade. Nevertheless, cement was recognized as having advantages when used as concrete. The French architect Philibert Delorme wrote in 1536:

> The excavations being made, whether for houses, harbors, bridges, or buildings in a marshy soil or even on land, and if, being deep and wide, stones of a large size cannot be used for the foundations, the best and surest method is to prepare a mortar composed of quicklime recently burnt, mixed with river sand which contains a quantity of pebbles of all sizes. . . . The composition thus executed hardens and solidifies so firmly in the foundations that, being heaped up in a mass and bound together, it becomes a uniform body or rock, such as nature forms, of a single block and so strong and hard that when dry it cannot be broken either by piles or any other instrument, nor can the pebbles be separated from it without breaking them to pieces.[5]

A variety of methods were used. According to George Godwin's essay on concrete, which won the medal of the Royal Institute of British Architects in 1836, for foundations a mixture of lime, water, and stone might be placed in a trench, layers of stone could be compacted with layers of grout poured over them, or the lime and stone could be thrown dry into a water-filled excavation.[6] Debates about critical innovations in the development of concrete were as common then as now, because instead of its being a new material it was developed gradually from the traditional methods of using grout in masonry. In fact, it was the late 1830s before the term "concrete" replaced "grouting" as the accepted reference to solid masses in which cement, sand, water, and stones were combined.[7]

Sir Robert Smirke, a Greek Revival architect best known for his design of the British Museum, exerted a strong influence on the popularization of concrete. It was Smirke's recognized skill in construction that led to his being asked to correct the faulty foundations of Millbank Penitentiary. The building had started as a project of Jeremy Bentham, the social reformer who had written on prison improvements, but a succession of architects of diminishing ability had tried to cope with the marshy site, in which foundations began to sink almost as soon as they were completed. In a complex operation, Smirke underlaid the penitentiary with a layer of large stone and gravel, pouring a thick covering of concrete over it.

Sixteen years later, when Smirke's British Museum was under construction, he followed his customary method for using concrete in new construction. Under load-bearing walls the concrete foundation's vertical dimension was 2½ feet, and the Museum's magnificent colonnade rested on concrete 6 feet thick.[8] The practice of the time was to make certain that each wheelbarrow of concrete was thrown into the excavation from a height well above the foundation. Smirke insisted on a height of at least 6 feet, and a specification cited by Godwin required a height not less than 9 feet.

At the Paris Exposition of 1855 there was exhibited a boat made six or seven years before of concrete (or cement mortar in present-day terms) and wire reinforcement by a Provençal landowner, Jean-Louis Lambot (fig. 7.2). The boat was almost 12 feet long and over 4 feet wide with sides around 1½ inches thick.[9] Soon after the Exposition, Lambot took out patents in France and Belgium, but his *ferciment* attracted little attention. In 1844 a Bristol physician named Henry Fox obtained an English patent for a floor system in which wood laths were fastened on the bottoms of cast-iron joists and the plaster troweled on them acted as a base for filling concrete between and above the joists.[10] Neither of these systems constituted reinforced concrete in which the compressive strength of concrete and the tensile qualities of metal worked together. Lambot's boat seems to have used wire to avoid the problems that would arise from cracks, and Fox's floors used concrete as a mass that spanned between the iron joists.

7.2 The remains of one of Jean-Louis Lambot's boats, as exhibited in a local museum. In 1955 a dry summer caused the water level in the lake on the Lambot estate to drop, and two concrete boats appeared in the mud, over a century after they had been fashioned out of *ferciment*. (*Concrete*, November 1967.)

7.3 A modern drawing of a Wilkinson floor system, built about 11 years after his patent, shows the square plaster forms and the very narrow beams between them. The upper drawing shows the center portion of a floor, and the lower drawing shows a supporting beam that spanned 9 feet. (*Structural Concrete*, July–August 1965.)

The essential nature of the combination of iron and concrete was understood by a plasterer, William Boutland Wilkinson, who manufactured plaster of paris and Roman cement in Newcastle upon Tyne.[11] In his 1854 patent Wilkinson described a shallow-arched floor slab in which iron strips cut as for barrel hoops would be set on edge at the crown of the arches "so that in this low position the strips may act with more power as tension rods." An alternative form of reinforcement was wire cable, which could easily be bought second-hand from mines. Patent drawings show the cable draped so that it was at the bottom of the slab at mid-span and at the top over supports, clear indication of an intuitive or reasoned understanding of the structural action required of a tensile material.

Even more indicative of Wilkinson's knowledge was the evidence discovered in 1954 in a residence he had built at his plasterworks around 1865. A careful inspection before the building was demolished showed that the floor of the upper story had been made by placing plaster molds upside down on formwork (fig. 7.3). The molds were about 22 inches square, and concrete was poured into the gaps between them to form a grid of beams. Concrete over the molds formed the floor.[12] Reinforcement for these small beams was wire cable, located at the bottom of the concrete but sloping upward at the ends. The slab that resulted spanned about 12 feet in each direction, and a grid of steel rods was placed at the bottom of the 1½-inch floor slab that covered the molds. In a larger beam the wire rope again rested at the bottom in the center of the span and sloped upward toward the ends.[13] The complementary actions of concrete and reinforcement had obviously been understood.

A different potential of concrete was explored when a patent for an "Improved Method of the Construction of Buildings" was taken out in 1875 by William Henry Lascelles, a general contractor in London. Before having his own construction company, Lascelles had worked with concrete in the construction of workers' housing for the Improved Industrial Dwelling Company, a philanthropic organiza-

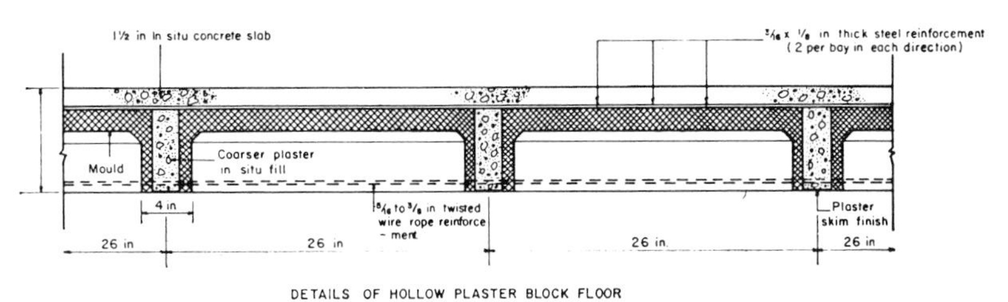

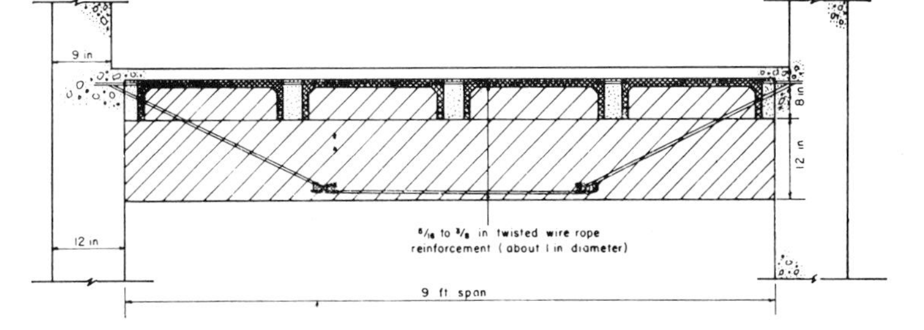

7.4 This two-bedroom cottage is one of the designs using Lascelles's system that were published by Richard Norman Shaw in 1878. The concrete panels are evident on the exterior, but they have no definite relationship to the location of interior partitions. (R. N. Shaw, *Sketches for Cottages and Other Buildings*, 1878.)

tion.[14] Lascelles's system was intended to be used in constructing inexpensive houses (fig. 7.4). A wood frame was erected, and on both sides of the studs and braces concrete panels, 2 feet by 3 feet and 1½ inches thick, were fastened with screws. The panels had simple butt joints at their vertical connections, which fell in the centers of the framing members, and Lascelles claimed that the walls admitted no water.[15] The joints of roof panels were filled with cement and covered with tiles. Lascelles experimented with concrete joists and rafters, but they proved to be much more expensive than wood. Many of the panels for exterior and interior walls and ceilings were molded in the patterns of traditional wood paneling; others for roofs and exterior walls bore a fish-scale pattern in imitation of roof tiles. The panels were reinforced by two half-inch-diameter iron rods placed diagonally, and floor panels had reinforcement of wire mesh.[16]

At the time that Lascelles patented his system for building low-cost cottages of concrete, he was acting as general contractor on projects under the supervision of Richard Norman Shaw, an architect particularly known for his tasteful and imaginative designs for residences. Around 1875

Shaw built a convent, for which the low budget encouraged use of concrete walls in much of the building, although portions of the concrete were covered with roof tiles. At the same time Lascelles was acting as contractor for a residence and studio that Shaw had designed for a popular genre painter, Marcus Stone. In Stone's house, molded concrete was used for the frames of three large windows that admitted light to the painter's studio on the top floor. The thin concrete frames, reinforced with a single rod in each mullion, permitted a lightness equal to that of wood, which the Building Act of 1774 had virtually prohibited on the face of buildings.

Although Lascelles built only a few of Shaw's projects between 1875 and 1878, there seems to have been a degree of collaboration in the development of Lascelles's patent. Two model cottages designed by Shaw were constructed on the grounds of the Paris Exhibition of 1878, and inside them Shaw's drawings of other designs using the patented system were displayed. At the same time in England a small booklet was published, *Sketches for cottages and other buildings: designed to be constructed in the patent cement slab system of W. H. Lascelles, . . . From sketches and notes by R. Norman Shaw.* The cottages were not among the best of Shaw's work. If they exhibited any of Shaw's principles, either in drawings or in those that were built, it was chiefly in their picturesque qualities and their color. Exterior panels were tinted by mixing the concrete with Spanish red, an iron oxide pigment that produced a color the public thought alarmingly vivid until time and soot had mellowed it to a rich terra-cotta tone. While the early hue shocked lay people and professionals alike, to Shaw any color was preferable to the pallid hues of classical architecture, and pigments that could provide lasting color in concrete were extremely rare. One significant change was made in Lascelles's system: concrete studs replaced the wood framing described in his 1876 presentation to the Royal Institute of British Architects, and metal members encased in concrete served as joists.

In the 1870s the most complete knowledge of the structural fundamentals of reinforced concrete seems to lie in the studies of Thaddeus Hyatt, a manufacturer of sidewalk gratings made of metal and glass that were employed to light basements.[17] Hyatt had begun manufacturing his "vault covers" in New York sometime during the 1840s. The business was well established by 1856, when Hyatt became active in support of the abolitionist settlers in Kansas. After he spent thirteen weeks in a Washington jail for his refusal to testify in Congress's investigation of the Harpers Ferry incident, the new Lincoln administration assigned him as American consul in the French port of La Rochelle, where he remained from 1861 to 1865. It seems safe to assume that Hyatt's curiosity would have led him to learn more of French experiments in concrete construction. When he left his consular post and returned to business, Hyatt immediately started experimenting in London with new ways of building sidewalk panels, and this led him to the study of concrete.[18]

Hyatt published his findings in 1877 as *An Account of Some Experiments with Portland Cement Concrete Combined with Iron as a Building Material.* All of Hyatt's tests, whether made by himself or in testing laboratories, are summarized to present the essentials on which the use of reinforced concrete is based today. He stated that the strength of iron does not withstand fire, and showed that all iron reinforcement should be completely sur-

7.5 It took four years to build W. E. Ward's concrete house in Port Chester, New York. Much of this time was due to caution. One year after a floor had been laid, Ward loaded it with 26 tons and left the load there through the winter. (*Transactions, American Society of Mechanical Engineers,* 1883.)

rounded with concrete, which resists fire. After a blazing fire had been kept under a concrete slab for ten hours, cold water was sprayed on the underside of the slab. No damage was detected, even after the slab had undergone a second heating and cooling.[19] Hyatt also discovered that the bond between iron and concrete was sufficiently strong to make reinforcement in the bottom of a beam act in conjunction with the concrete in the top, a fact that made a network of rods or straps more effective than the older method of embedding I-beams in concrete. Of the tests made on more than 50 beams by David Kirkaldy's laboratory, Hyatt wrote:

> A study of these [test results] will show, from the breaking of the metal, first, that all of the blades of iron were perfectly held while the beam was under strain; secondly, that the two materials worked in perfect harmony; and thirdly, that the proportionate power of the metal increased regularly as it became tie-metal . . . the higher the blades the more they lost as *tie*, and the more they gained as *compressive* material, the portion which came into compression being so much metal relatively lost, inasmuch as there was concrete enough to do this part of the work without it.[20]

Learning that the two materials expanded and contracted at almost the same rate under heat, Hyatt established the durability of the combination when subjected to fire or cold. For concrete columns Hyatt proposed a combination of longitudinal reinforcement and hoops. Furthermore, he patented precast planks of concrete that have a striking resemblance to many used today.

While Hyatt's reinforcement was straps of wrought iron placed on edge, he was aware that they would have been more efficient if placed flat and explained that he used them on edge only because the work was done more quickly that way. Iron straps afforded a larger surface area for bonding with the concrete than did round or square rods, and Hyatt favored the use of straps that had ridges on their surfaces to increase this bond.[21]

In the words of E. L. Ransome, the systematic and highly analytical investigations of Thaddeus Hyatt "ended the 'period of discovery' and put the theory of reinforced concrete construction on a rational basis."[22] Unfortunately, Hyatt's costly tests drew no investors and patent infringements eliminated his opportunities to profit from his discoveries.

A more profitable venture was that of a young French gardener, Joseph Monier, who began in 1849 to make tubs and pots of concrete. Finding the tubs brittle, he embedded in them a mesh of iron. Roughly contemporary with Lambot's boat, the plant containers built by Monier utilized very similar methods to combat an outward pressure, the opposite of pressures on a boat's hull. By 1867 Monier had advanced his method to the point of obtaining a patent and exhibiting his work at the Paris Exposition of that year. His first extension of the patent's function seems to have been a water tank constructed in 1872 to hold over 30,000 gallons.[23] Tanks built according to the Monier system were usually constructed by erecting a cylindrical cage of vertical and horizontal rods and attaching formwork on either the inside or outside of the rods. Concrete was then troweled on the reinforcing to attain the desired structural thickness.

Monier obtained a patent for the reinforcement of beams, a patent "backed up neither by theory nor by systematic experiment."[24] As a structural principle it did not greatly differ from the French floor systems that had been in common use for several decades. As an American engineer commented in 1900, "The Monier patent is so broad, embracing in a general way any iron parts enveloped by cement, that it is difficult to understand why all the other systems are not infringements."[25]

Little was done with Monier's patents until the German rights were purchased in 1884 and sold the following year to G. A. Wayss, a civil engineer and contractor.[26] Wayss quickly instituted thorough experimental studies, employing the services of a government engineer, Matthias Koenen. In 1886 the first fundamental analysis of reinforced concrete's structural action was published by Koenen in a German journal of construction management. The next year he collaborated with Wayss on a pamphlet, *Das System Monier*, that became the first German handbook of concrete fundamentals.[27] It is believed that the French gardener's contributions to the development of reinforced concrete were basically intuitive. In fact, the story has been told that when he was shown concrete with reinforcement placed at the bottom, Monier disagreed with Wayss about that placement of rods and closed the ensuing argument by asking, "Who is the inventor, you or I?"[28]

In 1873, William E. Ward, a mechanical engineer who had been a founder of the Russell, Burdsall and Ward Bolt and Screw Company, decided to build himself a house in Port Chester, New York, on a hill overlooking the river and the factory of which he was manager. Ward engaged the services of Robert Mook, a New York architect, but few have flattered the style of the building, Second Empire forms with two towers, crenellated and machicolated.[29] There is little wonder that in Port Chester the house was called "Ward's Castle."[30] In its structure, however, the building was revolutionary (fig. 7.5). The entire house is of concrete, except for doors and windows, and the concrete was moistened with only enough water to result in adhesion of the materials and permit its being tamped in the formwork. Floors were made of concrete with small reinforcing rods and I-beams embedded in it. Their construction followed the method used by Ward in preparing an experimental beam, 5 inches wide and 12 inches deep. First, a 1-inch layer of concrete was tamped in the bottom of wooden formwork and, after an iron beam 4 inches high had been placed on the

7.6 In 1901 Ernest L. Ransome constructed this four-story machine shop for a manufacturer in Greensburg, Pennsylvania. Floors and columns were molded in turn, leaving expanses of window to light the factory's work areas. (E. L. Ransome and A. Saurbrey, *Reinforced Concrete Buildings*, 1912.)

7.7 The first step toward Ernest L. Ransome's Unit System was his U.S. Patent no. 694,580. As each floor slab was poured, extensions above and below completed the wall, except for the insertion of windows.

concrete, more concrete was filled around the beam and to the top of the formwork, 7 inches above the top of the iron. About eight years later Ward explained: "The reason for placing the iron beam so near the bottom of the mold was to utilize its tensile quality for resisting the strain below the neutral axis when this composite beam was exposed to heavy loads, while the *béton* above this line was relied on for resisting compression from load strain."[31] The concrete beams varied in span and size, and above them the floor slab was formed with 1 inch of concrete beneath iron rods (5⁄16 of an inch in diameter, spaced 8 inches apart in each direction) and with 2 more inches of concrete above the rods.

An exterior finish was provided by painting the walls with a mixture of cement and sand before rubbing the surfaces with abrasive stones. Plaster was applied on inside wall surfaces. One of the building's towers contained two concrete tanks, one for rain water and the other filled with water pumped from a spring. Porch columns were hollow and carried off water from the roof. Experiments made by Ward showed that it was possible to build lightweight interior walls of thin panels, two thicknesses of 2½ inches of concrete being cast with 6 to 10 inches of space left between them. Every 2 or 3 feet connectors between the surfaces unified the wall, and the space between was utilized to provide an ingenious radiant heating system. From a heater in the basement hot air rose within interior partitions, moved through channels in the floors, and returned to the basement through the outside walls. It was the gravity warm air system of that period converted to provide radiant heat. Although delay is always a problem of radiant heating, Ward reported that it took only five hours to bring all spaces of the house to 68° F when temperatures outside were about 30° F.

While the major American contribution to the theoretical development of reinforced concrete was made by Hyatt's experiments, mostly executed in London, the major practical advancement in the United States was the work of Ernest L. Ransome, son of the English inventor of a rotary kiln for firing cement. Ransome came to the United States to promote Portland cement manufactured by his father's firm, and soon he found his way to San Francisco, where in 1870 he became superintendent of the Pacific Stone Company. The principal uses of concrete at that time were foundations and the floor arches poured between I-beams. Ransome replaced the I-beams by placing a rod at the bottom of the concrete, therewith making concrete beams between the floor arches. This change effected considerable savings, but the cost of rods was doubled by the need to thread the ends of round rods and attach nuts in order to tie them firmly in the concrete. A less expensive method, which Ransome said was discovered while he twisted a rubber band, required twisting the entire length of square rods. Improvising a machine for this purpose, Ransome found that he could inexpensively twist bars up to 2 inches thick.[32] This formed the basis of his 1884 patent, and it was the cornerstone of the Ransome system of reinforced concrete construction.

By 1888 Ransome had moved from the construction of what he called "small and unimportant structures" to entire buildings of considerable size. There were claims that by twisting bars he weakened the metal in them, and many doubted the strength of his concrete floors. A test and demonstration was carried out during construction of a building for the California

Academy of Sciences: "To satisfy all skeptics in regard to the strength, a section of the second floor 15′ × 22′ was uniformly loaded with gravel to 415 lbs. per square foot; the deflection was ⅛″. For the further satisfaction of

the doubtful, the load was left on for four weeks, but very few availed themselves of invitations to examine the work a second time."[33] An addition to a borax plant at Alameda, California, marked Ransome's first use of ribbed floor construction, in which an arched underside of the floor slab was replaced by T-beams. This system, which had been used three years earlier in construction of an Amsterdam library, freed concrete construction from the traditions of iron beams and plaster arches.

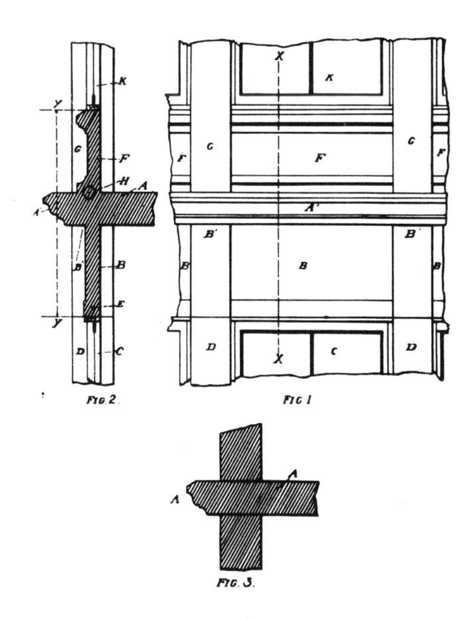

Through much of the 1890s Ransome spent his time promoting the Ransome system in competition with the other concrete systems then marketed in the United States, most of them distinguished from the others by a particular shape of reinforcement.

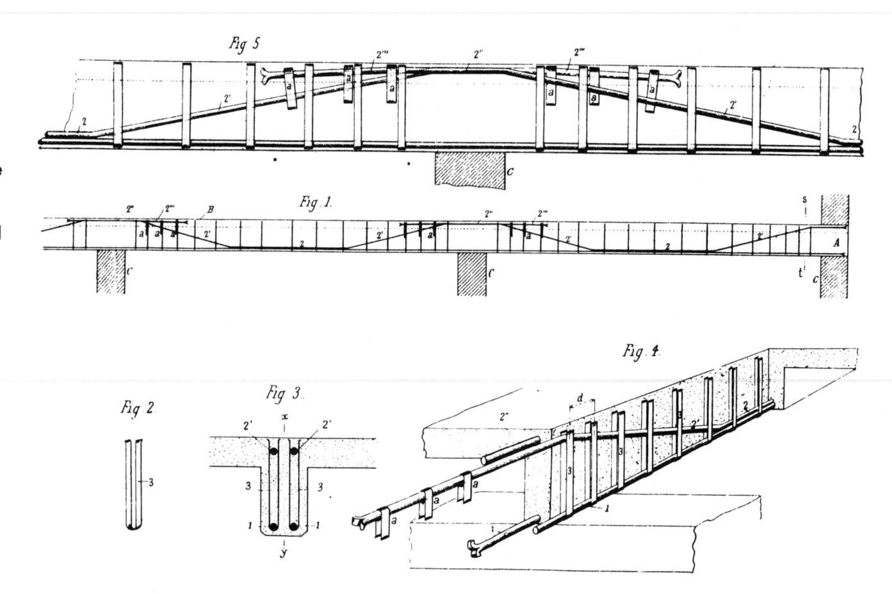

7.8 François Hennebique's patent "Improvements in the Construction of Joists, Girders and the Like of Cement." The vertical straps that serve as stirrups are correctly spaced more closely near the ends of spans. (British Patent no. 30,143.)

Moving his field of activity to the eastern states, he and his staff specialized in the construction of industrial buildings. In 1902 he patented a system in which his T-beam floors were combined with columns to form the concrete frame that was destined to dominate construction of factories and warehouse buildings (fig. 7.6). With the Kelly and Jones machine shop in Greensburg, Pennsylvania, Ransome ceased to treat concrete walls as substitutes for load-bearing masonry construction, and light curtain walls or broad expanses of windows were used to fill the bays of the concrete framework (fig. 7.7). Ransome's role was that of consulting engineer and licensor of others to use the Ransome system, which was principally characterized by the use of twisted square reinforcing bars.

Leadership in commercial development of reinforced concrete construction was first assumed by the German firm of Wayss and Freytag, which had bought and advanced the Monier system. Their system dominated Germany and Austria, and at first they also completed a surprising number of projects in France. Subsequently the organization of François Hennebique, begun eight years after Wayss and Freytag obtained the Monier patents, rapidly advanced use of the new material. Hennebique had apprenticed as a stonemason, and by 1867 he had established himself as a contractor with special capability and interest in the restoration of churches. His first reinforced concrete structure was a house in Belgium, for which Hennebique had planned to use Henry Fox's system that filled concrete between and over iron beams. While the plans were in preparation, a house of that construction burned near the site intended for the building, and Hennebique's client became justifiably determined that his house should not be prey to fire damage. Hennebique's solution was to dispense with iron beams and replace them with steel rods near the bottom of the slab, but covered with enough concrete to prevent their being weakened by intense heat.[34]

In 1892, when he was 50 years old, Hennebique took out patents that covered more than a decade of his experimentation, and he forthwith closed his contracting business. An enviable organization was quickly put in place. Establishing his office in the role of professional consultant, Hennebique chose only the most reliable contractors as licensees. By building a technical staff and appointing agents in different parts of the world, he developed a marketing organization that permitted rapid expansion at the same time that close control could be maintained to assure quality. In its first seven years the Hennebique organization carried out over 3,000 projects, and it averaged about 100 bridges per year. Starting in 1892 with a single office and two engineer-draftsmen, five years later Hennebique had 17 offices, 56 engineer-draftsmen, and 55 licensees. By 1909 he had 62 offices, 43 of them in Europe, 12 in the United States, and the remainder in Africa and Asia.[35] A large part of the company's success may have been due to the remarkably advanced business techniques with which Hennebique operated his far-flung enterprise. Its magazine *Le Béton Armé* appeared monthly with informative articles for the Hennebique staff, as well as reports that advertised the company's accomplishments. Engineers joining the firm underwent a training program. Banquets celebrated every thousandth contract fulfilled, and the company slogan *Plus d'Incendies Désastreux* was boldly stamped on every drawing that the company issued or approved. One of Hennebique's most effective advertisements was the firm's offices built in Paris on Rue Danton (1898), a stone-looking building in which the thinness of concrete construction had made it possible to gain an additional story within the city regulations on height and some 90 square feet of floor area at every level.[36]

The fundamentals of Hennebique concrete design were incorporated in his patent for the reinforcement of beams (fig. 7.8). Bars were set near the bottom of the concrete, some extending the full length of the beam and others bent to the top as they neared the supports at the ends of the beam. These two sorts of bars were fastened together by slender U-shapes of strap iron (stirrups, as they are now called), which counteracted horizontal shearing forces toward the beam's ends, and the bars at the top of the beam resisted tension there when the beam extended across supports. This patent summarized the soundest discoveries that had been made in the previous two decades, and there are many who have attributed to Hennebique more eclectic and selective ability than inventive or theoretical insight. However, as the organizer of a corporation that could furnish professional services and extend the use of a new material, Hennebique had no significant competitors.

As there was inevitable competition among the commercial proponents of different systems of reinforced concrete construction, particularly those spawned by early activity in France, there frequently were concomitant battles over patent rights. Around the turn of the century Simon Boussiron and Paul Piketty built some structures in Paris that used methods resembling those introduced by Hennebique. Thereupon Hennebique sued, charging infringement of his patent of 1892 and the supplement he had filed the following year, which had introduced stirrups. The defendants replied with an overall challenge to the validity of Hennebique's patent. They pointed out that in an 1878 addition to his basic patent, Monier had described his system of reinforcing beams, and the drawing submitted had shown some rods bent in a U-shape and placed upright. Although Hennebique claimed his patent offered the advantage of anchoring longitudinal rods to the stirrups, the patent court was unmoved. Its decision stated that the patent of Buissiron was invalid because it was not sufficiently different from that of Hennebique, but the court furthermore declared Hennebique's patent void because it duplicated significant features of the Monier patent and its attachments.[37]

In 1898 Thomas A. Edison took a look at the cement plant he had established a few years before as an offshoot of his ore-milling investments. By 1902 the Edison Portland Cement Company had spent 1½ million dollars on its newly opened New Jersey plant. Three years later the investment had doubled and it was reported by accountants that the sizable plant, covering an area of 80 acres, could only hope to be profitable and efficient in operation when orders were more numerous and prices were high. Edison's reaction was to search for a way to increase the demand for cement, and from this came his interest in the design of concrete housing.[38] A series of house designs were developed during his experiments.

The system that Edison pursued in all cases involved the assembly of metal formwork to make a mold for the entire structure, doing away with the piecemeal, level-by-level process of placing formwork and pouring concrete (figs. 7.9, 7.10). The problem was the development of a mixture and a method by which concrete could successfully make its way into the far corners of a form so complex, without there being voids and weaknesses that might affect the soundness of the structure. As in many of Edison's projects of this period, claims and doubts have produced an intricate web of narrative, and one of the simplest evaluations was that made collaboratively by two civil engineers who were sent by the journal *Cement Age* to visit Edison's laboratory at West Orange, New Jersey. One engineer represented the National Association of Cement Users, the other the Association of American Portland Cement Manufacturers. The engineers' reports, as published in 1908, describe many of the characteristics of the scheme, but some features are not defined, because of the changeable nature of the design, which was still in a very formative stage.[39]

The form for the house was to be made of cast-iron plates, varying from ½ inch to ⅞ inch in thickness. Inside surfaces of these plates were nickel-plated or surfaced with bronze where particularly fine detail was desired. Within the mold, reinforcing rods ½

7.9 Edison's patent for concrete construction was a simple proposal. A metal mold for the entire structure would be assembled and concrete poured in at the top. (*Cement and Engineering News*, January 1924.)

7.10 This is one of several designs for Edison's concrete houses. Other designs had more pronounced delineations of shapes, and this one may have been a relatively conservative effort to simplify the flow of concrete in the form. (*Scientific American Supplement*, 18 April 1908.)

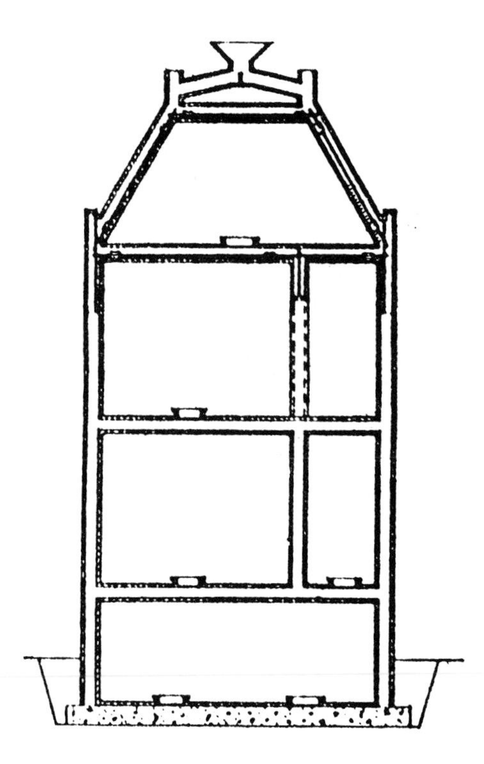

and ⅝ inch in diameter were placed, and all gravel was screened through a half-inch sieve so that there would be less likelihood of clumps of gravel preventing concrete's penetrating all parts of the form. Because the concrete mix was necessarily liquid, Edison added a colloid—in this case a clay—to prevent separation of the ingredients.[40] The flow of concrete had been tested in several small-scale molds, and at the time of the engineers' inspection a larger test was being made with a U-shaped mold that had a 4-inch-square cross-section throughout, a horizontal run of 26 feet, and vertical legs 10 feet high. When Edison poured his concrete into the top at one end of the mold, it flowed across the horizontal and rose 4½ feet in the other vertical.

The engineers were unable to examine the concrete and determine its quality, since forms had not been removed at the time of their visit. Judgments of both engineers very strongly questioned the optimistically low estimate of cost that had been provided by Edison, and both were concerned about the separation of ingredients as the concrete flowed through the mold. One of them concluded that it was "entirely unlikely that in such a mixture a concrete of 1:3:5 proportions in flat slab construction would even be self-supporting at the end of six days," and rapid removal and reassembly of formwork was the key factor in the low cost that Edison predicted for houses built as a group.[41] The project did not achieve Edison's goal and was abandoned.

Advances in the theory and practice of reinforced concrete construction meant little until technical information reached engineers in a form that could be readily employed. At the Ecole des

7.11 With a steam engine driving the equipment, the mixing of concrete in the 1860s made little provision for accurately measuring ingredients. (F. Beckwith, *Report on Béton-Coignet*, 1868.)

Ponts et Chausées, still one of France's principal schools of engineering, formal instruction in the design of reinforced concrete was begun in 1897 by Charles Rabut; but, on the whole, publications kept engineers informed long before professors did.

An 1871 report on the Coignet system of *béton aggloméré* was published in Washington by Major General Q. A. Gillmore, and little appeared through the next 15 years except Thaddeus Hyatt's privately published book of 1877. The formulas developed by Koenen for the German owners of Monier's patents were published in a periodical in 1886 and as a pamphlet the following year. In Germany and Austria this publication, *Das System Monier*, encouraged many engineers to investigate particular aspects of reinforced concrete design. The publications of Wayss and Company continued through the 1890s with other reports on the Monier system. Few significant publications on the subject appeared in France until late 1893 when Paul Planat, editor of the weekly journal *La Construction Moderne*, began a series of detailed articles on concrete engineering that appeared through almost a year.[42] At about the same time the British weekly *The Builder* printed a series of articles on reinforced concrete.

By the turn of the century, books on concrete engineering appeared more frequently. Many were routine repetitions of previous publications, but others were so influential that they ran through several editions and were translated into other languages. Paul Christophe, a former Hennebique engineer, published articles (later reprinted as a book) in a Belgian journal, showing detailed drawings of reinforcement placement and providing precise descriptions of a large number of projects.[43] The publications of the firms that owned patents on specific systems customarily showed finished structure without revealing details that were the firm's stock-in-trade, but Christophe's book withheld little.

A rapid succession of journals devoted to the subject of concrete construction were initiated in this period. The French organization of Portland cement manufacturers in 1896 began publication of *Le Ciment*, which combined engineering information with news on the manufacture of cement. Two years later the organ of the Hennebique system, *Le Béton Armé*, was begun. A number of American magazines (*Cement*, *Cement Era*, *Concrete*, and *Cement Age*) appeared in the first four years of the new century, and during that period two influential journals were started in Germany. Great Britain's first significant journal, *Concrete and Constructional Engineering*, did not commence publication until 1906, the delay a result of Britain's late activity in reinforced concrete construction.

Each country had its own standard books and manuals, although translation soon came to those that proved most useful. In 1902 Wayss and Freytag, concerned about the spread of so many theories other than their own, published *Der Eisenbeton*, written by the company's technical director, Emil Mörsch. Six years later it was translated and became a standby in English-speaking countries. Buel and Hill's *Reinforced Concrete*, published in New York, became popular for its middle section, which gave structural details of about 200 projects ranging from foundations to smokestacks. Charles F. Marsh's British book compared the different systems then in common use, focusing largely on Continental work.[44] Each handbook had its own virtues, but none could be complete because of the wide range of systems, theories, and practices at that time.

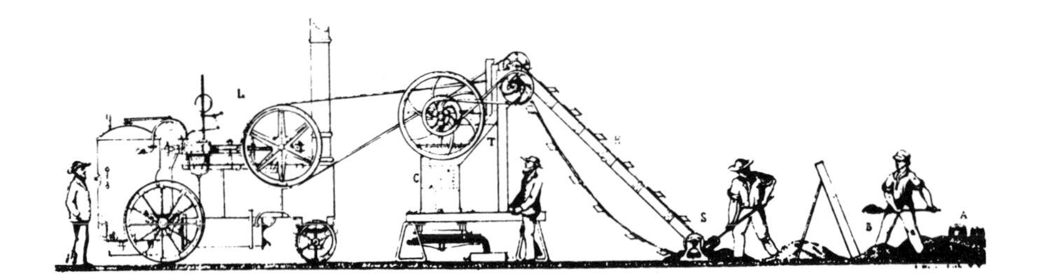

Loading tests and laboratory analyses were able to reassure many skeptical engineers and building officials about the strength of reinforced concrete construction, but the material's durability could only be truly substantiated by time. There were relatively few examples dated before 1880, and at the turn of the century there were many questions unanswered, particularly those about the possible damage that could result if reinforcement rusted. A 1905 report on advances in the use of concrete in Holland summarized the skepticism of a Dutch military engineer who opposed the use of reinforced concrete: "If the tension of the iron bars is allowed to surpass 200 or 300 kg. per sq. cm., fissures will appear in the concrete, which are sure to bring down the construction in the long run, when the iron has decayed by rust. . . . on the other hand, when less than 300 kg. of tensional resistance per sq. cm. is utilized from the iron skeleton, its presence is not to be defended from an economical point of view, and unarmed concrete might do as well."[45] The economic problems vanished as cement decreased in cost, but the fear of rust remained. Stories abounded about shovels, hammers, and muskets found unrusted in old concrete. In an informal discussion held at the 1902 convention of the American Society of Civil Engineers, there were proposals that reinforcing steel be sandblasted, painted with various waterproofing substances, or electroplated with copper or aluminum.[46] About 10 years later, members of the Science Committee of Britain's Concrete Institute visited France, where a roof slab that had been laid by François Coignet without waterproofing was opened by his son Edmond. The iron I-beams in the slab were found to be sound and unrusted after 60 years.[47]

Prior to the introduction of reinforced concrete, almost all construction procedures involved establishing physical relationships between objects that were supplied to the building location. But in concrete the material itself was formulated by workmen at the construction site; thereby the construction industry undertook "manufacturing" in addition to its customary procedures of assembly (fig. 7.11). With reinforced concrete, success depended on both the preparation of concrete and the accurate placement of the reinforcing metal—always assuming, of course, that the engineer's instructions were satisfactory. In 1903 the *Engineering News* stated its editorial opinion: "There is no doubt that a great deal of concrete work is built by men who are really not competent to undertake it, and that much work is done without sufficiently strict and continuous expert supervision to ensure the best and safest results. In view of the enormous increase in the use of concrete, and in the variety of purposes for which it is used, it is well for engineers to bear these facts in mind."[48] Fifteen years later the American Railway Engineering Association published a study of failures of concrete structures and classified them according to the most frequent causes: improper design, poor materials or workmanship, removing formwork too soon, and such external factors as fire and the uneven settlement of foundations.

At the 1900 Paris Exposition much attention was given to the concrete work of François Hennebique and Edmond Coignet, but newspapers around the world also carried news of the failure of a concrete footbridge on the grounds of the Exposition. The following year one wing of a five-story hotel in Switzerland collapsed while under construction. Workmen had poured concrete piers beneath a previously built concrete beam without supporting the beam adequately, but an investigatory board also found that the first-story columns had been designed too small, unwashed sand and gravel had been used, and other faults of construction had been permitted. When a two-story building being built for the Eastman Kodak Company in Rochester, New York, failed in 1906, careful examination showed that, although the collapse could be attributed to removing formwork before the concrete was ready, there were inescapable indications that a poor quality of cement had been used and wood chips, leaves, and other obvious impurities were found in the concrete.[49]

Because of these and many other failures, it was not uncommon for structures to be submitted to the most direct method of testing, the actual loading of slabs and beams. The Superintendent of Buildings for the Borough of Manhattan issued regulations in 1903 that required: "The contractor must be prepared to make load tests on any portion of a concrete-steel construction, within a reasonable time after erection, as often as may be required by the Superintendent of Buildings. The tests must show that the construction will sustain a load of

7.12 A Hennebique publication early in this century showed grain elevators in Tunis, which had tilted as much as 25° from the vertical because of unequal settlement. It was claimed that by loading the high side of floors and excavating under the higher parts of foundations the buildings were returned intact to a vertical position. (*The Hennebique Armored Concrete System*, circa 1908.)

7.13 In a relatively short period all of these theories attempted to explain the role that concrete played in a reinforced concrete beam. At the top of each graph the compressive action of the concrete is indicated, and at the bottom its response in tension. These remain arguable points in structural theory. (S. B. Hamilton, *A Note on the History of Reinforced Concrete in Buildings*, 1956. Building Research Establishment, U.K.: Crown copyright 1956.)

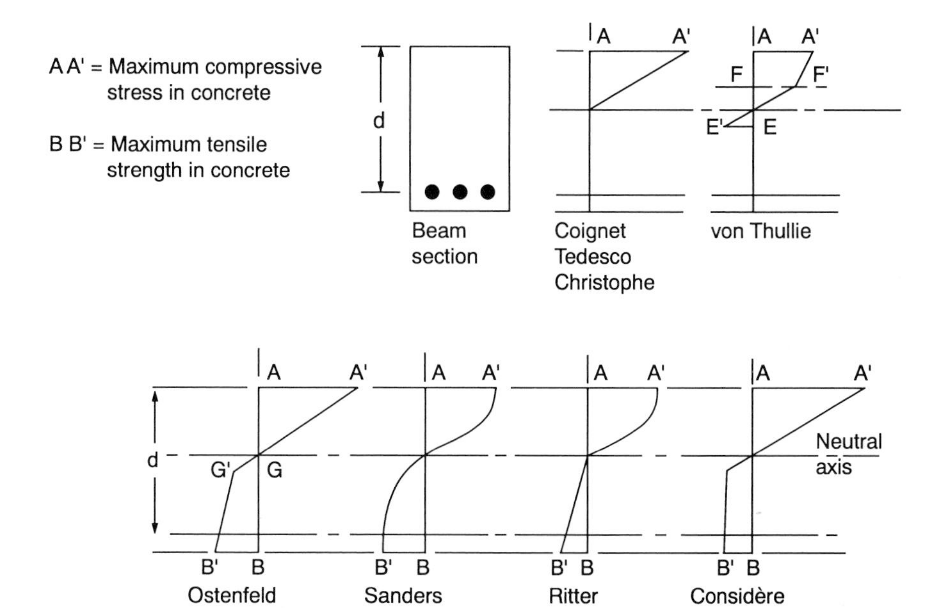

three times that for which it is designed without any sign of failure."[50]

As the use of reinforced concrete increased, professional organizations and governmental agencies moved to bring order to the extraordinary variety of theories, formulas, and practices that were employed (fig. 7.13). The German organization of architects and engineers, along with the German Concrete Association, in 1904 issued a preliminary draft of rules to govern the design, execution, and testing of reinforced concrete construction, and these rules became the basis of regulations that were promulgated shortly afterward by the Prussian government.[51] The regulations mandated examination of projects by building authorities, and their reports were to be utilized two years later in a review of the regulations. French regulations, as imposed two years later, were liberal, with the expressed desire of avoiding discouragement of experimentation and advances of technique. Maximum stresses allowed for steel, iron, and different qualities of concrete were stipulated, and other constants were specified—all conservative values, even for that time. In reviewing parts of the proposal, the editor of *Le Ciment* commented:

[Under these rules] the calculated bars in compression [in a beam] may be found often more important than those in tension, which appears absurd. In fact, applying the same rule to pillars, nothing hinders the provision of a metal section sufficient for resisting the load without any assistance from the concrete. . . . The men of science who elaborated these rules did not forget that some builders had already created their art, both practically and theoretically, a long time ago, and they wisely decided to limit their stresses, and in availing themselves of the knowledge [gained] by the leading constructors, to present the teaching of practical experience in scientific form, in order to prepare easy methods of controlling the works designed and carried out by less skilled men.[52]

7.14 A panoramic photograph of the construction of the State Normal School, San Jose, California, shows five towers and seven metal chutes used to bring concrete to all parts of the project. Chutes of this sort often extended well over 400 feet. The concrete was hoisted to the top of a tower, and there it was dumped into a tube through which it flowed to the point from which wheelbarrows would take it to the formwork. (*Architect and Engineer of California*, April 1910.)

In the same year the Royal Institute of British Architects appointed a Reinforced Concrete Committee which, with the support of other organizations and governmental agencies, was to investigate requirements for concrete construction. The British Fire Prevention Committee also designated a committee on the subject, and these groups were joined by reinforced concrete firms to form the Concrete Institute. From members of this organization there was assembled a Joint Committee, which issued its first report in 1907 and continued to revise it through the four years that followed. In many ways the stipulations of the report followed the example of the French regulations of 1906, which had been in several parts influenced by the publications of Armande Considère and Paul Christophe. When the London County Council in 1915 enacted regulations for the metropolitan area, there were but a few significant variations from the Joint Committee's report.

Standards for the use of reinforced concrete were developed in the United States by a Joint Committee, which included representatives of the American Society for Testing and Materials and the national organizations of civil engineers, railway engineers, and cement manufacturers. Taking a less sanguine view of the situation than their European counterparts, the U. S. Joint Committee found the results and interpretations of tests often inconclusive and decided to institute a program of fundamental research, marshaling the resources of 11 universities.[53] In 1903 there began seven years of laboratory tests, followed by five years of testing actual buildings. When it was released in 1917, the final report received strong criticism. Still, when a new code was issued eight years later, the 1917 regulations were remembered favorably. "The old Joint Committee Report was conservative in some respects but in [its] breadth of view it was a remarkable document as compared to the present report."[54]

Until almost the turn of the century concrete was usually used as a mixture so dry that, even after it had been dropped from a height, workmen tamped it to form a solid and unified mass. Traditional formulas, stated as a ratio of cement, sand, and stone, used a 1:2:4 mix. Several factors influenced the choice of proportions: a stiff mix with more stone was usually less costly; a mix with more water and cement filled the formwork more easily; and a denser mix was stronger. It was the accepted theory that sand should fill all voids between stones and that cement should only fill voids remaining between the grains of sand, and handbooks provided lengthy instructions on the gradation of stone sizes. As late as 1904 a specification assumed a stiff mixture, saying that it should "be of such consistency that when dumped in place it will not require much tamping."[55] In 1892 and 1897 a French government laboratory published the results of tests in which it was found that, when other factors were kept constant, the compressive strength of concrete varied with the square of the ratio between the amounts of cement and water.

Because of the honeycombing that resulted from inadequately compacted dry-mix concrete, contractors and engineers began employing a more liquid mixture. The rising costs of labor encouraged contractors to find a mechanical means of distributing mixed concrete to its formwork, and the growing importance of hoists and

derricks on construction sites led to the use of chutes and tubes. With a hoist tower at a central point where concrete was mixed, the concrete could be lifted to the top of a tower and there poured into chutes that extended as far as 500 feet from the hoist. If several chutes were provided, concrete could be directed alternately to several places about the project, from which wheelbarrows would take it the remaining distance to the formwork (fig. 7.14). A California maker of such equipment in 1910 boasted: "The old wheelbarrow method of mixing and distributing the cubic yard is from 60 cents to $1; by the use of carts, runs from 40 cents to 75 cents; by the use of 'Gravity System' it costs from 15 cents to 25 cents."[56] Such use of extremely fluid mixes appears to have been very much a phenomenon of American practice, and seldom seen in Europe. Wetter mixes offered a more watertight concrete and seemed advantageous in a period when there was great controversy about whether reinforcement rusted within concrete. Although stone tended to settle to the bottom of wet-mix concrete, producing a mass of uneven quality, American engineers and contractors strongly favored the generous use of water, particularly during the "wet-mix era" between 1905 and 1910.

Studies published in 1912 showed that excessive amounts of water reduced the density of concrete, thus lowering its strength. Six years later, when Duff Abrams issued his report on almost 50,000 tests, the matter seemed to have been settled, although the conclusion was perhaps overstated:

> The size and grading of the aggregate [stone] and the quantity of cement are no longer of any importance except insofar as these factors influence the quantity of water required to produce a workable mix. This gives us an entirely new conception of the function of the constituent materials entering into a concrete mix and is the most basic principle which has been brought out in our studies of concrete. . . . The equation expresses the law of strength of concrete so far as the proportions of materials is concerned. It is seen that for given concrete materials the strength depends on only one factor—the ratio of water to cement.[57]

By the time the water-cement ratio had been fully accepted as a critical factor, it was recognized that, while that ratio determined the fundamental

7.15 The quality of concrete mixes was often much more carefully controlled after the introduction of ready-mixed concrete in the 1920s. Here hydraulic dump trucks wait to load. (*Pictorial History of the Ready Mixed Concrete Industry*, 1964.)

7.16 The Cottancin system, used in this Paris factory, was one of about 21 reinforced concrete systems being applied in France around 1909. Cottancin did not believe the bond between concrete and steel was reliable, and therefore his system employed very narrow beams that were reinforced with a carefully woven mesh of wire. (*Concrete and Constructional Engineering*, July 1906.)

7.17 Anatole de Baudot proposed this design for a government building. His Gothic leanings are clearly indicated by the ribbed columns in the foreground, but in the upper structure he converts the Cottancin system into what he called "the solution through tangency." (A. de Baudot, *L'architecture, le passé—le present*, 1916.)

strength of concrete, the grading of stone and proportioning of ingredients provided a means of utilizing the ratio in an economical and workable mixture.

The most dramatic architectural development of early concrete construction was displayed in the church of St. Jean de Montmartre, designed by the architect Anatole de Baudot using the Cottancin system, still another patent that was challenged by Hennebique (fig. 7.16). This system was perhaps not really concrete, because large aggregate could not be used in beams so narrow. The whole web of the beams in the Cottancin system was reinforced by a woven network of wire or small round rods. Beams ran both in the customary rectangular pattern and as diagonals, and usually the diagonals were more numerous. A pattern of that nature was somewhat medieval in appearance and it is not surprising that Baudot, student and associate of Eugène-Emmanuel Viollet-le-Duc, the leading French Gothic Revivalist of the time, should find it challenging as a means of achieving large spans without the height required in arcuated systems (fig. 7.17).

Another very important consideration that is quite favorable to reinforced concrete is the ease with which it allows supports to be spread farther apart, a solution that is almost unavoidable today. Certainly we are able, like the Romans and Byzantines, to erect vaults and domes with great spans, to increase them greatly, thanks to assemblies of light-weight materials and reinforced concrete.[58]

Building authorities viewed Baudot's design with serious misgivings and, because of delays in granting the necessary approvals, the church of St. Jean de Montmartre was completed a full seven years after design was begun in 1897. When a British deputation from the Concrete Institute visited Paris around 1910 to inspect buildings in which reinforced concrete had been used, their tour included the church, and the committee found no fault with the condition of the concrete.[59]

In 1903, the year before St. Jean de Montmartre was at last completed, the first skyscraper of reinforced con-

crete, 16 stories high, was built (fig. 7.19, 7.20). The Ingalls Building in Cincinnati, Ohio, was designed by the architectural firm of Elzner and Anderson, and engineering of the concrete frame was done by a Ransome licensee. Most of the building's floor area was framed in rectangular panels 16 by 32 feet, subdivided into two squares by an intermediate beam. Instead of the ribbed slab used in most of Ransome's work, the engineer made floors of concrete 3 to 5 inches thick and reinforced with twisted square rods.[60] The exterior finish of the Ingalls Building was set against concrete walls 8 inches thick, except for party walls that were 3 to 4 inches thick. The bottom three floors were faced with marble, the top story and cornice with white terra-cotta, and the stories between with gray brick.

7.18 Franz Visintini, an Austrian working in Switzerland, produced precast lattice beams, either 8 or 12 inches wide. The thickness of the bottom and top surfaces could also be varied. Larger spans and heavier loads were carried by stacking the beams two deep. (*Journal, Western Society of Engineers,* November–December 1904.)

7.19 The Ingalls Building (Cincinnati, Ohio: Elzner and Anderson, architects) occupied a site only 100 by 50 feet and its 16 floors did not achieve a spectacular height for the time. Nevertheless, it demonstrated that all the characteristics of the skyscraper, which had been developed in steel-frame construction, could be present in a concrete-frame building. (*Engineer,* February 1906.)

7.20 Attaching a terra-cotta cornice to a concrete building frame was much the same as attaching it to a steel frame. On the Ingalls Building, Cincinnati, Ohio, the walls below this cornice were of masonry. Brick was supported at each floor by a 3-inch projection of the floor slab. (*Engineering Record,* 23 May 1903.)

However, the architect considered such coverings to be superficial:

> It is not incumbent upon us to face the concrete with marble, or brick and terra-cotta, as was done in the Ingalls Building, for reasons of momentary expediency, for as the state of the art advances, the architectural forms, moldings and what not, will be incorporated with the molds for the structural work, and upon removing the formwork, the surface of the exposed concrete, will be given the desired finish of rubbing or tooling, as the case may be.[61]

With the Ingalls Building the experience with fireproofed steel office buildings was transliterated into the techniques of reinforced concrete.

The Ingalls Building looked like any other office building of that time. On the outside, masonry concealed every line of the concrete frame; on the inside, plaster covered the structure. At the same time that the Ingalls Building was under construction in Cincinnati, Auguste Perret was building his first reinforced concrete frame structure, an apartment building at

25b Rue Franklin, Paris, which was to be occupied by Perret's parents and offices of the family construction firm. Like Hennebique's offices on Rue Danton, this project was required to make the most of a small site, and reinforced concrete seemed a logical way to avoid thick walls of masonry that would occupy precious floor area. Perret in 1936 explained his treatment of the exterior: "It was the first house built of reinforced concrete and exposing its skeleton, as it is done today. At that time I thought that in order to preserve the steel it was necessary to cover it with a facing. The proper material seemed to me to be tile, but I used the tile in different patterns, depending on whether it was used on columns or walls, outlining in this way the skeleton."[62] Flat tile covered the concrete frame, and the masonry curtain walls bore slightly darker tiles stamped with a floral pattern (fig. 7.21). Although the facade angles in and out in order to capture as much light as possible from the narrow frontage, there is a classical severity in the design, except for the hint of Art Nouveau in the floral pattern of some tile. Perret and his family were engaged in construction, but for this project the work in concrete was executed by a subcontractor who was familiar with the Hennebique system.[63] From that time Perret's work, including auditoriums and industrial buildings, explored the structural and visual characteristics of reinforced concrete construction. The material suited his interests in structure—"the architect's mother tongue," he called it. Two decades after his apartment building on Rue Franklin, Perret designed his crowning achievement in the use of concrete, a church in Le Raincy, an industrial suburb of Paris. There the framework was exposed, even showing the pattern of planks used in the formwork. Peter Collins

has pointed out the contrast between the opinions of an architectural periodical which believed the church was "the most significant building since Henry Labrouste's reading room in the Bibliothèque Nationale" and a civil engineering journal which insisted that the concrete should have been concealed behind some veneer of more proper materials.[64]

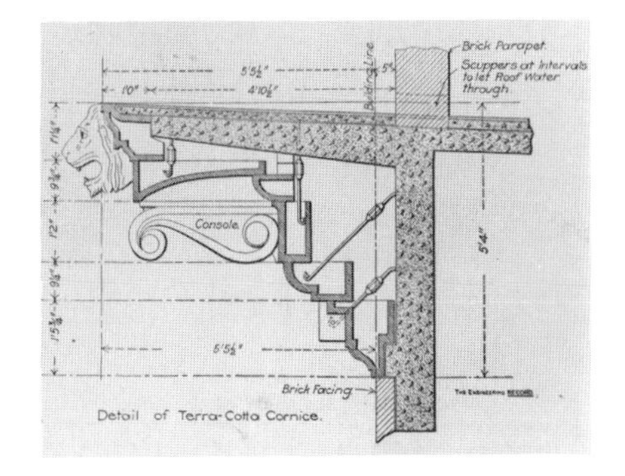

7.21 Auguste Perret covered the concrete frame of the apartment building at 25b Rue Franklin (Paris, 1903) with tile and used tile with a floral pattern on the wall panels. Because of his need to gain the maximum amount of light through the narrow facade, windows are more evident than the patterned walls. (Courtesy of Architectural Library, North Dakota State University.)

7.22 Thaddeus Hyatt's 1878 patent shows his grid of reinforcement and some of the precast units that were included in the patent. Besides reducing a floor's weight, Hyatt explained that the holes in some of these floor systems could serve as "longitudinal flues or ventilating spaces." (U.S. Patent no. 206,112.)

7.23 Two kiln houses for the Edison Portland Cement Company were constructed in 1907 with precast columns, roof slabs, and girders 50 feet long. Rotary kilns for cement making are shown inside. (*Engineering News*, 4 July 1907.)

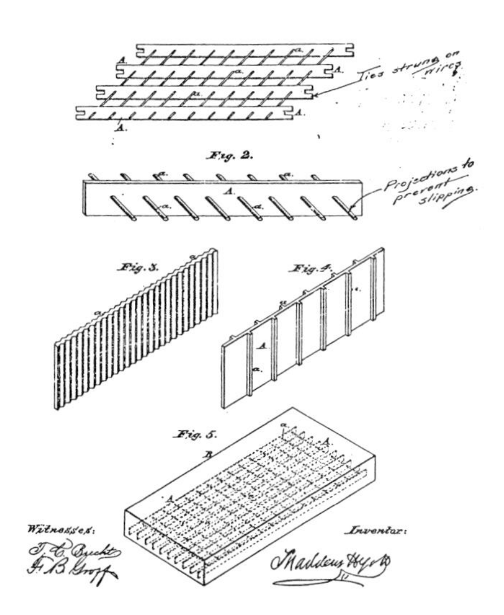

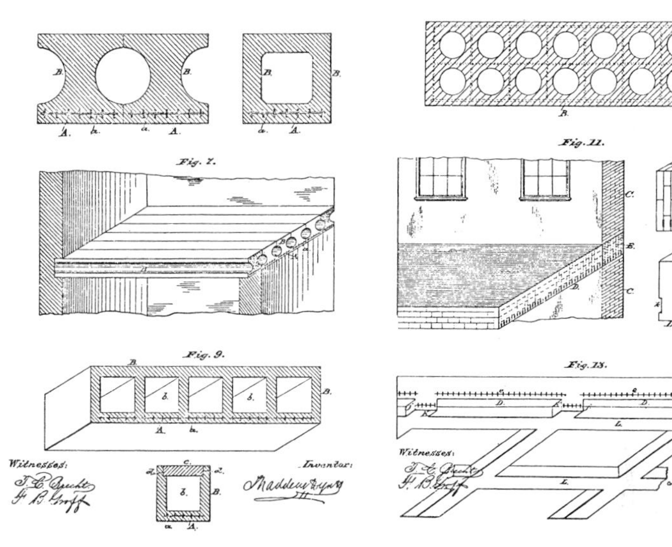

The works of Auguste Perret were not structurally adventurous and his designs, no matter how daring in other ways, had a traditional air. His contemporary, Tony Garnier, as architect for major projects of the socialist city government of Lyons, moved closer to the forms and methods of the future. Together, Perret and Garnier established reinforced concrete as the dominant medium of modern architecture in France.

Beams, columns, and floor slabs of precast concrete were as elementary to construction as the traditional stone blocks and heavy timbers. Hyatt's patents of the 1870s show precast planks and beams spanning between walls—not surprising since his business was manufacturing sidewalk panels (fig. 7.22).[65] In 1900 a stable was built in Brooklyn, New York,

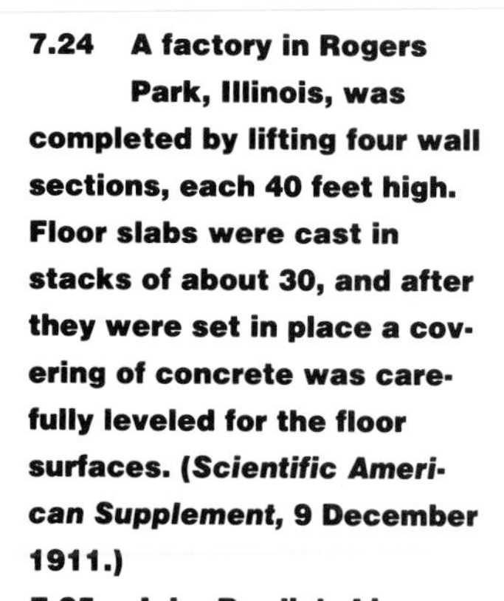

7.24 A factory in Rogers Park, Illinois, was completed by lifting four wall sections, each 40 feet high. Floor slabs were cast in stacks of about 30, and after they were set in place a covering of concrete was carefully leveled for the floor surfaces. (*Scientific American Supplement*, 9 December 1911.)

7.25 John Brodie's Liverpool housing employed a variety of panels with their notched edges dovetailed together. The floors had an inlaid finish of wood, and the walls were prefinished, some having color in the concrete. (*Journal, Royal Institute of British Architects*, September 1969.)

using precast roof decking 2 inches thick. Concrete slabs, 4 feet wide and 17 feet long, were secured to steel roof trusses, and similar panels served as partitions within the stable.[66] A more complete use of precast concrete was made in a four-story factory building constructed five years later in Reading, Pennsylvania. The factory's walls were brick and concrete columns were cast in place, but the roof and floors were carried by precast girders 24 feet long holding beams that spanned 12 feet. To reduce their weight, both girders and beams were perforated in a trusslike pattern.

Little formwork was needed when slabs were to be precast. The roof slabs employed in 1907 for the Edison Portland Cement Company, New Village, New York, were molded one on another with layers of paper as separation (fig. 7.23). In other cases, a coating of liquid soap was painted between to keep slabs from bonding to the ones beneath. By 1910 there had even been methods developed for casting panels in which the two surfaces of concrete were separated by an air space. In the United States the procedure used for constructing a cavity wall as a single panel involved first pouring a slab of concrete about 4 inches thick; then adding a layer of paper, 4 inches of sand, and another layer of paper; and finally adding 4 inches more of concrete (fig. 7.24). Connections between the two concrete surfaces were made with metal brackets or by removing portions of the sand layer.[67] As such panels were tilted to their intended position, rods were employed to make sand flow from the center.

In 1903 John Brodie, city engineer of Liverpool, presented his scheme for building less costly housing for the poor to the city's Housing Committee.[68] Eighteen months earlier Brodie had begun experiments with concrete containing clinkers from the furnaces of the Liverpool waste disposal system. As shown in his 1901 patent, Brodie employed precast concrete panels, whether walls or floors, with notched edges, the tenons arranged so that four walls could interlock along the same line (fig. 7.25). In this way the intersections of panels formed stiff verticals, eliminating the need for columns. The underside of floor panels had ribs that curved to an increased depth at the center of the panel.[69] Several tenements, stables, and an exhibition cottage were built between 1904 and 1906 using this system. Although the system was short-lived, Brodie's successes were widely known and influenced similar systems that were developed in Holland and the United States.

"Unit system" became the general term for construction methods using precast elements during the period before World War I. Systems for housing proliferated, although none were to affect significantly the accepted manners of constructing residences. Ernest L. Ransome developed the "Ransome Unit System," a method of constructing factory buildings and warehouses with precast concrete. For a four-story building erected in 1911 for the United Shoe Machinery Company of Beverly, Massachusetts, Ransome designed concrete columns, girders, beams, and wall panels that were all precast. Floor and roof surfaces were poured in place, and they did much to unify the assembled parts.

A more involved system was based on the many patents taken out by John E. Conzelman in the years from 1910 to 1916 (fig. 7.26). The "Unit Structural Concrete Method," as developed by his Unit Construction Company of St. Louis, Missouri, was principally utilized for industrial and railroad buildings. A major early (1911) application of this unit system was the five-story building of the National Lead Company, St. Louis, Missouri.[70] The structure was entirely assembled of precast elements, and it was designed to withstand floor loads of 500 pounds per square foot. Five years later a far different project was undertaken by Conzelman's firm for the Youngstown Sheet and Tube Company in the wake of fires and riots cause by workers in that Ohio city. In tardy response to their workers' housing problems, the company constructed 281 residential units with Conzelman's unit system. Part of this housing remained in use over 60 years after its construction.[71]

Most structural uses of reinforced concrete had followed the traditions of construction in other materials—columns, girders, beams, joists, and decking in a succession of elements that were a step-by-step reduction of the structural task in dimension and difficulty. An early departure from tradition was flat slab construction that dispensed with girders, beams, and joists, leaving only the column and a horizontal slab. In 1902 O. W. Norcross, the contractor on many of H. H. Richardson's buildings, was granted three patents related to a flat slab system.[72] These patents he sold to another company, which successfully defended them against a series of court challenges instituted by C. A. P. Turner, an engineer in Minneapolis, Minnesota. An article written by Turner in 1905 showed a concrete column with a slightly flared capital supporting a slab 7 inches thick (fig. 7.27).[73] The advantage most emphasized in Turner's article was the economy of eliminating the costly formwork required for beams and joists.

The following year Turner built the Bovey-Johnson building, five stories of flat slab construction. Minneapolis building officials insisted on testing the structure and found that almost three times the design load caused a deflection of only one-fourth of an inch. The new system proved to be popular for buildings with heavy loading, and it is reported that in the next seven years 80 percent of all buildings in the United States designed for loads of 100 or more pounds per square foot of floor area employed flat slab construction.[74]

In Europe the Swiss engineer Robert Maillart experimented with mushroom columns through 1908 and 1909. His first use of the mushroom column was a five-story industrial building erected in 1910. There were two clear differences between the approaches of Turner and Maillart: in placing grids of reinforcement in the slab, the Swiss engineer followed the two directions along which the columns were aligned, while the American put equal amounts of reinforcement in those directions and along the diagonals; and while Maillart insisted on a subtle curvature blending the undersurface of the slab into the column's shape, Turner used a more abrupt juncture of the two.

By the time both Turner and Maillart were building flat slabs and mushroom columns there was an obvious need for an accepted theoretical basis for the engineering design of such

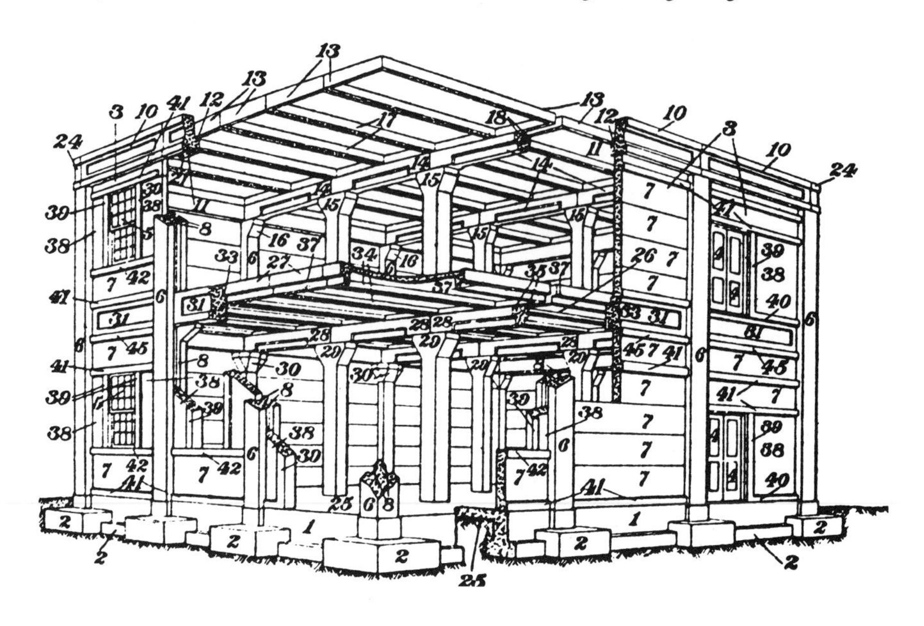

7.26 The Unit System developed by John E. Conzelman of St. Louis made use of more than 51 U.S. patents he obtained from 1910 to 1916. Applications of the system were mostly industrial and railroad buildings, although in 1914 about 50 shelters were constructed for passengers along the Pacific Electric Railway in Los Angeles. (*Journal, American Concrete Institute*, February 1954.)

7.27 In C. A. P. Turner's publication of his flat slab system of construction, a large circle of reinforcement was placed above each column, forming the basis of a four-directional pattern of slab reinforcement. (H. T. Eddy and C. A. P. Turner, *Reinforced Concrete Buildings*, 1914.)

structures. Franz Grashof had published in 1866 the first explanation of a plate's action when supported by equally spaced columns, and this was the basis of several theories. Others were derived from an assumption that circles of slab around the columns acted as cantilevers and the remainder of the slab was suspended from the ends of the cantilevers. One skeptic assumed a single set of conditions and calculated slab thickness and the reinforcement according to the formulas of the six most popular theories. For all but one formula the required slab thickness was the same, but the calculated amount of required reinforcement varied as much as 400 percent.[75] The smallest amount of steel resulted from the formula used by Turner. Arguments among theorists grew bitter, with phrases such as "multifarious absurdities" and "a riotous license of figuring" hurled at opposing factions. The theorists argued vociferously, but in the previous seven years Turner had completed buildings and bridges worth $200,000,000 and cautious clients and building authorities had required the testing of many of those projects. Against practical experience the fine points of theory had little chance to survive.

The first concrete shell was a railroad station at Paris-Bercy designed in 1910 by Simon Boussiron, the engineer of several major bridges and canal projects. The vaults of the station's freight-receiving room spanned between concrete beams that were themselves large enough to span the distance and support the vaults. In a similar freight room, built the following year, it was recognized that the vaults themselves had considerable spanning capability, and the beams on which they rested were reduced to little more than the concrete necessary to adequately cover reinforcing rods. Just before World War I, concrete air-

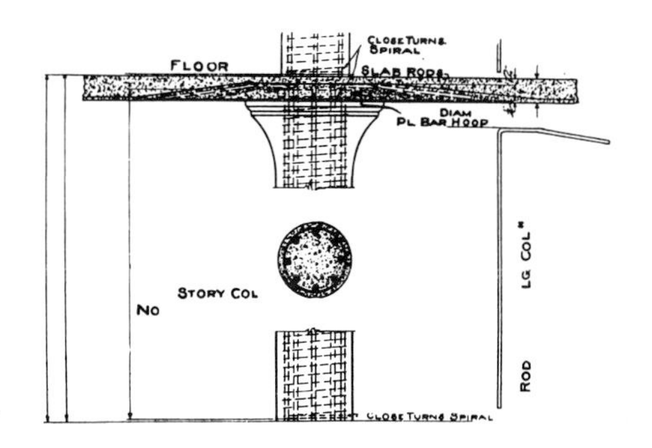

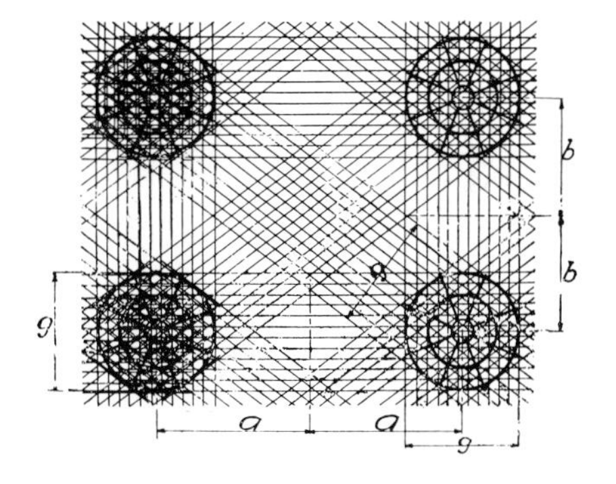

craft hangars were proposed to the French government, and the idea was revived in 1916, when eight were built of thin barrel vaults with ribs on the outside. These structures could be built quickly, and military authorities launched construction of many more designed in much the same way. The largest of this type, built in 1919, used three parallel vaults connected by a transverse vault and thus providing an unobstructed space about 395 feet by 148 feet. After the war, prospects of transatlantic transportation by dirigibles led to the construction of two hangars at the Orly airport outside Paris. The designer, Eugène Freyssinet, spanned 320 feet with a parabolic vault made of angular corrugations from 10 to 18 feet in depth. The concrete was not quite 4 inches thick.[76]

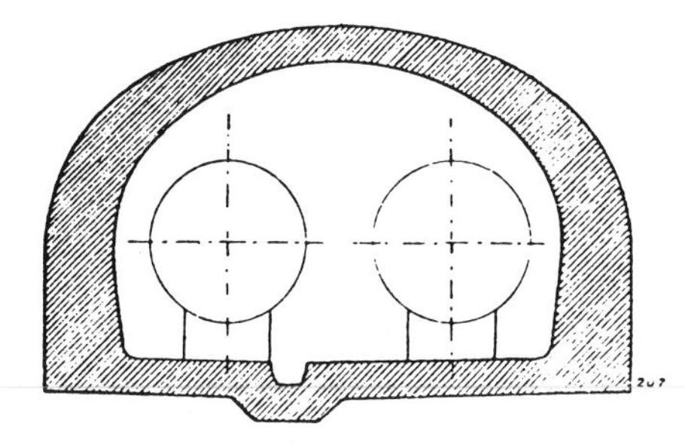

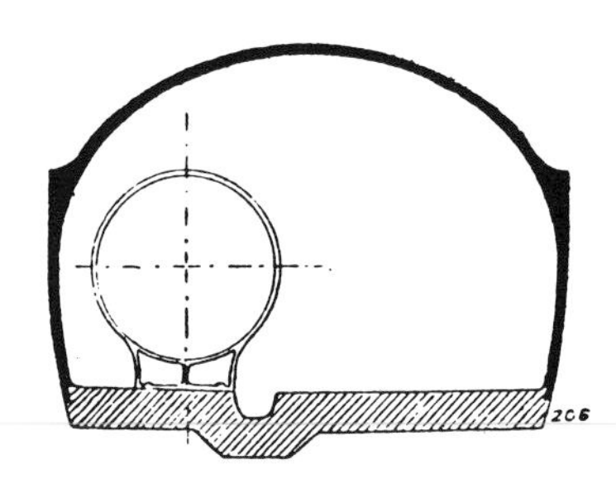

7.28 In 1892 Paris began construction of a new drainage system, with about 1½ miles of tunnel, 17 feet wide, connecting the system to fields outside the city. Masonry construction (shown at left) was first proposed, but officials chose to try the reinforced concrete design presented by Edmond Coignet, a vault slightly more than 3 inches in thickness. Success of the project did much to heighten interest in reinforced concrete in France. (*Concrete and Constructional Engineering*, July 1906.)

7.29 This hemispherical framework of metal bars was sprayed with concrete to form the first Z-D dome of thin-shell concrete. In later projects edge conditions and deflections led the engineers to build sounder formwork, which was removed after the concrete shell was completed. (W. Bauersfeld, *Projection Planetarium and Shell Construction*, 1957. Reprinted by permission of the Council of the Institution of Mechanical Engineers.)

A prominent German engineer-industrialist approached the Zeiss optical works to construct the first planetarium projector for installation at the Deutsches Museum in Munich. As the projector neared completion in 1922, it became necessary that the technicians be provided with a hemispherical space in which the projector could be tested and adjusted.[77] On the roof of a Zeiss factory building at Jena, a hemispherical network slightly over 50 feet in diameter was fabricated of steel bars about 2 feet long (fig. 7.29). It had been intended to cover the dome with wire mesh and trowel on gypsum, but that water-susceptible material was discarded when a covering of sprayed concrete was suggested by an engineer associated with the construction firm of Dyckerhoff and Widmann, contractors for several buildings in the Zeiss factory complex. The frame was sprayed with concrete, with an accurately curved form held against the inside of that portion of the dome being sprayed. Between the steel bars the concrete was about 1.2 inches thick.[78]

This system was developed through collaboration of Walther Bauersfeld of the Zeiss corporation and Franz Dischinger, an engineer with Dyckerhoff and Widmann. The patented system was known in England as "Shell-D" and in the United States as "Z-D" (an abbreviation of Zeiss-Dywidag, the last part being made of the first syllables of the contracting firm's name and the German equivalent of "Inc."). Their second project was larger, a shallow dome 131 feet in span and 2⅓ inches thick. As the complex mathematics of shells was studied, similarities were found between the actions of domes and barrel vaults. The first Z-D vaulting built commercially was for a fair in Düsseldorf, barrel vaults 2 inches thick and arching almost 40 feet, with a span of 75 feet.[79] By 1927 the processes of designing and constructing thin shell structures had advanced to the point where it was possible to build the Market Hall at Frankfurt am Main, which spanned about 120 feet with concrete barrel vaults three to four inches thick.

In 1932 Dyckerhoff and Widmann allied itself with Roberts and Schaeffer, an established Illinois engineering firm, for the promotion of the Z-D system of thin-shell construction in the United States. Appropriately, their first project was a planetarium, the Hayden Planetarium of the Museum of Natural History, New York. The concrete shell was sprayed over formwork covered with sheets of cork. Inside, a white-painted ceiling was suspended, perforated sheets of stainless steel providing some protection against the acoustical problems that plague concave shapes.[80]

The cost relationship of labor and material in the United States was quite different from that of Europe, and the design of shell structures had to be adapted in recognition of that difference. However, because shells used little of strategic materials, they were used during World War II for factories and aircraft hangars. By the 1960s shells were seldom employed, unless esthetic considerations called for a dramatic form.

Unlike steel, which in most buildings must be fireproofed, reinforced concrete is in itself fireproof, eliminating the necessity for a protective covering. In most parts of the western world the cost of concrete construction is about equally divided between the concrete itself, reinforcing steel, and the formwork in which the beams, columns, and slabs are poured. (Costs of formwork generally vary according to local labor costs.) In multistory building construction a steel framework, once it is protected from fire, has dimensions very near to those of a concrete framework. Therefore, the choice between the two materials for structural purposes is principally based on cost and speed of assembly.

The use of precast concrete advanced in the 1930s and 1940s when its use eliminated the mixing of concrete on the site. This advantage vanished when transit-mixed concrete came into use, and consequently the demand for precast concrete declined.[81] At present the majority of structural reinforced concrete is cast in place, developing a monolithic rigidity in the construction. Precast concrete, which for its rigidity relies on careful detailing of connections, has not yet effectively challenged the preponderant method of casting buildings *in situ*, and the Roman system of constructing molds in the form of the building has continued.

Systems

From its very beginning, mechanization caused workers to move to cities, seeking employment in factories and mills. In England, where machines first significantly influenced national life, an increase of rural poverty and of the birth rate during the eighteenth century further encouraged the movement of population toward cities. Only about half lived in rural areas at the middle of the nineteenth century. The most industrialized English cities became such crowded and sooty tangles that they were sometimes referred to as "the Black Indies," so different from the rest of the country that they seemed strange colonial islands.

During these centuries England's shipping activity and foreign trade blossomed, almost doubling during the nineteenth century. This growth was accompanied by increasing numbers in the social level between the gentry and laborers. At the upper reaches of this middle class were major businessmen, who had wealth without station. At the lower reaches were clerks and merchants. In London between 1851 and 1891 the number of "commercial clerks" increased about fivefold, while the city's population as a whole did not even double. Members of the middle class desired to better their living conditions, both at home and at work. Their expenditures may not have been so lavish as those of the nobility or the richest bourgeoisie, but the middle class became multitudinous, a ready market for building improvements that offered comfort and convenience. Manufacturers found that working conditions influenced the output of their workers, knowledge that was gained through the leadership of a few philanthropic factory owners, the adoption of legislation, and the insistence of labor unions. A stronger incentive to improvement was the fact that in warmer and better-lighted factories work could continue through longer hours and an owner's investment in the mill and machines would thereby produce greater profits. From such industrial applications came the elementary forms of devices that were later used in buildings of all sorts.

The comfort-producing systems that were developed had as their purpose the production of an environment in which one was not threatened by actual discomfort and was, perhaps, even somewhat cosseted. For a middle-class society that frowned on excessive luxury, it was difficult to determine what was a beneficial degree of comfort. In a period of radical change in the principles of medicine, it was even difficult to determine those conditions that were healthful. Until Pasteur's germ theory won out at the end of the nineteenth century, sea bathing, mineral waters, "fresh-air cures," and sunshine—the latter being particularly for the romantic wasting away that was tuberculosis—were among the many cures and nostrums that were briefly but seriously adopted.

In the case of cataclysmic events, such as fire or lightning, the owners or occupants of buildings could do little to prevent their occurrence. Lightning most often caused fires, and fires and structural failures threatened owners' investments in buildings. With the development of insurance this fact could be viewed as little more than an individual's wager on the likelihood of the problem arising, but from medieval times it had been recognized that this gamble also risked the safety of other structures and the people in them. Since there were no certainties involved, the process of regulation moved slowly, accelerated by catastrophes and slowed by the objections of influential owners of buildings.

The provision of comfort required the insertion of tubes, wires, fixtures, and equipment into the fabric of buildings. This work became increasingly specialized as time went by, until each type of system developed its own engineers, draftsmen, and contractors. No longer were the "natural philosophers" and toolmakers of the eighteenth century capable of applying their general knowledge of scientific principles to a variety of tasks, and tests and measurements were developed to provide more exact consideration of the systems involved. These provided much of the information on which standards and regulations were based.

Still, these systems (except for elevators, which demanded visibility) were largely unseen and certainly not displayed. Architectural styles of the eighteenth and nineteenth centuries were on the whole committed to presenting an illusion of the past, or at least a recognizable evocation of historical forms. Sir Charles Barry's Reform Club in London (1840) prompted the French critic César Daly to write:

> [It is] almost a living body, with its complex circulation systems; for in its walls which appear as immobile, there circulate gases, vapors, and fluids through the hidden ducts and wires. These latter constitute the arteries, veins and nerves of a new organized being, heat is conveyed by them in winter, fresh air in summer, and in every season they supply light, hot and cold water, food, and all those numerous accessories which an advanced civilization requires.[1]

While Daly admired the systems within the Reform Club's building, he seems to have been equally impressed by the cunning with which they were concealed. Architectural revivalism, particularly in the late nineteenth century when buildings' dimensions did not always allow precise historicism, left sufficient latitude to accommodate the necessary elements of comfort and convenience. Although fireplaces were included in many of the new tall commercial buildings, they were at one time recommended for ventilation purposes and often might not be at all suited for burning a fire. Lighting, long principally a transportable piece of equipment, had its system embedded in the fabric of the building, but the actual lamps were visible imitation of the lamps of the past.

When Le Corbusier in 1923 wrote "a house is a machine for living in," he not only proclaimed an esthetic principle, but at the same time recognized a fact regarding all buildings of that time. Wires and pipes might be hidden, but lamps, switches, grilles, and steam radiators were visible. By the middle of the twentieth century the systems within a building might account for a quarter or even half of its total cost. The function of architecture had acquired an increasing responsibility for the ease and well-being of the buildings' occupants.

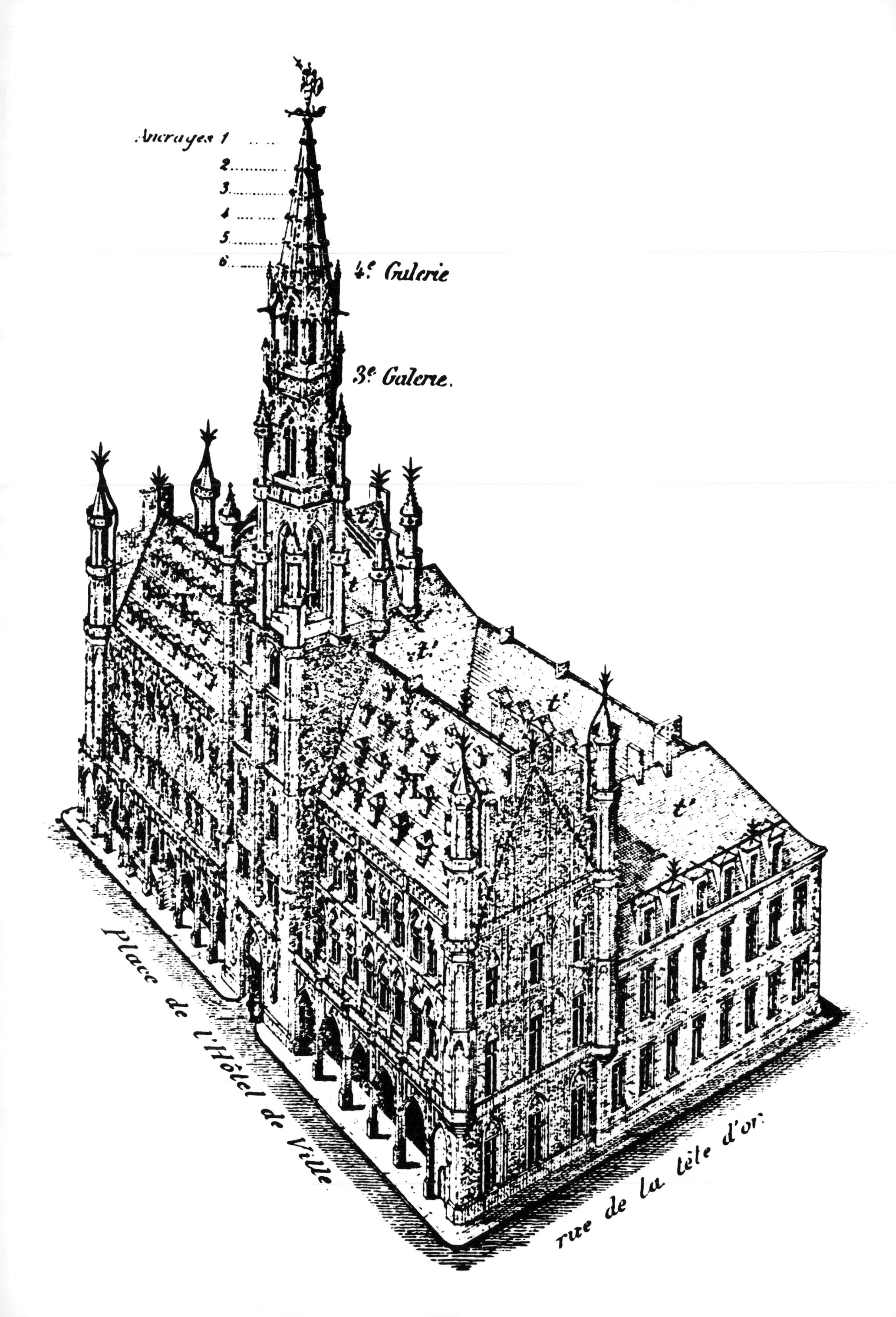
Ancrages 1
2
3
4
5
6
4e Galerie
3e Galerie
Place de l'Hôtel de Ville
rue de la tête d'or

8 Lightning Protection

1752 Lightning rod described in Benjamin Franklin's *Poor Richard's (Improved) Almanac*

1754 Prokop Divis erects a lightning tower in his Czech village

1755 Boston earthquake blamed on the use of lightning rods

1777 George III overrides the advice of Franklin and the Royal Society for protecting the Purfleet magazines

1782 Robespierre defends a French advocate of lightning rods against his neighbors' complaints

1823 Académie des Sciences supports J. A. C. Charles's "double cone" of lightning protection

8.1 The frequency of thunderstorm days varies as indicated by this map, developed from information of the World Meteorological Organization. The need for lightning protection depends also on the intensity of the thunderstorms that may occur. (G. P. McKinnon, *Fire Protection Handbook*, 1976. Reprinted with permission. Copyright 1986 National Fire Protection Association, Quincy, Massachusetts.)

8.2 Franklin explained his theory of points in a paper he published in 1749. Small dashes indicate the charge ("electrical atmosphere") around the metal shape. In Franklin's words, at the point "the quantity is largest, and the surface to attract and keep it back the least," and therefore "you can get it away still more easily." (*Journal of the Franklin Institute*, April 1906.)

In ancient times it was believed that sleeping bodies were protected from lightning because in a relaxed state they offered no resistance to it, and obeying another myth at least one Roman emperor donned a wreath of laurel whenever a storm threatened.[1] On the whole, the most popular explanations of lightning and defenses against it were based on the assumption that it was divinely instigated and directed. It became a custom to ring church bells in order to dissipate the force of lightning, but there remained a theological embarrassment in the frequency with which the tall towers of churches were struck and demolished by bolts of lightning.

Peasant practices regarding protection from lightning often related to the bonfires built in some parts of Europe for the midsummer celebration of the Eve of St. John. Brands from the bonfire, when taken home and carefully stored in a cupboard, were believed in one district to protect the house from "lightning, conflagration, and certain maladies and spells."[2] In other locales, defense against lightning was provided by children throwing garlands of flowers on the cottage roof or by a householder's hanging on a wall a bunch of mountain arnica, an herb to which medicinal qualities were also attributed. A surer protection in some areas was said to lie in the fact that lightning would never strike a house in which a crossbill finch was kept in a cage.[3]

Churches and army powder magazines seem particularly to have focused public and official attention on the use of lightning rods. These subjects were combined when in 1767 lightning struck a church in Venice where city authorities had stored 100 tons of gunpowder in vaults they assumed to be protected by divine authority. A large area surrounding the church was razed and many were killed in the explosion.[4] There were bitter arguments between faith and science, but during the lifetime of Benjamin Franklin lightning rods were mounted on many of the major churches in Europe and in countries elsewhere.

No other name is so strongly associated with lightning protection as Franklin's. His interest in electrical experimentation began at a demonstration presented in Boston by Dr. Archibald Spencer, a Scotsman who started his career as a male midwife, ended as a preacher, and at a time between those occupations lectured on electricity. A year later Spencer played an engagement in Philadelphia and sold all of his demonstration equipment to Franklin. In the meantime one of Franklin's English friends sent the Library Company of Philadelphia an "electric tube" in which could be stored the electrical charge produced by friction, along with pamphlets on European experiments with electricity and his own instructions. The gift was well-timed. At the end of 1748 Franklin retired from his printing business and had "no other tasks than such as I shall like to give myself and of enjoying what I look upon as a great happiness: leisure to read, study, make experiments, and converse at large."[5] Within six months of his retirement Franklin had conducted a sufficient number of experiments to write of them to Peter Collison and others of his many correspondents. The circle of Franklin's friends, the "Junto" organized in 1727 and later renamed the American Philosophical Society, provided both collaborators and an enthusiastic audience for demonstrations of the mysterious workings of electricity.

One of Franklin's early findings was that sharp-pointed conductors drew off electrical charges far more

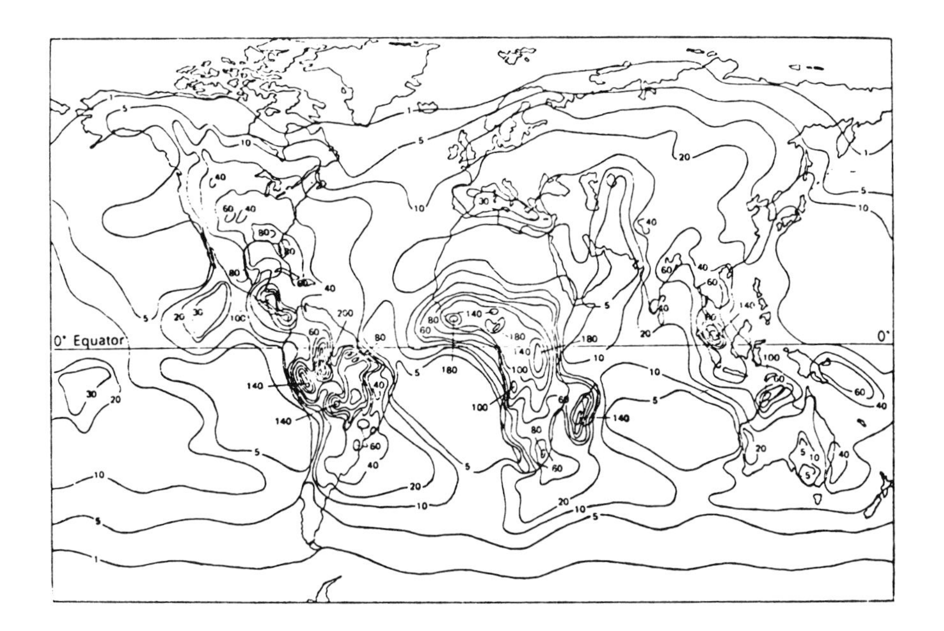

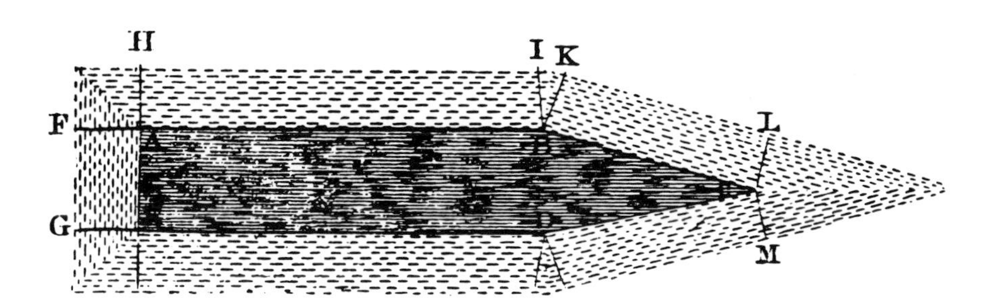

readily than blunt-ended ones (fig. 8.2). Since Franklin had concluded that lightning was much the same as the sparks that leaped between points in his experimental paraphernalia, he hungered to conduct experiments with thunderclouds overhead. He proposed to place "a kind of sentry-box" on a high tower with a pointed iron rod extending 20 or 30 feet into the air. Inside the sentry box a man would stand on a cake of wax to provide insulation and would hold a wire with which sparks could be drawn from the iron rod. But in Philadelphia there were no steeples or spires that offered sufficient height, and the famous kite trick, the "Philadelphia experiment," was devised:

To the top of the upright stick of the [kite] cross is to be fixed a very sharp pointed wire, rising a foot or more above the wood. . . . As soon as any of the thunder clouds come over the kite, the pointed wire will draw the electric fire from them. . . . And when the rain has wet the kite and twine . . . you will find it streams out plentifully from the key on the approach of your knuckle.[6]

be six or eight Feet above the Highest part of the Building. To the upper End of the Rod fasten about a Foot of Brass Wire, the size of a common Knitting-needle, sharpened to a fine Point; the Rod may be secured to the House by a few small Staples. If the House or Barn be long, there may be a Rod and Point at each End, and a middling Wire along the Ridge from one to the other. A House thus furnished will not be damaged by Lightning, it being attracted by the Points and passing thro the Metal into the Ground without hurting any Thing.[7]

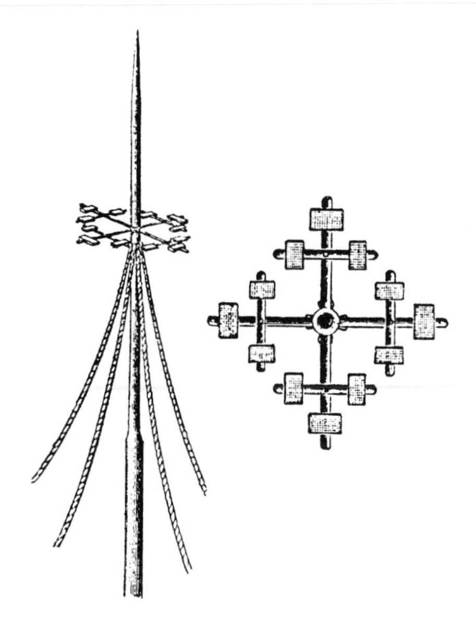

8.3 The "weather machine" of Prokop Divis had boxes of needles on all four branches, providing a multitude of points. It was mounted on a wooden shaft, but chains also connected it with the earth. (*Popular Science Monthly*, January 1893.)

This experiment led to Franklin's invention of the lightning rod. Scholars, as is their inclination, have questioned whether Franklin's kite experiment preceded his invention of the lightning rod. In fact, some doubt that he performed the experiment.

In September 1752, Franklin attached to his house an iron rod with two bells attached. The rod's purpose was experimental and the bells' ringing informed him when a charge had been accumulated for his use. Within a month of this project, there was published in *Poor Richard's (Improved) Almanac* a description of the means by which buildings could be protected from damage by lightning. Written by Franklin under the pseudonym of Richard Saunders, the instructions are simple:

Provide a small Iron Rod . . . but of such a Length that one End being three or four Feet in the moist Ground the other may

By 1755, a Boston minister could complain that there were "more [lightning rods] erected [here] than any where else in New England" (he blamed the lightning rods for overcharging the earth with electricity and thereby causing a recent earthquake in the city).[8]

There is at least one other person sometimes given credit for the invention of the lightning rod. Prokop Divis was a priest in Czechoslovakia, where in 1736 he was appointed pastor in Primetice, a village in southern Moravia.[9] Although relatively isolated, Divis kept in touch with scientific progress through visits to the region's seats of learning, reading published reports, and correspondence. The duties of his parish appear to have required little of Divis's attention, and he devoted much of his time to experimentation.

After a preliminary study of electrical phenomena, Divis focused his thought on atmospheric charges of electricity. In 1753 he wrote an explanation of why a Russian physicist had been killed while conducting experiments with lightning. A copy of this paper was sent to Leonhard Euler, president of the Berlin Academy of Sciences, and in his letter Divis implied his intention to construct a device that would diminish the threat of lightning over a relatively large area. In June of the following year this conductor was erected in a garden plot near Divis's rectory. Convinced that lightning was an electrical discharge and that metal points would attract it, he designed a lightning rod that bore a multitude of points (fig. 8.3). The central rod was supported by a wooden upright 48 feet high at first, and later said to have been increased to 132 feet, which was guyed with iron chains. At its top the iron rod bore two crossbars, each end of them having a smaller crossbar. In this way the rods provided 13 points, though blunt, but in addition each of the 12 horizontal points held a box filled with iron filings in which 27 brass needles stood erect. If one ignored the ends of the rods as points, there were altogether 324 brass needles to attract electrical charges.

For two years Divis experimented with his lightning rod. He proposed to his government the location of similar towers throughout the kingdom, but ranking scientists squelched the plan. The summer of 1756 was extremely dry in Moravia, and the farmers around Primetice blamed Divis's lightning rod with drawing off the moisture of clouds as well as their electrical charge. Anger grew and a throng of farmers tore down the tower.

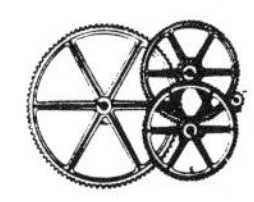

In France, too, experimentation and controversy thrived. According to a description of Paris published in Amsterdam during 1782, lightning rods were rarer in the capital city than in large provincial towns.[10] Abbé Bertholon, a priest of the Lazarist order and a professor of experimental physics, had installed lightning rods in Lyons, his native city, and in other parts of southern France. Bertholon was a friend of Benjamin Franklin and followed his principles, except for the French inclination to count height as the paramount consideration in the design of lightning rods. In Paris a *paratonnerre* 185 feet high was installed at the Hôtel de Chârost in Faubourg Saint-Honoré, its lower end extending 28 feet below ground level into the water of a well. On the other side of Paris, a convent had a rod 3 feet higher that went 90 feet into the ground.

Elsewhere in France, M. de Vissery de Bois-Vale, a retired lawyer, was an amateur experimenter who dabbled in the scientific paraphernalia that fascinated so many gentlemen of that period. He corresponded with other experimenters, including Franklin. A few months after he wrote Franklin in 1782, Vissery put a lightning rod of his own devising on his house at Saint-Omer.[11] Neighbors gawked and gossiped, soon reaching the conclusion that the apparatus might attract bolts of lightning and thus endanger their own properties and families. They were deaf to Vissery's explanations. When his neighbors took their complaints to the sheriff, Vissery's theories failed to sway that official, and an order was issued to force removal of the lightning rod.

In defense of scientific truth, the amateur appealed to the Council of Artois, which sat in the regional capitol. Vissery's lawyer gathered scientific opinions from France's most famous experts, including Abbé Bertholon. It was at that time customary for established lawyers to prepare the substance of a case, but assign its oratorical presentation to someone younger. The lawyer chosen to argue M. Vissery's case was Maximilien F. M. I. de Robespierre, ten years later to mastermind the Reign of Terror before he was himself guillotined. When offered the case, Robespierre had been a member of the bar for only two years, and the case of Vissery provided a rare opportunity to advance his career at a time when legal oratory was inclined to lead to political oratory. Using the older lawyer's compilation of theories and opinions, young Robespierre began his argument by speaking of Galileo and others who had been oppressed because of their unpopular scientific views. He went on to review the electrical principles known at that time—although his own scientific education was negligible—and summarized the short history of lightning rods. He closed by invoking royal support through a description of several electrical experiments that Louis XVI himself had caused to be conducted. "If there were any doubts about the effects of these devices, the scientists involved would never have experimented on such a dear and such a sacred head." At Robespierre's request Vissery financed the publication of the speech, and Robespierre sent copies to friends, associates, charming young ladies, and even Benjamin Franklin.[12]

8.4 The Continental method of protecting buildings at the start of this century required tall vertical rods spaced about the roof. At corners the rods were usually projected at an angle of about 60°. This installation at the Vatican used iron rods instead of copper, apparently seeking to lower the cost of protecting over seven acres of buildings. (K. Hedges, *Modern Lightning Conductors*, 1905.)

Franklin's first statement on lightning as a form of electricity had been presented to the British Royal Society of Arts in 1750 by John Mitchell, physician, cartographer of the American colonies, and a member of the Society. Although Franklin's *Autobiography* reports that the paper "was laught at by the Connoisseurs," it was published and attracted the attention of scientists. Three years later the Society awarded Franklin the Sir Godfrey Copley Gold Medal for his electrical experiments, and after another three years he was elected to membership. When the dean and chapter of St. Paul's Cathedral in 1769 asked the Royal Society for advice on lightning protection for the church's dome and cupola, it was Franklin who advised them regarding the placement of rods, points, and wires. In 1772 the Board of Ordnance consulted Franklin, who was once more in England, about protection of its principal installation of powder magazines at Purfleet. He visited the location and recommended that pointed lightning rods be installed. The problem was then taken to the Royal Society, which appointed a committee. The members of that committee agreed with Franklin that pointed rods should be placed on the magazines, and it was he that wrote their final report.

There was one dissent among the five members of the committee. Benjamin Wilson, an experimenter and writer on electricity, was also a painter and had completed a portrait of Franklin about a dozen years before. It was Wilson's contention that pointed rods would attract lightning to the magazines, causing a greater likelihood of damage than would result from blunt rods. Wilson published pamphlets that angrily presented his argument.

Years later, after the American Revolution had begun, lightning struck at Purfleet, although none of the magazines was damaged. The controversy was then revived with George III avidly supporting Wilson's point of view and patriotic undercurrents strongly influencing opinion. The king ordered blunt-ended conductors placed on the Purfleet magazines and the royal palace. The Royal Society was thus placed on the defensive, since George III had publicly contradicted its scientific advice.[13] The

president of the Society, Sir John Pringle, court physician but for more than 20 years a close friend of Franklin, refused to be persuaded that blunt ends were superior because of royal preference, and in the end he resigned the presidency of the Royal Society.

Lightning protection had come to be accepted, but not without a residue of doubt. Sir William Thomson, later Lord Kelvin, commented, "If I urge Glasgow manufacturers to put up lightning rods they say it is cheaper to insure than to do so."[14] This attitude was not peculiar to frugal Scots, for many articles on lightning rods published in the late nineteenth century advised that the cost of rods and conductors should be carefully compared with insurance premiums. At the same time, insurance companies sometimes assumed a portion of the cost of placing rods on buildings for which they held policies. Some German companies paid a small percentage of the cost, but the East Prussian Fire Insurance Company bore half the cost for its customers.[15] Other companies reduced premiums for protected property. In Schleswig-Holstein some fire insurance companies reduced their rates on houses roofed with slate or tile by 5 percent if lightning rods were installed, and they lowered rates on thatched houses and windmills by 20 percent.

Another debate about lightning protection occurred in the mid-nineteenth century between two prominent British scientists. Michael Faraday had a long and distinguished record in electrochemical experimentation when he was asked in 1853 to propose the proper protection for Eddystone Lighthouse, perched on a rock outside Plymouth harbor. Faraday's opponent, William Snow Harris, had begun his study of lightning in 1822 by proposing that rods and conductors be installed on British navy vessels. The subject continued to hold his attention, and in the 1850s Harris was recommended by the architect, Sir Charles Barry, to provide a protective system for the new Houses of Parliament.

The Faraday-Harris controversy centered on the form of conductors that should be employed in an installation of lightning rods. Faraday viewed them as electrical conductors and insisted that their cross-sectional area determined the degree of protection that they afforded. Harris considered as critical the amount of conductor surface along which a streak of lightning might travel to the ground, and he employed metal ribbons or hollow tubes in which the extent of surface was very great in relation to the amount of metal. Years later a British admiral recalled the vehemence of the two scientists: "Each told me that the other 'knew nothing about it.' "[16] In 1892 Oliver Lodge, then the ranking British expert on lightning, concluded that Harris was correct in placing primary value on the surface area of conductors, but in the 1970s the lightning protection regulations of several countries (Australia, Germany, Great Britain, and the United States) made no distinction among strips, rods, and wire cables as conductors.

Although Franklin has assumed that his points attracted lightning, others portrayed their action as that of effectively shielding a specific area. J. A. C. Charles, a French physicist, proposed that each lightning rod be considered as protecting everything within a cone having its point at the tip of the rod and its base a circle with a radius twice the height of the rod, hence the "double cone of Charles." The conical shape was suggested by observations that lightning often approached grounded objects in an upward movement. Thus, if one could define an area of protection, it seemed reasonable that it would be generally conical, although the exact proportions were an arbitrary assumption (fig. 8.5).

If lightning rods were to be effectively placed for protecting the towers of cathedrals and chimneys of mansions, some geometric principle seemed necessary to guide their placement, a principle much more specific than Franklin's speaking of his points attracting the electrical charge of clouds. One of the first rules accepted was the "double cone of Charles," which the French Académie des Sciences espoused in 1823.[17] A "single cone" with its base's radius equal to the height was theorized by Joseph Louis Gay-Lussac, professor of physics at the Sorbonne. As the notion of cones of protection came to be more firmly implanted, other proportions were proposed. The steepest form considered through the years was a cone with its radius half its height, and there were unusual variations in which the slope of the cone's sides was a concave curve or the zone of protection was cylindrical. In 1854 a French committee, formed to assess the current knowledge of lightning protection, firmly stated that cones of space should not be considered as describing the extent of protection by lightning rods, and through the remainder of the nineteenth century this admonition was repeated by other committees and several individual scientists. Nevertheless, in lieu of any other specific rules that could be used to determine the placement of points, the notion of cones of protection has persisted.[18]

The French Ministry of the Interior in 1822 had ordered that lightning protection be provided for all public buildings, and the following year a distinguished committee of the Acadé-

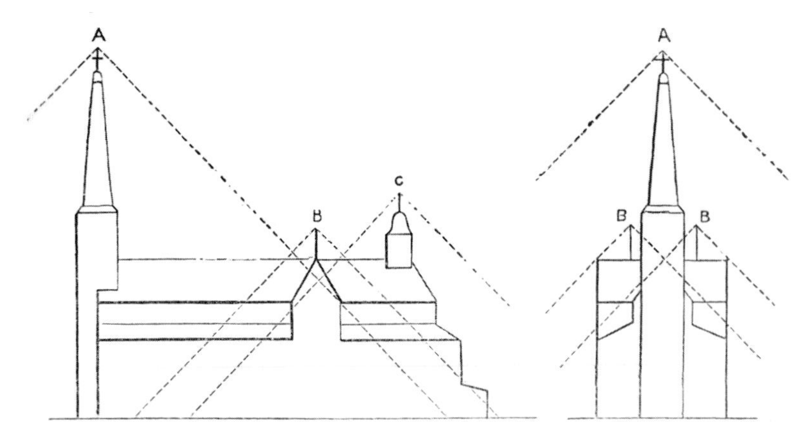

8.5 Although different experts and various national agencies recommended other proportions, English military authorities and most English practitioners employed a cone of lightning protection with its height equal to the radius of the base, a 45° slope on the side. Rods were placed so that such cones enveloped the entire form of the structure. (G. J. Symonds, *Lightning Rod Conference*, 1882.)

mie des Sciences presented its recommendations. From large buildings an iron rod should extend upward 23 to 30 feet, they said, including a top 22 inches of copper point, either gilded at the end or tipped with a bit of platinum. An iron rod, held away from roofs and walls by brackets, should extend to the ground, where it could be taken to a well or a pit filled with charcoal. The "double cone" of protection was favored by the committee, but the "single cone" was accepted for installation on church towers. The committee strongly urged that all metal on the surface of a building—gutters, tie rods, or roof coverings—be connected to the system, so that such metal would not divert lightning from the conductor's path.

At midcentury there began a succession of French committees charged with the study of lightning protection. As a beginning, the committee in 1854 supported the recommendations of 30 years before but pointed out two new considerations. First, because of an inquiry made by the company building the Palais d'Industrie for the next year's exposition, the problem of metal-framed buildings was approached. With the construction of great glass and iron buildings for international fairs and railroad stations, the building itself became a conductor. The question could have raised doubts about the wisdom of constructing metal buildings, but the committee considered lightning rods imperative on metal buildings and required that they be connected to the building's frame.[19] Second, with regard to new additions to the Louvre, the committee stressed that if conductors could not be extended into a well they should have two branches, one extending into the earth to reach a stable level of moisture and the other kept near the ground, where surface moisture from a thundershower might provide a ground.

A subsequent French committee in 1867 gave much of its attention to the placement of rods on powder magazines, because the year before several French magazines had been struck by lightning although provided with lightning rods. At the same time the committee reduced the recommended height of rods by 13 feet, thus lowering the high rods that had long been the earmark of French installations.

If the principles of lightning protection were debatable among scientists, they were even more bewildering to ordinary citizens. Lightning rod salesmen became the equal of patent medicine hawkers in the United States, and the stories of their trickery multiplied. Mark Twain in 1870 wrote of being persuaded by a glib peddler to install lightning rods on his house and being sold a much greater number of rods than was needed and "3,211 feet of the best quality zinc-plated spiral-twist" conductor wire.[20] Twain's hyperbole was deserved by the fly-by-night door-to-door salesmen who bilked householders, and the genuine danger was inflated by an endless store of tales about buildings and trees having been wrenched apart by bolts of lightning and balls of fire blazing erratic paths through bedrooms while the occupants cowered in bed. Like patent medicines, the sale of lightning rods was very much a trade based on fear and fabrication.

In 1875 a brilliant Scottish scientist, James Clerk Maxwell, proposed a variation on the usual methods of lightning protection, which assumed that rods should gather electrical charges from the sky and drain them into the ground (fig. 8.6). Experiments almost a century before had shown that there was no electrostatic charge in the center of a hollow conductor, and on that basis Maxwell proposed the “bird cage” method of protecting buildings from lightning. He recommended an enveloping network of conductors, with connecting rods on the top of a building, leading to the ground in several directions, and connecting conductors all around the base of the building. The Germans to some extent adopted this idea for a period, but one British expert shrugged, “It would be unpleasant, when you reached home out of a storm, to find it so highly charged as to knock you down directly as you tried to go in.”[21]

8.6 The founder of electromagnetic theory, James Clerk Maxwell, designed a system of lightning protection that did not make use of metal rods jutting into the sky. Instead, conductors (shown here by dashed lines) ran along the edges and corners of the structure. The system was usually referred to as being like bird cages or the wire meat safes of that day. (O. J. Lodge, *Lightning Conductors and Lightning Guards*, 1892.)

8.7 Professor Melsens's treatment of the Hôtel de Ville, Brussels, set *aigrettes* of metal points around the figure at the top of the tower, on each of the turrets, and at the gables of the building's many dormers. (L. H. F. Melsens, *Des Paratonnerres*, 1877.)

Protecting the splendid Gothic town hall of Brussels became a preoccupation of Professor L. H. F. Melsens, a Belgian physicist and chemist. His system for that building was begun in 1865, and some believed it the best protected building in the world. The town hall's tall tower was particularly susceptible to lightning discharges, so eight conductors led from it to the ground.[22] Melsens did not use the tall rods favored by the French (figs. 8.7, 8.8). Instead he installed 37 “point arrangements,” some of which included several pointed short rods plus *aigrettes*, which were made of eight wires fanned out.[23] Beyond any scientific basis, this method had the distinct advantage of not seeming to clutter the intricate ornamentation of the tower. In spite of such elaborate precautions, in 1888 the town hall was struck by lightning and set on fire. After the roof was examined it was concluded that a horizontal bar of metal had not been connected to Melsens's network, and from it sparks had ignited gas.

If Lord Kelvin had difficulty persuading Glasgow investors to provide lightning protection for their buildings, greater economic resistance was to come. Reports of governmental committees represented the views of respected scientists and consequently became the principal basis for regulations issued by city governments, governmental agencies, and the military forces. Such rules were necessarily written in a general manner, intended to provide sound protection for all buildings, in spite of the extreme variety of sites and forms that the structures included. At the end of the nineteenth century, a German building official pointed out that if all buildings were protected according to such regulations, the cost of installing the rods and conductors would be greater than the cost of damage that could normally be expected from lightning.

Discussions of lightning protection at the turn of the century increasingly focused on rural structures. An early statistical report on an area of central Germany concluded that the danger from lightning was nearly four times as great in the country as it was in urban areas.[24] The case was best stated in a 1901 study of lightning fatalities in the United States.

It is not surprising that so few lightning strokes fall in the cities. The modern city building, with its metallic roof and steel frame, is a fairly good conductor of electricity, and is in much less danger of receiving a damaging stroke of lightning than an isolated dwelling in the open country. The multiplication of telegraph, telephone, and electric light wires in cities also adds to the effectiveness of silent discharges in relieving the electrical tension during a thunderstorm; but should a cloud with a tremendous store of energy quickly approach, all the wires in ten cities would not prevent it from discharging right and left until its store of energy had been dissipated.[25]

Spires, towers, smokestacks, water towers, and skyscrapers might certainly be subject to lightning even in the densest urban environment, but at the same time they could serve as protection for lower structures.

Skyscrapers were a new problem. Towers and chimneys in the midst of cities had often been damaged by lightning. How would the tall steel-framed buildings survive? By 1902 it was reported that in the United States the quality of lightning rod installation had so greatly and generally deteriorated that architects were dropping that portion from their specifications for new construction. Hence, many of the finest and highest buildings in the centers of cities were without protection. With insurance policies covering any loss, financial risk was lessened.[26] It was the conclusion of Britain's 1905 Lightning Research Committee that steel-framed buildings provided themselves with a high degree of protection, much like the "bird cage" system that had been expounded by Maxwell about 30 years before. If small rods were placed about the roof and properly connected to the metal frame, the building itself would con-

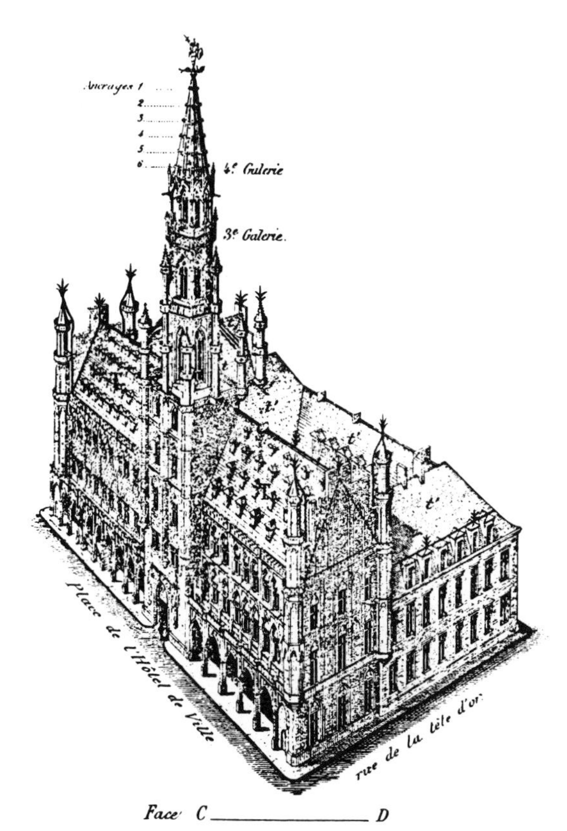

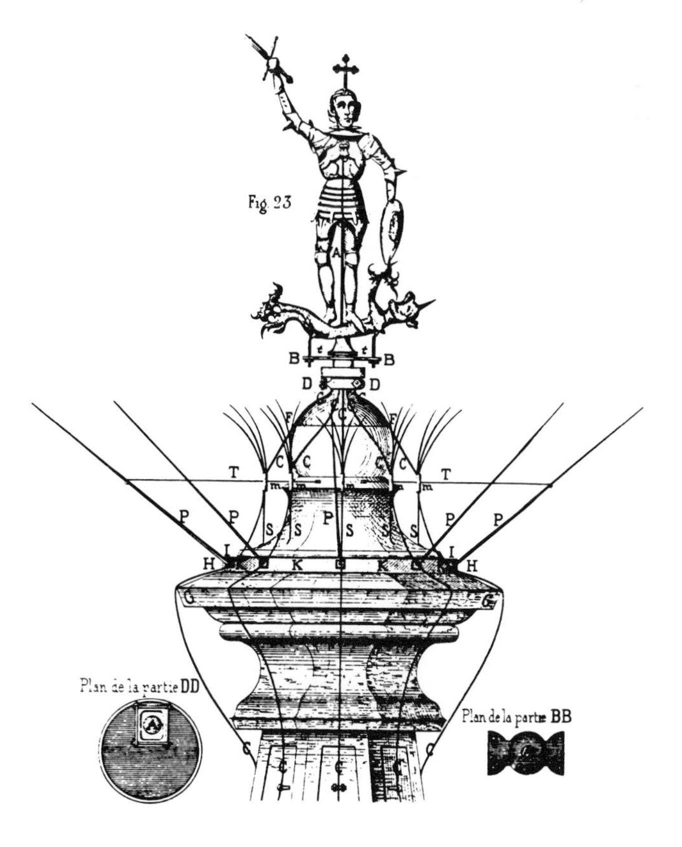

8.8 **In 1882 Melsens laid out the lightning protection of a Belgian farmhouse with 37 *aigrettes* spaced along the roofs. Conductors were grounded to wells (*P, P′*) and the farm pond. (L. H. F. Melsens, *Des Paratonnerres*, 1877.)**

8.9 **This diagram from the 1882 report of a British Lightning Rod Conference indicates the terminology of the time. Methods shown for grounding the system (G, G[1], G[2], G[3]) include most that were popular at the time, although a well was usually conceded to be best. (G. J. Symonds, *Lightning Rod Conference*, 1882.)**

8.10 **The Royal Society was consulted in 1769 about the protection of St. Paul's. Early in the present century, inspection of the cathedral found that the effectiveness of the installations that had been made through many decades was limited due to drainage, paving, and pipe-laying projects that had altered the moistness of the ground. Connections were then made to water and hydraulic mains, and many small grounds were provided by metal plates or perforated pipes buried in the earth. (K. Hedges, *Modern Lightning Conductors*, 1905).**

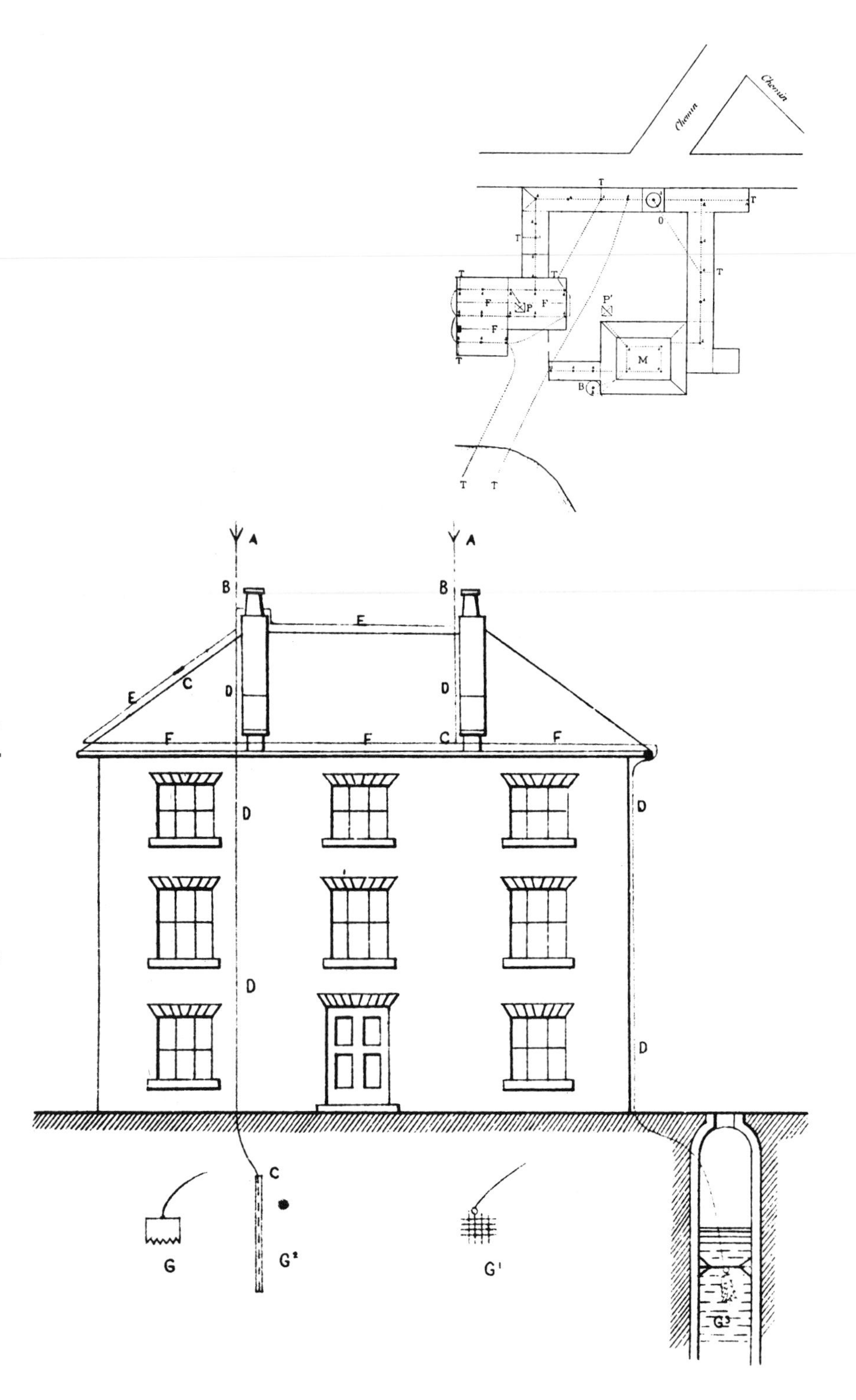

stitute a network of conductors. The principal fault found was the frequent failure to provide adequate grounding for the columns, which ended on foundations of stone or concrete without buried conductors leading to moist earth or, better yet, a deep well.[27]

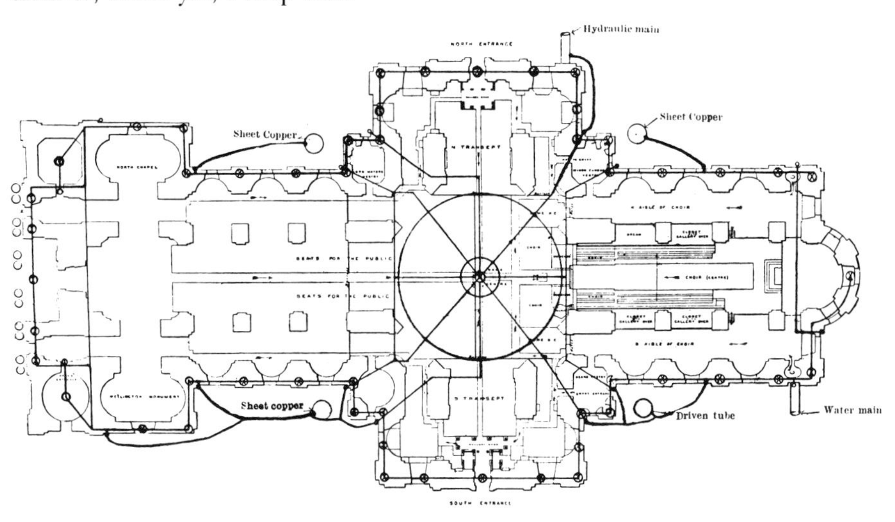

Curiously enough, these same statements made in 1905 correspond to the views of the 1976 *Fire Protection Handbook* as published by the U.S. National Fire Protection Association, which also describes cones of protection in terms very much like those used in the early part of the nineteenth century.

In tall buildings the steel framework or the bars within concrete (when linked by wires or clamps) afford protection by distributing the current of lightning. But despite the tradition of cones of protection, it has been considered wise to install conductors on the walls to contend with lightning strikes against the sides of a building.[28] With the increased amount of metal within buildings, mechanical and electrical systems, it has been necessary to isolate or even shield the metal parts of interior functions from the network of lightning protection. Today lightning conductors may serve as much to defend computer installations and the information stored in them as to protect the fabric of a building.

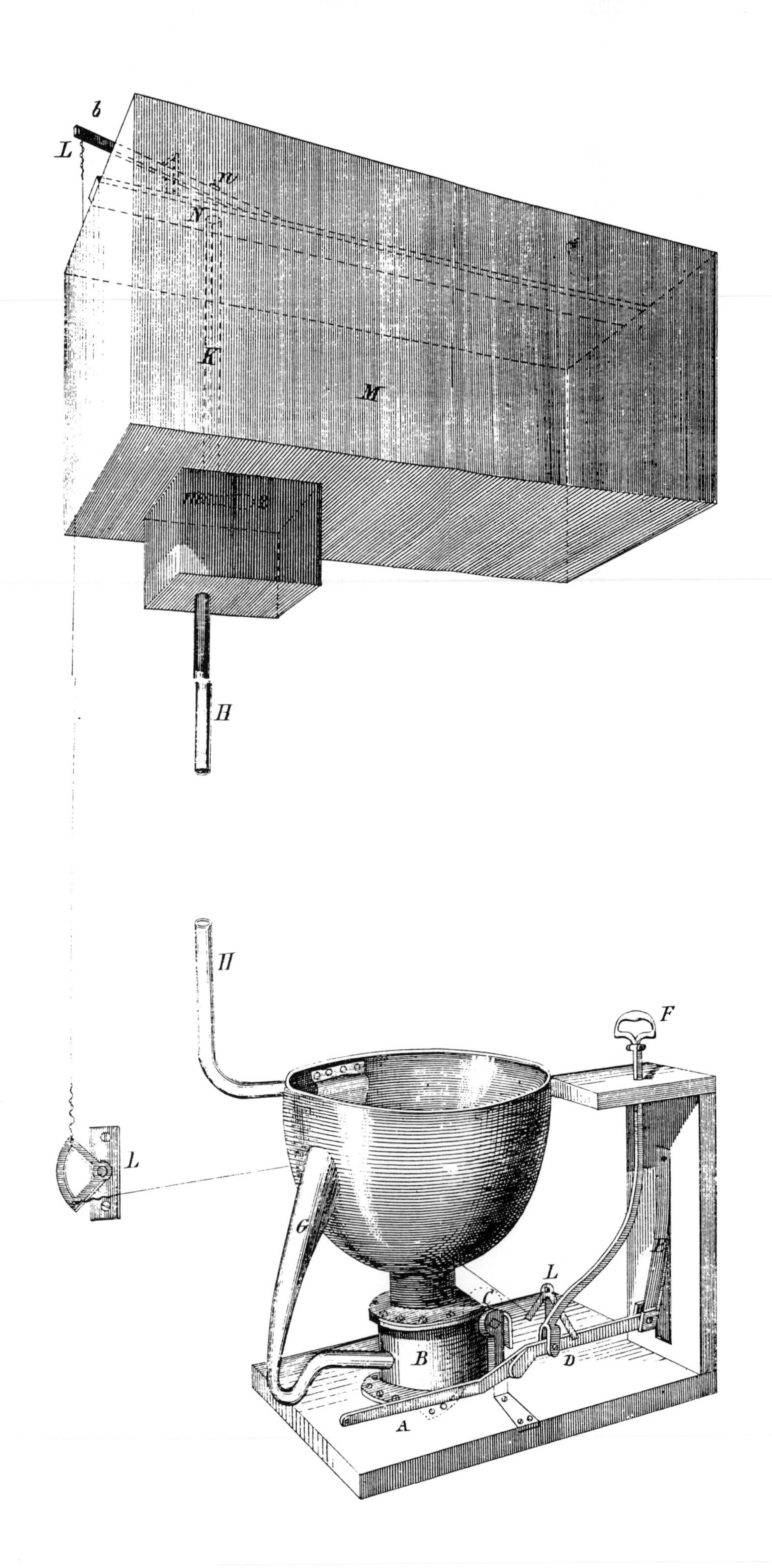
b
L
N
K
M
H
H
F
L
G
E
L
C
B
D
A

9

Sanitation

1596 Sir John Harington describes the water closet in *Metamorphosis of Ajax*

1775 Valve water closet patented (Britain) by Alexander Cumming, a London watchmaker

1778 Joseph Bramah patents (Britain) his improvement on the valve water closet

1860 The earth closet introduced by Henry Moule, an English clergyman

1866 New York's Metropolitan Health Law enacted following a survey of the city's sanitary conditions

Because of the remarkable engineering skill of its empire, the city of Rome in the fourth century A.D. could provide 11 public baths, 144 public latrines, 865 private baths, and some 1,352 public fountains and cisterns.[1] These were supplied with water from 13 aqueducts, and around 70 aqueducts are known to have been built in other parts of the empire. (Repairing such aqueducts was often a medieval expedient for improving a city's supply of water.) Some public latrines had flowing water beneath seats of stone or wood. Containers were placed at certain street corners, and Vespasian instituted a charge for their use and a system by which the urine collected was sold for use by fullers of cloth.[2] Within Roman houses, chamber pots were used, being usually emptied into the street. But water was plentiful, and drainage was well provided in most Roman cities. All of these advantages declined with the empire.

As late as the middle of the seventeenth century even the nobility of England were little concerned with sanitation. After the court of Charles II visited Oxford to escape the plague, a local diarist wrote: "Though they were neat and gay in their apparell, yet they were very nasty and beastly, leaving at their departure their excrements in every corner, in chimneys, studies, cole houses, cellers. Rude, rough, whoremongers; vaine, empty, careless."[3] Medieval traditions had made city streets almost as much sewers as thoroughfares. With drainage in the middle of the paving, the pedestrian was forced at night to choose between walking in filth or passing within the reach of any thieves who might lurk in dark doorways. In Edinburgh the danger of having a chamber pot emptied over one was only slightly diminished by the custom of householders shouting "Gardy-loo" from upper-story windows a moment before hurling sewage into the street.[4] From the street such ordure drained into streams and was swept to the sea and the cities downstream. Englishmen on their Grand Tour complained that garments laundered in Rome stank forever of the Tiber's waters; just as some German visitors to London claimed that clothing washed in water from the Thames never lost traces of that river's sickening stench.[5] Visitors to Paris claimed they could smell the city's filth two miles outside its gates, and fastidious Parisians who ventured into the streets in the time of Louis XIV covered their noses with hands in perfumed gloves.[6] In smaller towns the problems were much the same. Viollet-le-Duc mentions that in towns of central France a stream through the center would customarily be called *merderel*, named for its foul contents.

Castles and monasteries often had privies projecting over their moats or streams. Built within towers or buttresses, these "garderobes" (wardrobe being a euphemism) emptied through shafts hollowed in the masonry. Where there was neither stream nor moat, cesspits were dug. Additional comfort was provided its users when the garderobe was warmed by being built into the masonry mass of a fireplace. For defecation peasants had the custom of retiring to a distance "a bow's shot" away from their dwellings. At the country houses of well-to-do Englishmen the gardens served a similar purpose, and leaving a social gathering "to pluck a rose" was another of the numerous euphemisms, meaning either a visit to a "bog-house" at the back of the garden or a more casual solution.

Chamber pots were provided inside houses of the prosperous, and the task of emptying them ranked high in servants' complaints.[7] Still at Versailles the ceremonies of Louis XIV's daily awakening included two *porte-chaise d'affaires*, attendants dressed in black velvet, who had the honor of removing the royal pot from beneath the king's *chaise-percée*. Those of noble birth or considerable wealth had such seats, ornately decorated, richly padded with velvet, and having a hole in the center of the seat cushion.

The lavish public baths of the Roman empire were not primarily intended for cleanliness. When a gong announced the daily opening of a bath, Romans, including citizens, slaves, freemen, and children, thronged there to watch sporting events, take pleasure in the luxurious decoration, hobnob with acquaintances of their own station, and perhaps bathe.[8] Smaller public baths were provided by the government or operated by private owners, and the houses and country villas of the rich were usually outfitted with a few small rooms that were the domestic bath, perhaps heated by circulating hot air beneath the stone floor as were the public baths.

Public baths continued to exist in much of Europe during the medieval period. The stews—as medieval baths, both public and private, were called—by the seventeenth century had become little more than brothels, although they were sometimes used for the steaming that might be prescribed as a cure for illnesses. Churchmen attacking the stews as centers of sin may have had some effect, but the decline of public baths was probably as much influenced by the increasing price of soap and fuel, as well as the years of plagues. Nobility continued to bathe splendidly. Tubs were hollowed out of marble for kings and dukes, although they sometimes required submerged sheets and cushions to protect royal rumps from the cold stone, a problem avoided in the wooden tubs that were used by people less grand. It is said that Louis XIV had at Versailles at least 100 bathrooms to serve the multitude of residents, visitors, and attendants, as compared with an inventory of 264 *chaise-percées*. Marie Antoinette bathed daily, but used only one tub rather than the pair considered proper, one for cleansing and one for rinsing. Except for sponge baths, the queen's subjects in Paris seldom took more than two baths per year, those during the summer and in public bathing places in the Seine. In 1800 there was not a single public bath in London and, when a lady of fashion was chided about her grimy hands, she laughingly replied, "If you think that dirty, you should see my feet!"[9]

When Queen Victoria assumed the throne in 1837 there was not a single bathroom in Buckingham Palace. For the most part, as in other houses of the wealthy, portable tubs were brought into bedrooms or dressing rooms, servants rushing up the back stairs bearing containers of hot and cool water and spreading many sheets around the tub to catch the splashing. Copper was the most desired material for making these tubs, but its expense frequently led to the substitution of tin. In simpler households a water seller might be summoned to bring a tub and the hot water required. In 1838 there were 1,013 water sellers offering that service to citizens of Paris, and there were only 2,224 tubs that were fixed in place.[10]

In 1596 the high sheriff of Somerset and a godson of Queen Elizabeth, Sir John Harington, published his design of a water closet. As a courtier, Harington had become known as a poet and man of wit, if not of impeccable taste. When he circulated about the court his translation of a salacious portion of Ariosto's *Orlando Furioso*, Elizabeth ordered him to depart from the court and before returning to translate the remainder of the poem, the literary work for which he is most remembered. Several years later Harington was again banished from court when he published *A New Discourse of a Stale Subject, Called the Metamorphosis of Ajax*. ("Ajax" is a pun on "jakes," a colloquial term for a privy.) Although filled with roguish humor and loaded with literary allusions, this and subsequent tracts on the subject give a clear description of the water closet that Harington had constructed for his country seat at Kelston near Bath (fig. 9.1)

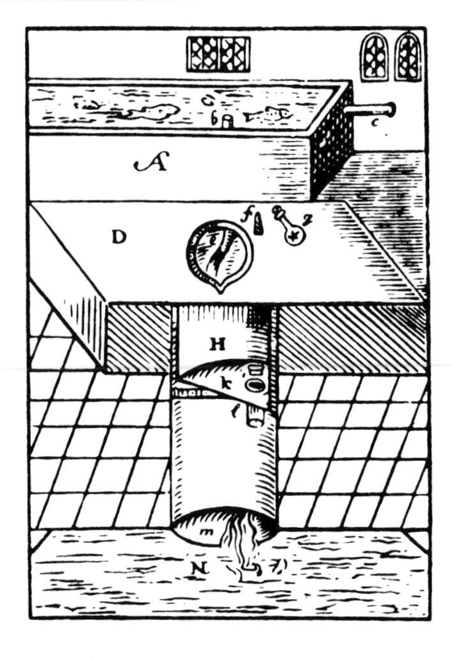

9.1 In Harington's drawing of his water closet, water cleans the bowl and an iron rod extends from the seat surface (*f*) to the stopper (*k*) beneath. If water were scarce, it was recommended that the "stool pot" (*H*) be emptied at least once a day, and after being drained it was supposed to be filled with 6 inches of water. (Fish are drawn in the tank merely to indicate that it is filled with water.) (J. Harington, *Metamorphosis of Ajax*, 1596.)

9.2 In Alexander Cumming's water closet, an upward pull on the handle (*O*) activated a mechanism that simultaneously moved the "slider" at the bottom of the bowl and opened a valve that allowed water to flow into the bowl. (British Patent no. 1105.)

In the privy that annoys you, first cause a cistern, . . . to be placed either in the room or above it, from whence the water may, by a small pipe of lead of an inch be conveyed under the seat in the hinder part thereof (but quite out of sight); to which pipe you must have a cock or washer, to yield water with some pretty strength when you would let it in.

Next make a vessel of an oval form . . . two feet deep, one foot broad, sixteen inches long; place this very close to your seat, like the pot of a close-stool; let the oval incline to the right hand.

This vessel may be brick, stone, or lead; . . . the bottom and sides all smooth, and dressed with pitch, rosin, and wax; which will keep it from tainting with the urine.

In the lowest part of the vessel which will be on the right hand, you must fasten the sluice or washer of brass, with solder or cement; the concavity, or hollow thereof, must be two inches and a half.

To the washers stopple must be a stem of iron as big as a curtain rod; . . . with a strong screw at the top of it; to which you must have a hollow key with a worm to fit that screw. . . .

These things thus placed, all about your vessel and elsewhere, must be passing close plastered with good lime and hair, that no air come up from the vault, but only at your sluice, which stands closed stopped; and it must be left, after it is voided, half a foot deep in clean water.[11]

A water closet of Harington's design was installed for the Queen at Richmond Palace, but this first example of the valve closet was not widely adopted and seems to have disappeared from use until almost two centuries later, when it was reinvented and introduced once again.

A British patent for a valve water closet was issued to Alexander Cumming, a London watchmaker, in 1775.

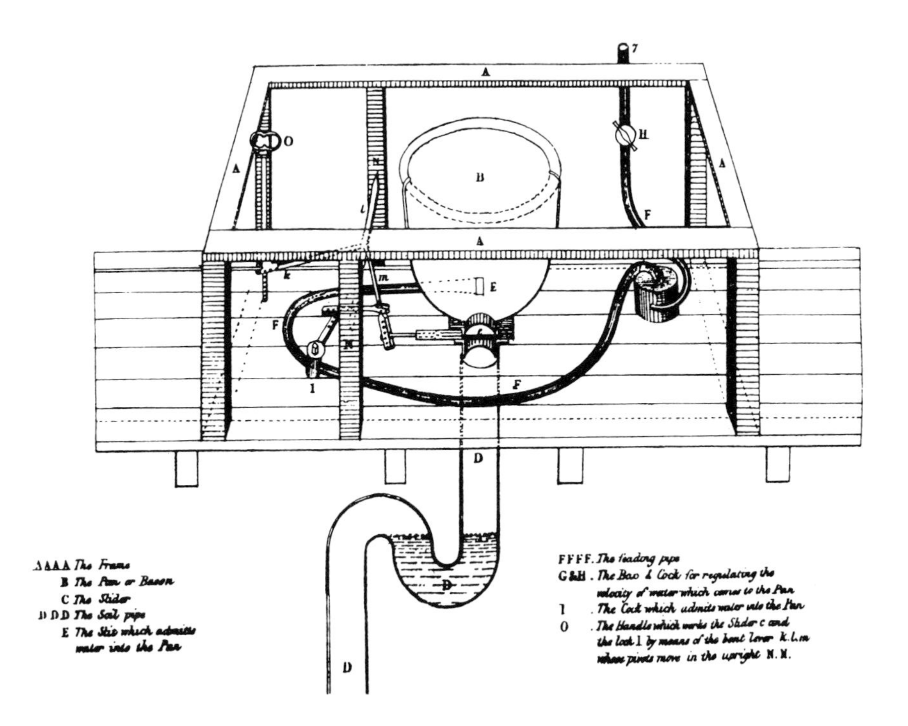

The principle was the same: a lever beside the seat operated a leather-covered valve at the bottom of the bowl, and at the same time water from an overhead tank was released through an opening in the side of the bowl.[12] Beneath the water closet itself the waste drained through the "stink-trap," an S-shaped bend in the waste pipe that retained sufficient water to seal the pipe and prevent odors entering the toilet room (fig. 9.2). Until rubber valve seals were introduced, the fit of valves quickly deteriorated, and there was little to prevent entrance of smells from the system.

Various adaptations of the basic design were developed. The most influential of these was the first patent of Joseph Bramah, a London cabinetmaker who later invented locks and many other devices (fig. 9.3). Bramah fabricated cabinets to surround valve water closets, and he turned his mind to improving the closets. Three improvements contributed to the popularity of Bramah's design, as compared to that patented by Cumming only three years before.[13] First, a metal flange was set in front of the opening through which water entered the bowl, deflecting the flow to cover much of the bowl's interior surface. Second, an overflow pipe bypassed the valve in case of stoppage. The third improvement was in the valve itself. Cumming's valve, the "slider," had moved horizontally to close or open the bottom of the bowl; Bramah's was hinged and pivoted when the lever was pulled. Thousands of such water closets were manufactured by Bramah. For about a century the Bramah valve closet was in general use, until in 1870 the British firm of Dent and Hellyer introduced their "Optimus" improved valve closet. Almost a hundred years later this model was much like Bramah's design, except for its having a flushing rim and improved mechanics that quieted its noise and made chain-pulling less frustrating. In alphabetical order from "Optimus A" to "Optimus P," models of this closet were produced until World War II.[14]

The pan water closet was probably introduced almost as early as eighteenth-century valve closets, and the difference between them was slight. Instead of the flat face of the valve closing the opening at the bottom of the bowl, in the pan closet a shallow metal dish was hinged beneath the opening. An 1829 British patent provided a leather ring to seal the edges of the pan against a flange of the bowl, and water in the pan further sealed against odors.[15] Leather gaskets were not usual in the many pan closets available during the middle half of the nineteenth century, but one would assume that water seals at the pan and in the "stink-trap" would have been sufficient. Nevertheless, leaks in the entire assembly of pipes and fixture often defeated the purpose of these water seals. Pan closets were widely used, but there were constant complaints about them. Even with the addition of a flushing rim around the upper edge of the bowl, the design was not considered satisfactory. A nineteenth-century plumber commented: "I consider this closet a very unsanitary piece of mechanism, and totally unfit for its intended purpose, inasmuch as in a short time the internal parts become besmeared, and consequently become offensive."[16] The pan closet persisted as a basis of negative comparison for other designs, "as bad, or worse, than the pan-closet."[17] Toward the end of the nineteenth century the pan closet was sometimes still in use in Chicago, although it was not admired:

> The pan-closet of twenty-five years ago was identical with the pan-closet of today. The old pan cistern-closet was defective in structure in two particular things: First, the supply pipe to the closet bowl was far too small for a proper flush; one-half inch pipe was not sufficient. Second, the closet bowl, especially the French [round] bowl, was defective in principle. The swinging of the water around the bowl in a whirlpool-like shape was not enough to cleanse the bowl nor remove the soil from the trap, or clean the trunk of the closet or wash out the soil pipe.[18]

Not much better was the hopper closet, which dispensed with valve or pan. It consisted of the hopper, an inverted cone placed directly over the

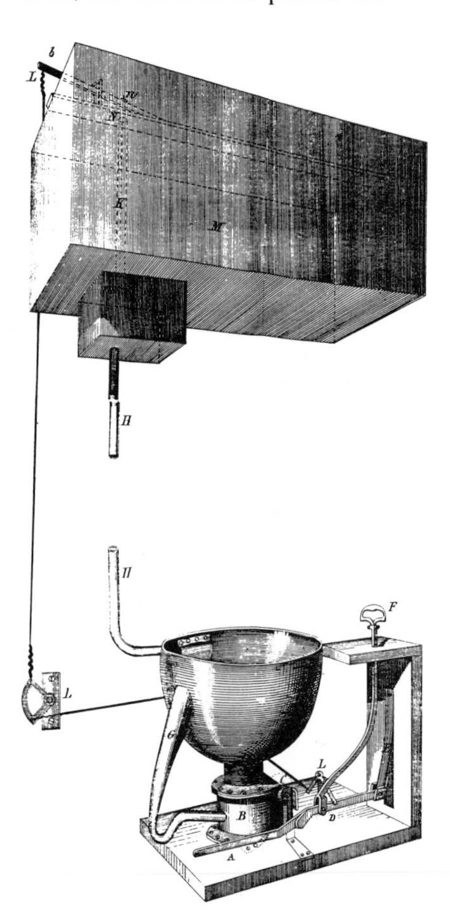

9.3 The Bramah valve closet was operated by pulling the handle (*F*), which opened a hinged valve in the chamber (*B*) and pulled the wire (*L*) that opened a valve at the bottom of the water tank. As with all valve closets, a tight-fitting valve was needed if any water were to be retained in the bowl. (British Patent no. 1177.)

"stink-trap," and an ineffective trickle of water that spun around the sides of the bowl. The sole advantage of the hopper closet was its extremely low price. In 1849, 18 hopper closets were shipped from New York to St. Louis for installation in the Planters' House Hotel, but a few years later the installation was improved when the hopper closets were replaced with the repugnant pan closets.[19] In general, it could be assumed that hopper closets would be provided for servants and workers, because they cost about a fifth as much as more sophisticated and efficient models. At any time every country used a variety of designs, cost being the major factor in selection. So many hopper water closets were manufactured so cheaply that an English sanitary reformer suggested abandoning use of the hopper and selling the unused earthenware cones to gardeners, who could use them to protect rhubarb from spring frosts.[20]

Although the plunger water closet was used at the same time as valve, pan, and hopper closets, few improvements in its design were patented until the latter part of the nineteenth century. The mechanism was simple, though clumsy. At the bottom of the bowl, instead of a valve or pan, there was a plunger, which, when lifted by raising a handle, opened the passage between the bowl and the trap. Some plungers were simply cylinders, but others were hollow and contained a valve that opened under pressure to prevent the bowl's overflowing. Two major problems of the plunger closet were the tendency of the plunger to make an imperfect seal, even when provided with a rubber gasket, and the fact that concealed portions of the mechanism were quickly fouled. An advantage was the quantity of water retained in the bowl, about half its capacity, and this feature was kept in many later designs of water closets.

The valve, pan, and plunger water closets had provided a bowl within which feces might fall into water and a means of conveying the feces and water into a drain in which there was a trap, but the more fastidious classes required a closet that was more sanitary, although it might also be much more expensive. The washout water closet had its outlet located at the back side of a shallow bowl. Waste and water flowed through this opening into the trap, leaving an inch or two of water in the bottom of the bowl. Its principal fault lay in the fact that the water seldom had enough force to cleanse the bowl properly and eject the wastes. A variety of designs attempted to make improvements by changes of the flushing rim and the surface of the bowl, but they were, on the whole, unsuccessful.

The washdown water closet was little more than a hopper closet with its S-trap at a level so high that the water in it filled much of the water-closet's bowl. A variation of the washdown closet was the siphonic closet, which used different water inlets and amounts of water to cause a siphon action with two or three sequential flushing actions.

Earlier water closets, like the other bathroom fixtures, had required cabinetwork to conceal the inelegant connections between parts made of earthenware, cast iron, and enameled iron. Makers of the ceramic portions of fixtures had long resented the fact that their portion of the final product was priced far below the pipes, traps, and valves that were made of metals. After the introduction of the flushing rim, the skill in manufacturing extremely intricate shapes of earthenware grew year by year (fig. 9.6). One of the advantages of the washout and washdown water closets was the pos-

9.4 **The basic types of water closets in the nineteenth century: *A*, pan closet; *B*, valve closet; *C*, plunger closet; *D*, long hopper closet; *E*, short hopper closet; and *F*, a washdown closet. (W. P. Gerhard, *House Drainage and Sanitary Plumbing*, 1882.)**

9.5 **Later water closet designs: left, the washout closet; center, the washdown closet; and right, the siphon closet. (W. P. Gerhard, *Water Supply, Sewerage and Plumbing of Modern City Buildings*, 1910.)**

sibility of making them from a single smooth material, earthenware, using complex shapes to replace the complicated mechanisms of earlier models. The shapes of the fixtures were made visible in bathrooms, where they shone bright and clean. Decoration could be added to their surfaces, plants and animals shaped in bas-relief and colored patterns added with sanitary glaze.

After cholera first struck England in 1831, a physician in the industrial city of Leeds began compiling studies showing that mortality rates were higher and epidemic disease spread more rapidly in the working-class areas of the city. A later investigation proved that most of those who died in London from the cholera epidemic of 1849 had drunk water from a single public pump. With repeated epidemics in Europe and the United States, cholera came to be respected as "the great sanitary inspector of nature."

Typhus, typhoid fever, and smallpox added to the death rate in nineteenth-century slums, but the frequency of children's deaths probably was largely the result of their increased susceptibility due to malnutrition. A London laboring-class family of five in 1841 is reported to have been fed for a typical week with five pounds of meat, twenty of bread, forty of potatoes, and little else.[21] Although the dispersed squalor of farmworkers' hovels was little better, the slums of large cities attracted the attention of reformers. Friedrich Engels in 1845 described *The Condition of the Working Class in England*, and 16 years later Henry Mayhew in *London Labour and the London Poor* dramatically recorded the lives of the city's impoverished. Prior to both these books, a commissioner of the English poor laws, Edwin Chadwick, published the *Report on the Sanitary Condition of the Labouring Population of Great Britain*. A hard-working and cantankerous man, Chadwick combined a massive compilation of questionnaires to physicians, police, builders, and others with the tabulation and mapping of the vital

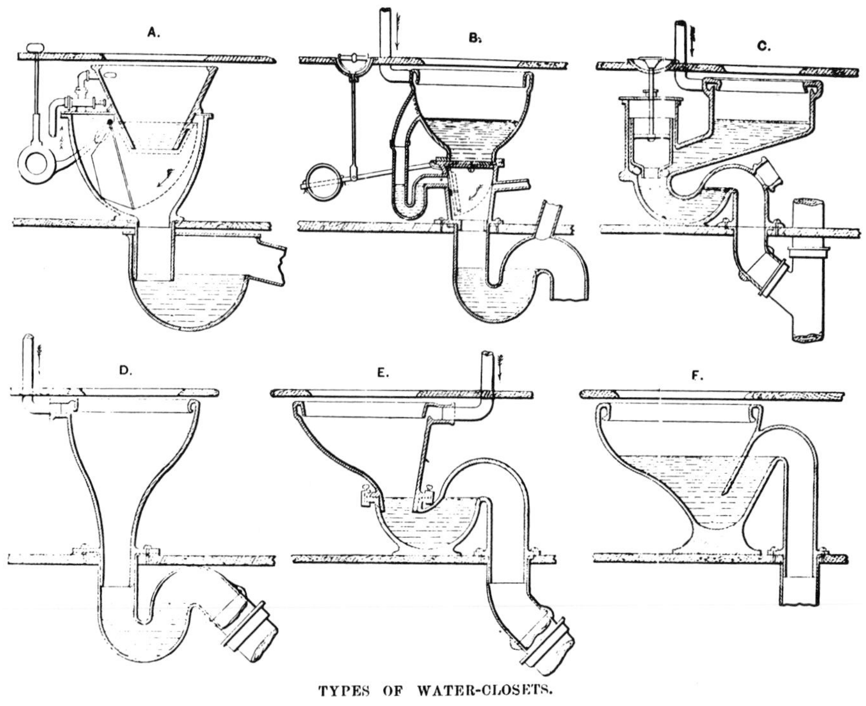

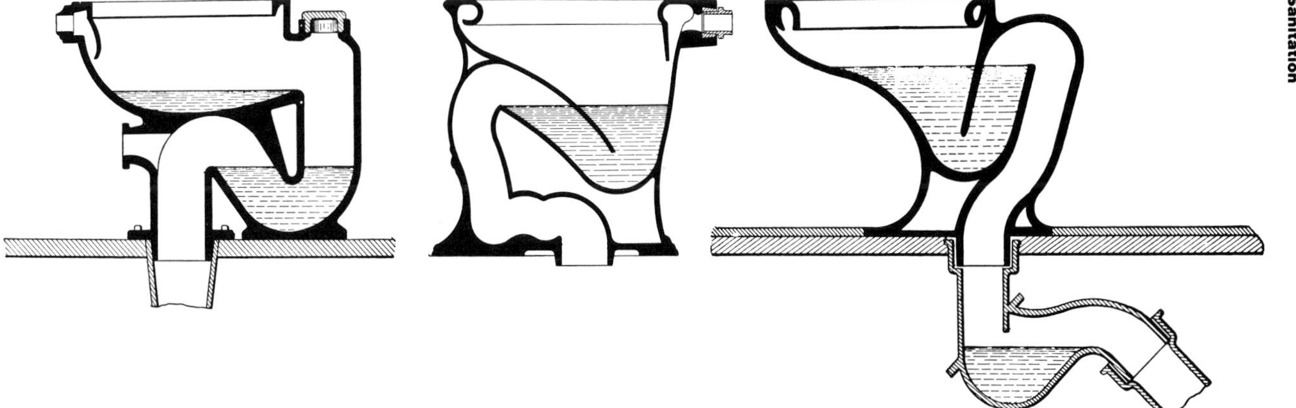

statistics available at the time. Over 10,000 copies of "Mr. Chadwick's Report" were sold, an extraordinary demand for anything published as a government document.[22] A series of boards and commissions contributed to the preponderant evidence that insanitary conditions were closely associated with the occurrence of disease and early death. Although not everyone was alarmed by the facts as they were reported, the righteous impulses of Victorian England were aroused. An exception was Thomas Carlyle, who declared that "if paupers are made miserable, paupers will needs decline in multitude."[23] Unfortunately Hamburg, Paris, and other cities were to make major improvements in their sanitary conditions before the political and business interests in London would permit such construction.

By the middle of the nineteenth century an increased portion of housing for English workmen was built with an alleyway behind, supplanting the back-to-back construction that had opened only at the front. Privies, often shared by several houses, were situated in rear yards, and the alleys permitted nocturnal removal of the privies' contents, which was sold to farmers as fertilizer. In some cases, instead of cesspits being emptied, other holes were dug and the previous pit merely covered, a practice that greatly increased the likelihood of polluting wells. Later the increased number of water closets installed resulted in larger amounts of fecal material entering streams, because it was no longer available for agricultural use.

Problems of building sanitation, while more prevalent in slums, were forthrightly disrespectful of rank. In 1844, 53 overflowing cesspits were found beneath Windsor Castle, providing an explanation of the frequent ailments suffered by the royal family's servants at the castle. The Prince Consort ordered the installation of water closets and a sewage system. Seventeen years later Prince Albert died of typhoid fever, and a decade afterward the Prince of Wales fell ill of the same disease. Such events brought the attention of the English public to bear upon the need for improved household sanitation. The

Prince of Wales, after recovering from his nearly fatal fever, was said to have declared, "If I were not a prince, I would be a plumber."[24] But progress was slow. In 1882 the Duchess of Connaught, wife of Victoria's third son, was dangerously ill, and the ailment was attributed to her newly built residence, Bagshot Park. The duchess's doctor and a civil engineer inspected the building and reported: "Offensive smells had long been perceived about the house, and had been a common topic of conversation; but no one had suspected their origin or had realized the dangers they were likely to cause. Many of the inmates, however, had suffered from various forms of indisposition, such as sore throats, diarrhea, and a general sense of heaviness and malaise, and these generally affect new comers."[25] *Lancet*, a medical journal, commented that the plumbing system of the building appeared to have been designed "by a Machiavellian policy which would seem to be the pastime of modern builders."[26]

In Great Britain the first legislation on sanitation, enacted in 1848, had little effect because it gave permission for improvements without requiring that any action be taken. A succession of later acts established standards and required compliance with them. As legislation advanced, the training of inspectors improved and their numbers increased. In 1864 Dublin had only one sanitary officer, but about 30 years later the city employed a corps of 50.[27]

A reform movement in New York, led by concerned businessmen, in 1864 conducted a survey of sanitary conditions within the city, using procedures that had been developed in England. Their report of shocking conditions and the threat of cholera led to passage of the Metropolitan Health Law two years later. Used as a precedent for many sanitation laws of other cities in the United States, the Metropolitan Health Law and subsequent stricter legislation were particularly directed toward the control of tenement construction.

9.6 Bathrooms of the well-to-do in the United States: During the 1880s (left) it was deemed necessary to encase all fixtures in wooden paneling, a requirement no longer present in the 1890s (center) when open fixtures were given decorative treatments. By the start of the twentieth century (right) a shower over the bathtub was standard. The presence of sitz baths and bidets in these catalog illustrations indicates the enthusiasm of the manufacturers of plumbing fixtures more than the common practice of the times. (A. M. Maddock, *The Polished Earth*, 1962.)

One of the most important factors in protecting the public health was the provision of a safe water supply. In 1829 sand filter beds were installed by the Chelsea Waterworks in London, and about 20 years later all water companies in the city were required to do the same. At the end of the century there were in all almost 15 acres of sand filter beds in London. At the same time techniques of filtering water in the United States proved so effective that sanitary engineers recommended that municipal authorities invest only in water treatment, because, as one engineer put it, a dollar spent on water treatment "would do as much as ten dollars spent in sewage purification."[28] There was, of course, a sharp difference of opinion between the citizens of a small downstream town and those of the upstream metropolis. However, it was discovered in Lawrence, Massachusetts, that sand filtration reduced the incidence of typhoid fever by 79 percent.

Early pipes, those of the ancient and medieval periods, were made of earthenware, lead, or wood. For drains made of fired clay it was simple to taper or enlarge tubular shapes in order to provide a firm joint between lengths of pipe. Sheet lead was obtained by spreading the molten metal on a sand-covered surface and beating the sheet until the desired thickness was attained. An oval cross section resulted from bending the sheet lead and making a soldered joint; a teardrop cross section resulted when the metal was folded and crimped for a joint. Wooden pipes have been found in several early medieval English installations, notably monastic complexes. Early in the seventeenth century, when the first large-scale water supply system was available in London, water was brought from the north into the city by an open stream 38 miles long, and from a reservoir it was distributed to users through mains made of elm trunks and fastened together with iron bands. Branches of the trees were often used to form Y's and T's needed for water mains. In spite of the development of metal pipe, the use of wooden mains for the supply of water and gas persisted into the last decades of the nineteenth century. The American Pipe Company of San Francisco in 1878 advertised sections of fir and pine 8 feet long, which were bored, steam-treated, wrapped with iron bands, and coated with asphalt.[29] Wooden pipes must have done their work well. In one of the company's advertising brochures a testimonial letter reported the excellent condition of wooden water mains in Manhattan that had survived more than 75 years of use.

Metal pipes for water supply were later made by the same methods employed in manufacturing gun barrels. In fact, many early installations for water supply, gas lighting, and steam heating utilized inexpensive gun barrels, those that were a manufacturer's surplus or were judged unsatisfactory for the sudden pressure of firing a rifle.

The principal concern in plumbing systems was the piping of drainage systems that carried away the waste, and tell-tale odors were ample indication that the drains were not satisfactory. A major factor in the public's anxiety about drains was the fact that the "miasma theory" of disease's ori-

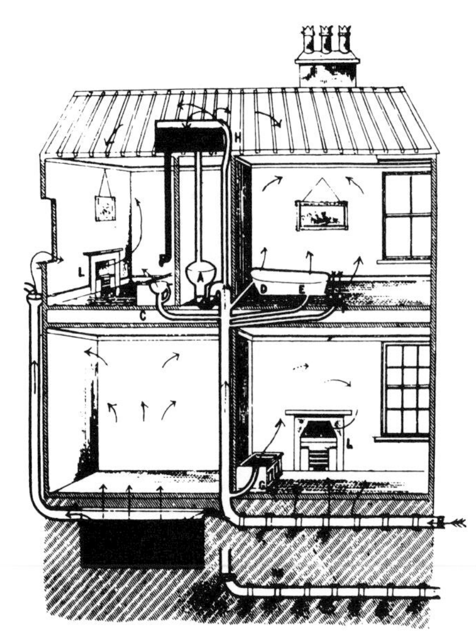

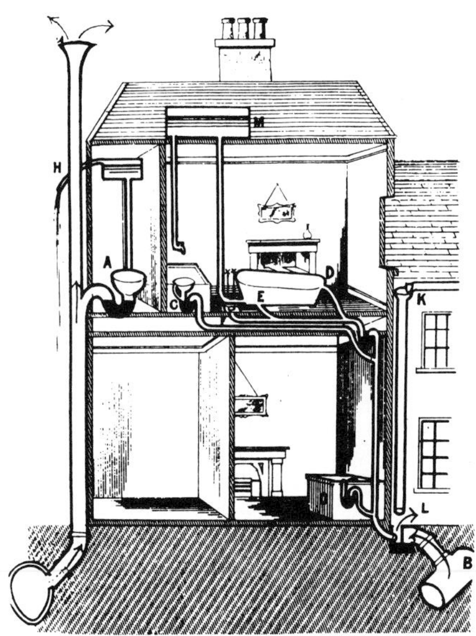

9.7 A publication of 1879 shows "a house with every sanitary arrangement faulty" (left) and "a house with faulty arrangements avoided" (right). Arrows within the dwelling at left show the possible circulation of sewer gas leaking from fixtures and seeping from beneath the floors. (T. P. Teale, *Dangers to Health*, 1879.)

gins and transmission persisted in the minds of the public and many physicians long after the conclusive experimental determination of the "germ theory." Unseen microorganisms were recognized as the source of disease by the 1880s, but only among those of advanced scientific background. Most of the population continued to associate the spread of sickness with the presence of identifiable foul odors.

Sewer gas leaked into toilet rooms primarily through early water closet installations (fig. 9.7). "The odor of their memory, or the memory of their odor, still lingers with us."[30] An almost endless number of ailments were attributed to the presence of sewer gas. In addition to its supposedly causing specific sicknesses, there was a more general claim that sewer gas deprived "men of ambition and women of beauty." As late as 1901 William Paul Gerhard, a leading sanitarian in the United States, told a gathering of knowledgeable health officers: "Modern German sanitarians are nearly united in being opposed to the so-called 'sewer-gas theory;' they claim that the researches of Von Pettenkofer, of Pasteur, of Dr. Koch and others have established, almost beyond a doubt, the fact that every infectious or zymotic disease requires the presence of a specific microörganism or pathogenic bacterium to cause it, and that the gases of putrefaction *per se* cannot cause the disease."[31] Another authority considered that sewer gas, while not a cause of specific disease, made one susceptible because it produced a condition of general ill health.

The primary defense against sewer gas entering a building was the trap, which was intended to seal drains by means of the water retained in a downward bend of the pipe. There was a danger, however, of water in a trap evaporating when buildings were closed for a long period, but hardly so great a danger as imagined by Chicago plumbers who in an 1888 issue of the *Sanitary News* recommended filling such traps in vacant buildings with glycerine or oil, or even draining out the water and packing the trap with salt.[32] A more realistic danger was that downrushing quantities of water in the main drains might create a vacuum and draw water out of a trap. The solution proposed was venting each trap by extending a pipe from the down side of the trap through the roof, and so providing air to relieve any vacuum (fig. 9.9). A Chicago plumber remembered that he first saw a major vertical drain extended through the roof in structures after the fire of 1871.[33] In the previous decade he had seen vents installed in connection with water closet traps, but they had been meant to avoid pressure developing within the major drains rather than to protect the seal of traps. By the end of the century regulations often called for back-venting traps. This requirement, still in force, was debated with great fervor. The opposition was formidable and outspoken.

J. Pickering Putnam, a Boston architect: "One of the most unfortunate and burdensome building-laws ever inflicted upon the people, and an imposition upon the public."

George E. Waring, the famous New York City sanitarian: "Does more harm than good, that is to say, that a trap is more likely to lose its seal if it is back-vented than if it is not."

An "English expert on drainage:" "A diagram of house-plumbing, protected by ventilation-pipes as prescribed by most American authorities, a bewildering nightmare of complicated ingenuity."[34]

In 1892 the *Northwestern Architect*, a Minneapolis publication, queried

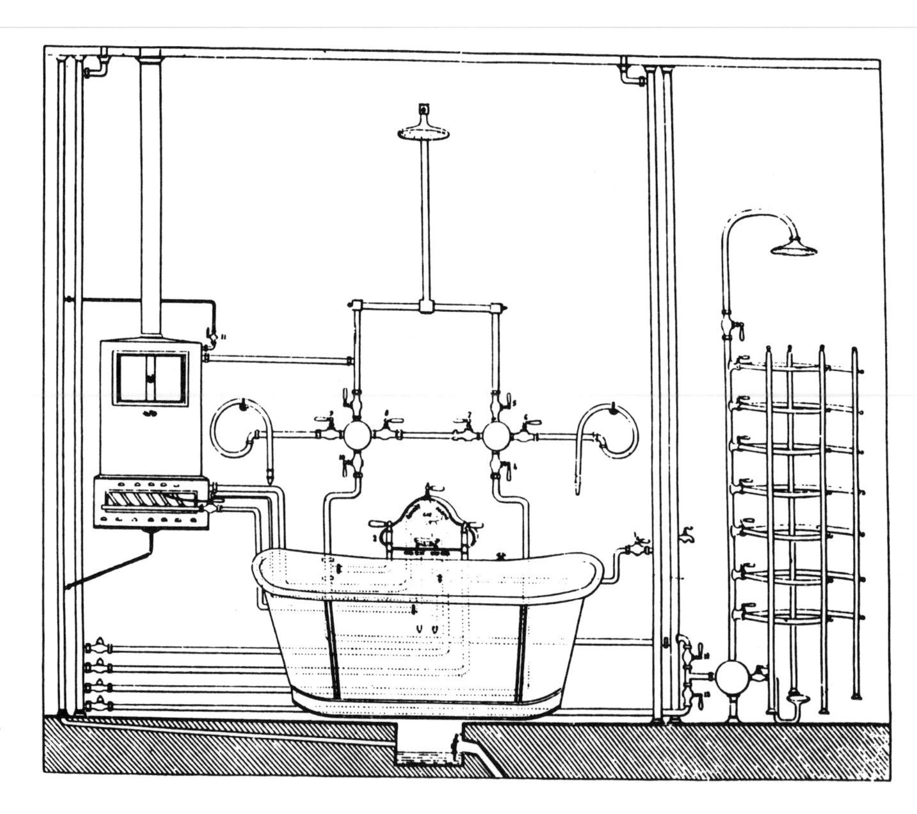

9.8 **In commenting on this 1890 bath arrangement displayed at the Palais de l'Industrie, *La Construction Moderne* said, "This system is perhaps extremely ingenious but less than practical." The journal's illustration bore 21 lines of caption explaining the ways of achieving different bathing effects. (*Construction Moderne*, 20 December 1890.)**

9.9 **Diagrams at the turn of the century suggested eliminating "back air pipes" by care in the provision of special traps. Such precautions permitted simplification of the piping required. (R. Sturgis, *Dictionary of Architecture and Building*, 1901–1902, vol. 2.)**

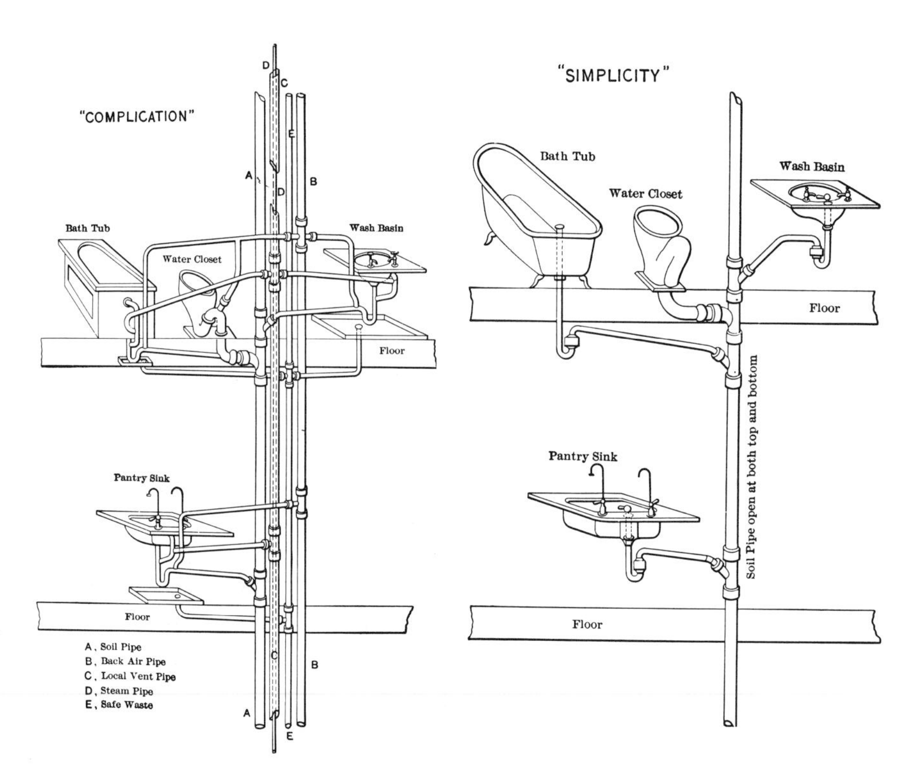

architects and plumbing contractors on the matter. Plumbers were about evenly divided on the subject of back-venting. The architects gave varying replies, but all were convinced that traps of greater depth than customary would provide as much protection as back-venting. This point of view was echoed by Dankmar Adler, the respected Chicago architect, who added, "I am glad to see that you are moving in the cause of reform in a matter which is of vital interest to the entire community."[35]

Since ancient times the disposal of wastes has been accomplished with the use of water. Water closets, by definition, consumed large amounts of water, and even the filth thrown into streets was washed into rivers and lakes. There were few exceptions. The "night-men" who gathered the contents of cesspools, privy pits, barrels, and buckets sold them to farmers when cities were so small that a short trip outside the city would be profitable. When the distance was too great, the excrement was usually deposited in a single large pit or, more likely, taken to the river. The construction of metropolitan sewage systems and the increased number of water closets unfortunately meant only that an even greater portion of the waste went into the river.

"No more pipes, no more cesspits, no more unhealthy odors, absolute cleanliness." When a writer for *La Construction Moderne* read those claims for the earth closet invented in 1860 by the Rev. Henry Moule, he accepted the first and second claims, questioned the third, and rejected the last.[36] Lime, ashes, and sand had long been periodically dusted over the contents of privies (fig. 9.10). In the Moule earth closet, fine, dry earth or sand was kept in a hopper above and behind the seat. After each use a lever released a layer of earth from the hopper and, if ashes were used, gratings retained the cinders. One version of the earth closet automatically activated the hopper whenever a user rose from the seat, but the contraption appears to be have been noisy. "In sick rooms, this method of distribution of earth may be found objectionable, as more or less vibration follows the rising and this is apt to disturb the nerves of a patient."[37] The earth closet enjoyed a limited popularity, and there were nine designs patented in the United States between 1872 and 1882. Much of this interest in what now seems an improbable solution to the problem came from cold regions in which water pipes were liable to freeze. In addition, it should be recalled that the earth closet may have offered a relatively inexpensive alternative at a time when about 95 percent of New York City's population still depended on privies or chamber pots.[38]

The earth closet was a short-lived invention. Chemical toilets containing liquids of high alkalinity have become the most common solution for toilets in temporary locations, and pit privies remain a characteristic of isolated and primitive locations. The discussion of methods for disposing of human wastes has in recent years been related to discouraging forecasts of the future water supply. Toilets employing bacterial decomposition, incineration, and oil as a flushing agent have received attention, but almost all modern systems are fundamentally based on Harington's design made at the end of the sixteenth century. The water closet of that time has been improved in operation and appearance, but little has been done to confront the fact that water-flush systems are manifestly inappropriate in our time.

9.10 Earth closets scattered soil, sand, or ashes over the wastes deposited in a box beneath the seat. (G. E. Waring, *Earth Closets and Earth Sewage,* 1870.)

9.11 Systems of sanitation that avoided the use of water were particularly fitting for northern areas, where frozen pipes were a constant problem. For the Smead Dry Closet and Cremation System, a chamber was constructed beneath the floor of a school toilet with air going over a heater at one side (right) and up a flue on the other side (left). According to an 1889 catalog, over a hundred of these systems at that time had been installed in schools in Minnesota and Wisconsin. (Catalog of Ruttan Warming and Ventilating Co., 1889.)

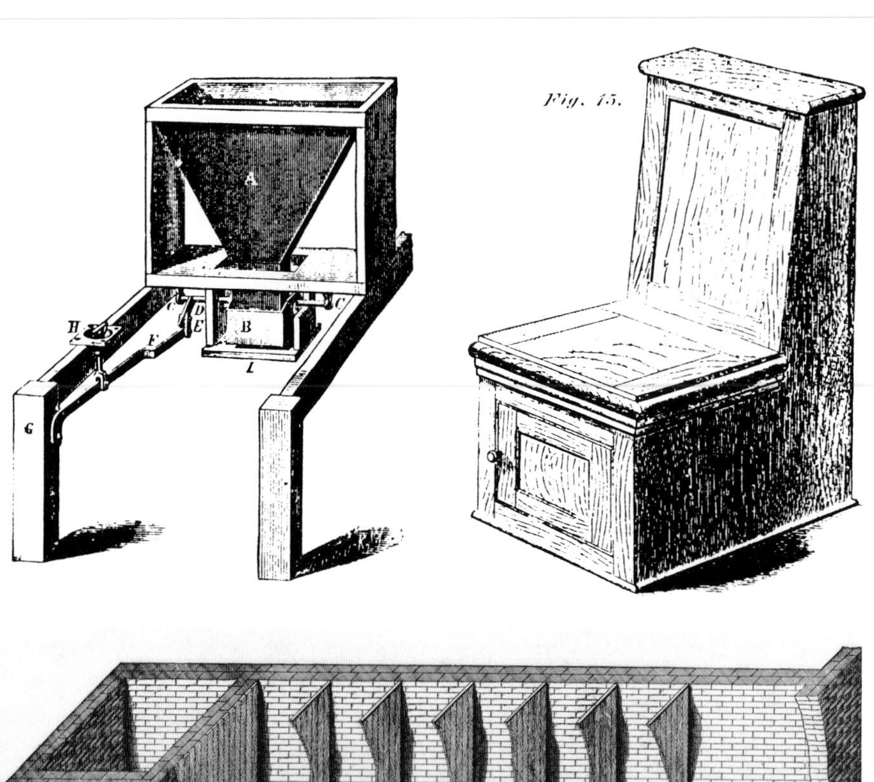

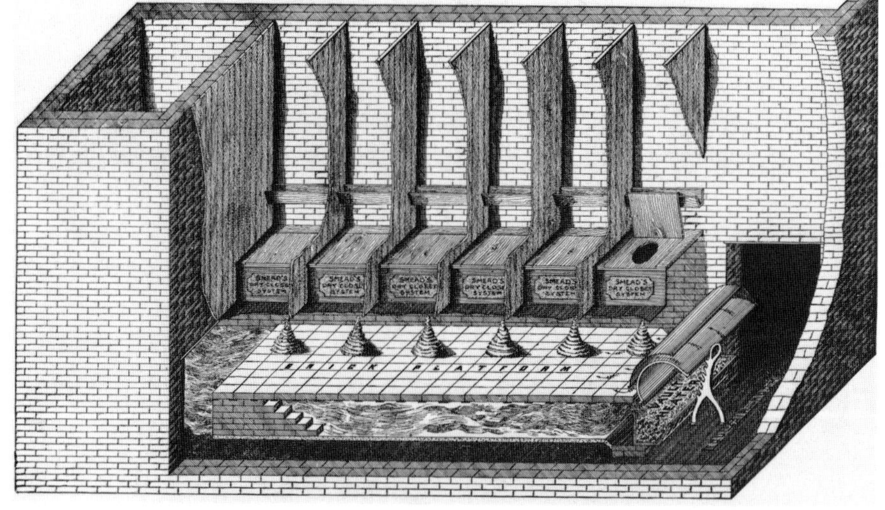

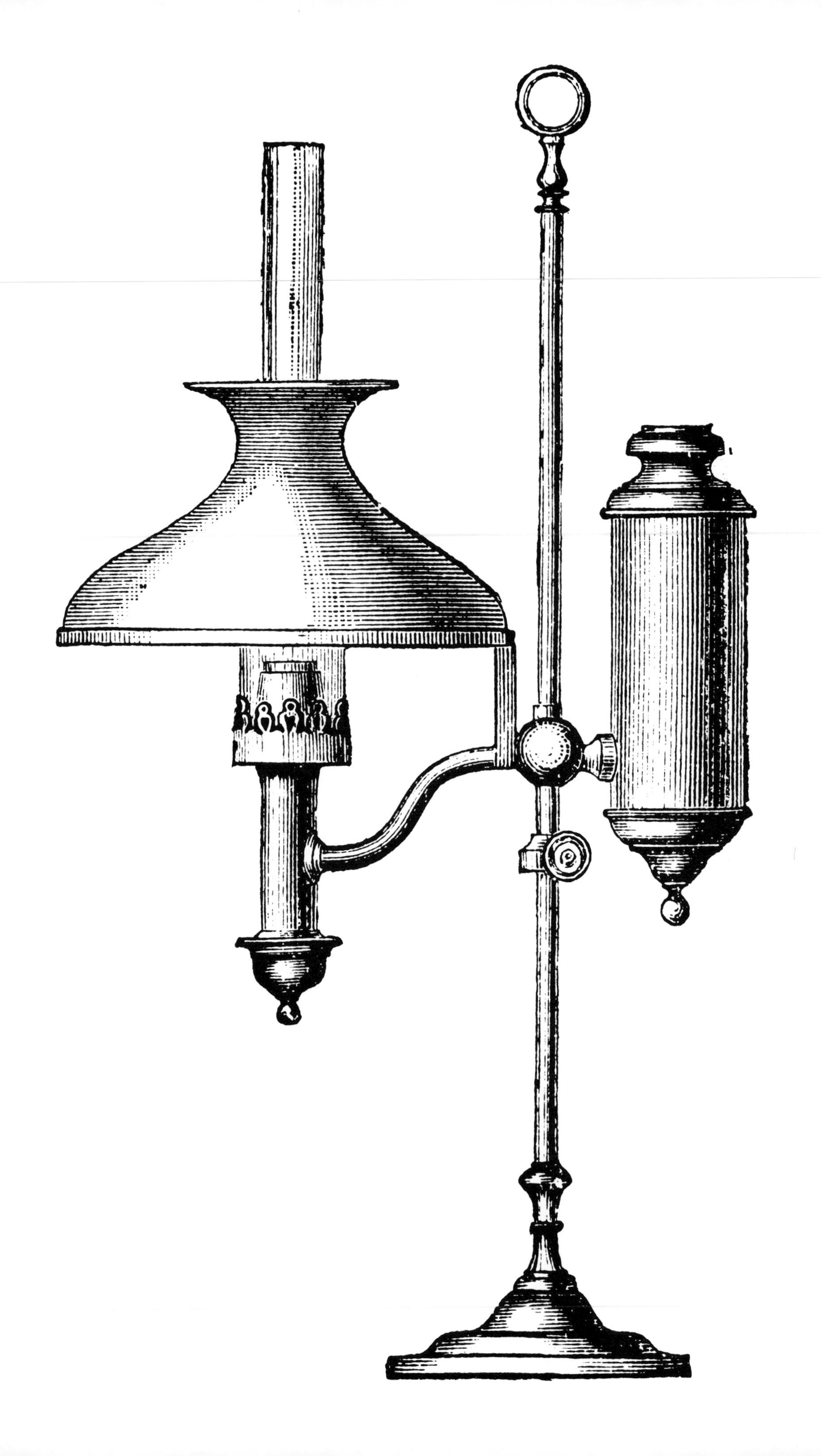

10

Lighting

1660s Experiments conducted with gas gathered from springs

1783 A circular burner for oil lamps invented by Genevois Aimé Argand

1798 Philippe Lebon moves to Paris to develop his "Thermolampe"

1802 Peace of Amiens celebrated by William Murdock's gaslight display at the Boulton and Watt factory, Birmingham

1829 Limelight perfected by Thomas Drummond (Britain)

1845 Patent (Britain) awarded for incandescent lamp of J. W. Starr

1859 Moses G. Farmer lights his home in Salem, Massachusetts, with an incandescent lamp

1870s Continuous electrical current provided by the Gramme dynamo

1876 The Jablochkoff arc lamp introduced in Paris

First demonstration of Charles Francis Brush's arc lamp in Cleveland, Ohio

1879 Thomas A. Edison obtains his principal U.S. patent for the incandescent lamp

Joseph Swan exhibits his incandescent lamp before the Newcastle-upon-Tyne Chemical Society

1893 Commercially practical incandescent mantles for gas burners developed by Carl Auer von Welsbach

1895 D. McFarlan Moore displays his first vapor lamp

1907 Tungsten filament incandescent lamps introduced

1939 Planners of the New York World's Fair obtain fluorescent lamps for use in fair buildings

10.1 This popular kerosene-burning student's lamp had a reservoir (at the right) that released fuel to the base of a cylindrical Argand wick. An ingenious valve controlled the kerosene's flow and made it possible to place the tank of fuel at the side, so that it did not prevent ample light falling downward, where it would be needed on a student's reading table. (*Appleton's Cyclopedia*, 1880, 2:225.)

The primitive firebrand taken from a campfire appears to have been the root from which a minor industry had grown by the seventeenth century. Sticks of pine and other resinous woods were shaped for lighting, and stands, brackets, and ash pans in considerable variety were manufactured to hold such torches. Other vegetable matter was burned for light, such as the oily nuts of the candleberry tree in Southeast Asia, and in some areas that had little plant life the oily carcasses of certain fishes and birds were burned. Through time each of these alternatives acquired its own fixtures and accessories.

From Roman times candles were often made of beeswax (plentiful at a time when honey was the usual sweetener), and tallow candles were made by rendering animal fats. There was also the relatively inexpensive light available from the rushlight, which was made by dipping the pith of rush stalks in tallow. The outside material of the stalk was left on one side of the pith as a stiffener.

Tallow candles flickered and smoked, but wax candles were many times more expensive. A "composite" candle, made by mixing tallow and wax, was more reasonable in price and long served as a popular compromise. In the early years of the eighteenth century, whalers turned their attention to the sperm whale, which had body tissue rich in oil and in the front of its brain had a cavity that contained oil and spermaceti. The oil and spermaceti mixed together at the body temperature of the whale, but when cooled the spermaceti separated and set as a white wax. Of this were made the finest candles, the standard against which later lighting systems were compared.

As candlewicks charred, the amount of light delivered by the flame was reduced. This problem was so great that in seventeenth-century theaters the chandeliers were lowered between acts, and the actors—still costumed as heroes or buffoons—came forth to crop "snuffs," the charred ends of wicks, so that there might be brighter light for the act that was to follow. Starting around 1810 braided wicks were soaked in boracic acid, and after this treatment the wicks would burn away more completely, improving the candle's light.[1] It was estimated that a typical Philadelphia family of five members at the first of the eighteenth century would have had two oil lamps for those rooms most used, perhaps the parlor and the kitchen, and five candlesticks for errands into other rooms or going off to bed.[2] As faint as its light might be, a candle was safer than a lamp, for lamps could spill their fuel when overturned.

The earliest lamps were little more than shallow dishes with tubular spouts in which lay a wick. With a thick wick and limited air supplied to the flame, such oil lamps could provide little light. The first fuels used were olive oil in the south of Europe and animal oils, including that of whales, in the north. In the middle of the seventeenth century the introduction of colza oil, made from rape seed, reduced lighting costs to such an extent that many poor families could afford light better than had been provided by their lamps filled with the animal grease they saved.[3]

In spite of changes in the form of wicks and the lamp itself, the oil lamp gave off a weak and wavering light. A major improvement was made in 1783 by a young Swiss scientist, Genevois Aimé Argand. Argand demonstrated his lamp to the police administrator responsible for lighting Paris streets; meeting a cool response, he traveled on to England, where he found an opportunity to present George III with a model of his lamp. With the king's encouragement, Argand soon was introduced to Matthew Boulton, James Watt's partner, and they patented the lamp and set up a factory to manufacture it.

The Argand lamp recognized the need to provide ample air to a flame. The reservoir of oil was shaped in a ring with an opening through its center. Around this opening a hollow tubular wick stood in the circle of fuel. Argand's design permitted air to rise around and in the center of the flames, establishing a draft like that of a chimney. (The same form was later used in Argand burners for gas lighting.)

In 1785, feeling assured of his invention's success in England, Argand returned to France to try once more for official recognition of his lamp. To Argand's amazement, both the court and Paris were filled with talk about the marvelously brilliant lamps that had lighted a performance of the *Marriage of Figaro* at the Comédie-Française. Although he had not known that they were to be used, newspaper accounts awarded Argand credit for inventing the lamps. At the same time credit for the use of glass cylinders around the flames was given to Quinquet and Lange, a pharmacist and a grocer who had been part of Argand's circle of scientific enthusiasts when he lived in Lyons. Argand protested loudly, and when he applied for a French patent he met with resistance from his former friends. Waged in newspapers, at scientific conferences, and even before the Parlement, the debate boiled down to this: Lange had displayed a lamp with a glass chimney to the Académie des Sciences more than four months before Argand's application for an English patent was filed, and while Argand's English patent included a chimney, there was no stipulation that it be of a transparent material. In the end the French authorities granted Argand exclusive rights to manufacture the lamps, and Lange was given exclusive rights to market them. Quinquet, on the other hand, through some private arrangements with the other two, displayed and sold a small number of the lamps in his apothecary shop.[4]

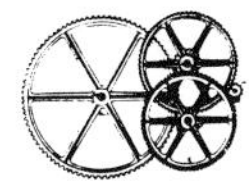

Near Wigan, a Lancashire town just west of Manchester, the pools of a small spring bubbled with strong-smelling vapors. Natives believed that the water was flammable, but Thomas Shirley tested it and found that the water was merely water. It was the foul air that floated on the surface of the water and the rising bubbles that caught fire from a torch. This he reported to the Royal Society in 1667. At about the same time Rev. John Clayton wrote the prominent scientist Robert Boyle describing similar stud-

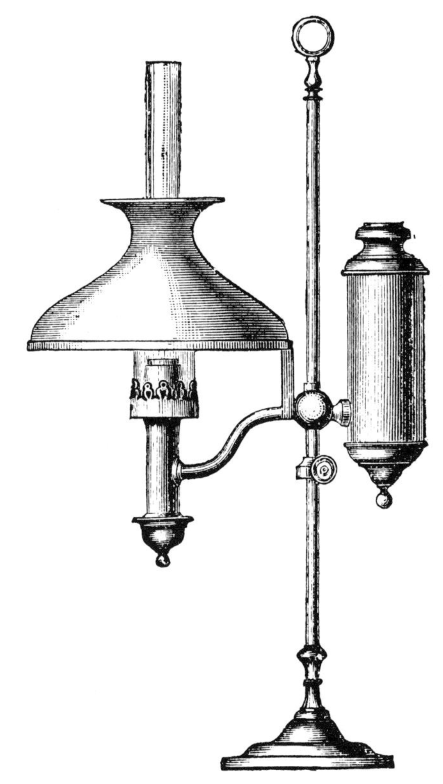

ies he had made. Conjecturing that the vapor in such streams must seep up from the strata of coal that underlay much of the region, Clayton had heated coal in a closed vessel.

At first there came over only Phlegm [water], afterwards a black Oil [coal tar], and then likewise a Spirit arose, which I could in no ways condense, . . . I observed that the Spirit which issued out caught fire at the Flame of the Candle, and continued burning with Violence as it issued out in a stream, which I blew out, and lighted again, alternately, for several times.[5]

To entertain friends Clayton inflated bladders with the gas, then pricked the bladder with a pin and lit the thin stream of gas that spurted out as he squeezed the bladder.

In all such early experiments no proposal seems to have been made for the use of gas for lighting or heating. It seems to have been viewed as a curious phenomenon without particular utility, until 1765 when the English town of Whitehaven briefly considered the possibility of piping gas from a nearby coal mine to light the town.[6]

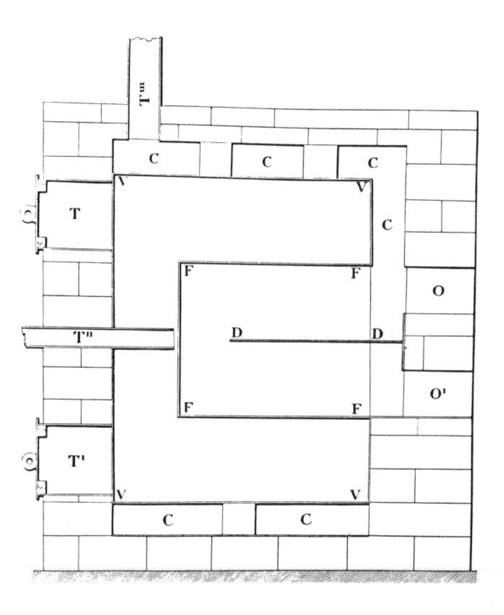

10.2 The gas-producing system patented by Philippe Lebon is shown in cross section. The material from which gas is to be extracted is inserted through one door (*T*) and removed through another (*T′*). Within the retort (*V*) is a firebox (*F*) inside which the fuel rests on an iron plate (*D*), its heat reaching the flue (*T‴*) through channels (C) around the retort. The "Thermo-lampe" was a refinement of this scheme with gas-burning lamps attached to it. (C. Hunt, *History of the Introduction of Gas Lighting,* 1907.)

Experiments in the middle of the seventeenth century initiated more than a century of further study without appreciable consideration of gas's large-scale practical application. One investigator discovered that when coal gas was passed through tubes immersed in water some matter would condense, while the gas retained its flammability. A bishop in Wales investigated the gases coming from different kinds of coal and wood when they were heated in a closed retort. Jean Pierre Minkelers, a Belgian professor of science, was paid in 1784 to explore ways of producing light gases for use in balloons, only a year after Montgolfier's first hot-air balloon ascension. One of the first of the substances considered in his investigation was coal gas and, while this proved not to be suited for use in balloon ascensions, it was used to light Minkelers's laboratory.

At the end of the eighteenth century two men, a Frenchman and a Scot, took steps toward commercialization of gas lighting. The Frenchman, Philippe Lebon, was an engineer and chemist who experimented with the distillation of wood in an iron retort and realized that the gas could be useful for heating and lighting (fig. 10.2). In 1798 he moved to Paris and continued his experimentation. After obtaining a patent, Lebon sought to interest the French government in using gas to light the public buildings of Paris, but in the rising flood of Napoleonic ambitions the proposal was passed over. Lebon's "Thermo-lampe" was a self-contained unit that both produced and burned the gas. A few gardens and a few houses in Paris were lighted by this system, but after Lebon's early death the system failed to become popular, in spite of his widow's vigorous efforts to market it.

The Scot was William Murdock,

son of a millwright who taught him that craft. In 1777 he was hired to work in the Boulton and Watt factory where steam engines were built, and two years later he was sent to the county of Cornwall to assume responsibility for installing and maintaining the pumping engines used in mines there.

It is now [1806] nearly sixteen years since, in a course of experiments I was making at Redruth, in Cornwall, upon the quantities and qualities of the gases produced by distillation from different mineral and vegetable substances, I was induced . . . to try the combustible properties of the gases produced from it [coal], as well as from peat, wood, and other inflammable substances. And being struck with great quantities of gas which they afforded, and as well with the brilliancy of the light and the facility of its production, I instituted several experiments with a view of ascertaining the cost at which it might be obtained, compared with that of equal quantities of light yielded by oils and tallows. My apparatus consisted of an iron retort with tinned copper and iron tubes, through which the gas was conducted. . . . The gas was also washed with water, and other means were employed to purify it. In the year 1798 I removed from Cornwall to Messrs. Boulton, Watt and Co.'s works for the manufactory of steam engines at the Soho Foundry [Birmingham], and there I constructed an apparatus upon a larger scale, which during many successive nights was applied to the lighting of their principal buildings. . . . These experiments were continued with some interruptions until the peace of 1802, when a public display of this light was made by me in the illumination of Mr. Boulton's manufactory at Soho. . . . Since that period I have, under the sanction of Messrs. Boulton, Watt, and Co., extended the apparatus at Soho Foundry so as to give light to all the principal shops.[7]

Many years later the Scottish engineer William Fairbairn recalled a walk through an unlighted part of Manchester as Murdock pressed on a gas-filled bladder, as if it were a bagpipe. Their way through the night was lighted by a flame at the end of a clay pipestem that Murdock had attached to the bladder.

When Murdock returned to the Soho factory, a brief period of experimentation with gas was followed by a hiatus of about two years, during which the system appeared to be unpatentable because of litigation. Efforts resumed in late 1801 when one of Watt's sons wrote giving news of Lebon's activities in Paris. A few months later the Peace of Amiens provided the opportunity Murdock described for introducing gaslight to the public. A retort was set in a fireplace of the major factory building and tubes led gas from there to copper vases that served as lanterns at the ends of the building.

George Augustus Lee, a partner in Phillips and Lee, a leading cotton-spinning concern, arranged in 1804 for a gas system to be installed in his house, and the project expanded to include two rooms of the mill and the company's accounting offices (fig. 10.3). When the gas system later grew to encompass all the mill building, there were a total of 271 Argand burners and 633 cockspur burners which, according to Murdock's calculations, cost only a fraction of what would have been spent for tallow candles. In addition, there were some savings in insurance rates, for underwriters viewed the highly combustible, lint-filled mill buildings as very high risks.[8]

Although Murdock stated in 1808 that he was "unacquainted with the circumstances of the gas from coal having been observed by others to be

10.3 **William Murdock provided lighting for England's largest textile mill by installing six iron retorts of this type, each 5½ feet high. Coal was lowered into the retorts in metal baskets. The gas was piped to an air-cooled condenser from which tar was drained. (C. Hunt, *History of the Introduction of Gas Lighting*, 1907.)**

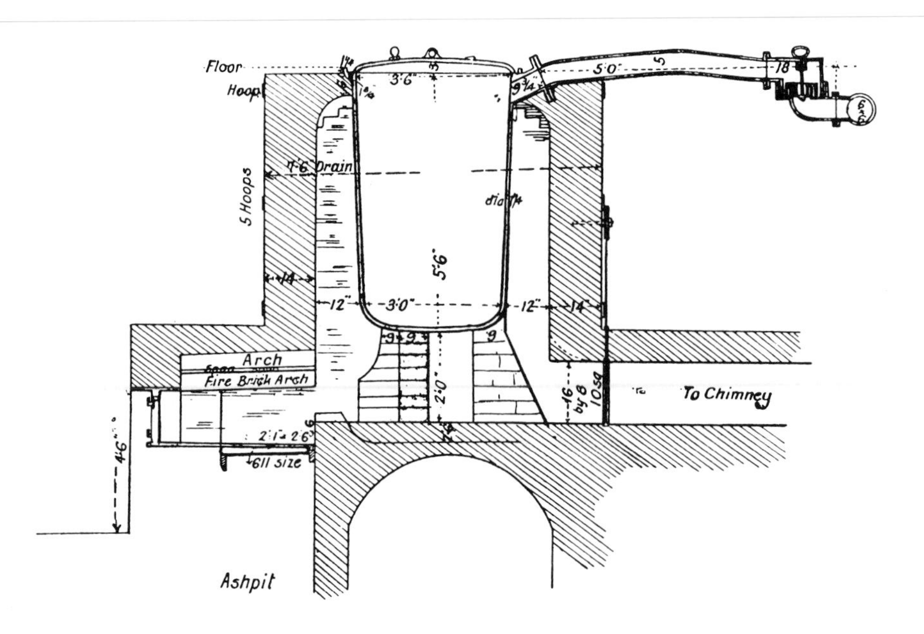

10.4 **Early in the twentieth century, English country houses could escape from the use of candles and lamps by installing their own gas systems, such as the "Country Petrol-Air Gas Plant." The air-pumping bell (right) was suspended with counterbalancing weights and provided pressure. The movement of this bell operated a pump (G) that supplied small amounts of gasoline to the upper part of the large tank (*I*) from which a gasoline-air mixture of between 2 and 6 percent went to the lighting system. It was said that 100 to 120 gas burners could be operated with one gallon of gasoline per hour. (*Nature*, 6 January 1916.)**

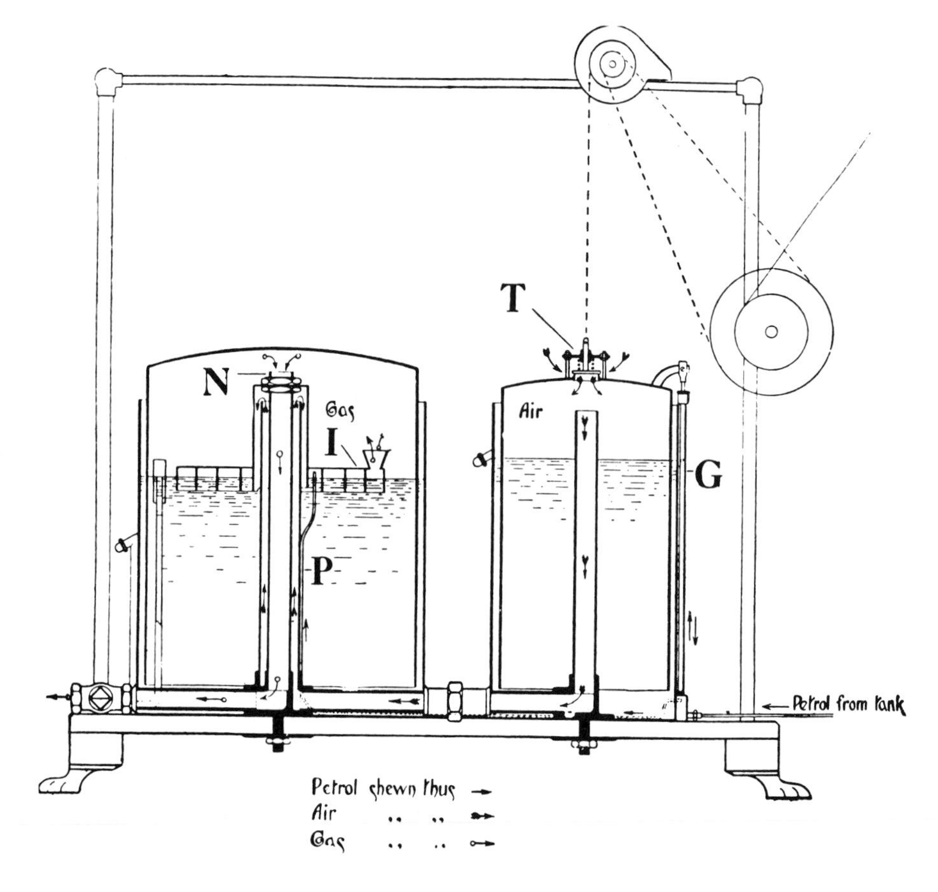

capable of combustion," when the Gas Light and Coke Company applied to Parliament for incorporation, Murdock's enterprise with Boulton and Watt was not judged to be singular, in spite of the firm's vigorous battle in the hearings. James Watt the younger led the fight, with George Augustus Lee giving testimony and the case being presented by Henry Brougham, later to become a political reformer of governmental practices. After competition had been widened in the field of gas lighting, Murdock lost interest in the subject, and in 1830 he retired from Boulton and Watt to his house nearby, which was lighted with gas from the Soho plant.

When Murdock completed his installation at the Phillips and Lee mills, he reported that there was "no Soho stink," which makes it clear that at Soho any purification had been primitive and inadequate. The early displays of gas lighting were most often outdoors and a majority of early installations were in large spaces such as theaters and mills, because the fumes from gas lamps were found to soil furniture, to damage goods in shops, to cause a room's occupants to have headaches, and to have an extremely foul smell. One of the early schemes for lighting residences with gas proposed placing lamps outside windows with reflectors to direct light into the house. Thus light would enter while odor was kept out, and gaslights would not add to the monumental task of spring housecleaning, a tradition required by lamps and open fires.

10.5 In 1815 the first gas supply in London started with 15 miles of mains. Thirteen years later the area to which gas was supplied was much larger, and this gasworks with two halls containing retorts loomed near the Regent's Canal. (Nineteenth-century engraving. Reprinted from D. Chandler, *Outline of History of Lighting by Gas*, 1936.)

Once gas lighting began to spread in popularity, methods of purification demanded and received attention, perhaps more than was given the basic procedures of gas production. Removal of impurities was first accomplished by passing the gas through masses of moistened coke, through tanks of water, or through fine sprays of water. These practices drew off some by-products but did little to improve the gas. Samuel Clegg, who had been Murdock's assistant at Soho, added caustic lime to water and forced the gas to bubble through this mixture. However, the pressure resulting from this treatment caused problems in the early stages of production. A later method of purification, long used, was accomplished by placing two or three inches of moist lime on trays and forcing the gas through several of them. After it had been used, the lime was extraordinar-

10.6 Although overturned oil lamps had always presented a danger, the explosion of gas was more feared. There were many tales of asphyxiation, and some experts recommended that gas not be piped to bedrooms. In residences, one was constantly reminded of these risks by the loud pop when a burner was lit and by the faint pervasive odor of illuminating gas that leaked from pipes and burners. (F. L. Collins, *Consolidated Gas Company of New York: A History*, 1934.)

10.7 In Samuel Clegg's systems for factory gas lighting, coal was put in an iron retort (*A*) beneath which the fire was built. A pipe (*D*) led the gas to a tank of water in which a gas holder was suspended by counterbalances. (C. Hunt, *History of the Introduction of Gas Lighting*, 1907.)

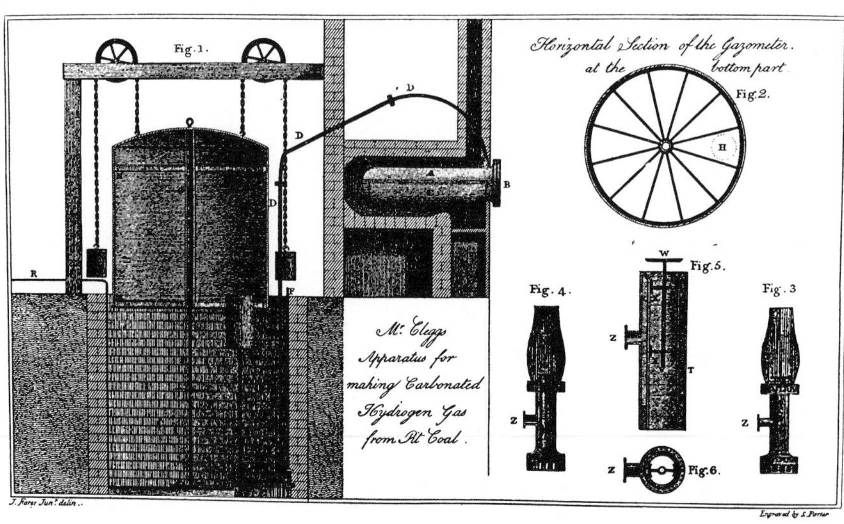

ily odoriferous, but it could be sold to farmers as fertilizer. London's Gas Light and Coke Company evaporated the stinking liquid that came from the lime process and carted it away at night, during the hours when the cleaners of cesspools also scheduled their work.[9]

As gas became tolerable, attention focused on the efficiency with which it might be burned, particularly on devices that could increase the amount of light gained from a jet of gas. The first form of gas burner, a single small hole in a tip applied to the end of a gas pipe, soon came to be used only in places that required little light or merited only the cheapest burner (fig. 10.8). The cockspur burner, in Murdock's words, consisted of "a conical end, having three circular apertures . . . one at the top of the cone, and two lateral ones." He described the shape of the flame as resembling the fleur-de-lis, but the flames' similarity to a fowl's footprint caused cockspur to be the common name for the burner. The cockscomb burner was an extension of the same design, with the tip a flattened fan-shaped piece on which the outside edge had as many as ten holes. The batwing burner had a hemispherical tip with a narrow slit from which the gas issued. Its flame spread to such an extent that glass chimneys, which increased the draft and protected the flame from wind, were seldom used because the width of the flame could cause the glass to break.[10]

All of those were elementary variations on simple openings. The union-jet or fishtail burner used a different course. The tip was flat or slightly concave and two holes were drilled at about 45-degree angles so that the jets of gas would strike against each other. This spread the gas and produced a flat flame like that of the batwing burner, but longer and narrower.

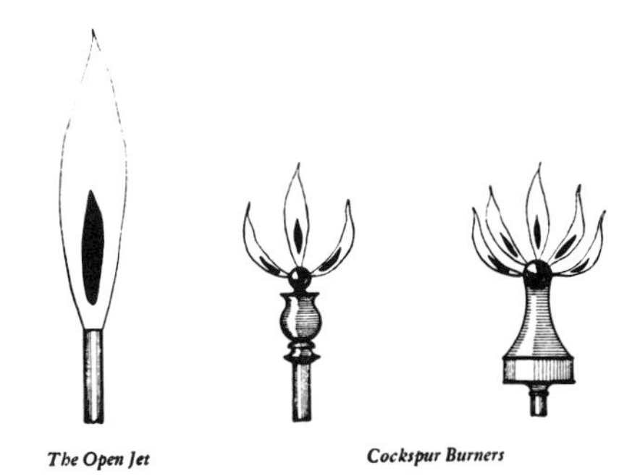

10.8 These standard types of gas burners were developed between 1808 and 1820. The corrosion that troubled burners made of iron and steel was eliminated by using enamel, soapstone, porcelain, and other materials. This change provided a slightly brighter light, for such burners did not draw heat away from the flame so rapidly. (F. L. Collins, *Consolidated Gas Company of New York: A History,* 1934.)

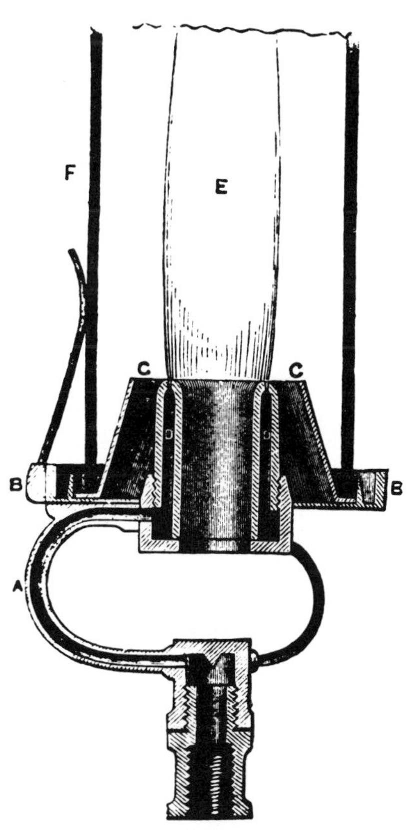

10.9 Argand gas burners derived their name and design from the cylindrical-wick oil lamp invented late in the eighteenth century. Air moved within and around the ring of many gas jets. It was found that a small number of holes increased velocity of the gas and reduced the amount of light given off. The height of the glass chimney could be varied to compensate for gas of low quality. This is the 1869 Government Standard Burner, serving in England as a means by which the quality of gas could be determined. (D. Chandler, *Outline of History of Lighting by Gas,* 1936.)

Although the batwing burner gave off slightly more light, the fishtail burner was used in most cases because it was desirable to use a glass chimney in order to eliminate the smoke given off when a slight movement of air caused the flame to flicker.

The same principle used in Argand oil lamps was applied in the Argand gas burner, which had a horizontal circle of tube with many perforations around the top (fig. 10.9). Because of the many small flames and the danger of their being extinguished, Argand burners were always enclosed by chimneys. Although it was almost twice as efficient as the batwing burner (in terms of energy radiated as light versus total energy consumed), at the end of the nineteenth century the Argand burner was only slightly more efficient than a good oil lamp.[11]

When electric lighting was in its infancy, it was urgent that gas lighting reach higher levels of illumination in order to meet the challenge of electric lighting. It was not always practical for suppliers to provide a richer gas mixture, so one short-lived invention, the albo-carbon burner, attempted to enrich the gas at the burner. A metal container to be kept filled with naphthalene (a component of coal tar) received the gas pipe, and on its way to the burner the gas took with it the fumes of the naphthalene, providing a richer fuel for the flame. A metal rod extended from the naphthalene container to a point over the flame, conducting some of the flame's heat to the naphthalene so that it would evaporate more readily.[12]

The most effective burner was one that took heat from the end point of the lamp system and returned it to the beginning of the process, a principle that had been used in furnaces for making iron and glass. These regenerative burners had complex methods of placing glass chimneys, ceramic cylinders, or metal tubes so that they were heated by the flames. By sealing off other channels, the air on its way to the flame was forced to pass through the heated passageway. At the same time, there was in most cases a method by which heat from the flame was conducted to the pipe in which the gas approached the flame. With both gas and air preheated, combustion and illumination became more efficient.[13] For a very brief period the regenerative burner staved off early advances of electric arc lighting. Like the arc lamp, the regenerative burner was a cumbersome apparatus and found little application, except for large halls, stores, and streets.

Another means of increasing the light radiating from a flame was based on the incandescent properties of certain materials, the way in which they glowed when heated. By using the heat of a flame, rather than its brilliance, incandescence became the method through which gas lighting was advanced during its last decades. During his experiments in the 1820s with various substances, Goldsworthy Gurney, a physician who came to London to pursue a career in science, discovered that the oxy-hydrogen blowpipe that he had invented produced an intensely white light when the flames played on a mass of solid lime. (Lime is calcium oxide, and the oxides of some elements produce strong light when heated, the intensity of the light increasing as the temperature rises.) From this the limelight was born, but, beyond demonstrations for fellow scientists and visiting nobility, Gurney did little with the discovery.

In 1820 young Thomas Drummond joined the military group charged with completing the ordnance survey of the British Isles. One of the greatest problems of the survey was the foggy and rainy climate of the sur-

10.10 Almost to the end of the nineteenth century, limelights persisted as theatrical spotlights and for magic lantern projectors. In this piece of theater equipment, a cylinder of lime was set on the spindle beside the gas jet at the right. As the lime burned away, intensity of the light could be sustained by rotating it with the knob at left. (*The Magic Lantern*, 1890.)

10.11 Lamp governors contained diaphragms that restrained sudden surges of gas pressure, which might extinguish flames or cause noise in some types of burners. The governor at the far right used a pellet of soapstone instead of a diaphragm. (D. Chandler, *Outline of History of Lighting by Gas*, 1936.)

vey area, and this was especially true of the work in Ireland where Drummond went in 1824. To assist in the essential long-distance sighting that established the framework within which detailed cartography was determined, Drummond invented the heliostat, a mirror that pivoted with the sun's movement to maintain a constant beam of reflected light, and from the experiments of Gurney he developed what became known as the limelight or Drummond light. Its intense white glow served well as a target for sighting over long distances through rain and mists. Sir John Herschel, the English astronomer, described a demonstration of the Drummond light:

> The common Argand burner and parabolic reflector of a British lighthouse were first exhibited, the room being darkened, and with considerable effect. Fresnel's superb lamp was next disclosed, at whose superior effect the other seemed to dwindle, and showed in a manner quite subordinate. But when the gas began to play, the lime being brought now to its full ignition and the screen suddenly removed, a glare shone forth, overpowering, and as it were annihilating, both its predecessors, which appeared by its side, the one as a feeble gleam which it required attention to see, the other like a mere plate of heated metal. A shout of triumph and of admiration burst from all present.[14]

In 1829 Thomas Drummond began development of the limelight for use in lighthouses and concerned himself with methods that might reduce the cost of manufacturing the lights. However, the limelight did not become a commercial enterprise for Drummond because he soon entered into government service and became a leader in the Irish government.

The limelight was often fitted with a reflector and lens to act as a powerful spotlight. It was this form that was employed in theaters to provide a light much stronger than that possible from candles, lanterns, or gas flame (fig. 10.10). (From this use "limelight" came to mean the glare of public attention.) For less spectacular purposes the lamp's white brilliance was not always thought to be desirable. When New York's American Museum on Broadway mounted a Drummond light outside its door, it was described as "sending a livid ghastly glare for a mile up the street."[15]

In the 1860s "Scholl's platinum light perfecter" was introduced and had brief popularity. It was a thin band of platinum that was within a fishtail burner at the point where the two jets of gas met. The brilliantly glowing metal did increase the burner's light, particularly when it was poorly adjusted. However, the perfecter's assistance was appreciable

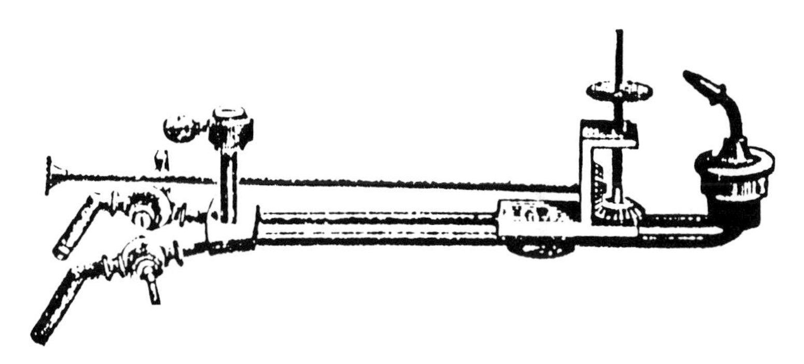

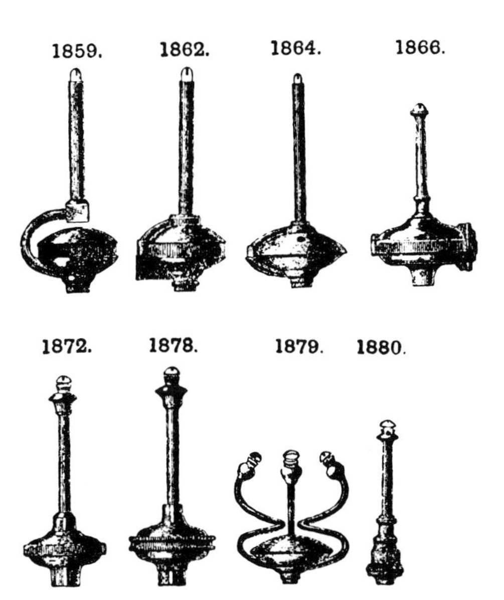

eral substances.[17] Even when placed around the relatively low temperature of a bunsen flame, the mantle produced a brilliant incandescence.

One of the first substances used by Welsbach was erbium oxide, which gave off an unpleasant greenish light. He found a white light was produced by magnesia and the oxides of lanthanum, yttrium, and zirconium, while neodym zircon emitted a yellow light. But these mantles crumbled in a few days. By the time Welsbach took out his second patent in 1886, he had found that a mantle of thorium oxide provided a strong light and a durable mantle, and the addition of very small amounts of cerium altered the green-tinged light of thorium to an acceptable warm white tone. Because early mantles were easily damaged in shipment, they were coated with collodion, which was burned off by the consumer after the mantles were mounted on burners.

Welsbach made extravagant announcements when his first mantles were introduced, but it was 1893 before a commercially plausible mantle was produced. That mantle still required several improvements. Cotton did not have fibers long enough, and short fibers produced fuzzy mantles that burned up rapidly. Ramie, an Asiatic plant also called Indian hemp and Chinese grass, provided longer fibers, and the process of mercerizing, a treatment with caustic soda, increased the strength of ramie fibers. Artificial fibers, made by squirting a cellulose solution through small openings, came to be one of the last solutions.[18] The improvement of the fibers of which mantles were made followed much the same pattern as that of the development of filaments for electric incandescent lamps, experimentation with natural materials leading to methods of fabricating materials to match the requirements of the device.

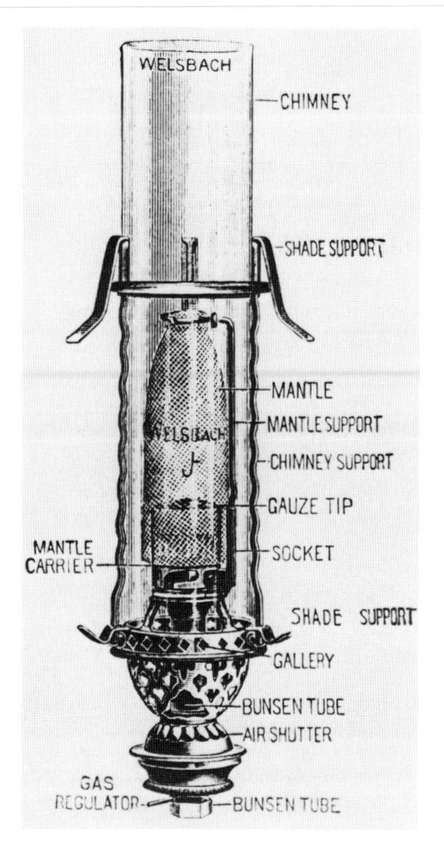

10.12 At the turn of the century the Welsbach burner had reached this form. A major difficulty was the greenish tint of its light, as compared with the yellowish gas flame and the bluish cast of the electric arc lamp. (*Electrical World and Engineer*, 10 November 1900.)

10.13 An inherent fault of most lamps was that the lamp itself blocked the light from being cast downward, where it was most often needed. This problem was eliminated by the inverted gas mantle lamp, as shown by these diagrams of light distribution. (L. Gaster and J. S. Dow, *Modern Illuminants and Illuminating Engineering*, 1919.)

only when it was new, and in use it soon diminished to insignificance.[16]

Alexander Cruikshanks in 1839 combined these prior notions and surrounded a gas flame with a mesh of platinum wire coated with lime. Almost a half century later, the next step was taken by Carl Auer von Welsbach, a Heidelberg student of Robert Wilhelm Bunsen, who had studied those oxides of metals that are known as rare earths. Welsbach's invention in its first form was a cotton mesh cylinder that had been soaked in a solution of one of the rare earths (fig. 10.12). The mantle, as it was called, was heated to burn out the organic fibers of the original cotton shape, leaving a delicate shell of min-

The Welsbach mantle was developed at about the same time that the electric arc lamp came into use. Jablochkoff's "candles" and Brush's arc lamps had faltering starts and reached maturity in 1893, the year Welsbach introduced the mantle that had resulted from his ten years of trials and errors. The efficiency of electric arc lamps and incandescent gas mantles was roughly the same; both produced light that equaled about 13 percent of the total energy they consumed.[19]

Welsbach's first patent on an incandescent mantle for gas lamps was granted by the German patent office in 1885, and he obtained three related patents in the six years that followed. Once the mixture of 99 percent thorium oxide and 1 percent cerium oxide had been settled on, the manufacture of incandescent mantles became a thriving and competitive industry, for almost anyone who could buy cotton, a knitting machine, and thorium nitrate was prepared to set up business. In several countries, lawsuits were brought against companies that infringed the Welsbach patents. A suit to stop infringement in Germany was decided in 1896 and Welsbach's original 1885 patent was upheld. But that patent did not include the use of thorium oxide, the key ingredient that had been included in a supplementary patent the following year. The portion of the later patent that dealt with thorium oxide was not allowed by the judges of the German Patent Bureau, because use of thoria was not original with Welsbach. Competitors were thus unrestrained by the patent deci-

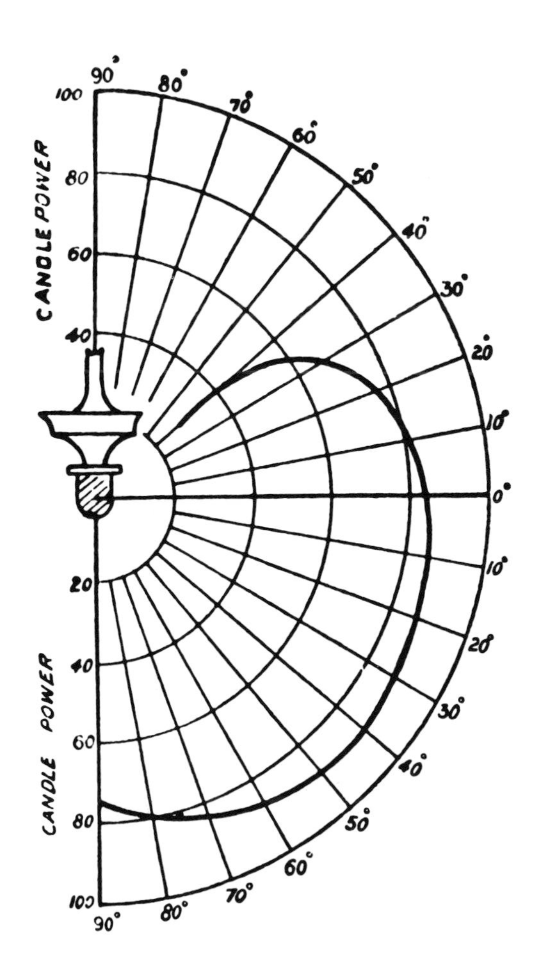

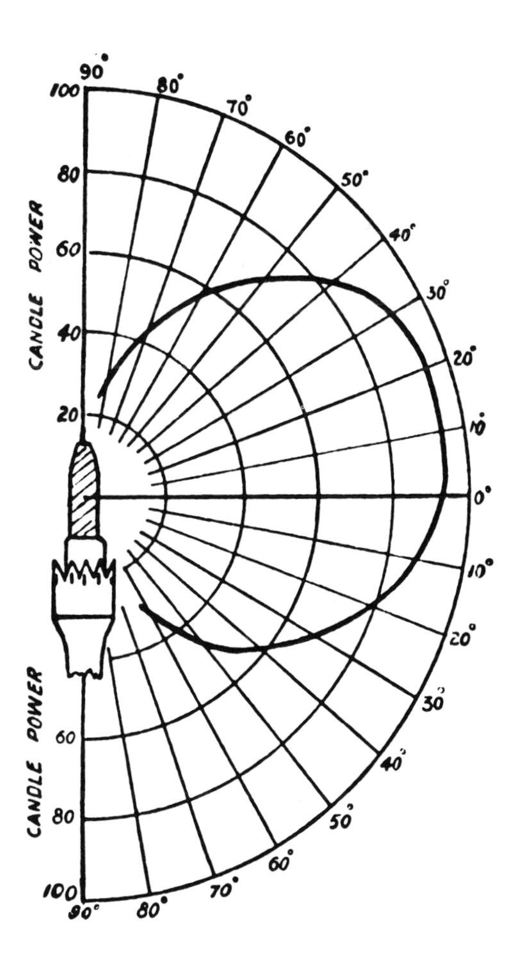

sion, and the manufacturers licensed under Welsbach patents were forced to lower their price drastically in order to meet their competitors' prices.[20]

Welsbach's British patent had been taken out in 1893, but the Welsbach company seems to have been reluctant to test its patent in the courts. At the turn of the century actions for infringement were taken against sellers, not manufacturers, of competing mantles. This delay permitted the organization of a group of competitors willing to share the cost of any litigation that the Welsbach company might initiate, and 27 firms were represented at the initial and exploratory meeting of that group.[21] But this was not the major concern of the Welsbach company. Failing to provide its stockholders with a dividend, the management was investigated by a committee of stockholders in 1902 and judged incompetent. The company was in a precarious state, for only about 15 percent of British gas users employed incandescent mantles.[22]

10.14 Until a satisfactory dry gas meter was devised around the middle of the nineteenth century, meters contained a liquid that sealed the measuring mechanism. Although water was used elsewhere, in the U.S. harsh winters required that the water in a meter be laced with alcohol. Here the installer of meters, carrying his day's supply, is followed by the company representative who filled the meters with alcohol. (L. Stoltz and A. Jamison, *History of the Gas Industry*, 1938.)

Early in the nineteenth century young Humphry Davy was deep in the study of electricity. He gave an account of one of his experiments:

> Pieces of charcoal about one-sixteenth of an inch in diameter whose ends are connected to a source of electric current, are brought near each other (within 1/32 or 1/40 part of an inch), a bright spark is produced, and by withdrawing the points from each other a constant discharge takes place through the air, producing a most brilliant ascending arch of light, broad and conical in form in the middle; hence the term "arch light," commonly called "arc light."[23]

Little was done with this discovery until the 1840s, when Jean Bernard Léon Foucault, a French science journalist and experimenter, replaced Davy's electrodes of wood charcoal with rods made of the carbon from retorts in which illuminating gas was produced. Shaped and baked with syrup as a binder, the hard carbon rods brought a significant improvement in the amount of light produced, but the "arc light" had to be frequently adjusted by hand and was more a laboratory display than commercial competition for the gas burners and oil lamps of the time. All of the arc lamps devised in the middle of the nineteenth century had little practical purpose. The electrodes eroded as the lamp burned, and any durable material for making them had also to be capable of producing large amounts of light. To provide light for a long period of time, it was necessary to find a practical means of maintaining the proper distance between the electrodes, and a system for supplying electric current at lower cost was essential.

Regulation of the electrodes attracted much attention from inventors (fig. 10.15). One design diminished the erosion of electrodes by using two large slabs of carbon so that the arc danced along the narrow foot-long gap between the slabs. Another used five disks of carbon, which were slowly rotated by clockwork. Some inventors adjusted electrodes with the action of springs. Others used weights, clockwork, or magnets. On the whole, the designs were more remarkable for their ingenuity than for practicality. Through the 1850s the more successful arc lamps improved sufficiently to increase the duration of the light in laboratory conditions from 1½ hours to about 12 hours.[24] Arc lamps were installed in a few English and French lighthouses,

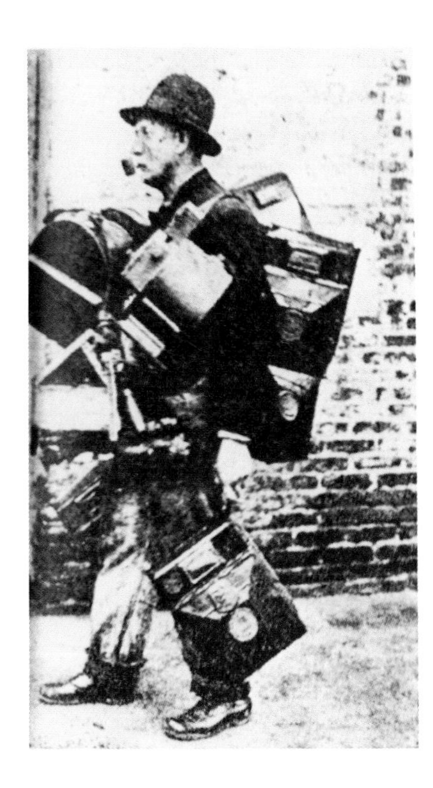

and there were uses in the burgeoning field of photography. Still, electricity, which then came from extremely inefficient batteries, was too costly, and the arc lamp mechanisms were far too complex. For over a decade, 1860 to 1872, experimentation with arc lamps was at a virtual standstill. Only after the invention of a practical dynamo did the development of arc lamps resume and progress.

In the 1820s three discoveries opened the way for electricity to become more than a laboratory curiosity. It was found that current flowing through a wire created a magnetic field. Then André Ampère discovered that current in a coil of wire caused magnetic properties, and that when a bar of iron was placed within the coil of wire an electromagnet was created. Within a few years there were experiments to reverse the process, to derive electric current from a magnet. More than 30 years of experiments in electromagnetic induction were required to produce laboratory models of primitive dynamos, and around 1867 several experimenters found methods for using a portion of their machines' output to energize the electromagnets within them. Improvements, principally focused on the methods of winding wires within the mechanism, led to the perfection of a practical dynamo that could supply inexpensive electric power. After 50 years of development and scores of experimental discoveries, the costly current produced chemically in batteries was replaced by electricity from a mechanical source.

In the early 1870s the only significant applications of electricity were telegraphy and electroplating, but the public's rapid acceptance of the telephone soon brought it to a position of leadership among those industries based on the use of electric current. The 1870s saw a revival of interest in arc lamps, and local companies were quickly organized to promote this new method of lighting. Finding an adequate power source was no longer an obstacle, and the principal problem remaining was the means of maintaining the distance between electrodes. In 1876 Paul Jablochkoff, a Russian telegraphy engineer living in France, introduced his "electric candle" (fig. 10.16). Instead of arranging carbon rods point to point, he placed them parallel, with a band of ceramic clay or plaster between them. The arc was started by a small piece of carbon between the points. As the carbon rods burned, clay melted or the plaster crumbled, and the arc moved toward the bottom of the candle. The first of Jablochkoff's candles burned for about 1½ hours, giving off an unsteady yellowish light. Two years later he introduced a lamp that would burn through an entire night, six candles connected so that as one burned out, the next was automatically lighted. Standards for street lighting were low at the time; the typical gas burner along downtown thoroughfares provided less light than today's 25-watt bulb.[25] The Jablochkoff arc lamp, many times brighter, was quickly adopted for street lighting in Paris and London. In theaters and large halls the arc lamp did not give off so much heat as gas burners, but its light was harshly brilliant and the arc flickered (fig. 10.17). The Jablochkoff candle made sputtering sounds, wasted much of its light upward, and did not have significant economic advantages over gas lighting. The Reading Room of the British Museum was provided with Jablochkoff candles in 1879, an improvement that doubled the hours during which patrons could use the Reading Room during winter months, but within about 18 months these lamps were replaced by a different design.[26] Seldom did installations of Jablochkoff candles remain more than a few years, usually being supplanted by the complex mechanisms of earlier lamps.

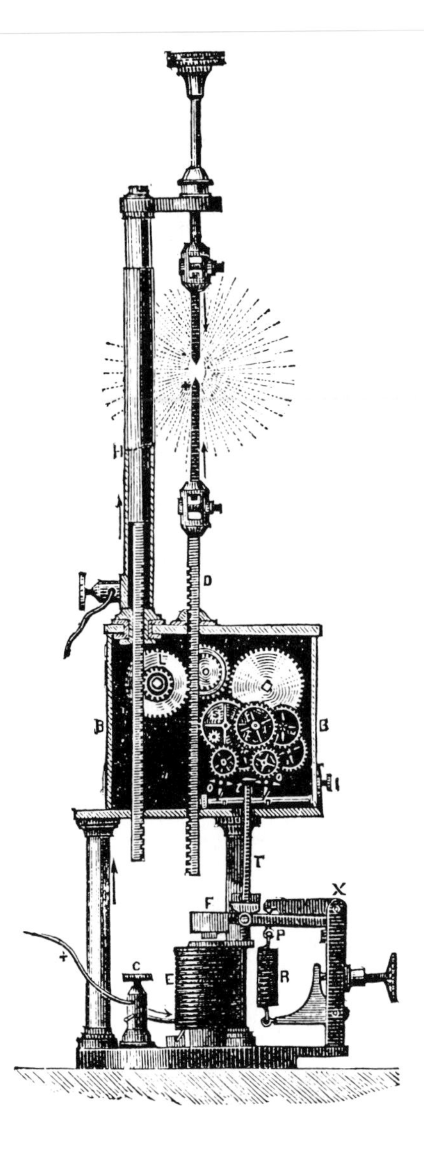

10.15 One of the first mechanizations of the arc lamp was the Foucault Regulator. In order to maintain the desired distance between carbon rods, electromagnetic controls caused the carbons to move closer together when the current carried across the arc decreased, and stopped the movement when the current increased. The movement itself was produced by a spring in the mechanism. (*Appleton's Cyclopedia*, 1880, 1:545.)

10.16 The Jablochkoff candle had plaster or clay between the two rods, which were about 10 inches in length. If the dividing material melted more quickly than the carbons or crumbled, the light became reddish, and carbons had to be replaced every time the light was extinguished. A typical installation of Jablochkoff candles required a steam engine to operate a Gramme dynamo, which provided power for 16 arc lamps. (E. Alglave and J. Boulard, *The Electric Light*, 1884.)

Arc lighting moved from its experimental phase into commercial exploitation with the work of Charles Francis Brush, a graduate mining

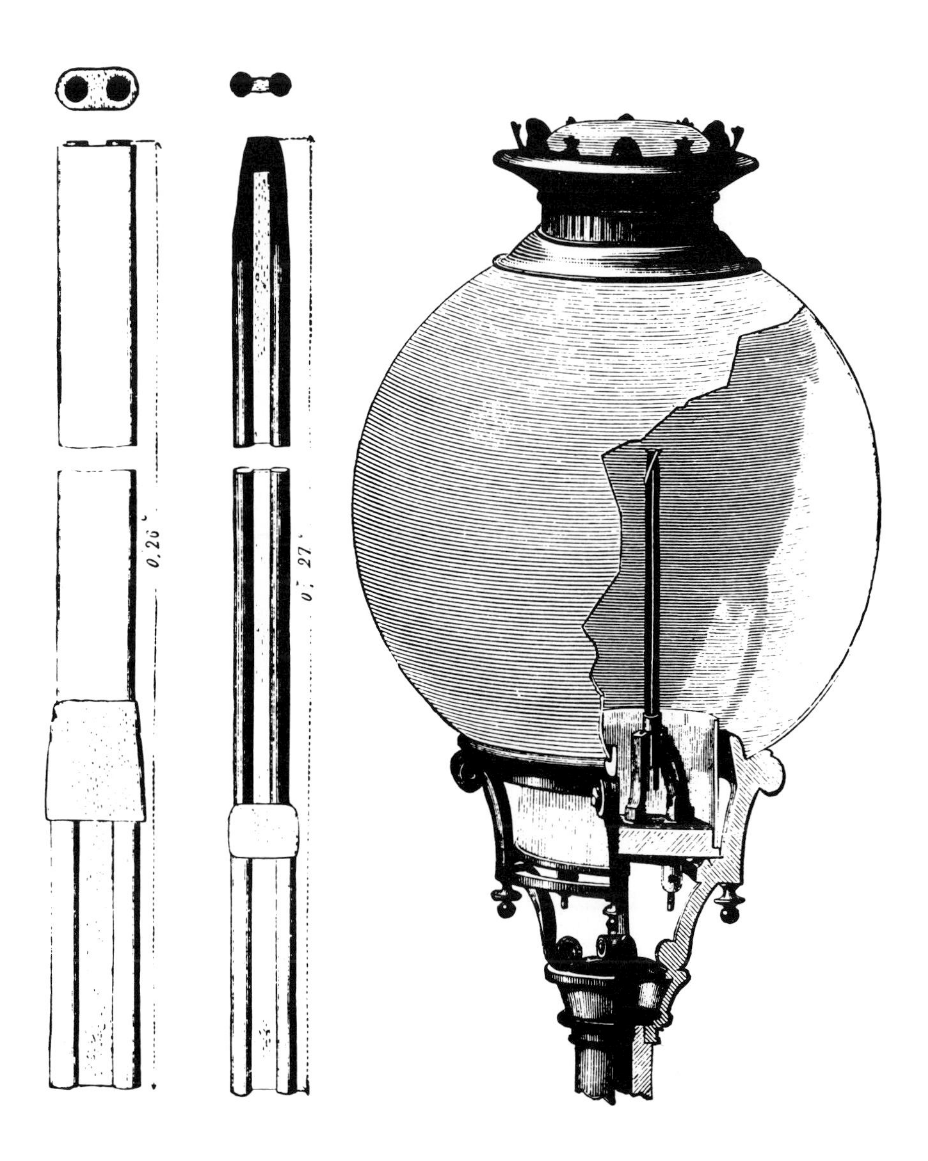

engineer and a dealer in iron and steel in the city of Cleveland, Ohio. His contribution began in 1876 with an improved dynamo that could provide current for several arc lamps. The following year he developed a lamp in which differences in the distance between the carbon rods, and the consequent variation of current passing through the arc, caused variation in the amount of current in a solenoid that, in response, repositioned the carbons.[27] The first demonstration of Brush's system took place in Cleveland's Public Square. The lamp was located in a second-story window of the Telegraph Supply and Manufacturing Company and current came from a dynamo about a block away. Brush described the event:

The light was a very small one, of course, but it was concentrated by a parabolic reflector. The occasion was one of a parade

10.17 The brilliance of arc lighting limited its use to locations out of doors and large spaces, such as department stores and factories. This music hall in Paris combined gas chandeliers, Jablochkoff candles in ovoid globes, and mirrored walls and columns. (E. Alglave and J. Boulard, *The Electric Light*, 1884.)

of horsemen and foot soldiers, and all that sort of thing, and the light was thrown in their faces as they came up the street. I remember how the eyes of the horses looked like green balls of light. I do not know how well the horses liked it. They did not seem to care for it very much. After a while a big policeman came up and said, "Put out that damn light!"[28]

A few years later, when Brush lights were first turned on in another Cleveland park, the public brought dark glasses and pieces of smoked glass to protect their eyes from the glare.

The advantage of Brush's work was his attention to the entire system by which light was produced. A Brush lighting plant included the Brush dynamo (which could be operated by a building's steam engine), the arc lamps, and carbon rods that had been copper-plated. The amount and quality of the light produced by the Brush system and others of the period were acceptable for street lighting, railroad stations, department stores, and factories. Wanamaker's department store in Philadelphia installed a Brush plant operating 20 lamps, and a Rhode Island textile mill purchased a system of 80 lamps in 1879, at that time the largest electric lighting plant in the world.[29]

The period from 1873 through 1875 saw a world-wide depression due to excessive speculation, but after it ended practical dynamos were ready at the same time that investment capital was once more available. There followed a startling surge of activity in the development of systems to provide lighting by electricity. The arc lamp gave off little heat, as compared with gas lighting, and it provided light more cheaply than other systems available at the time. But a textbook published in 1881 protested: "Its brilliancy is painfully and even dangerously intense, being liable to injure the eyes and produce headaches. Its small size detracts from its illuminating power—*it dazzles rather than illuminates*—and it cannot be produced on a sufficiently small scale for ordinary purpose of convenience. There is no mean between the absence of light and a light of overpowering intensity."[30] The problem of providing small lamps came to be known as the "subdivision of electric light," and arc lamps proved to be incapable of economically producing small amounts of light. As the lamps were made smaller, their efficiency was greatly reduced, and at the same time there was an increase in the cost and inconvenience of replacing carbons and the care required for the mechanisms that positioned the carbons.

Prior to the 1890s, the carbon rods used in arc lamps had not lasted much longer than 7 hours, and for street lighting to last the night it was necessary to provide a second pair of carbons that would light automatically when the first set was finished. Another attempt to increase the duration of arc lamps centered on the composition of the carbon electrodes. Longer and thicker electrodes were tried. Electroplating carbons with a thin coating of copper or zinc was found to increase their life by 25 to 50 percent. A successful device was a metal framework resting on the point of the lower carbon and holding a small tube of fireclay around the point of the upper carbon. By impeding the circulation of air, the fireclay slowed combustion of the carbon.[31]

Through the 1880s many experimenters tried to lengthen the life of carbons by enclosing arc lamps in glass. Their principal problems were the flickering light given off and black deposits that quickly formed on the globes. In the early 1890s several U.S. patents were granted for systems that discovered a simple solution: when carbons were enclosed in a globe that restricted the movement of air, a current smaller than customary would produce a steady light, and carbons would have a life of about 150 hours. In spite of the resistance of lamp trimmers, whose livelihood depended on the short life of carbons used in street lights, enclosed arc lamps soon replaced open arcs. In cities such as Cleveland and Detroit, street lighting was provided by erecting widely spaced towers as tall as 250 feet and raising racks of arc lights as one would run up a flag.

A further improvement came from the development of carbons that replaced the sickly purplish light of the enclosed arc lamp with a yellowish hue rather like firelight. By combining carbon with compounds of magnesium, strontium, fluorine, and other metals, the color of the light was improved. With the electrodes arranged as a V, no shadow obstructed the downward spread of light. The flame arc lamp, as this development was called, had the great advantage of giving off about a fifth the heat of a customary carbon arc lamp, and less than a fiftieth that of equivalent Welsbach gas lights.[32] For each watt of electrical current, the enclosed flame arc lamp gave almost three times as much light as an enclosed lamp of ordinary carbons. A "midget" arc lamp was introduced, but it did not prove to be economical in operation. An enthusiast said in 1911 that "electric lighting by means of powerful sources is inferior to none in the ability to produce tasteful and magnificent effects in immense buildings," but even in the most immense buildings it was most often recommended that the arc lamps be located between a skylight and a ceiling of ribbed glass.[33] In the show windows of department stores a translucent ceiling was usually set between the arc lamp and the display. Not even the "midget" arc lamp was bearable in a room of ordinary dimensions, and consequently the use of arc lamps remained greatly limited.

A British patent for "Improvements in Obtaining Light by Electricity" was granted in 1845 to Edward Augustin King, the leader of a group of English investors in the experiments of J. W. Starr, a young American working in England. A colleague spoke of Starr as "one of the ablest experimental investigators," but the inventors did little with the invention after Starr's sudden death.[34]

Although the Starr incandescent lamp was an unsuccessful solution, it contained the elements of later designs and confronted the problems that were to trouble others during more than three decades of research that would follow. The incandescent elements described in the patent (elements that glowed when electrical current passed through them) were to be either thin sheets of platinum, a metal of high electrical resistance, or rods of carbon. This incandescent material and the framework that held it were enclosed in a glass tube that had one end sealed. A vacuum (necessary to prevent the incandescent material from being soon consumed by oxidation) was obtained by filling the tube with mercury and inverting it with the open end in a vessel of mercury. Electrical current entered from a copper wire through the mercury and a platinum wire sealed in the glass at the end of the tube. Many of the patent's details were altered as Starr undertook his developmental experiments. Platinum was discarded, and his attention focused on small carbon rods, about ½ inch long and ⅒ of an inch square in cross section. Obtaining and maintaining a vacuum was difficult, for as the temperature rose within the glass container the materials inside gave off gases. Above all, Starr's lamp was a laboratory device and could not be manufactured in quantity and easily installed.

In the design of incandescent electric lamps during the decades that followed Starr's work, the incandescent element was commonly a rod or strip of carbon, although many experimenters used coils of platinum wire. Platinum was expensive, it soon disintegrated when used in air or an imperfect vacuum, and the temperature at which it provided a strong white light was only slightly below its melting point. Iridium had a melting point somewhat higher, but in both cases temperatures were usually kept safely low in order to avoid destruction of the wires. "Lamps with metallic wire filaments were condemned to remain at comparatively low temperatures, emitting light decidedly yellow in color, often looking more like red-hot hairpins than practical artificial light sources."[35] Carbon had a high melting point as well as a high resistance to electrical current. A major problem was carbon's predisposition to vaporize or combine with whatever gases might be present. With imperfect vacuums, carbon elements seldom lasted long, and many designs of lamps made provisions for the replacement of carbon filaments. The extreme case was a Russian lamp, patented in 1875. It contained five carbon rods with provisions for them to be lighted in sequence. A description of this lamp said: "The first carbon of a lamp never lasts for less time than a quarter of an hour; sometimes it breaks at the end of thirty to thirty-five minutes, but that is very rarely; its average duration is twenty-one minutes. The succeeding carbons last upon an average for two hours, so long as the luminous intensity does not exceed 40 [gas] burners, in which case the average duration is only half an hour."[36] It was intended that the first carbon would exhaust the oxygen that might be within the lamp, providing a more desirable atmosphere for the succeeding carbons. The automatic series of elements was a method then being used in arc lamps. In fact, it is apparent that much experimentation in incandescent lighting was strongly influenced by practices in arc lighting.

Most of the incandescent lamps invented during this period were the work of experimenters who were largely self-taught in the science of electricity. Progress made in the middle of the nineteenth century can be fairly represented by the work of Moses G. Farmer, a New England schoolmaster, who had left that occupation for a position with a telegraph company. That job he left in 1851 to manufacture an electric fire alarm system he had invented, and as he approached the age of 40 he shifted his attention to the problem of the incandescent lamp. He developed a lamp with a filament of platinum wire and current taken from wet cell batteries, and in 1859 he lighted the parlor of his home in Salem, Massachusetts:

> When we shall see the electric light distributed in our dwellings it may prove a source of pride to Salem to call to mind that this boon met its first success in that city, where a parlor, in Pearl Street, was lighted every evening during the month of July, 1859, by the electric light, and was undoubtedly the first private dwelling house ever lighted by electricity. A galvanic battery furnished the electric current, which was conveyed by conducting wires to the mantelpiece of the parlor, where were located two electric lamps. Either lamp could be lighted at pleasure, or both at once, by simply turning a little button. The light was soft, mild, agreeable to the eye, and more delightful to read or sew by than any light ever seen before. It was discontinued, for the reason that the acids and zinc consumed in the battery made the light cost about four times as much as an equivalent amount of gas light.[37]

Almost a decade later Farmer patented a lamp with a carbon element, and he lighted a residence in Cambridge, Massachusetts, with 40 lamps of this design. He filled the lamp's globe with nitrogen, although he believed that "a vacuum is, perhaps, better, were it not for the difficulty of maintaining it." The repeated heating and cooling of a lamp made most methods of sealing lamp globes unreliable, especially in designs that were stoppered bottles of the sort Farmer used.

The introduction of the Gramme dynamo in the mid-1870s provided a practical source of continuous electric current and stirred new activity in the efforts to develop an incandescent electric lamp. Between March 1878 and the following spring, two similar patents were developed by the team of William E. Sawyer and Albon Man, experimenter and entrepreneur respectively, in their New York shop. The lamp was derived from earlier European designs (fig. 10.18). Electrical current passed through a carbon pencil inside a globe filled with nitrogen, which deterred the deterioration of the carbon. Because the life of such carbon rods was short, a long pencil

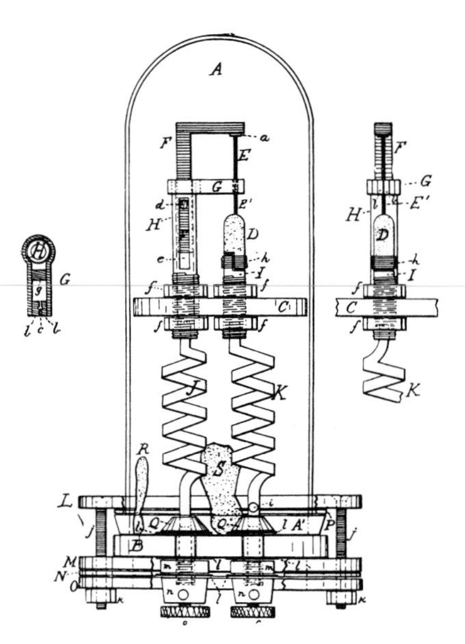

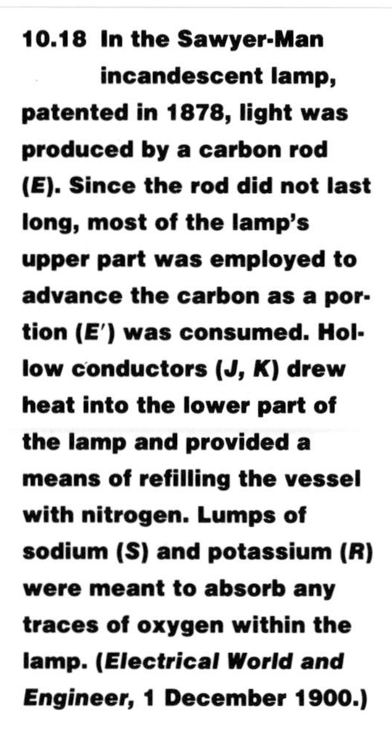

10.18 In the Sawyer-Man incandescent lamp, patented in 1878, light was produced by a carbon rod (*E*). Since the rod did not last long, most of the lamp's upper part was employed to advance the carbon as a portion (*E′*) was consumed. Hollow conductors (*J, K*) drew heat into the lower part of the lamp and provided a means of refilling the vessel with nitrogen. Lumps of sodium (*S*) and potassium (*R*) were meant to absorb any traces of oxygen within the lamp. (*Electrical World and Engineer*, 1 December 1900.)

was used and current was passed only through a segment of its length. As the lamp burned, more of the carbon pencil moved into the light-producing portion of the lamp. By having the pressure of the nitrogen about the same as atmospheric pressure outside the globe, it was hoped that the entrance of air could be prevented. In use, however, the heating and cooling of the lamp caused pressure to vary, and air usually entered when the lamp was turned off. After Sawyer was discharged by the company's investors (an action attributed principally to his drunkenness), he started another company, but he seems to have focused his attention more on competition with his former employers than on the improvement of his lamp. The Sawyer-Man Electric Company, organized in 1886, had its name more from the ownership of patent rights than any close association with the inventors, and within two years it was bought by Westinghouse.

Joseph Swan, an English chemist, experimented with incandescent lamps in the 1860s, but the difficulties arising from imperfectly maintained vacuums caused him to halt his studies of the problem. For several years he shifted his inventive attention to the development of photographic plates and papers. Learning of successes using the Sprengel mercury vacuum pump for other fields of experimentation, Swan resumed his study of the incandescent lamp (fig. 10.20). The lamp that Swan exhibited to the Newcastle-upon-Tyne Chemical Society at the end of 1879 was a sealed glass container, abandoning the notion of replacing incandescent units during the life of the lamp, and it contained a thin carbon rod held between platinum wires. In operation the lamp globe was blackened by vapors until Swan began the practice of passing current through the lamp during the process of pumping out air.

Late in 1878, work on incandescent lighting was begun by Thomas A. Edison, already known as the "Wizard of Menlo Park" for his invention of the mimeograph, phonograph, and numerous telegraphic devices. The first patents of Moses G. Farmer and William Sawyer were dated that same year, but it should be remembered that these designs were in many ways crude and impractical and were rapidly surpassed by Edison's efforts. Within a few months the Edison Electric Light Company was formed, with several major financiers included, and Edison informed the New York press that he had solved the problem of incandescent lighting. The announcement was overly optimistic, for at that time Edison was experimenting with systems and incandescent materials that others had already tested without success. There was, however, a significant difference in Edison's approach to the problem of electric lighting. At the very outset he employed a scientist-mathematician to provide a theoretical point of view (something Edison had not always welcomed)

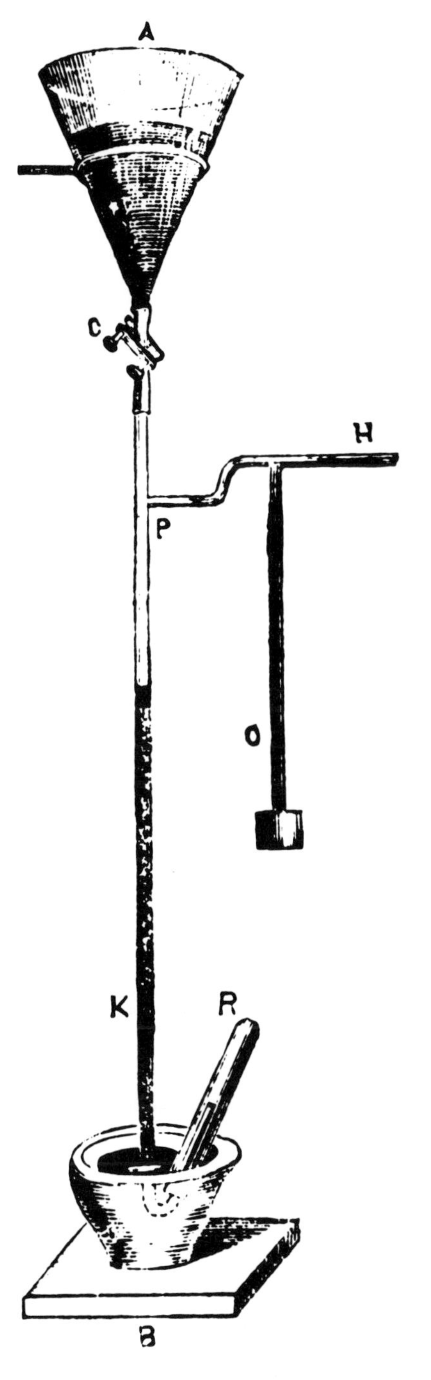

and, as the first lamps were tested, attention was also directed to generators and means of distributing current. As an experienced inventor, Edison viewed the prospect of an efficient incandescent lamp in relation to an entire system of electrical supply that would create a market for the lamp. He quickly reasoned that lamps needed to have a brilliance about the same as that of gas jets, that if the lamps had high resistance it would reduce the cost of copper wires in the distribution system, and that the choice of filaments depended largely on finding a material that could be heated to incandescence without reaching its melting point.

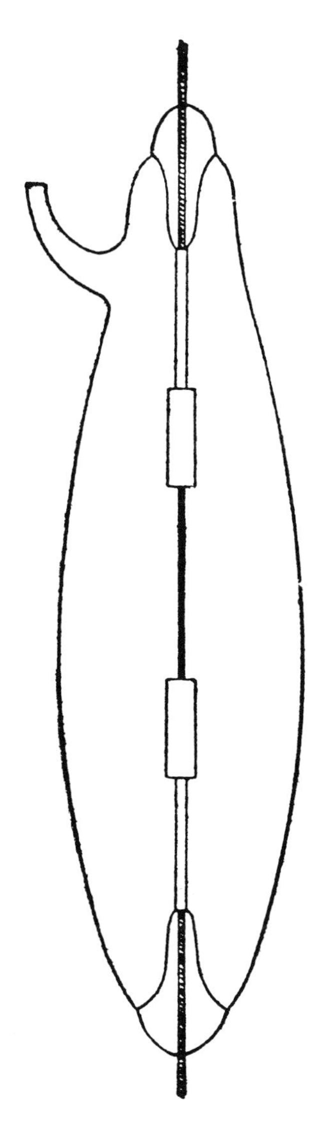

10.19 Invention of the Sprengel air pump greatly improved the vacuum that experimenters could obtain within their lamp globes. Drops of mercury entered (*A*), rolled down a tube (*C-K*), and took along air in falling to the drain (*B*). In the tube (*H*) a small amount of phosphoric anhydride absorbed any moisture that might be present. The bulb to be evacuated would be fastened at the end of the tube (*H*) along which a barometer (O) measured pressure. (*Engineer,* 8 March 1867.)

10.20 The lamp displayed by Joseph Swan in 1879 had a thin carbon rod as its incandescent element. Current entered through wires of platinum, the only metal that would not expand and crack the glass. The lamp shone only briefly and deposited carbon particles on the inside surface of the glass. The following year Swan began experimenting with other materials as filaments. (Reprinted with permission of Macmillan Publishing Company from Arthur A. Bright, *The Electric-Lamp Industry.* Copyright 1949 by Macmillan Publishing Company, renewed 1976 by Evelyn F. Hitchcock.)

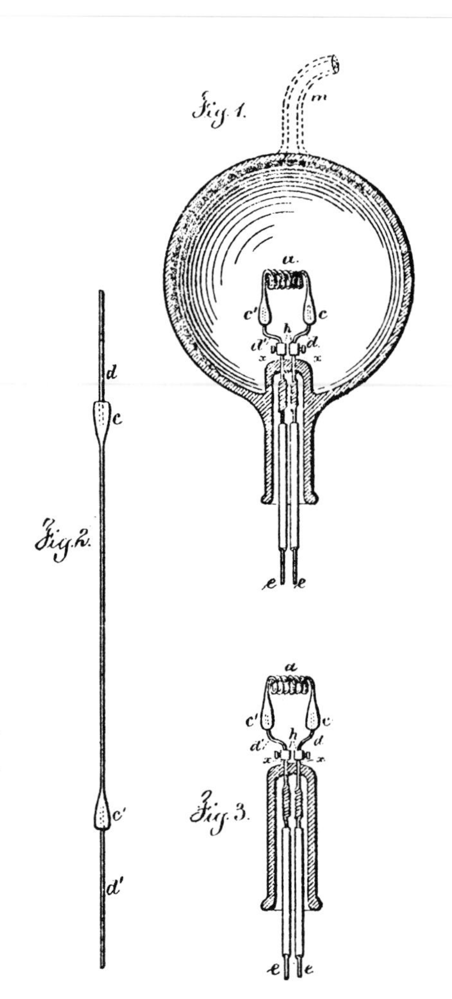

10.21 In his 1880 patent, Thomas Edison was vague regarding the filament: "I have carbonized and used cotton and linen thread, wood splints, papers coiled in various forms, mixed with tar and kneaded so that the same may be rolled out into wires of various lengths and diameters." The spiral filament shown in the patent proved difficult to make and was succeeded by a simple horseshoe shape. (U.S. Patent no. 223,898.)

10.22 Between 1881 (left) and 1905 (right), the Edison lamp changed considerably: (*A*) The hairpin of carbonized bamboo gave way to a double loop of carbonized artificial cellulose fiber. (*B*) Platinum lead-in wires were shortened. (C) The stem of the glass tube became smaller. (*D*) Copper wire and copper-plating used to secure the filament to lead-in wires were replaced by a carbon paste. (*E*) "Free-blown" bulbs, which varied slightly in size were followed by bulbs that were machine-blown into molds. (*F*) The screw base was simplified and reduced in size. (*Journal of the Franklin Institute*, July 1905.)

Early incandescing elements were usually made of platinum or iridium (two metals with high melting points) or carbon (which had a melting point higher than that of any metal). The brightness of the light increased with the temperature of any filament, and it was a relatively narrow range of temperature—between glowing heat and melting—that drew the attention of experimenters. In the 1870s, when enthusiasm for the study of incandescent lighting increased because adequate sources of electrical current became available, and for many years after, carbon filaments were the most satisfactory available.

Carbon filaments were made by cutting fine strips of a cellulose material and slowly subjecting them to high temperatures while in a closed crucible. Hydrogen and oxygen were thus driven from the filaments, leaving a skeleton of carbon, hard and dense like coal. Joseph Swan in 1860 had used carbonized strips of paper and cardboard, and Edison's first commercial lamps had filaments of carbonized paper, but they proved to be too fragile (figs. 10.21, 10.22). Edison had tested a variety of materials (6,000 according to some stores), including monkey grass, jute, coconut palm, and hair from the beards of his laboratory workers.[38] When he tested bamboo in 1880, it produced a filament that was firm and hard. Agents were soon dispatched to the Amazon, China, Japan, and India with instructions to obtain samples of every kind of bamboo they encountered and send them back to New Jersey. A Japanese bamboo, madake, was found to be the best, and it was used by Edison until 1894, although agents continued to travel in search of a variety still better.[39]

In 1894 the General Electric Company began marketing incandescent lamps in the United States with "squirt" filaments. These filaments, first used by Swan over a decade earlier, were made by dissolving cotton and squirting the syrupy mixture through a die into alcohol. The threads produced by this process—cellulose, like the materials that had been used previously—were then carbonized, and those filaments were better than any made from bamboo. Carbonized "squirt" filaments became commercially available in incandescent lamps at about the same time that the first "artificial silk" made by a similar process was sold in France.

Platinum and iridium had been found to have melting points too near their temperature of incandescence. Carbon filaments, no matter what form of cellulose had been carbonized, had the disadvantage of containing minute portions of impurities that, when the filament was lit, caused the carbon to vaporize and be deposited on the surface of the glass bulb. Gases were inserted to inhibit vaporization and reduce blackening of the bulb, and "getters" (air-absorbing chemicals) were placed in the lamps to insure a complete vacuum. By 1905 many manufacturers were treating filaments by subjecting them to temperatures much higher than those of the carbonizing process. This method of "metallizing" the carbon filament resulted in improvements of around 25 percent in efficiency, but by that time the development of wire filaments had begun.[40]

The Nernst lamp was introduced in 1897, but its incandescent element was a rod of rare oxides that had to be preheated with an electric coil, a system requiring a wait of almost 15 seconds before the light came on.[41] Welsbach, developer of the gas mantle, experimented with wire filaments of osmium in the same period, but his lamps were so expensive and the filaments so fragile that his lamps were rented rather than sold.

The tungsten filament was introduced commercially in the United States in 1907 after the General Electric Company purchased Austrian patents (fig. 10.23). Tungsten is a plentiful metal, extremely heavy and hard (the name is Swedish for "heavy stone"), with a high melting point and a low electrical resistance that increases as the metal is heated. But tungsten is not ductile, and the early filaments made of the metal were produced by mixing a fine tungsten powder with sugar and gum arabic. This paste was squirted through diamond dies, and those threads were heated to burn away the binding ingredients. Lamps with such filaments were very efficient, but the tungsten wire was fragile. It required several years of concentrated experimentation to find a method by which lamp manufacturers could draw tungsten wire, overcoming the stubborn metal and providing sturdy filaments for incandescent lamps.

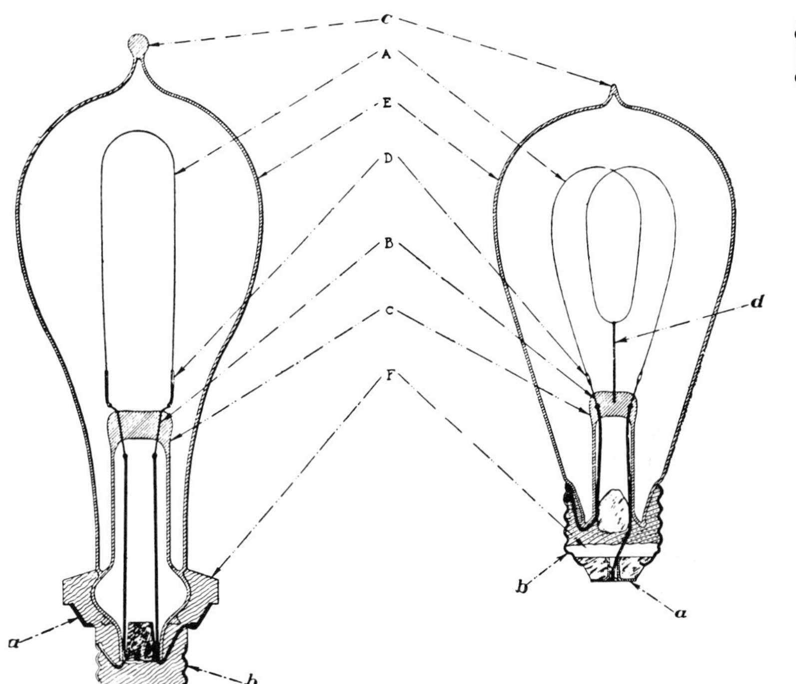

After the introduction of the durable ductile tungsten filament of drawn wire, three major improvements were made in the incandescent lamp (fig. 10.25). Even with tungsten filaments, bulbs blackened and efficiency was reduced; almost 95 percent of the electrical energy employed in a vacuum tungsten lamp was squandered in heat.[42] It was found that filling bulbs with nitrogen eliminated blackening and allowed a higher temperature in the filament, a change that increased a lamp's efficiency. In some cases it was necessary to place a mica disk in the neck of the bulb to prevent overheating the base of the lamp. Frosting of bulbs was recognized as desirable to diffuse the light and eliminate the glare of visible filaments, but it proved difficult to achieve. Opal glass and external frostings were unsatisfactory, and years passed before a method was found to etch the inside of bulbs while retaining the strength of the glass. The remaining improvement was accomplished by altering the form of the filament. If a tungsten filament were tightly twisted into a screw-shaped coil, the heat loss was determined by the diameter of the coil, not the diameter of the wire. In order to further increase the effective diameter, after 1926 the coiled filament was coiled a second time, resulting in the coiled coil or double-coiled filament that reduced heat losses. Drawn wire filaments and these three improvements caused the efficiency of incandescent lamps, as measured in units of light per watt of electrical energy, to roughly double in the period from 1910 to 1940.

Although gas companies and those producing electricity were clearly competing for lighting customers, at first there seemed a possibility that gas and electric lighting might coexist with a reasonable level of profit for both. In 1891 the gas companies of Massachusetts, for instance, produced and sold 20 percent more gas than they had in 1889, and in that same period the state's arc lighting increased by 33 percent and incandescent electric lighting by 117 percent. One writer pointed out in 1892 that "though the world has absorbed in vast amount the new illuminant known as electricity, it does not follow that the use of gas has decreased to a corresponding degree."[43] His argument followed a simple, though fragile, logic. In order to attract customers, stores were lighted electrically above the required levels for visibility, and merchants who grew accustomed to that level of light insisted on as much from gas burners at home. Furthermore, he reasoned, the shadows of skyscrapers darkened the lower stories of buildings nearby and thus created a need for more light. Supporters of each method of illumination published intricate calculations comparing the cost of equal amounts of light as produced from gas and electricity. Some partisan exaggeration must be assumed, but at the turn of the century it is evident that a unit of light was more expensive if provided by electricity than by gas, and the comparative investment in electrical generating equipment was even higher. Since gas-producing plants were in place and electric stations had to be built, investment costs were a decisive factor. One expert in 1900 compared light sources according to the percentage of energy radiated as light (the rest expended as heat); see table 10.1[44] The calculations were even more staggering when the radiated light energy was compared with the energy in a given amount of coal used in its production (table 10.2).[45]

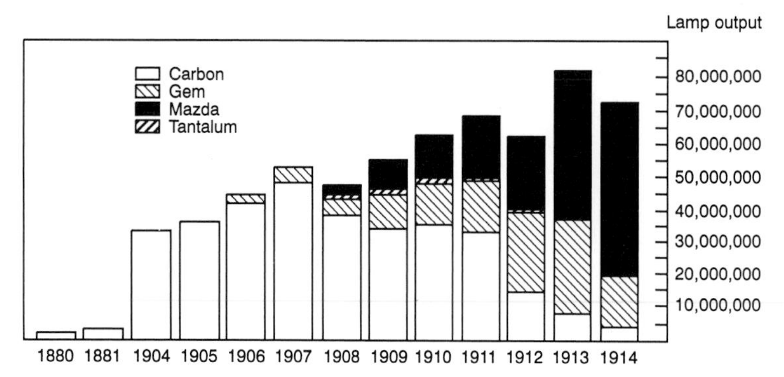

10.23 Electric lamp production in the United States, 1880 to 1914, by number and type, shows the decline of carbon filament lamps with the introduction of Gem (metallized carbon) and Mazda (tungsten) lamps. Between 1907 and 1914 the light produced by a 60-watt Mazda lamp increased almost 30 percent and the price was reduced by about 75 percent. (Redrawn from *Electrical World*, 16 October 1915.)

Table 10.1 Energy Efficiency of Various Light Sources

Light Source	Percent of Energy Emitted as Light
Batwing gas burner	1.3
Candle	1.5
Oil lamp	2.0
Argand gas burner	2.4
Incandescent "glow" lamp	5.0
Limelight	9 to 14
Welsbach gas mantle	13
Arc lamp	13

Table 10.2 Energy Efficiency of Various Light Sources

Light Source	Percent of Coal Used as Light Energy
Batwing gas burner	0.23
Argand gas burner	0.43
Incandescent "glow" lamp	0.5
Arc lamp	1.3
Welsbach gas mantle	2.3

The incandescent gas burner, particularly Welsbach's mantle, gave new hope to the gas industry. At a meeting in England in 1905, the manager of Manchester's city-operated gas plant reported that in his city equal quantities of light were eight times as costly with electricity as with gas.[46] In the same year, stockholders of London's South Metropolitan Gas Company were told that the company's new customers had resulted in less than two-thirds the predicted increase in the amount of gas used, a disappointment attributed to the fact that incandescent gas burners required about two-thirds the amount of gas used by previous burners to provide the same amount of light.[47]

Cost, however, was not the deciding factor. The Victorian desire for comfort and convenience (which favored electricity) was strong among the middle and upper classes, and the lower classes still used lamps and lanterns. In the first decade of the twentieth century, the pneumatic switch for gas lighting was introduced in an effort to gain some of the operational ease to be found in electrical installations. In this device, a push-button sent a surge of air pressure through a tiny metal tube that ran along walls and ceilings to the gas fixture. There the pressure opened a gas valve, and the flame was ignited from a pilot light that was incorporated in the fixture. As late as 1929 a British firm introduced a switch that operated electrically from a small battery that heated a wire until it ignited the gas.

A boost to gas consumption was increased attention to lower-income users. The well-to-do switched to electric lighting, retaining gas as a cooking fuel. The poor kept to coal or oil, which could be purchased in small amounts as money was available. The average workman's life and income were too uncertain to set aside money for quarterly gas bills and a deposit to the company; and their landlords were reluctant to invest in improving the illumination of their properties. Oil for lighting and coal for cooking were the most accessible fuel supplies until the introduction of the penny-in-the-slot gas meter in the 1890s. Coin prepayment meters were tried in London in 1888, but did not succeed until introduced by the Liverpool Gas Company four years later.[48] Eyeing that success, London's South Metropolitan Gas Company again ordered and advertised penny-in-the-slot meters. During a period when that company added 1,063 ordinary customers and 3,654 cooking ranges, it added 17,055 coin-metered customers

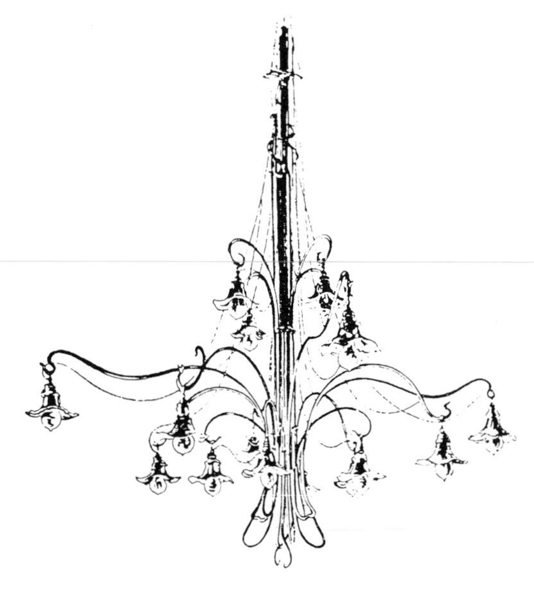

10.24 The transparent glass globes and the wires of electric lighting gave impetus to the development of new fixtures. Art Nouveau designers were particularly inclined to relate the lamps to flowers, as in this Belgian chandelier for a boudoir. (*Art et Décoration*, January 1897.)

and 18,865 cooking ranges metered in that manner. In the United States, meters were designed to receive a twenty-five-cent piece, and before World War I they had become at least one-third of the total number of meters installed annually.[49]

The provision of electrical power was obviously a promising investment in the United States. A group of prominent businessmen in Buffalo, New York, in 1887 offered a prize of $100,000 to the inventor who would find a way to make the tremendous force of Niagara Falls work for their city. For this purpose the Niagara Falls Power Company was formed, but in 1889 its stock was purchased by a coalition of New York bankers, who had formed the Cataract Construction Company with the intention of building a hydroelectric project of much larger scope. The Swiss had made turbines of the size needed to harness Niagara, and therefore the generation of power was judged to be feasible. But how should that power be transmitted the 20 miles to Buffalo? This led to the question of whether alternating current or direct current should be used.

Alternating current could easily be raised to high voltage, which was transmitted with much less loss than direct current. Furthermore, alternating current generators and transformers were superior in efficiency to those of a direct current system. From the point of view of producing and transmitting electrical power, alternating current was unquestionably preferable. On the other hand, at that time electric lighting constituted the main use of the power, and therefore the amount of energy to be used fluctuated greatly through the day and night. For a direct current system, a combination of several generators mounted in parallel and a group of storage batteries could compensate for variations in load by operating more generators as loads increased and

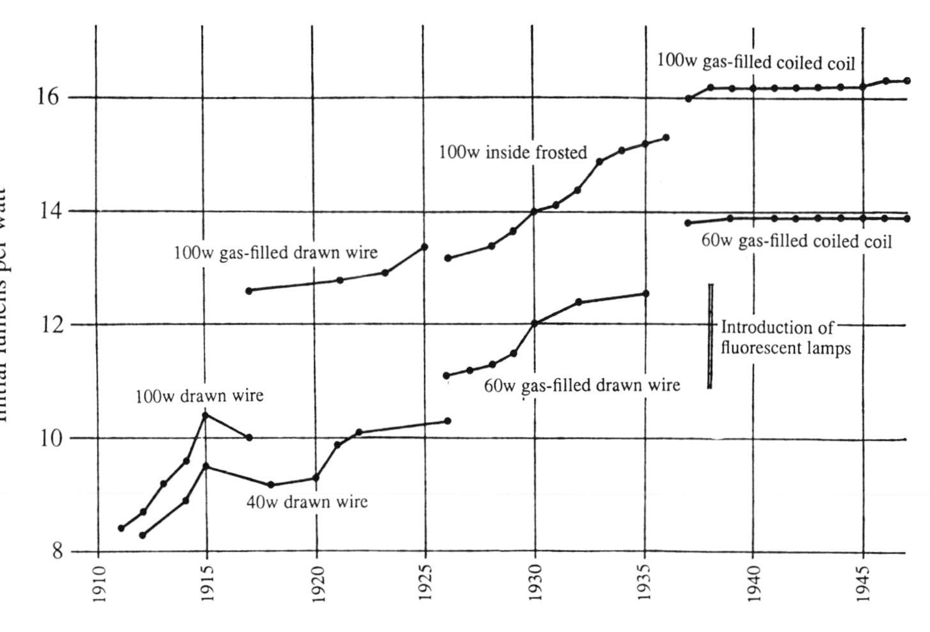

10.25 Improvements in incandescent lamps about doubled their efficiency in the period from 1910 to 1940. However, when introduced in 1938 the 30-watt fluorescent tube had an efficiency of 35 lumens per watt, twice that of the 100-watt incandescent bulb of that time. (Graph developed from data in A. A. Bright, Jr., *The Electric-Lamp Industry*, 1949.)

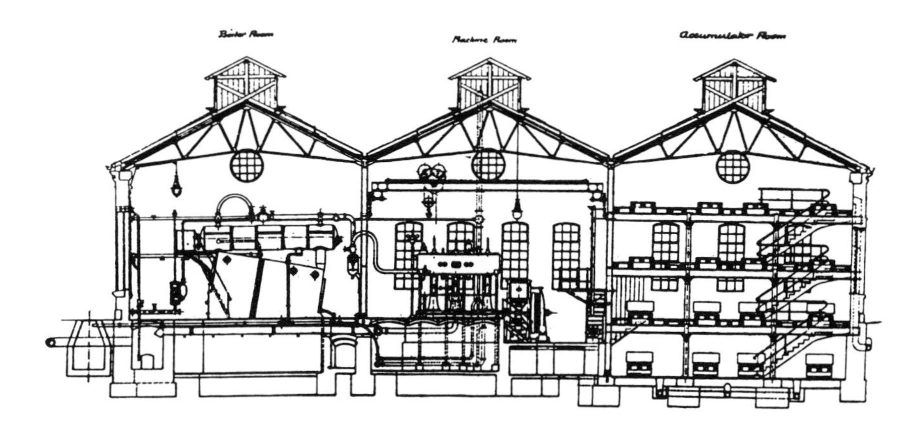

10.26 Early direct current electrical plants used accumulators (storage batteries) to supply power when generators were not running. This European plant, built in the 1890s, had three sections (left to right): the boiler room, in which steam was produced; the machine room, in which steam-driven generators supplied electricity; and the accumulator room, in which four levels of lead-acid batteries stored some of the electricity. (K. Hedges, *Continental Electric Light Central Stations*, 1892.)

drawing from batteries during the short period of peak load (fig 10.26).

The "battle of the systems" raged during the period in which the first large generating plants were being built. In England, the Deptford station was in operation in 1891, using alternating current. Debate over the two systems continued among British engineers for many years, and in some areas of England direct current persisted into the 1950s. In the United States the argument was bitter, but its resolution was decisive. Edison advocated the use of direct current and opposed alternating current as presenting a greater danger to the user. George Westinghouse replied angrily in support of alternating current. Because high-voltage transmission was recognized as being advantageous and the technique of constructing high-voltage generators was more firmly established, the contract for the Niagara Falls project was awarded to Westinghouse. Thus alternating current became standard in the United States, but competition continued, much of it focusing on inventions and their patents.[50]

By the end of the eighteenth century, patents were no longer royal prerogatives, monopolies bestowed at the pleasure of the government, but instead had become recognized as rights of the inventors. The U.S. Constitution authorized Congress "to promote the Progress of Science and the useful Arts by securing for limited Times to Authors and Inventors the exclusive Right to their respective Writings and Discoveries." It was assumed that a monopoly for a limited time would serve as an incentive to inventors, afford them a reasonable opportunity to develop production of their inventions, make them known to the public, and reap fitting rewards before they became available for others to manufacture. This monopoly was in some ways contradictory to the antitrust legislation of the late nineteenth century, but both had the objective of increasing and improving the goods available. However, these objectives were often neglected when extensions of patents were granted, additions were recognized, and closely related improvements were patented by others. A common practice—one

10.27 In little more than a decade U.S. manufacture of lighting became concentrated in two giant corporations, although a few smaller companies remained. This diagram shows the development of the Westinghouse Company and the intricate lineage of the General Electric Company. (Reprinted with permission of Macmillan Publishing Company from Arthur A. Bright, *The Electric-Lamp Industry*. Copyright 1949 by Macmillan Publishing Company, renewed 1976 by Evelyn F. Hitchcock.)

used in the case of electric incandescent lamps—involved the owners of major competing patents agreeing to pool their patents. In some cases, the agreements by which other companies were licensed to use those patents included restrictions, limiting production and controlling the retail price at which the product could be sold. There were sometimes, in fact, requirements that any improvements in the product or process patented by licensees should be made available to the entire patent pool. With patents granted for periods from 15 to 20 years, it was possible for the owner of an initial patent to establish a mercantile dominance that could be maintained for several decades. Nevertheless, there was often a reluctance to institute legal action against those who infringed a patent. Pursuit of an infringement case often required exposing details of manufacturing the product, details much more specific than those revealed in the patent. More important, the courts in most countries considered both the evidence of infringement and the validity of the patent itself. Many of the advantages, perils, and frustrations of patent suits are present in the case of Thomas A. Edison's incandescent lamp patent.

Edison filed application for his major U.S. patent for incandescent lighting (no. 223,898) in early November 1879. It was granted on 27 January 1880, with patents also awarded by Canada, Belgium, Italy, and France between those dates. Many of the problems came from the wording of these patents. The German Supreme Court declared the German patent "extraordinarily obscure"; the English Court of Appeals suspected that the patent it considered was "studiously and willfully obscure." A U.S. judge commented, "The language of the patent itself. That is the greatest cloud that hung over it."[51] In 1882 the Edison Electric Light Company and Swan United Electric Light Company charged each other with infringement of basic British patents. This litigation was settled out of court by a merger of the two firms into the Edison and Swan Electric Light Company, commonly referred to as "Ediswan." In France the clash between Edison and Swan companies was also solved by a merger, producing the Compagnie Générale des Lampes Incandescentes. There was no merger in Germany, the courts deciding in 1891 that the Edison patents were valid but that Swan patents did not infringe.

When the Brush company in 1882 obtained rights for the use of Swan patents in the United States, Edison's company was threatened. The Edison Electric Light Company did not very actively defend its patent rights until after the first five years of the patent's 17-year term had passed. At that time Edison interests may have been encouraged by the German courts' decisions favoring their 1880 patent, but the U.S. patent examiners rejected Edison's efforts to amend his patent. The Edison Electric Light Company in 1885 brought a suit against the U.S. Electric Lighting Company that was used as the test case on the strength of the patent. Edison's lawyers began a delaying action, obtaining 23 continuances. In 1889, four years after the suit had been filed against the U.S. Electric Lighting Company, there was a change in the participants. George Westinghouse, newly interested in electrical goods after developing a company based on his invention of air brakes for railroads, bought the U.S. Lighting Company. When a decision favoring Edison was given in 1891, the latter's company rapidly brought suit against most of the small companies producing incandescent lamps.

Through the six years that remained of the patent's term, the Edison Electric Light Company continued a militant campaign of protecting its patents. The syndicate of financiers that controlled the Edison Electric Company arranged a merger with the Thomson-Houston Electric Company, its principal competitor (fig. 10.27). This new electrical firm, named General Electric Company, brought together the two largest groups of electrical patents. Their hold on the manufacture of electric lamps through the remainder of the patents' terms was so strong that when Westinghouse was given the contract to light the 1893 Columbian Exposition in Chicago, it was necessary quickly to obtain a patent that did not infringe on those held by General Electric.

Ten years before, Edison's Pearl Street Station had begun generating electrical current in New York. The previous year the International Electrical Exhibition and the first International Electrical Congress had been held in Paris. It was time for speculators to get in on the ground floor of a new industry's development. The establishment of electric power companies in the United States followed the patterns that had been established by gas companies and the streetcar investors. Before the Pearl Street Station was constructed, Edison took the precaution of inviting New York city aldermen to Menlo Park and providing them with a lavish buffet brought from Delmonico's fashionable restaurant. In other, smaller cities, a newly organized electrical service company most often had a list of investors that included influential local citizens, who hardly needed to curry the favor of city officials. In 1898 Chicago was the largest city in the United States to have its own electric lighting system, as did 58 other municipalities, but in almost all of these cases, there were also privately owned electric works operating within the city.[52]

In England, authorization of electric works was obtained at the national rather than municipal level. Instead of entertaining aldermen, British speculators invited the leaders of Parliament to dinner before that body acted on the law that was to regulate their company. At that high point of English optimism about investment in electricity, it was proposed that monopolies be granted to private companies for a period of seven years, for it was assumed that within that term a company would have achieved a reasonable return on its initial investment. But the law as finally passed by Par-

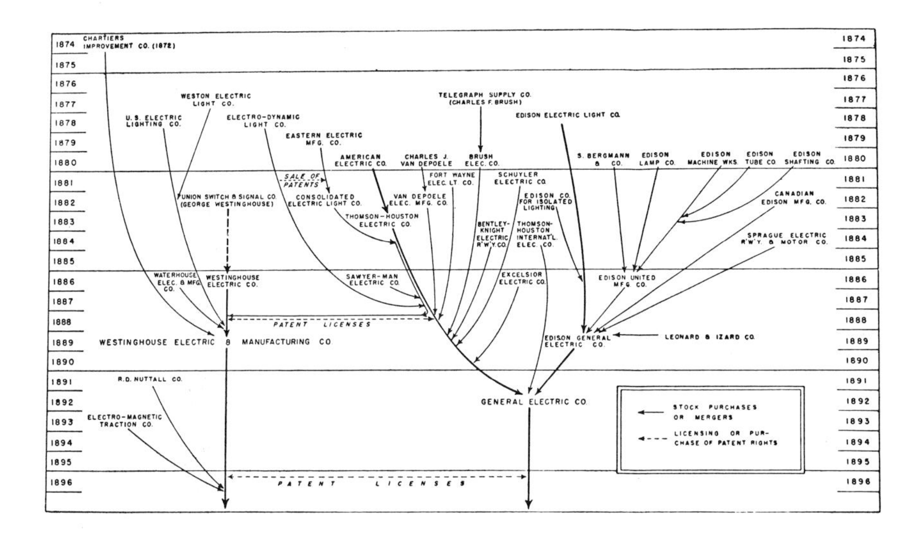

10.28 In 1904 this banking room was lighted with Moore tubes. The tube was 1¾ inches in diameter and 154 feet long, the longest lamp up to that time. (*Cassier's Magazine*, September 1904.)

liament was more cautious, providing a tenure of 21 years for private companies.[53]

With the enthusiasm raised by the Paris and London electrical exhibitions of 1881 and 1882, British investors were anxious to put their money in any form of electrical enterprise. Limits placed on the ownership of power stations by the Electric-Lighting Act of 1882 did not deter these investors, but the major companies' performances caused concern. The Anglo-American Brush Electric Light Corporation, formed to extend the parent company and its system into the British market, licensed many small companies ("Brush babies" as they were called), but it soon became evident that the Brush arc light system was no longer a leader in the field of lighting, and their patent on an incandescent lamp was challenged in courts. By the time the Edison and Swan United Electric Light Company, Ltd., was formed, enthusiasm had waned. In 1883, 69 charters were granted for British central stations and 62 of those were later withdrawn; in 1884 four were granted, all being withdrawn; and in 1885 none were granted. Ten years later the electric industry was referred to as one of "municipal enterprise," and speculation had ended.

Milan was the first city in continental Europe to build a central station for electric lighting. On the whole, Italian and French cities at the end of the century depended on privately owned power sources and, because few of the significant technological contributions had been made by natives of those countries, the spread of electric lighting was hampered by dependence on the importation of foreign equipment.[54] In Germany, Emil Rathenau purchased patent rights through the Compagnie Continentale Edison in Paris, and with them he established the Deutsche Edison Gesellschaft für angewandte Elektrizität. German cities were divided between municipal ownership and private, although some cities that had built their own plants leased them for operation by private companies.

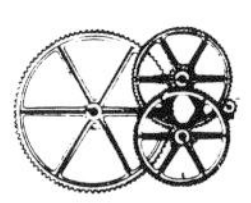

Heinrich Geissler, instrument maker at the University of Bonn, in 1854 fabricated small glass tubes filled with gas that glowed when current was passed through the platinum wires embedded in the ends of the tubes. Little was done to develop the Geissler tube as a light source until the experiments of D. McFarlan Moore. In 1895 Moore exhibited a bulb shaped much like those being used for incandescent filament lamps, but with an arrangement of wires that served as electrodes between which the gas was lighted.[55] By the following year, Moore had concluded that the larger surface area of a tube was necessary to produce sufficient light for commercial purposes. When the tube was filled with nitrogen, a pinkish light was given off; carbon dioxide produced white light, although less efficiently.[56] A major difficulty of the Moore tube was the absorption of gas as the electrodes deteriorated. By using a valve that was electromagnetically activated by decreases in the current in the tube, Moore was able to admit additional gas to the tube.

Although the valve and transformer required for Moore's tube lighting were somewhat cumbersome, reminiscent of the regulators needed for early arc lamps, a tube could be made long enough to light an entire room (figs. 10.28, 10.29). By airtight connections between lengths of glass, a tube 154 feet long and 1¾ inches in diameter was assembled and suspended around the perimeter of a banking room. The ends of the tube met near the point where a transformer and gas replenishment system were recessed in the wall.[57] Moore's "artificial daylight" was installed in prominent places during the period following 1900. According to studies of the installations at the Savoy Hotel in London and the Palace of Ice in Berlin, the system's consumption of electricity was relatively economical.[58] When displayed in 1905 at the Electrical Show in New York's Madison Square Garden, it was declared to be "absolutely free from glare."[59] Most of the installations were in large rooms, where the glare of arc lamps would be offensive and maintenance of a multitude of small incandescent bulbs was expensive. The major competing lamp using gaseous discharge was the mercury vapor lamp of Peter Cooper Hewitt. The light provided by mercury vapor was lacking in the red portion of the spectrum, a light in which "a person's face looks green and the lips purple."[60] Attempts were made to add other gases in order to correct the color of the mercury light, but they were on the whole unsuccessful (fig. 10.30). Hewitt around 1910 devised a reflector with a coating that fluoresced in a pinkish color, and lamps were made combining the mercury vapor tubes with tungsten filament bulbs in an effort to balance their different hues. In spite of all efforts to ameliorate the inherent characteristics of mercury vapor lighting, the lamps were soon relegated to use in industrial installations and other occasions where large amounts of light were needed and there was little concern for color.

Early in the twentieth century Georges Claude began his work on the production of liquid air while employed by the French branch of the Thomson-Houston Electric Company. In the process of liquefying oxygen he was able to obtain the rare gases, including neon, that had been identified only a decade before. Searching for a light that might equal daylight, Claude experimented with mixtures of helium and neon, trying to balance the blue and red tints of the light produced by these gases in a Geissler tube. This avenue of research proving unproductive, Claude focused his attention on neon, in spite of its red light. The gas's light was dimmed appreciably if only 1 percent of nitrogen were added through deterioration of the electrodes. In an effort to

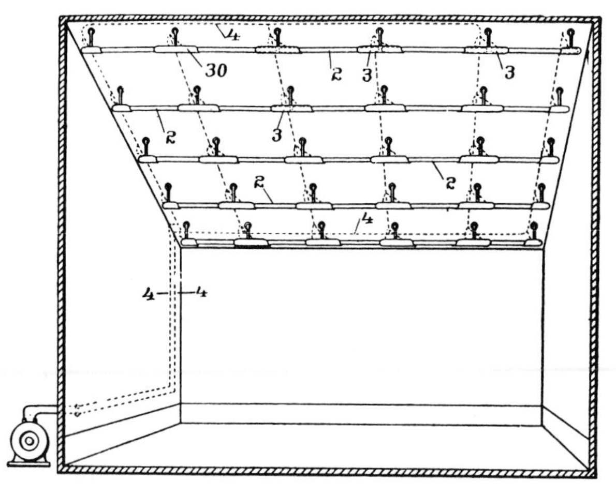

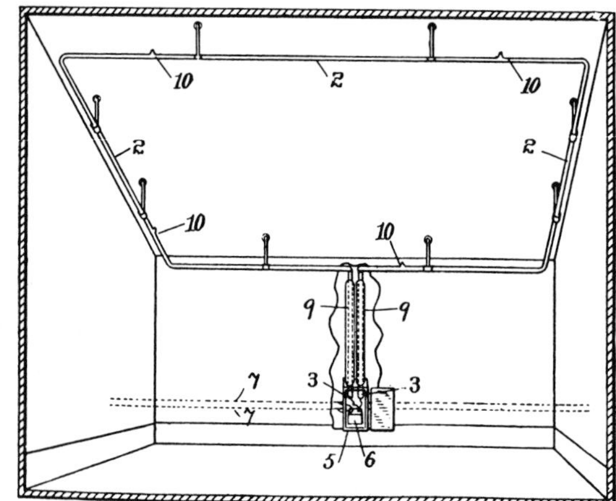

purify the neon, a bulb containing coconut charcoal was attached to the tube.[61] After the tube had been in operation some time and the nitrogen had been absorbed by the charcoal, the glass bulb was sealed off from the tube and removed. Decay of the electrodes was decreased by making them larger, which kept them cooler.[62]

For the Paris Motor Show in 1910, the Grand Palais was "immersed in a sea of bright golden light" from two neon tubes, about 1¾ inches in diameter and each 115 feet long.[63] One of these tubes succeeded in running almost 9 days before being accidentally broken. For advertising or decorative uses, the brilliance and efficiency of neon light proved useful, but few interior spaces could employ light that had little other than the red portions of the spectrum. A mixture of mercury vapor and neon gas had failed in Claude's early experiments, but the mixture of their light could produce a satisfactory color. In 1934 the General Electric Company built a small auditorium to display new lighting equipment in its New York office building. Wanting a general light similar to that of the sun, a combination of lamps was placed in coves around the ceiling: two neon lamps in clear tubes (red) for each two mercury vapor lamps in yellow glass tubes (green) and three mercury vapor lamps in blue tubes (blue). The design of the installation did not smoothly mix the colors, for descriptions comment: "There are areas of color corresponding to the colors of the lamps directly below them and areas of other colors between these, caused by the mingling of adjacent colors, all grading out . . . until they are lost in white near the edges of the ceiling. Then, too, there are varied enlivening colors in all the shadows in the room."[64]

In his book *La Lumière*, Alexandre Edmond Becquerel gave an account in 1867 of his experiments of placing granules of luminescent materials inside Geissler tubes. Although the glow of these materials was far from strong, the Versailles professor proposed obtaining light by coating the inside of a tube with powders. During the remainder of the nineteenth century there were fitful attempts to pursue Becquerel's suggestion. Inventors and scientists, including Thomas Edison and Sir John Ambrose Fleming, tested a broad range of solutions. Fluorescent materials were placed within tubes, on the outside of tubes, and mixed in the glass from which tubes were made. The investigation grew particularly intense at the start of the twentieth century, when D. McFarlan Moore's tubular gaseous discharge lamp and Peter Cooper Hewitt's mercury vapor lamp were introduced. In 1923 a French experimenter coated fluorescent powder on the outside of a tube containing neon and mercury vapor. A few years later he patented a lamp having the fluorescent material on the inside of the tube, but the varnish used to attach the powder produced a vapor that made the lamp's performance unsatisfactory.

European manufacturers around 1930 were much more active in advancing lamps that used mercury vapor, sodium, and neon than were manufacturers in the United States. A trio of German engineers patented a lamp in 1926 that included all of the essentials of the modern fluorescent lamp, but it was not pursued commercially. Although General Electric and the Westinghouse Corporation, virtually controlling the electric lamp industry in the United States, had conducted desultory investigations of fluorescent lighting before World War I, both manufacturers accepted incandescent lighting as completely satisfactory for general purposes of illumination and viewed gaseous-discharge and fluorescent lamps as suitable only for specialized functions. Then in 1933 General Electric's manager of development and engineering for its lamp department returned from Paris, where he had seen installations using fluorescent coatings on tubing much like that employed in neon lighting. His interest was increased the following year by a report from Arthur Compton, a recent winner of the Nobel prize for physics and a consultant for General Electric, who had seen cold-cathode fluorescent lighting in England and was convinced that it was a most promising direction for development. General Electric then began a rapid study of fluorescent lighting.

European installations of fluorescent lighting had concentrated on the use of long tubes custom-bent to suit the requirements peculiar to each installation. Because General Electric's production centered on broad marketing of mass-produced and standardized lamps, it was necessary to find ways in which the custom-made fixtures of European practice could be adapted to the characteristics of industry in the United States. Electric current in Europe was commonly distributed at voltages as high as twice the 110 volts used in the United States; work therefore began on the study of low-voltage fluorescent lights. General Electric's research budget for fluorescent lamps increased from a

10.29 After about six years of exhibiting tube lighting, D. McFarlan Moore received his definitive patents in 1902. Patent drawings indicated two patterns of installation. In one, the tube was shown as a continuous line suspended at the perimeter of the room; in the other, short tubes were connected in a pattern strongly resembling today's installations. (*Electrical World and Engineer*, 28 June 1902.)

10.30 The unpleasant colors of early mercury vapor lamps led to several years of unsuccessful experiments adding various gases to the tube and metals to the mercury. A brief trial by Peter Cooper Hewitt combined his mercury vapor lamp with a reflector above the tube. The reflector was coated with rhodamine or some other fluorescent dye that would be activated by the lamp and provide light in the red portion of the spectrum. (L. Gaster and J. S. Dow, *Modern Illuminants and Illuminating Engineering*, 1919.)

few hundred dollars in 1934 to $13,400 in the following year, and in two more years it was five times that amount.[65]

Westinghouse joined in the research. For three years the engineers of General Electric labored on the development of the standard low-voltage fluorescent lamp. Since the two companies cooperating in the research dominated the U.S. market in incandescent lamps, their research was both rapid and cautious. A new form of general illumination, especially one that required a different kind of fixture, could prosper only through replacing incandescent installations, new or existing. Commercial failure of the new fluorescent lamps could result in General Electric's losing a portion of its market in the sale of electric lamps. Sylvania Electrical Products, a minor participant in the market-sharing agreement that controlled incandescent lamp production, was developing its own patents for fluorescent lighting. Limited to about 5 percent of the market for incandescent lamps, Sylvania was freer to compete in the field of fluorescent lighting. It was expected that the introduction of fluorescent lighting would rapidly result in large sales. The new lighting was economical; the color of its light approached white; the large surface area of fluorescent tubes reduced glare problems; and the tubes generated less heat than incandescent lamps—a consideration that was to grow in significance with the increased use of air conditioning.

But caution soon had to be abandoned. The New York World's Fair and the Golden Gate International Exposition in San Francisco were to open in 1939. Designers and illumination engineers occupied with planning the buildings for both fairs were aware that the new method of lighting was near introduction. They made it known that they would employ European high-voltage custom-made lighting if low-voltage fluorescent lamps were not available for use in the fairs. Despite their plans to introduce the new lamps over a period of several years, General Electric and Westinghouse were forced to provide fluorescent lamps for the two fairs. The first year's production was mostly used for the two fairs, but in 1938 the two manufacturers listed fluorescent lamps in their catalogs. Four years later 33,600,000 fluorescent lamps were sold. In the first five years in which fluorescent lighting was on sale, the system was more often installed to raise the level of lighting than to reduce the amount of electricity consumed.[66]

In a modern office building about 40 percent of the cooling load for air conditioning results from the lighting system, and for other types of buildings the percentage may be about half that. Although the efficiency of fluorescent lamps has almost doubled since their introduction, less than a fifth of the energy used goes into light (still this is about twice as efficient as incandescent lighting). By current techniques the heat produced by lighting can be used as part of a building's heating system, but nevertheless heating by lamps is not the most efficient method and artificial lighting is most needed in those areas of a building needing the least heating. From the days of gas burners, illumination has always been hampered by its dependence on a form of heating as the source of light, whether oil, gas, or electric current.

IDEAL

11

Heating and Ventilation

1713 Ventilating fireplaces described in *Mécanique du feu* by Nicolas Gauger

1740 Benjamin Franklin designs an open-front firebox, the "Pennsylvania Fire-Place"

1796 Count Rumford (Benjamin Thompson) advances methods for improving fireplaces

1802 Steam heating installed in textile mill by Boulton and Watt

1806 The Derby Infirmary provided with William Strutt's improvement on the cockle stove

1830 Hot-water radiators installed by Joseph Bramah in Westminster Hospital, London

1830s Three complex stoves installed in the Long Room of the London Custom House

1833 Base-burner stove patented (U.S.) by Eliphalet Nott, president of Union College, Schenectady, New York

1839 David Boswell Reid chosen to plan heating and ventilation of the Houses of Parliament, London

1840–42 Pentonville Prison, London combines hot-air heating with extreme security requirements

1854 Reid's system for St. George's Hall, Liverpool, completed

1859–90 Theories of ventilation propose that specific gases are associated with certain diseases or that organic poisons are exhaled into the air

1876–89 Construction of Johns Hopkins Hospital, Baltimore, Maryland

1877 District heating introduced by Birdsill Holly in Lockport, New York

11.1 **To construct Benjamin Franklin's "Pennsylvania Fire-Place" one assembled the iron plates shown in this drawing to achieve the stove shown at the lower left. The ribbed plates (upper right) provide the channels at the back through which air circulated as it was heated. (B. Franklin, *An Account of the Newly Invented Pennsylvanian Fire-place*, 1744.)**

The ancient Romans had two systems of heating. For public baths after the first century B.C., for villas of the wealthy, and for some public buildings, heat was provided by hypocausts, systems of flues built within floors and walls through which passed the exhaust gases from furnaces. For other buildings braziers of charcoal were customary. At home on a chilly day, when the braziers did not suffice, they put on additional clothing or perhaps left for the nearest public bath.

In parts of Italy a central hearth was the usual source of heat until around 1400, when "French-style" wall fireplaces were adopted.[1] In France, fireplaces might be expected in certain rooms of the finer houses and braziers were carried into others. A French writer in the fifteenth century reported: "For the cold that comes to Germany in the winter they have stoves that heat in such a way that they are warm in their rooms, and in winter craftsmen do their work and keep their wives and children there and it takes very little wood to heat them."[2] Stoves were considered so much a German device that a French king, wishing them installed in his castles, found German workmen to perform the task.

When fires stood against walls, hoods were built over them and flues provided. The only fire in an English manor house of the sixteenth century might be in the central hall, sometimes then referred to as the "fire-house," although another would be provided if there were a separate kitchen.[3] The hood was high enough to permit a cook's stepping under it to tend a spit or stir a pot, and these early fireplaces functioned more as heated alcoves then as heating mechanisms. Almost all the air that was warmed went up the chimney, and fewer people clustered around the fire than could when it stood in the center of the hall. Air that left the room was replaced through cracks around windows and doors, and the latter were often provided with a vestibule of curtains or screens to divert the cold drafts entering.

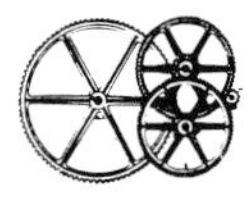

At the start of the nineteenth century, home life in the United States centered about the fireplace:

> Wood was cheap and labor scarce; therefore the fireplace was made capacious enough to contain a large backlog which lay in the ashes at the rear, and in front of which was the forestick, resting on andirons. The space between these two logs was filled with smaller wood. The living-room in which this fireplace was located served for both kitchen and dining-room, and at night high-backed settees were arranged in front of the fire to intercept the heat, and prevent cold draughts from behind.[4]

The area that could be heated by such fireplaces had been increased early in the eighteenth century by use of the jamb stove, a box made of five iron plates that was set in the rear wall of the fireplace and extended into the room behind it. Coals from the fireplace were shoveled into the jamb stove and ashes could be removed by raking them forward into the fireplace.

When a larger number of fireplaces were included in dwellings, they were smaller and remained equally inefficient. The problem was one of finding a way to evacuate smoke and gases without drawing too much air from the room or restricting the radiant intensity of the flames and the mass of the fireplace itself. In addition, downdrafts and sooty rooms were to be avoided. By the seventeenth century, attention was more sharply directed to problems of heating because of the soaring cost of fuel. Fireplace openings and flues were made somewhat narrower so that the rising gases moved faster and downdrafts were less likely. A Paris physician, Louis Savot, noticed a fireplace in a library of the Louvre, and the principle he observed came to be the basis for a profusion of designs after it was published in 1685.[5] The fire was built on an iron plate a few inches above the floor, and behind it was another iron plate with space left between it and the wall. Air was admitted between the floor and the bottom iron plate, and it rose between the wall and the hot surface of the back plate. This hot air was carried by tubes to the front of the mantel, where it returned to the room. The air required for combustion was completely separated from the air utilized for heating. A method had been found to capture and utilize additional heat from the fire by circulating air around the metal plates that held the flames.

A more complete study of the ventilating fireplace was presented in a book, *La mécanique du feu*, published in 1713 by Nicolas Gauger.[6] In Gauger's design of an improved fireplace, the sides were shaped in parabolic curves intended to reflect heat straight out into the room, a subtlety of questionable value. The dimensions of the flue were much smaller than was the usual practice at that time; in fact, they were nearly the dimensions common two centuries later. Fresh air drawn from outside the building was brought into the space between the iron walls of the firebox and the masonry walls of the building, baffles causing it to make several trips up and down before being released into the room. John Theophilus Desaguliers converted the design to the use of coal, at that time the most economical fuel in Britain, and a number of such installations were made in London until they were condemned by some scientists who claimed that such fireplaces "burnt the air, and that burnt air was fatal to animal life."[7]

Early in 1740 Benjamin Franklin designed an open-front cast-iron firebox.[8] His "Pennsylvania Fire-Place" was meant to stand well forward in an existing fireplace (fig. 11.1). Between the plates of cast iron, smoke and heat rose against the front surface of the "air-box," descended along its back surface, and from there rose to the chimney. Within the "air-box" fresh air from a channel beneath the floor passed through a maze of passages before entering the room. In the first half of the nineteenth century there were at least seventeen U.S. patents with titles that included the term "Franklin Stove," and descriptions of the "Pennsylvania Fire-Place" found their way to most European countries. Franklin did not patent his design, preferring to make the stove available

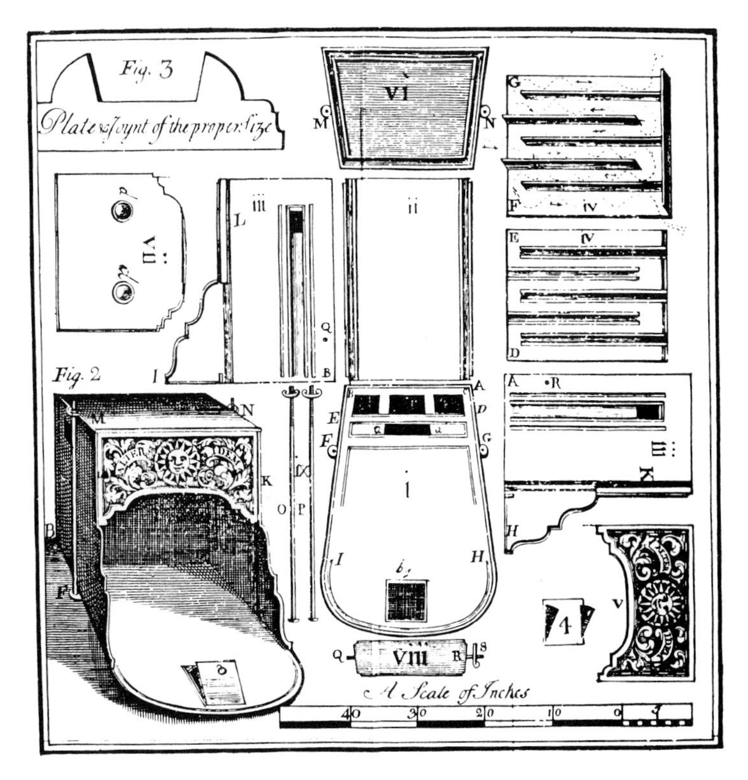

to all without payment of royalties, and he criticized others for wishing to profit from people's need for heat.

It was recognized that the smoke leaving chimneys and falling over sooty cities was unburned fuel. French and German experiments late in the seventeenth century developed cast-iron machines shaped very much like tobacco smoking pipes (fig. 11.2). By establishing a downdraft through the fire, smoke and gases were drawn through the glowing coals, where the gases were burned to release additional heat. Franklin's later "Pennsylvania Stove," an urn-shaped stove, was a more advanced form of downdraft heater.[9] His improvement consisted of channels at the bottom through which the hot gases meandered before entering the flue.

Some of the most intensive studies of heat-producing systems were conducted by Count Rumford, born Benjamin Thompson in Massachusetts.[10] A Tory spy during the American Revolution and knighted by George III, Thompson pursued much of his scientific work in Munich, where he was made Count Rumford by the Elector of Bavaria and was assigned duties that afforded an opportunity to apply his theories to the practical problems of heating and cooking for army barracks, prisons, and institutions that cared for the sick and poor. Rumford's improvement of institutional cooking facilities focused on the elimination of waste. Fires were insulated and served individual cooking utensils that were fitted into the heated mass of the range, eliminating the vast amount of heat that escaped around pots set on open flames. Rumford's fireplace improvements, proposed in 1796, were equally frugal and included practical methods for reshaping existing fireplaces (fig. 11.3). Rumford recommended that flues be narrowed above the fire, providing a "smoke shelf" (a ledge that discouraged downdrafts) at the base of the flue and leaving an opening only about 4 inches deep at the front of the flue. The width of the opening to the flue was allowed to vary according to the size of the fire or fireplace. The fire itself Rumford set well forward in a shallow recess with angled side walls.

Many fireplaces, particularly in England, were "Rumfordized," the most common complaints being that they provided too much heat after the alterations had been made. Some opponents pointed out that the separate elements of the Rumford design had been previously proposed by others. His scheme was so practical, requiring little more than a mason's filling portions of an existing fireplace, that there was great resentment among manufacturers who had flourished on the sale of cast-iron fireboxes. It was not long, however, before iron founders began to cast new fireboxes that followed the principles of Count Rumford.[11]

During the first half of the nineteenth century there appeared a number of inventions aimed at eliminating the frequent task of feeding fuel to fires. All were curious and a few became popular. By placing enough coal for a full day's heat in a box beneath or beside the fire, these mechanisms allowed the householder to

11.2 By establishing a downdraft, Leutmann's "Vulcanus Famulans," a so-called "smoke-consuming" stove, burned gases and unconsumed fuel that would otherwise have gone up the chimney. A description in 1681 stated that "matters which stink abominably when taken out of the fire, in this engine make no ill scent, neither do red herrings broiled thereon." (J. P. Putnam, *The Open Fireplace in All Ages*, 1882.)

11.3 The process of Rumfordizing fireplaces was essentially a matter of decreasing dimensions. A fireplace of the usual construction (figures 1 and 2) would have its front opening reduced and its back wall brought forward (figures 3 and 4). At the bottom of the chimney, these alterations would result in a smoke shelf and a much smaller opening into the chimney. (Count Rumford, *Essay IV, Of Chimney Fire-places*, 1797.)

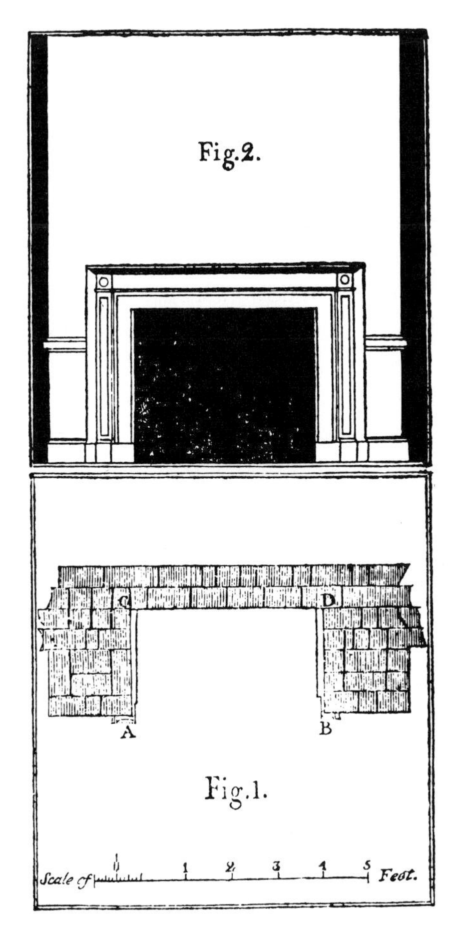

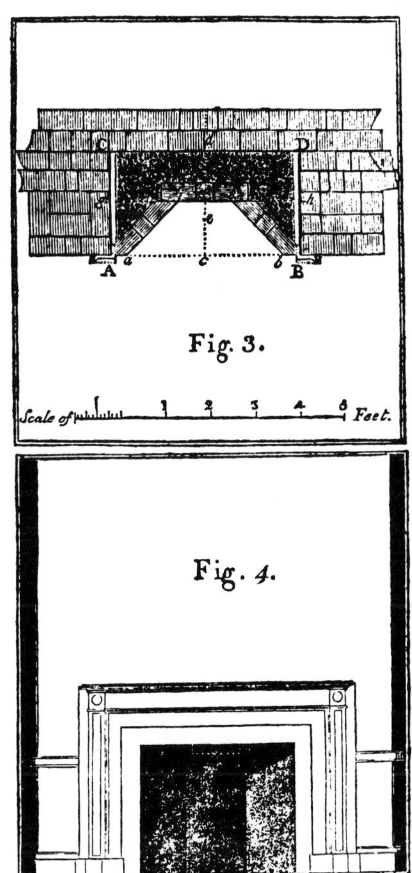

feed the fire periodically by manipulating levers, pistons, or plates that separated flames from the fuel supply.

In Europe, stoves were more often used in the northern countries, where large heaters were made of ceramic materials or metal and contained zigzagging flues and interconnected chambers that slowed the burning of fuel and made certain that additional heat was extracted from gases as they escaped to the chimney. The English, however, clung to their fireplaces, accepting only an occasional use of open-front cast-iron boxes as sources of heat for entrance halls. It was in the United States that stoves flourished and went through many transformations.

Most stoves stood out from a wall so that air could circulate and be warmed at all their surfaces, but the stovepipe between the fire and the flue was almost as effective as a heater. A study made early in the nineteenth century measured the duration of heat when different stoves were used to produce a fixed temperature with equal amounts of fuel (table 11.1)[12]

22 elliptical tubes, in cross section roughly 10 inches by 6 inches, circulated air upward from a space beneath the stove's platform. A tank within the jacket held water, which was reported to have evaporated at a rate of about eight gallons per day for each stove. At that time heat was usually controlled by regulation of the openings through which hot air was delivered, a method that wasted fuel as the fire continued to burn. Arnott's stove had a tube of mercury extending into the space through which air circulated, and a float at the opposite end of the tube regulated the entry of air to the fire according to the expansion of the mercury. The lid of the firebox was sealed to prevent leakage of gas by fitting its edge into a groove filled with fine sand. Stoves they were called, but many elements of a first-class warm-air system were embodied in the design.

11.4 By the late nineteenth century, circulating fireplaces were commonly available. Ornate grilles above the mantel released air into the room after it had been heated by the metal firebox and the "distributor," a network of flue pipes. In England the grille was usually placed high enough to avoid interfering with the tradition of gentlemen warming their rumps before the grate. (J. P. Putnam, *The Open Fireplace in All Ages*, 1882.)

11.5 At each end of the Long Room of the London Custom House there stood a square stove almost 12 feet high. Smoke was exhausted through a flue (c) beneath the stove. Air was heated by circulating through 22 vertical tubes (*b*). An additional stove, circular in shape, was situated in the center of the Long Room, and its flue carried smoke upward through the ceiling. (*Transactions, Royal Institute of British Architects*, 1842.)

Table 11.1 Duration of Heat for Various Stoves (see text)

Stove	Duration
Ordinary fireplace constructed for burning wood	**10 hours**
Open parlor grate for anthracite coal	**18 hours**
Open Franklin stove	**37 hours**
Sheet iron cylinder stove with 5 feet of horizontal flue pipe	**67 hours**
Ditto, with 42 feet of flue pipe having 17 elbow joints	**100 hours**

Obviously, a slow-burning stove with a large amount of heated surface was most efficient in supplying heat, and this was the goal of later inventions.

The heating system devised in the 1830s by Dr. Neil Arnott, a London physician, for the Long Room of the London Custom House consisted of three stoves set within the room, but the total organization of the system introduced several of the elements developed further in later heating designs (fig. 11.5). The Long Room was 186 feet in length, 64 feet wide, and 44 feet high, and it provided work space for 130 officials and clerks who served the many merchants doing business there.[13] Additional sashes were placed over the 1,100 square feet of windows in the room, but a skylight of almost the same area seems to have been too difficult to cover. Arnott spaced three stoves in the room, two cubical in shape with descending flues, one a center stove, circular with a flue through the ceiling. Between each stove and its jacket

Downdraft stoves in the Bank of England had been removed in 1787 as a result of workers' complaints, and there were many objections to those placed in the Custom House. Customs officials declared that air from the stoves caused "a sense of tension or fulness in the head, throbbing of the temples and vertigo, followed not infrequently by a confusion of ideas."[14] Stoves were popular in the United States, where stovemakers entered into intense competition with names such as "Morning Glory," "America," and "Oriental" for their products. Patents for stoves in the United States reached their highest number during the 1830s and 1840s. Cast iron, sheet metal, and soapstone were the more common materials, and coal-burning stoves were sometimes lined with brick to avoid excessive temperatures on their surfaces. Enameled metal permitted colorful decoration, and mica inserts allowed glimpses of the flames inside. A stan-

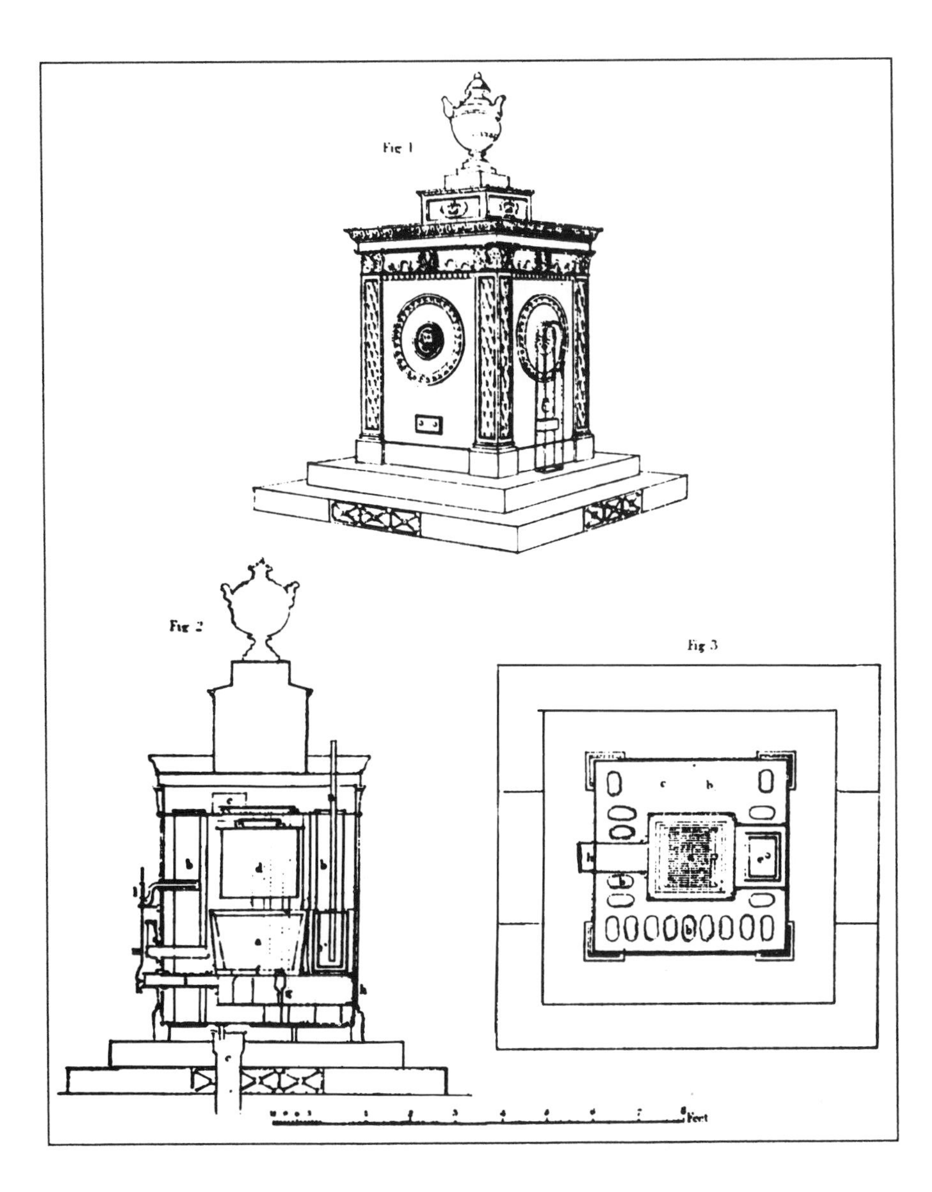
Fig 1
Fig 2
Fig 3
Feet

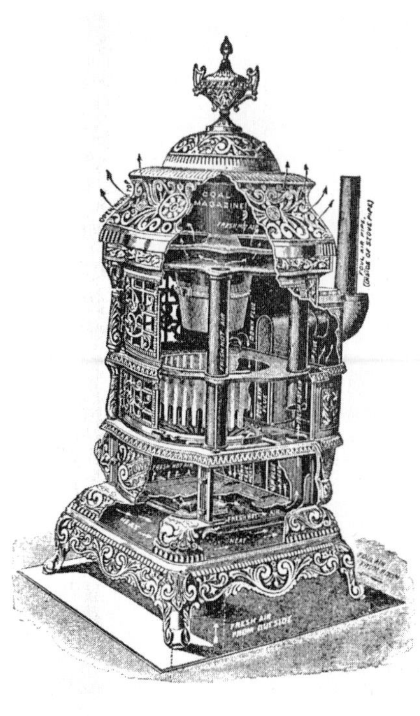

11.6 At Chicago's Columbian Exposition (1893) manufacturers eagerly displayed the stoves that were so popular in American homes. Open-grate stoves, such as the "Magic Ideal" (left), let the flames be seen. The "Cortland Howe Ventilator" (right) provided only flickering lights as seen through mica inserts, but received outside air through a floor grille. Both of these examples were crowned with decorative urns from which water evaporated. (*Report of the Committee on Awards of the World Columbian Commission*, 1901.)

11.7 This German design of a base-burner stove was said to hold a coal supply that lasted 10 or 12 hours. Air and flames moved horizontally through fuel, and ashes fell into the bottom of the stove. This model was meant to be installed in a wall, so that coal could be supplied and ashes removed in an adjacent room. (P. Planat, *Chauffage et Ventilation*, 1880.)

dard embellishment was an ornate metal vase at the top of the stove, large enough to hold a quantity of water that would add humidity to the hot dry air rising from the stove (fig. 11.6). In spite of the many technologically complex and heavily ornamented stoves that were available, the model most frequently seen was the globe stove, the "pot-bellied" stove of the American frontier. Dampers were the only significant technology of this simple design.

Base-burning stoves were designed to hold a sizable quantity of coal, with flames limited to a portion of the fuel supply (fig. 11.7). One of the best known of the base-burner stoves was patented in 1833 by Eliphalet Nott, the clergyman president of Union College in Schenectady, New York, who had previously designed a box stove for use in all the school's dormitory rooms. Coal was poured into a bin at the top of the Nott base-burner, and the fire burned in a receptacle beneath. A chamber over the flames received air from the room and released its heated air at the top of the stove. Coal was fed to the flames by a mechanically controlled aperture. In early models, as the stored coal became warmer, gas collected and might explode whenever the lid was raised to check on the fuel supply. A vent was added to allow that gas to escape to the flue, but if one forgot to close the vent the entire load of coal might burn.[15] The Nott stove never overcame this difficulty, but other models advanced the principle of the base-burner stove.

A problem of the westward extension of railways in the United States was the heating of railway cars. The English had seen no reason to heat travelers: "Their idea is that it is quite sufficient to keep the feet warm and not exhaust the lungs or stupefy the brain. Passengers are provided with cylinders of hot water, renewed as occasion requires, on which to place their feet."[16] Much more was needed for rail travel in the United States, where it required days to cross the plains and winter weather was severe. The rocking motion of trains often caused rickety stovepipes to fall and even overturned stoves. Perhaps the most ingenious attempt to solve this problem was a stove holding a supply of water that was supposed to extinguish the fire when the stove fell.

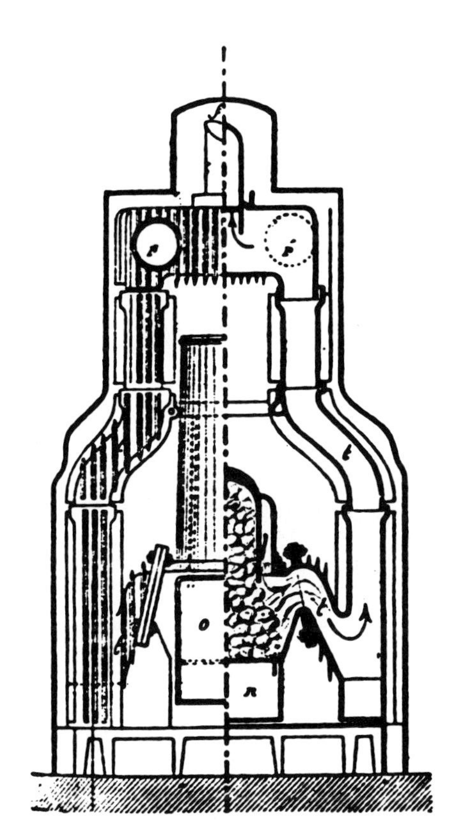

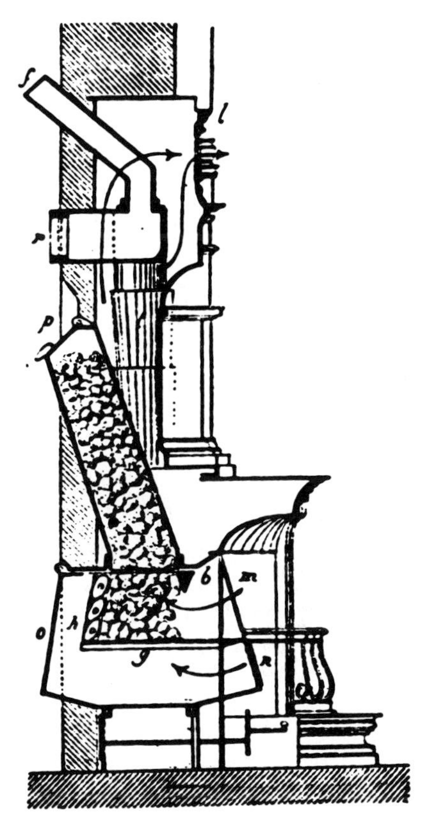

By the 1870s, gas stoves had begun to appear, offering an escape from the bother of supplying coal or wood and removing ashes. Reflective metal surfaces were common in the gas fires that fitted into fireplace openings, and they could be shaped behind the flames so that heat was reflected into rooms at the same time that rising air was heated by the metal surface. Even more effective were shapes of ceramic materials over which gas jets played. Once heated to a glowing red incandescence, the ceramic units afforded a large radiating surface and bore a slight resemblance to the flickering light of a fireplace.

As industrial processes increasingly made use of steam power, there was nothing remarkable about piping some of that heat to points where it could be utilized for other purposes. Hot water was proposed in 1610 as a means of drying gunpowder, and by the late eighteenth century many greenhouses were warmed with hot water.[17] A diagram published in England in 1745 showed steam pipes looping through all the rooms of a three-story building and valves in each room to control the flow.[18]

A much more sophisticated device invented by Bonnemain, a Paris physician, was reported to the French Académie des Sciences in 1782. Intended for the incubation of poultry eggs, the system recirculated hot water (fig. 11.8). Apparently tests had shown that bubbles from boiling water could block the pipes, so an outlet at the uppermost point of the piping let steam escape, collecting condensate in a small container. Automatic regulation of the temperature was provided by a long rod that expanded, thereby closing off the flow of air to the fire.[19] In 1812 the Marquis de Chabannes published an English pamphlet on the Bonnemain system, adding his own variations (fig. 11.9). One of his diagrams shows steam circulating through a three-story building and includes an air outlet, provisions for recirculating hot water, and a pipe that allowed for expansion of the water in the system.[20] While French heaters based on Bonnemain's scheme usually made use of pipes of small dimension, Chabannes encountered difficulties that caused him to employ pipes 3 or 4 inches in diameter.[21] This meant that his heating system was seldom installed in buildings of formal character.

James Watt in 1784 extended a steam pipe from the boilers into his small office at the factory where the firm of Boulton and Watt manufactured steam engines. The pipe ran between two tin-coated iron plates, 3½ feet by 2½ feet, fastened an inch apart. Watt also placed similar heaters in two rooms of his house. He was not well pleased with the heat provided by these radiators, perhaps because the shiny tinned surfaces were less efficient than bare iron

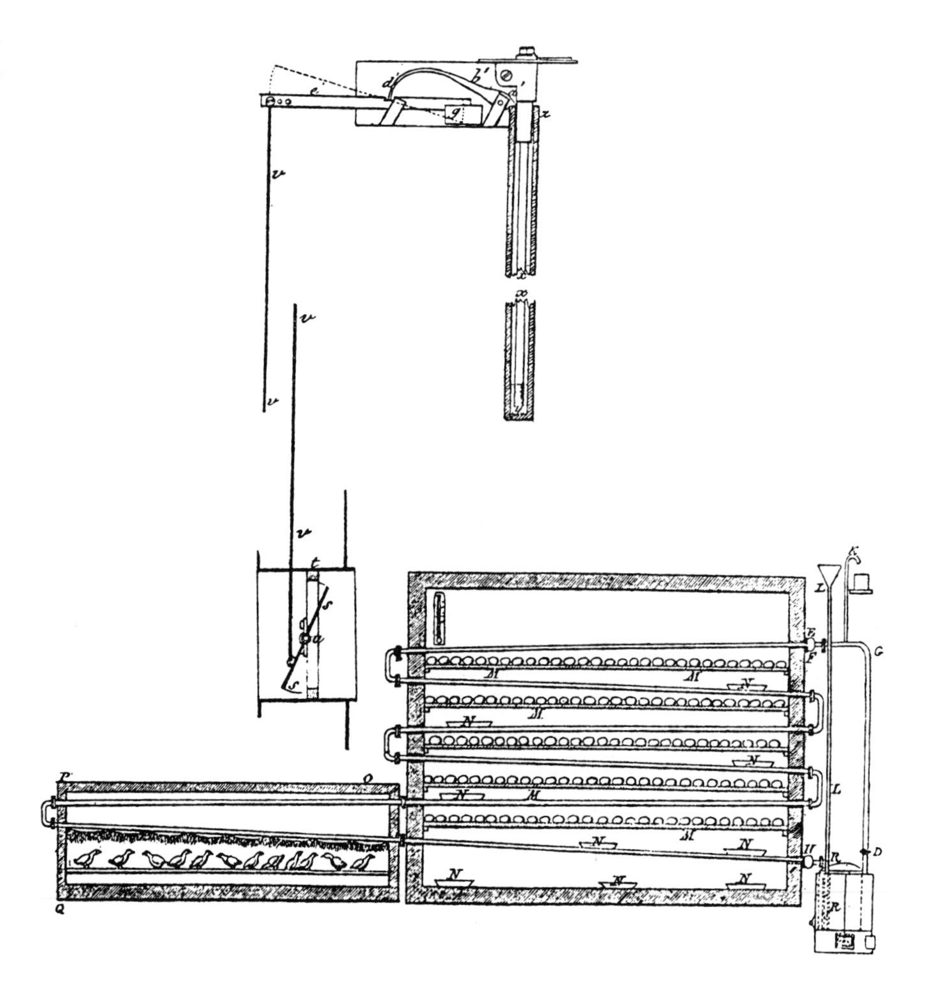

11.8 Bonnemain's hot-water incubator was displayed in the 1770s. As water circulated among the racks of eggs, one loop of the pipe warmed the chicks. At the top of the incubator, a funnel permitted addition of water and a valve was available to release air from the pipes. An enlarged diagram (top) shows the mechanism by which the expansion and contraction of a long rod (x) within the boiler controlled the damper(s) that admitted air to the fuel. (*Gesundheits-Ingenieur*, 31 August 1907.)

11.9 The hot-water heating system devised by the Marquis de Chabannes in the 1810s is said to have been installed in two London houses. Cold water (*c*, *n*) was fed from the roof to the furnace (*a*), and when heated, circulated to radiating vessels (*f*) in the rooms and to a bath (*m*). (*Gesundheits-Ingenieur*, 31 August 1907.)

plates would have been. In 1802 the firm of Boulton and Watt installed its first steam-heating system in a cotton mill. The design used hollow cast-iron columns to carry steam from floor to floor. Although there was an elementary logic to this system, heating by vertical pipes presented problems, whether columns or independent pipes were used. To evacuate air when heating was started and to keep the steam moving in the pipes, a valve was required at the top of each vertical, and condensation had to be drained at the bottom. In addition to the obvious danger of workers brushing against such columns, the building and piping connections were often damaged by expansion of the iron members at high temperatures. In one of Boulton and Watt's first projects, expansion was reported as "1/10 of an inch per 10 feet nearly."[22]

In the first mill installations using horizontal pipes, they were fed by a riser at one end of the building and ran along each ceiling to the other end of the building, where a valve allowed the escape of air, steam, and condensate. Later the ends of the heating pipes were connected to make a circulating system, and they were placed near the floor. It was necessary to slope horizontal pipes to prevent explosions when steam entered cool pipes and encountered water that had been trapped there.

For years Jacob Perkins, who came to England from the United States in 1818 in order to promote his many inventions, experimented with the use of high pressure steam in engines and boilers. When his son Angier March Perkins joined him in 1827, these findings were applied to heating systems. Large pipes were required for heating systems in order to avoid cooling too rapidly and to assure that circulation was maintained. By providing a sealed system of pipes, water could

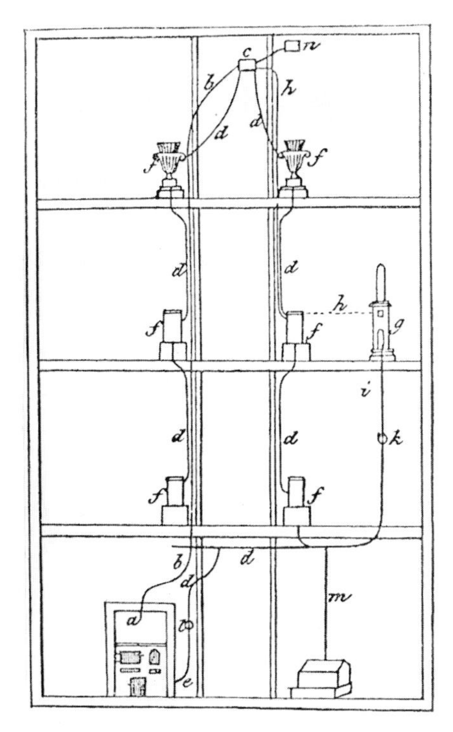

be heated above the normal temperature of boiling, and it would circulate through pipes of smaller diameter. In these systems the temperature of the water was usually between 220° F and 350° F and pipes were commonly one inch in diameter.[23]

The first installation of the Perkins system was for growing grapes in winter, but in 1832 the second heated the offices of a London insurance firm.[24] Angier March Perkins made improvements in the system during the next decade, and its use spread to the Continent and the United States. Jacob Perkins described his son's installation in the British Museum: "It would seem incredible, but it is a fact, that in this apparatus a stream of hot water which is not thicker than your little finger, carries the heat 700 feet, and then returns to the boiler at a temperature of 100 degrees."[25] The

Perkins system was occasionally used in residences and offices, but it and other steam systems were at that time more applicable to mill buildings and other industrial works, where exhaust steam could be utilized.

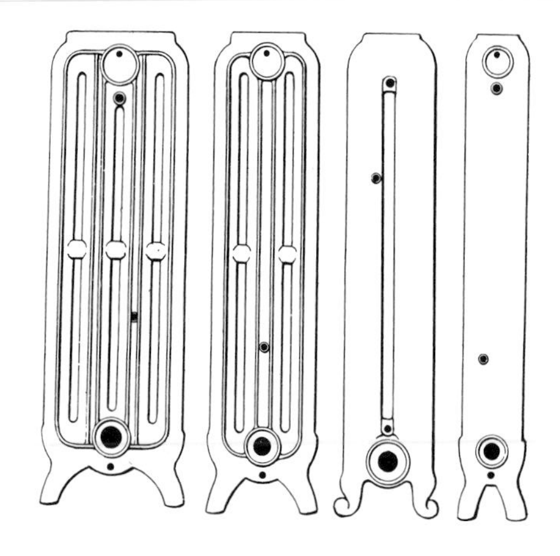

11.10 The manufacture of standardized radiator sections provided a great range of application. Any number of one-, two-, three-, or four-loop units could be combined to supply the required amount of radiation. Four-loop units were least efficient because the outer loops obstructed radiation from center loops. (*Engineering Magazine*, February 1898.)

11.11 The ease of casting metals and the flexibility of standardized radiator units encouraged production of an extraordinary variety of designs. Scrolls, frets, and other decorative additions abounded, and this dining room cabinet also warmed dishes and food brought from the kitchen. (Catalog of A. A. Griffing Iron Company, 1886.)

A handbook written in 1880 stated that "almost any respectable plumber can do hot water fittings, and they all know that one foot of four inch pipe warms 100 feet of air." This would mean that a room twenty feet long, fifteen feet wide, and ten feet high would require a large pipe running along most of two walls to make a total of thirty feet. For mills and greenhouses exposed pipes of such size might be acceptable, but they were not to be tolerated in Victorian parlors. When the heating system for Westminster Hospital was developed in 1830 by Joseph Bramah, the halls and stairways were warmed by radiators of a form somewhat more developed than those of James Watt. A series of closely spaced vertical pipes were supplied with hot water at the top and joined a return pipe at the bottom.[26] This arrangement was surrounded by a box, perforated at bottom and top to allow circulation of the air as it was heated. Radiators were most strongly developed in the United States, where extremes of temperature were far more challenging than those of the English climate (figs. 11.10, 11.11). In the late 1850s U.S. patents of radiators began to be filed, and about two decades later a radiator was introduced made by assembling a series of standard cast-iron units.

A heating engineer at the end of the nineteenth century compared U.S. practice with that of England. Because water in the British Isles often severely corroded wrought-iron pipes, cast-iron pipes of cumbersome dimensions were most often used.

> Within the last few years far more wrought iron pipe is being used for both the construction of wall coils, box coils, and mains; and in districts where the water is very bad, rain water is used in the heating apparatus.
>
> Wonderful strides toward improvement in installation of heating systems have been made in Great Britain within the last few years; the class of work is better, and far more radiators are used than ever before, although the average Britisher is still rather inclined to think them very ugly; but American radiator manufacturers are selling a large quantity of their goods, and are likewise manufacturing a class of goods to meet the requirements of the English market.[27]

In the United States there was a tendency toward decorative effects in the castings that made up sectional radiators, but many architects still preferred to conceal them behind paneling.

Steam and hot-water heating required piping that could withstand the pressures that were present within the systems. Gun barrels had been manufactured in quantity during the long years of the Napoleonic Wars. For this purpose a strip of metal was heated a few inches at a time and hammered around an iron rod, the edges being lapped and welded together. As war ended, the use of piping for illuminating gas increased.

At first much of this need for small pipes was answered by screwing together gun barrels, either those that flaws made unsatisfactory for use in firearms or the unexpected surplus of factories. An English manufacturer, recognizing that gas pressure did not require a weld so strong as that for a gun barrel, in 1825 simplified the work by introducing the use of butt welds. The shape was obtained by pressing a strip of heated metal between two semicircular dies, in this way bringing its edges together. Another system, patented the same year, employed bell-shaped dies through which the metal was drawn, shaping and welding the tube as it advanced. Later the pressure of steam used in engines and heating demanded a return to the stronger lap-welded seams. To accomplish this, strips of metal with tapered edges were drawn through dies, reheated, and shaped by passing between rollers with concave faces that corresponded to the desired diameter of the pipe.[28]

The manufacture of pipes and fittings was begun in the United States by a Philadelphia firm in the early 1830s and by Joseph Nason and his partner James J. Walworth of Boston in 1847.[29] After working five years for the Boston Gas Light Company, Nason went to London, where he was employed for seven years by Angier March Perkins. Upon his return, Nason's newly organized firm purchased the stock of pipe and fittings of the unsuccessful New York branch of the major English maker of pipes.[30] In the first installations of hot water and steam heating, there were problems of preparing new systems, and the task of devising the connections and preparing the pipes required patience and ingenuity. Until the middle of the nineteenth century most pipe was imported from England. As domestic manufacture of pipe increased, imports were reduced, and no small part of production was the tubes and valves required for gas lighting.

Once it was possible to pipe steam under high pressure with satisfactory regulation of pressure, public building

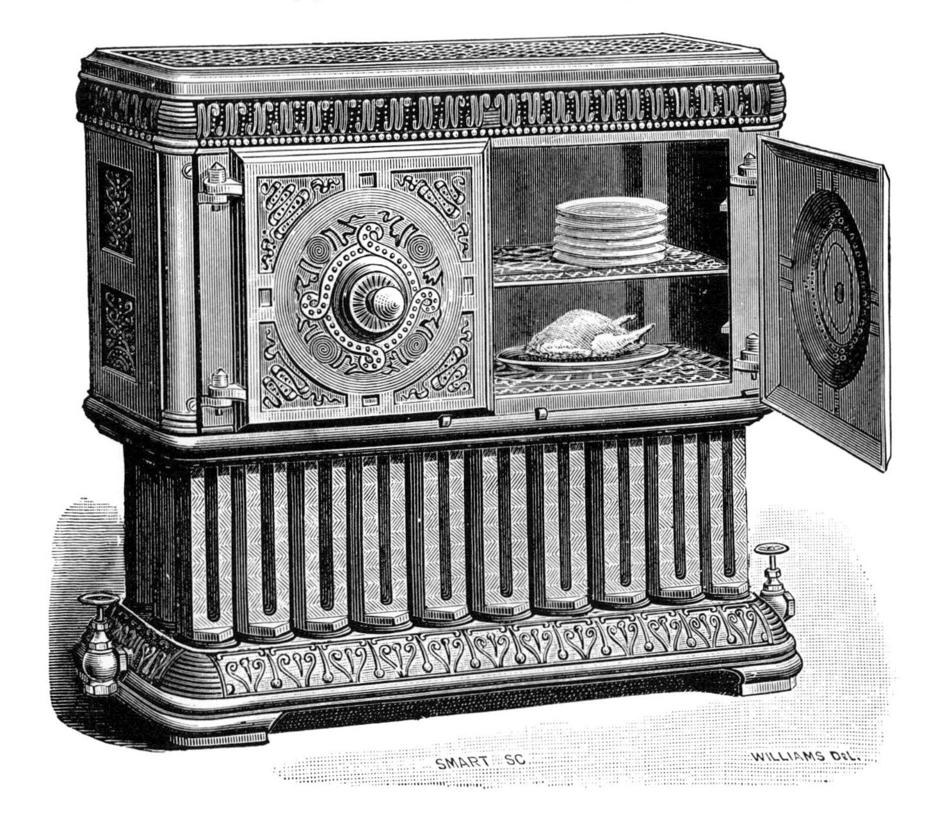

11.12 In the 1910s the Illinois Maintenance Company coordinated the steam plants of buildings in Chicago's Loop district. Underground lines were installed across streets and alleys to connect the boiler plants that were kept active (black) with nearby buildings (hachure) in which the operation of boilers had been discontinued. (S. M. Bushnell and F. B. Orr, *District Heating*, 1915.)

complexes began to be connected by tunnels that contained pipes carrying steam from a central plant, a system that had long been employed for industrial buildings. District heating, the commercial distribution of steam from a central plant, began when Birdsill Holly, inveterate inventor and owner of the Holly Manufacturing Company, moved to Lockport, New York, in 1877. Within the year Holly had run pipes from the boiler in the basement of his house on Chestnut Street to a residence about 500 feet away. Soon 4 miles of piping had been laid to serve houses, stores, and offices in Lockport. As Holly's system grew, the American District Steam Company was established.[31]

The New York Steam Company began service in 1882 and nine years later it was serving buildings, including New York's city hall, along 5½ miles of mains and return pipes from its downtown plant and 2⅓ miles from its uptown plant.[32] These plants had been erected for the sole purpose of providing steam for distribution. The former manager of the company stated that the only buildings not gaining from a steam supply company were large buildings located in areas where there were no legal restrictions on having steam boilers. Among such buildings, he said, many did not have sufficient basement area to install an adequate system.

One of the early followers in the move toward district heating was the Philadelphia Electric Company, which in 1887 ran a steam line to an adjacent building. In generating electricity great quantities of steam were exhausted and wasted. Since at that time electric stations were usually located at the fringe of a city's business district, they were near office buildings large enough to be profitable users of that exhaust steam. Philadelphia's system grew slowly. A line was extended to the company's offices about a block away, but it was 19 years before the Jefferson Hospital contracted to buy steam and two department stores joined the system.[33] Competition in providing electric service was keen. At one time there were 20 electric companies and four district steam systems operating in Philadelphia. The sale of waste steam was a profitable activity for electric companies, for replacing boilers in the basements of large buildings usually meant also supplying the electric power that had been generated by the building's own steam engine.

This was the nature of district heating until the start of the century when the Illinois Maintenance Company introduced a different service to owners of Chicago buildings. The company undertook the operation of existing plants in downtown buildings, connecting them with underground steam lines in order to close down the smaller and less efficient plants (fig. 11.12).[34] Soon similar companies commenced operation in Boston and New York, the latter adding equipment to manufacture ice during the summers. Seasonal differences in the use of steam were not so great as one would suppose. Records of a large Chicago office building indicate that during a cold day of winter the amount of steam used to generate electricity during office hours was almost equal to that required for heating.

District heating systems were established in France, Germany, and other north European locales, but Great Britain apparently did not have an installation until around 1911.[35] Where circumstances offered an economical source of heat, district heating was not limited to densely built metropolitan locations. Reykjavik, Iceland, employs natural hot springs. In Virginia, Minnesota, a town of about

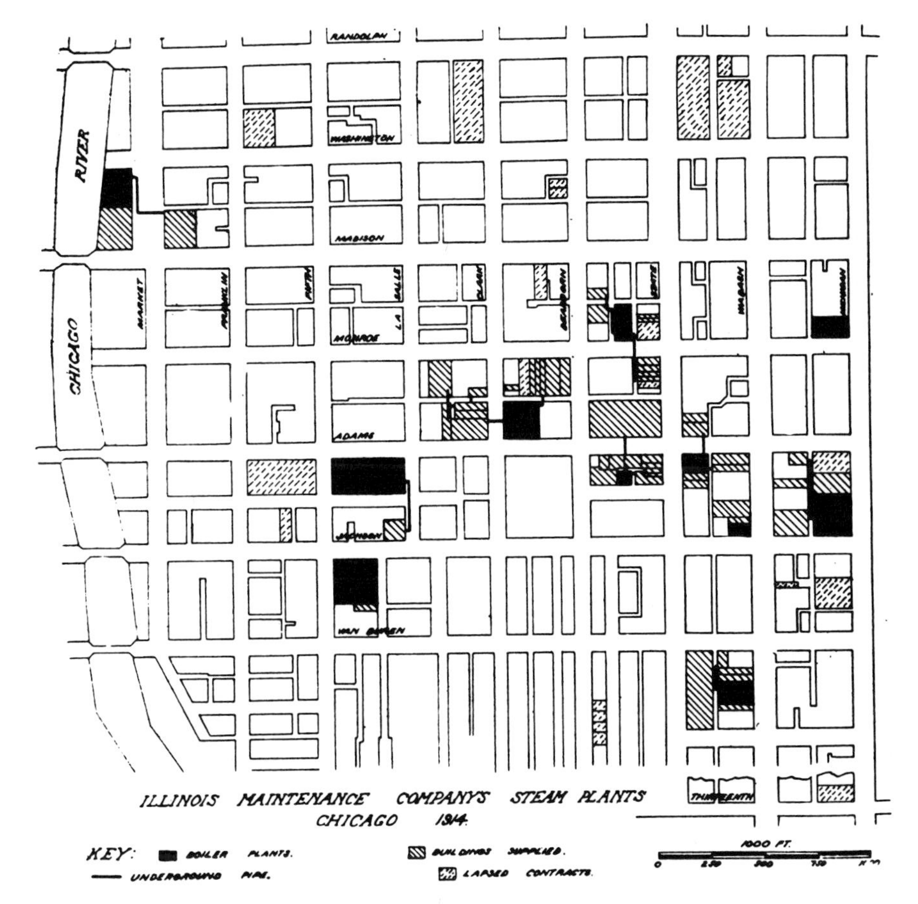

10,000 population with a climate similar to that of Moscow, the city system began in the nineteenth century with steam from a local sawmill and was augmented by steam from an electrical power company when that was established. Now the city has over 14 miles of mains operated by the city administration.

In addition to its legislative debates, the Houses of Parliament served for over 200 years as a battleground about the principles of heating and ventilation.[36] This single project illustrates the concerns and principles of heating and ventilation during that time, and it displays many features of technological endeavor when it is conducted in a public arena populated with strong-willed professionals and politicians. It all began in 1660, when Sir Christopher Wren attempted to relieve the heat from candles and MPs in the royal chapel, St. Stephen's, where the House of Commons had met for more than a century. In each corner of the room's ceiling, Wren cut a large square hole. In the room above he set over these openings decorative obelisks, around 7 feet high, through which the heated air could rise. Unfortunately, if the upper space were very cool a downdraft might reverse the intended movement of air. In 1723 this problem of ventilation was passed on to John Theophilus Desaguliers, a graduate of Oxford who lectured on science and presented frequent scientific papers before the Royal Society. His solution expanded on the work left by Wren. At each end of the upper room a small fireplace was built and connected to the two nearest obelisks by iron ducts near the floor. When fires were started

11.13 **The Edinburgh laboratory and lecture room of David Reid so impressed audiences that a committee engaged him to plan the heating and ventilation of the Houses of Parliament. Fires in the principal demonstration table were fed by air entering from above, and smoke traveled through ducts under the floor to reach the chimneys. (D. B. Reid, *Brief Outline Illustration of the Alterations in the House of Commons*, 1837.)**

in the fireplaces, their strong drafts drew air through the ducts and up from the House below. Although she was instructed to light the fires an hour before the House met, the action of this arrangement was defeated by the building's housekeeper, who resented the contraption's presence in rooms that were at the time assigned to her use. By waiting until a downdraft was well established, her lighting the fires added warm air from around the fireplaces to the mounting heat of the House. After Desaguliers invented a centrifugal fan for use in mines, one was installed above the chamber, and it was used there, turned by hand, for about 80 years. In 1791 the fan was moved to the center of the ceiling, and the chill of which MPs complained was corrected by installing a large stove beneath the floor of the House with warm air rising through a floor grating, 3 feet in diameter.

In the House of Lords there were complaints about cold and foul air. A solution was attempted in 1811 by Humphry Davy, the handsome and dynamic lecturer who had succeeded Rumford as leader of scientific studies at the Royal Institution. Davy provided fresh air through brick flues that were warmed by their running alongside chimneys. This fresh air entered the hall through a multitude of perforations in the floor. The peers' failure to compensate Davy for his efforts is recorded in a rhyme of the day:

For boring twenty thousand holes,
The lords gave nothing—damn their souls.[37]

Conditions in the House of Commons were improved several years later by the Marquis de Chabannes, who had used steam pipes to improve the Covent Garden Theater. Steam pipes were set beneath the members' seats, and hot air was drawn off at the ceiling by ducts that had coils of steam pipes at their bases.[38] The logic of this system was impeccable, but it proved to be difficult to adjust and control. Chabannes had been promised payment only if successful and, because the costs were half again his estimate, the Commons consented to pay him only a small fraction of the cost.[39]

On 16 October 1834 careless workmen used stoves beneath the House of Lords to burn two cartloads of refuse from the Exchequer Offices; the Houses of Parliament burned. Soon plans to rebuild were underway, and an architectural competition for the new building was won by Charles Barry, an ardent classicist, with the assistance of A. W. N. Pugin, a medievalist of remarkably narrow enthusiasms. This odd partnership was ordained by the competition's requiring a Gothic air about the design.

A few months before the fire, members of both houses had been among a group of visitors to the lecture room of David Boswell Reid during an Edinburgh meeting of the British Association for the Advancement of Science (fig. 11.13). When the University of Edinburgh failed to establish a chair of "practical chemistry," Reid had resigned his position there and set up his own program of lectures on chemistry and medicine. For purposes of demonstration, Reid's laboratory included a complex and efficient method of ventilation in which the downdrafts of various fires employed in experiments were led underground to flues elsewhere in the room. A year after the burning of the Houses of Parliament, a select committee concerned with plans for the heating and ventilation of the new building interviewed six experts on the subject, Reid among them. He was put in charge of installing a system for the quarters of the House of Commons. It was agreed that his work was to be controlled by the architect only in those aspects related to the style or soundness of the building.

Reid's scheme started with air intakes in two towers about 300 feet above ground level. Even this precaution was sometimes insufficient. Several brickyards operated in the vicinity, and through the years discomfort in the Houses was caused by taking in, at different heights, scents from passing bargeloads of manure, a nearby factory that made strawberry jam, and the floating scum of "bluebilly," an extremely malodorous byproduct of the manufacture of illuminating gas. From the towers, air moved to a basement space in which it was filtered and warmed. As in Davy's scheme some 20 years earlier, air entered the chamber through perforations in the floor, which was covered by a coarse carpet through which the air could pass (fig. 11.14). From ceiling openings, much larger in total area than those previously provided, ducts drew the air to a tower that served as exhaust for most of the systems within the building. Reid spoke of the system as a "pneumatic machine," but the power of the fans then available to bring air in and the suction of heated exhaust flues were not sufficient to operate his "machine" effectively.[40]

Relations between Barry and Reid deteriorated, each deeply resenting the authority given the other. Reid was certainly irascible and vague regarding his plans, but Barry had in 1836 predicted a construction period of six years and was only laying foundations when that period elapsed. An excuse for delay would have been useful, and the diary of Barry's son in 1846 speaks of his father's "arranging his weapons of attack toward Dr. Reid."[41] In 1846 the architect was officially assigned the task of heating and ventilating the House of Lords, while Reid continued his work on the House of Commons. The system installed by Barry took heated fresh air to the ceiling, and it entered the room at the sides of the ceiling, being extracted at the center of the ceiling. The public arguments between Barry and Reid

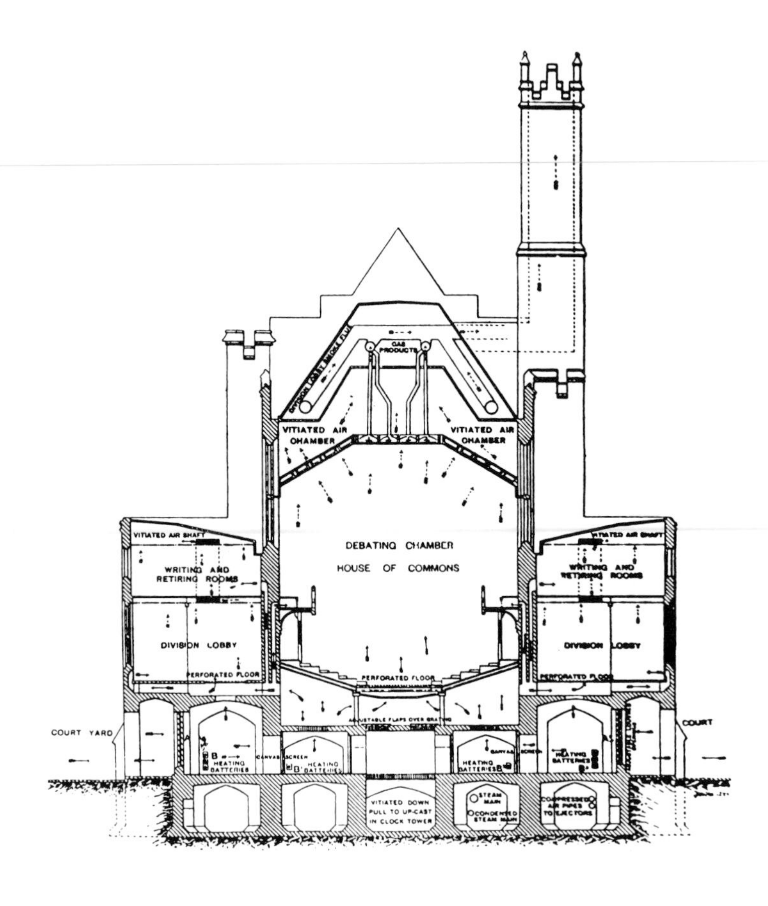

11.14 Until the end of 1891, the House of Commons kept many of the basic characteristics of Reid's "upcast" system of ventilation. Treated air rose through the perforated floor and left through the ceiling, taking with it the heat of the 64 gas burners that lit the chamber. (*Industries*, 4 November 1892.)

and the long investigatory hearings conducted by committees of Parliament serve principally to identify the volatile temperament of Reid. Each of these professionals found his supporters, usually along party lines, as indicated by an epigram printed in the *Times* of London (2 July 1845):

ON DR. REID'S BEING ALLOWED TO VENTILATE THE HOUSES OF PARLIAMENT BY ALTERNATE BLASTS OF HOT AND COLD AIR.

> Peel's patronage of Dr. Reid
> Is very natural indeed.
> For no one need be told.
> The worthy scientific man
> Is acting on the Premier's plan
> Of blowing hot and cold.

Faraday, who had been consulted by Barry, lectured at the Royal Institution in defense of the system used in the House of Lords, but there were still complaints about cold drafts down the peers' necks. Reid's lecture in reply was impressive, but his eloquence did not eliminate the dust that rose from his perforated floor. One angry MP proved this point by exhibiting a piece of gummy paper, grimy after having been fastened to a bench through an evening session.

By that time the architect and "ventilator" had fallen to accusing each other of being secretive and uncooperative. After dealing with an unmanageable tangle of supervision—committees, boards, and officials—Barry in the 1850s responded to questions about the date of completion and the cost by framing statements that were skillfully evasive. Benjamin Disraeli expressed one committee's frustration when he stated:

You hanged Admiral Byng, and the Navy increased in efficiency till we won Trafalgar. The disgrace of Whitelock was followed by the victory of Waterloo. We had decapitated Archbishop Laud and had thenceforward secured the responsibility of bishops. That principle we had never yet

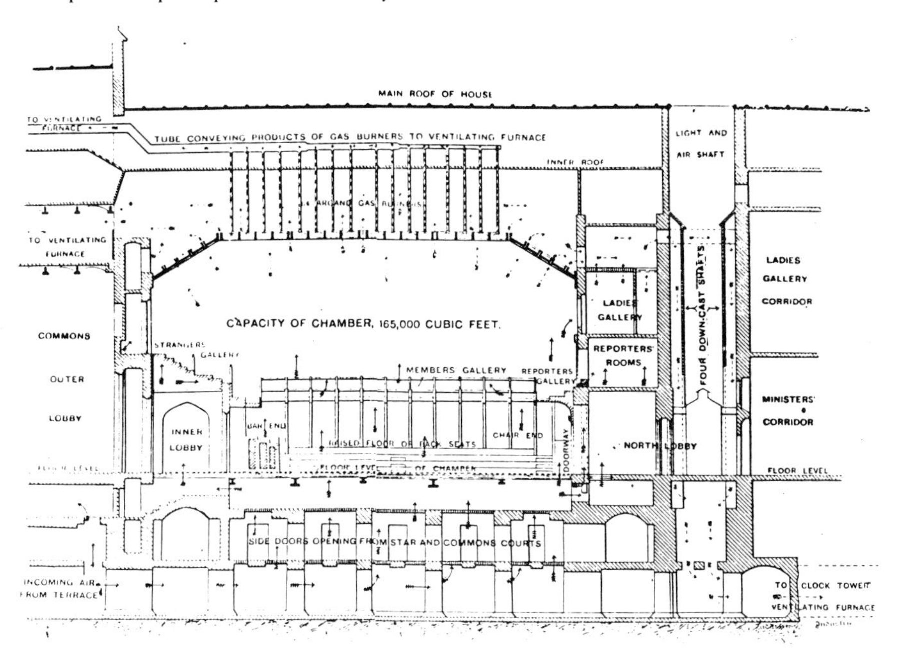

applied to architects; and when a member of that profession was called to execute a very simple task and utterly failed after a large expenditure of public money, it really became the Government to consider the case, and they might rest assured that if once they contemplated the possibility of hanging an architect, they would put a stop to such blunders in the future.[42]

But Disraeli suggested no method of forcing a legislative body to proceed in an orderly manner.

In 1852 Reid was discharged, and three years later he emigrated to the United States where, after a stay at the University of Wisconsin, he became an inspector of military hospitals during the Civil War.

After Charles Barry's engineer had made a few changes in the Houses, Goldsworthy Gurney was given the task of correcting air movement in the House of Lords. Gurney believed that Reid's system, which principally relied on air being drawn through rooms by the force of an exhaust system, lacked sufficient power, and he preferred to push fresh air into rooms. To this end he reversed the system, after a series of studies in which the movements of air were traced by smoke, produced from burning over 60 pounds of gunpowder and other materials. The two towers that Reid had used as air intakes were converted to exhaust flues, and air was drawn from ground level. This system proved an abject failure, for in addition to occasional specific sources of odor, the Thames itself was at that period little more than an open sewer. Tons of lime were scattered on the water during the hours that Parliament sat, but there was little effective relief until the city's drainage system was improved. Dr. John Percy, who succeeded Gurney, believed that fans could never successfully challenge the air-moving force of heated chimneys. He maintained the ground level intakes installed by his predecessor and made the sensible recommendation that the courtyard from which air was taken should be kept free of horse dung. Percy's principal changes were the addition of intakes on the side of the building toward the river and the installation of sprays of water and steam that could humidify, cool, or heat the air as it entered the system. Percy's revision of Gurney's scheme survived into the twentieth century. A great deal of extemporaneous adjustment was apparently required: "When for any reason less heat is required, the radiators are simply covered with cloths."[43]

There were periodic appointments of committees to investigate the ventilation systems of both Houses and search for improvement. The most curious, a bacterial study of the House of Commons in 1903, showed that, although the organisms floating in the room proved not to be associated with specific diseases, they also had little political discrimination, being most prevalent at the seats of the government's ministers and least present at the benches of the same party's MPs.[44] The House of Commons was destroyed during World War II, and when restored a modern air conditioning system was provided.

If David Boswell Reid fared poorly in his work on the Houses of Parliament, his design of the heating and ventilating system for St. George's Hall, Liverpool, was more successful.[45] The project began some years after Reid had begun his troubled collaboration with Charles Barry and ended in 1854, shortly before Reid emigrated to the United States. St. George's Hall was an imaginative agglomeration of Classicism by architect Harvey Lonsdale Elmes, and it solved the difficult problem of combining a large hall, two court rooms, a concert hall, and a smaller lecture room (figs. 11.15, 11.16). Collaboration began early, and the fact that Reid's system was viewed favorably over 50 years later indicates the cooperation of architect and engineer on the requirements of the design. The ventilation system was essentially what Reid had attempted in the House of Commons. Air entered at the bottom of rooms and was exhausted through the ceiling, taking with it the heat and odors of gas lighting.

11.15 In the original plans, St. George's Hall, Liverpool (left), was to exhaust air through a large ventilating tunnel leading to the tower of a Daily Courts building (background). When plans for the Daily Courts were abandoned, the ventilation scheme for St. George's Hall, which was already under construction, was changed to employ fans and vertical air shafts that would not be tall enough to harm the building's classical silhouette. (D. B. Reid, *Illustration of the Theory and Practice of Ventilation,* 1844.)

11.16 To ventilate St. George's Hall, Liverpool, four fans, each 10 feet in diameter, drew outside air into a basement passage, where it was heated by both steam coils and hot-water coils. All rooms, except the Central Hall, were exhausted through four shafts that extended to the uppermost parapet level. (*Civil Engineer and Architect's Journal,* 1 May 1864.)

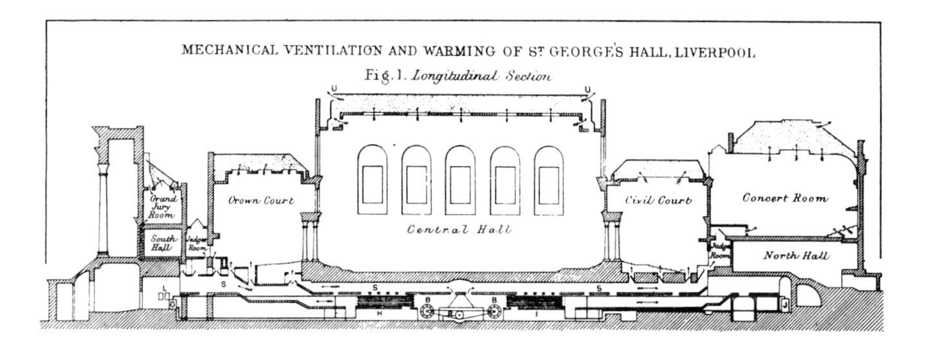

Fresh air was drawn into St. George's Hall down two large shafts. Once it had reached the basement levels, the air was heated by warm-water coils and supplementary steam coils or it was cooled by running water from city mains through the coils. Moisture could be added to the air by sprays of steam, and the shafts that admitted fresh air held water sprays that cleared particles from the air. Between the ceilings of rooms and the roof, horizontal flues separately gathered the exhaust air from major rooms and the smoke from the fireplaces that were used to ventilate small offices and meeting rooms in the building. The dictates of classical architecture did not permit towers so tall as those allowed by the somewhat medieval style of the Houses of Parliament, so the four vertical shafts that expelled smoke and foul air from St. George's Hall were fitted with metal louvers to prevent downdrafts. The ceiling of the central hall, 169 feet long by 74 feet wide, was so near the top of the towers that they could not effectively serve to evacuate air from that space. Louvers at the ends of the central hall's attic permitted air to be exhausted, the north or south louvers being opened according to the direction of the wind.

Thermometers were scattered about the system and read regularly. Strands of thread and small paper cylinders were hung at intake and delivery registers to provide evidence of the system's activity. Although St. George's Hall lacked most of the automatic controls that were later developed, only minor adjustments were made through the first half-century the building was used. This was the largest and most complete installation designed by David Boswell Reid, in contrast to the political muddle of the House of Commons. When St. George's Hall was completed, its system of heating and ventilation was overblown and excessively complicated; however, the size and complexity of the building made the scheme a remarkable exhibition of the scientific and practical knowledge of that time. In fact, during construction the building committee had found it necessary to dampen Reid's enthusiasm by reminding him that "the building was not intended as a mere shell in which to exhibit the principles of perfect ventilation."[46]

The eighteenth century has sometimes been called the Age of the Greenhouse, because it became fashionable for the wealthy in northern Europe to indulge themselves with the luxury of out-of-season fruits and flowers. Many greenhouses were heated by flues running along a back wall or through the floor as they made their way from a stove to a chimney. Not only was it difficult to maintain a draft for a horizontal distance long enough to be gen-

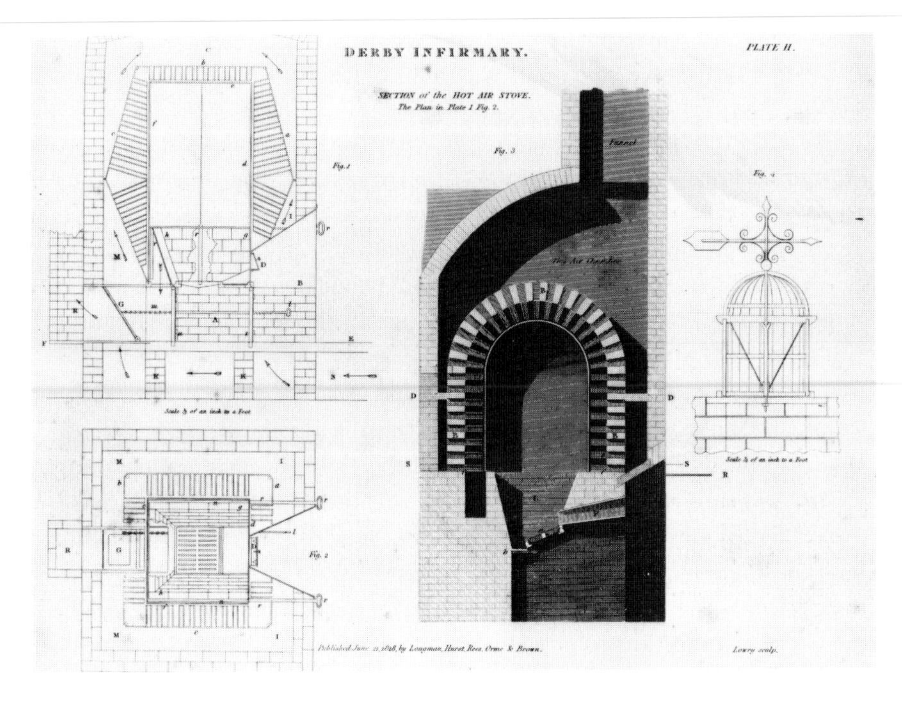

11.17 Devised by William Strutt and developed by Charles Sylvester, the cockle furnace was installed in 1806 in the Derby Infirmary. Around the iron jacket of the cockle a perforated shell of brick heated fresh air that rose from an underground passage. (C. Sylvester, *The Philosophy of Domestic Economy*, 1819.)

11.18 In planning Pentonville Prison (London, 1841–1842), Colonel Joshua Jebb, Surveyor-General of Prisons, saw a close connection between construction of a prison and the discipline under which it might operate. Although he insisted that there be no possibility of sound traveling between cells, he also insisted on healthful conditions. It was reported that each of the cells received between two and three air changes per hour. (J. Jebb, *Report on Pentonville Prison*, 1844.)

uinely useful in heating a greenhouse, but any leak in a flue might release coal gas that could rapidly damage the plants. A method of flue heating for buildings was found in what was called the "English smoke pipe system." Its flue and heating surface was a cast-iron pipe, extending upward through floors of a building and entering the chimney at the attic level. Although the danger of fire was inherent in the smoke pipe system, it seems to have been briefly popular around the 1820s, even being used with horizontal runs on the different floors of a textile factory.

A safer and more reliable method of heating provided two separated air flows, one to feed the flames and carry away smoke, the other to heat air and distribute it to the rooms of a building. (This problem is little more than a large-scale restatement of the various early systems in which iron fireboxes were set in fireplaces.) The simplest method of doing this was to locate a stove in a small chamber from which warm air could rise into the room above, where it was desired. The thirteenth-century city hall in the German city of Lüneburg had three stoves in a basement room below its principal meeting hall. Fresh air entered the basement and heated air rose under the councilmen's seats, where iron lids permitted each of the officials to regulate his own comfort.[47] In one German castle, heat from a tile stove rose through a perforated floor. None of these early warm-air heating systems did much more than relegate the stove and the bother of caring for fires to a space adjacent to the room that was to be heated and, in many cases, provide for the stove to be tended by servants without bothering their masters.

English textile mills in much of the eighteenth century were heated by open fires placed about each floor of the building. Profiting from trials made by others, William Strutt in 1792 developed a safer heating system for Belper Mills. Many mills had changed to the use of cockles, a form of closed stove, but cockles often overheated, which was dangerous in the highly flammable environment of a mill.[48] In Strutt's design for the Belper Mills, the cockle stove was surrounded by a brick enclosure into which fresh air was admitted (fig. 11.17). Another brick shell, arching only a short distance from the iron surface of the cockle, had a pattern of perforations in its brickwork. This jacket increased the heating surface and so assured a higher temperature in the air that rose to the upper floors of the factory. By 1806, when Strutt planned the heating system for the Derby Infirmary, his design had been refined. Here the brick covering of the cockle had greater mass, and less than an inch of space was left between the iron and brick surfaces. Fresh air was admitted through a slit surrounding the bottom of the brick shell. The air was drawn inward to the surface of the cockle through openings in the brick, moved upward along the iron surface, and once more passed through openings in the brickwork to reach the ducts that carried air to the rooms. Strutt's respected position among mill owners and his many scientific acquaintances assured that his ideas were applied in many buildings executed by others. Strutt's warm-air heating system was further developed by his friend Charles Sylvester, who had been tutor of Strutt's son. In his book *The Philosophy of Domestic Economy*, Sylvester explained the Strutt system. As a consultant on building projects he developed applications, and as a manufacturer of stoves he made the system available generally. By 1817 Sylvester had found it difficult to make certain that the brick shell was built correctly around the cockle without providing close personal supervision of the work. As a substitute for such intricate brickwork, he devised an iron jacket perforated by tubes that led to points near the surface of the cockle. Little supervision was required when workmen assembled the simple parts that came from the factory.

One of the first systems to confront the problem of providing warm air to a large number of rooms from a single source was that of the model prison at Pentonville, built in the 1840s under the supervision of Joshua Jebb, England's Surveyor-General of Prisons.[49] Fresh air was heated in a basement furnace room and left there through a passage built beneath the ground-level corridors (fig. 11.18). Individual ducts rose within the walls to release air above the door of each cell in the three stories of the building. In the outer wall of each cell an opening near

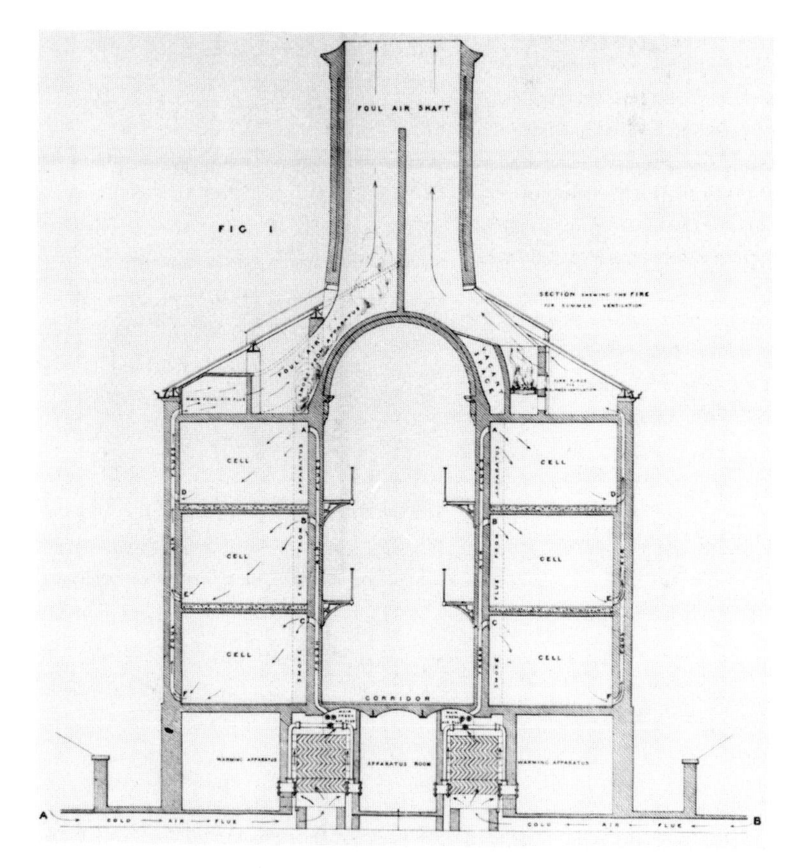

the floor let air escape into a vertical channel that led to a large horizontal duct in the attic, which in turn emptied into a vertical shaft that extended 20 to 25 feet above the roof.[50] In winter the flue from the furnace rose inside this exhaust shaft and guaranteed a strong draft; in the summer a fire was maintained for that purpose. This system was long an example to be followed in the construction of new prisons, and one visitor is said to have thought its air as fresh and invigorating as that in the Crystal Palace.[51]

At the end of the nineteenth century, warm-air heating was generally limited to residences or to buildings not much larger than residences and to assembly places where the crowds caused ventilation and cooling to be a major concern. For the smaller buildings, warm-air heating was little more than that provided for the burghers of Lüneburg five centuries before. In a basement the cast-iron furnace was surrounded by brick walls or a sheet-metal jacket (fig. 11.19). Furnace gases went directly to a chimney and fresh or recirculating air passed between the furnace and its surround before rising through chases or sheet-metal ducts to the rooms above. In some designs the surfaces of the furnace were shaped in ribs or corrugations to increase the amount of surface radiating heat. As the air cooled in rooms it usually returned to the basement through grilles or down a central stairwell (fig. 11.20). In larger installations air was drawn through racks of pipes containing steam or hot water and distributed in a similar manner.

Steam heating was more popular for large buildings, particularly if their purpose required division into small spaces. Steam radiators placed under each window and provided with individual controls allowed for any replanning of the building that might be executed in the future. As a writer reported in the *North American Review* in the 1880s: "Even for ordinary dwellings, [steam heating] is being gradually extended; while for larger blocks, such as apartment houses, hotels, manufacturing establishments, school-houses, and other large structures, the system is so greatly in demand as to have given rise to new and extensive industries devoted to the manufacture and installment of the necessary apparatus."[52] A system of radiant heating by coils of hot-water pipes distributed in concrete floor slabs and plastered walls was used in the United States during the 1940s and 1950s. It provided well-distributed heat, but was slow to respond to sudden needs.

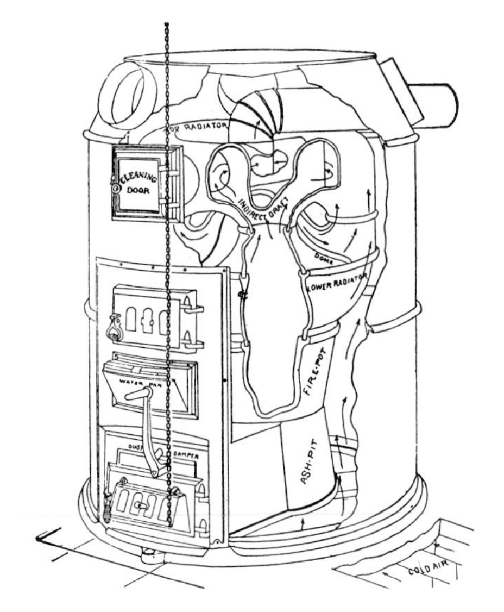

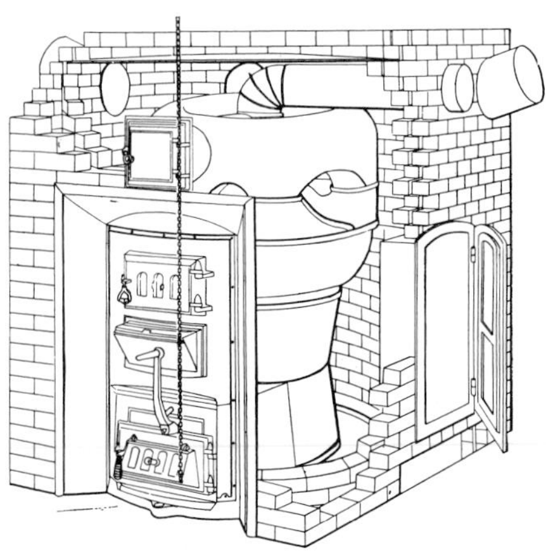

11.19 Residential hot-air heating systems at the start of the twentieth century commonly had round furnaces, little more than cast-iron stoves, as their central element. The surrounds were of two sorts: a "portable setting" (left) that consisted of a casing made of one or two layers of sheet metal, or a "brick setting" (right) that was a small chamber with masonry walls. (*Engineering Magazine*, June 1898.)

11.20 This diagram illustrates a residential heating system of the 1880s in which ducts were provided for both the supply of warm air and its return to the furnace. The advertiser pointed out that "in a majority of the houses already built, boxes in their basements can be constructed." An inset at the top shows a register that can be adjusted to either return air to the furnace or admit outside air. (*Useful Information Pertaining to the True Systems of Heating and Ventilation of Buildings*, 1885.)

Steam heating and hot-water heating were subject to all the long-standing complaints about stoves and the parched air they produced. Even when radiators were concealed in the most elaborate paneling and grillwork, it was common to find atop them a shallow pan of water that was intended to increase humidity. Steam and hot water were the most efficient methods of moving heat for long distances, for when hot-air ducts became too long their efficiency diminished and the difficulty of equalizing the flow of air increased. In addition, the acceptance of air conditioning required delivery by air, because of the condensation that would form on radiators filled with chilled water. Warm-air systems afforded the best opportunity to provide ventilation, control humidity, and in the summer utilize the same installation for cooling.

The automatic regulation of heating systems developed as a result of the discovery that materials expand or

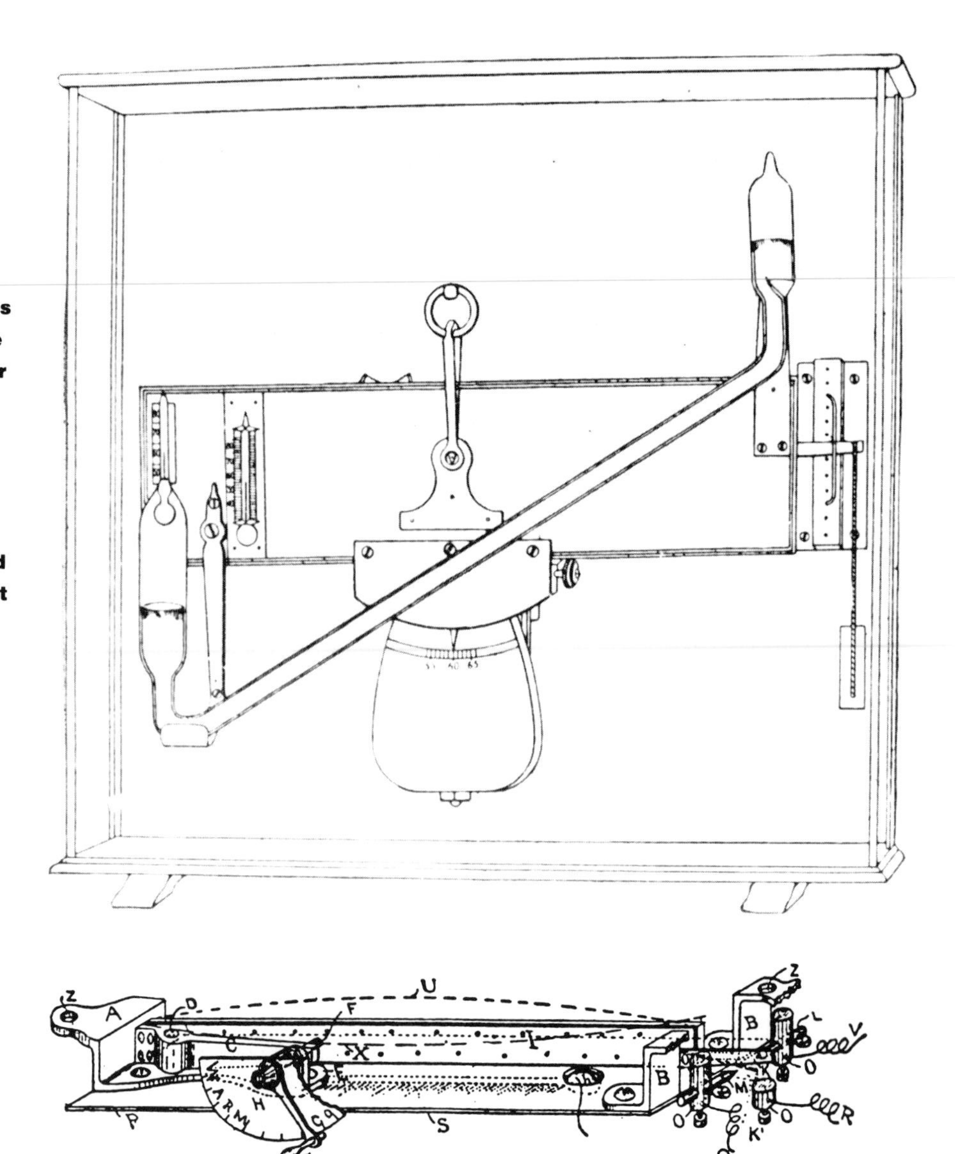

11.21 Thermostatic controls were available by the 1890s, although often rather complex in nature. Above: Mercury was poured into a balanced glass tube until bulbs at the ends were half-filled; the lower bulb held ether over the mercury. Cold caused the ether to contract rapidly, drawing more mercury into its bulb; the tube then tilted and thus activated the furnace controls. Desired temperatures could be set by adjusting the fulcrum on which the tube balanced. Below: A bar was fabricated by riveting together strips of brass and hard rubber. Because of the two materials' unequal expansion, the bar would bend with temperature changes, making electrical contact with the heating controls. (J. S. Billings, *Ventilation and Heating*, 1893.)

contract when their temperatures rise or fall. This knowledge was utilized in John Harrison's development of compensators that would adjust his marine clock for changes of temperature. His first device, made in 1726, relied on the lengthening of metal rods, the principle used some 50 years later to regulate the fire in Bonnemain's incubator. Harrison's second device was based on a bimetallic strip, two metals with very different coefficients of expansion bonded together so that heat would cause the strip to bend to one side. Whether they involved the use of floats on expanding liquid, the expansion of rods, or bimetallic strips, such devices converted the thermal expansion of materials into mechanical movements that regulated valves or dampers to control the flow of air to the fire or the flow of heated air or water from the furnace. Nevertheless, during sessions of Parliament at the end of the nineteenth century, attendants trudged about the Houses carrying thermometers and inspecting others that were fastened on the walls.

The need to control heat or refrigeration in industrial processes spurred the development of thermostats whose principles were often applicable to the heating systems of buildings. In addition to the principles employed in previous thermostats, in the 1910s devices were introduced that

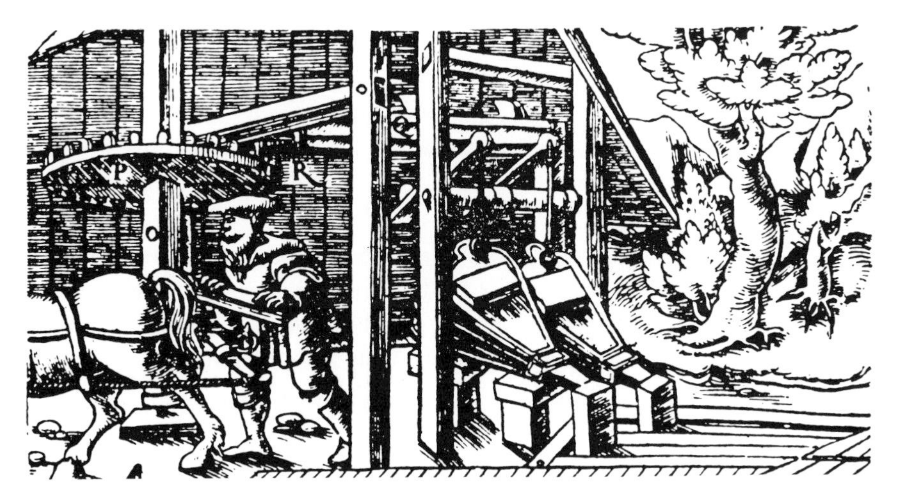

depended on the expansion of certain fluids in metal bellows, thin-walled containers that were corrugated around their circumferences.[53] The regulation of furnaces by thermostats located in rooms of buildings was made possible by electrical connections of later types (fig. 11.21).

The required capacity of heating systems could be determined with reasonable accuracy after the publication of handbooks, such as that of Thomas Box (1876), which included formulas for heat loss in buildings and factors for estimating the transmission of heat through walls of masonry and other materials.[54]

Any regulation of heating systems required an effective method by which air could be supplied to and removed from the spaces that were to be heated. In *De re metallica*, the sixteenth-century German metallurgist Agricola had referred to the use of fans and other mechanical means of drawing air and gases from mine shafts (fig. 11.22). Nevertheless, such machines were seldom used until the start of the nineteenth century when air pumps with strokes as long as 8 feet were introduced.[55] It was the 1830s before fans were commonly used in mining.

When Desaguliers was busy finding a way to ventilate the House of Commons, he introduced a fan in which air was admitted at the hub of a paddle wheel 7 feet in diameter and was driven out in a tangential direction (fig. 11.23). As they were turned by hand, the 12 blades of the fan forced air out of the fan's jacket. The Admiralty requested a demonstration of Desaguliers's machine, but the Sur-

11.22 The earliest methods of forced ventilation were often designed for the protection of miners. This system of horse-powered bellows was shown in a 1555 edition of Agricola's *De re metallica*. (*Refrigerating Engineering*, March 1934.)

11.23 After devising a machine to draw foul air from the mines operated by the Earl of Westmoreland, Desaguliers developed this "fanner" in 1734 for use in hospitals and prisons. The wheel was 7 feet in diameter and 1 foot wide. Air entered near the axle, and cranking by hand forced air through the outlet. (W. Bernan, *On the History and Art of Warming and Ventilating Rooms and Buildings*, 1845.)

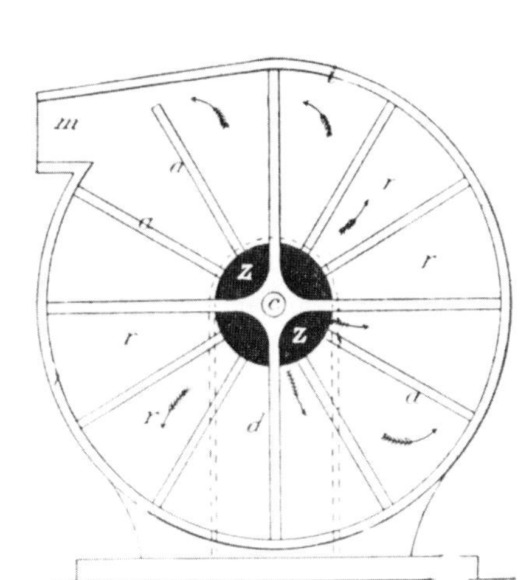

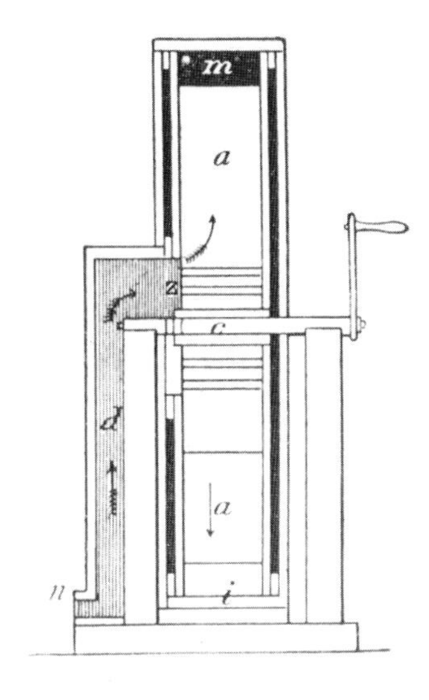

veyor of the Navy insisted that sailors first demonstrate the efficacy of "wind sails," a traditional method of forcing air below decks, and he left before witnessing the performance of Desaguliers's invention.[56] Few significant advances in the construction of fans were made in the century that followed the work of Desaguliers. The major improvement was to change the casing of the fan from circular to spiral, gradually enlarging the space between the blades and casing as air neared the outlet. Curved blades were experimented with, and formulas were developed for the relationship of fan size and the amount of air that could be moved. In the 1880s builders of an English tunnel made use of a fan 18 feet in diameter.[57]

By the 1870s, fans, which had been seldom used outside factories, began to be installed in schools and auditoriums. Electric motors contributed much to expanding the use of fans, but the heated chimney remained a common method for moving air. When members of the American Society of Mechanical Engineers, meeting in 1897, discussed the relative efficiency of heated chimneys and fans for exhausting air, a speaker presented data that proved a fan to be at least three times as efficient as a chimney; but one listener insisted that fans were "poor devices" and that chimneys were almost 14 times as efficient.[58]

For many years *Principles of Warming and Ventilating Public Buildings*, published in 1824, was the English-language handbook for problems of ventilation, and its stipulations were frequently borrowed by other writers on the subject. The book's author, Thomas Tredgold, had studied under an architect before setting up as an engineer and a prolific writer of books on the practical aspects of building construction. According to one of the rules he presented, every person in a room should be provided with 4 cubic feet of fresh air every minute. Others thought Tredgold's recommendations for ventilation excessive. Arnott, who favored supplying 2–3 cubic feet per person per minute, claimed: "There are, in England, many persons who, under all circumstances, call out for open fires and open windows, and by the cold currents and other concomitants of a ventilation more than necessary, prodigiously waste fuel and injure or kill their children and friends by catarrh, rheumatism, etc."[59] A few decades later, a dissent from the United States was to be found in *The American Woman's Home*, by Catherine E. Beecher and Harriet Beecher Stowe: "The great majority of the American people, owing to sheer ignorance, are, for want of pure air, being poisoned and starved; the results being weakened constitutions, frequent disease, and shortened life."[60] In spite of the fervor of this statement, it was accompanied by the requirement of one hogshead of fresh air per person per hour—only about 8½ cubic feet. Much of the debate on ventilation and temperature requirements appears to have related to personal and, even more, national perspectives. Around 1840 an Irishman in the United States wrote: "Casual visitors are nearly suffocated, and constant occupiers killed. An enormous furnace in the cellar sends up, day and night, streams of hot air. . . . It meets you the moment the street-door is opened to let you in, rushes after you when you emerge again, half-stewed and parboiled.[61]

In the 1770s Antoine Laurent Lavoisier, the founder of modern chemistry, disproved the phlogiston theory and showed that air consisted of two gases, one combustible and the other not. His experiments in combustion and oxidation were extended to include the behavior of animals in air, oxygen, and nitrogen. It was Lavoisier's conclusion that harmful atmospheres resulted from the increase of carbon dioxide, and that this inhibited the body's absorption of oxygen. Thus the problem of ventilation was not the depletion of oxygen in a room but the increase of carbon dioxide. The theory was incorrect, but it persisted for a century. A later experiment showed that an animal could recover from 45 minutes in an atmosphere that was 30 percent carbon dioxide. At the same time, it was found that the increased amount of carbon dioxide that occurred in crowded rooms was insignificant. Max Joseph von Pettenkofer, a professor of medicine at the University of Munich, in 1863 found that an atmosphere having 1 percent carbon dioxide could be breathed for long periods without ill effects. (Normally, air contains about 0.03 percent carbon dioxide, and one expert in 1928 said that "the worst ventilated room" reached only 0.5 percent carbon dioxide.)[62] This meant that even the worst conditions of crowded rooms could not be harmful, at least not because of the presence of carbon dioxide.[63] However, Pettenkofer, who did not subscribe to the germ theory of disease, was convinced that the amount of injurious organic impurities breathed into the air were indicated by the amount of carbon dioxide present. This relationship encouraged the continuation of the popular belief that the gas itself was harmful.

A remarkable connection was made in the middle of the nineteenth century when the theory of poisonous "organic effluvia" led to a scientific association of gaseous characteristics with specific ailments (fig. 11.24). One writer stated this chemical extension of the miasmic theory of disease thus: "On the whole, it would appear that sulpheret of ammonia is the morbific agent exciting typhus fever, sulpheretted hydrogen being the pestilential virus [meaning "causative agent"] producing yellow fever and the bilious remittents and agues of tropical climates."[64] Another scientist was cited as having proved the relationship between hydrogen and malaria. It must be remembered that at this time Pasteur was engrossed in the studies that would soon result in germs being identified as the cause of disease. Until Pasteur's findings had been generally accepted around the end of the century, it was commonly assumed that diseases were spread by vapors from infected or putrid materials. In that context it is not surprising that any discussion of ventilation might consider a specific relationship between different gases and illnesses.

11.24 With the advance of public education, schools became a prime interest of those concerned about ventilation standards. Here ghostly flying monsters, bearing the names of specific diseases, flit over the heads of drowsy students. (Bettmann Archive.)

The notion of organic poisons was advanced by the experiments of Charles Brown-Sequard and Jacques d'Arsonval in 1887. In one study, air entered the first of four boxes containing healthy rabbits, and air from that box supplied the next, and so on until the last box was provided only with air that had passed through all three others. It was reported that the fourth rabbit died first, the third died next, and the first two survived. When the air entering the fourth box was passed through sulphuric acid, the fourth rabbit survived, but the third died.[65] Findings of this sort greatly strengthened the belief in there being organic poisons exhaled into the air. As late as 1911 an experimental group reported animals made ill by injections of material from the breath of other animals, but subsequent experiments did not corroborate those results.

Carbon dioxide was branded the culprit, either as a poisonous gas or a vehicle of injurious substances, until the 1880s when a Dutch experimenter theorized that the effects of poorly ventilated rooms were more a matter of physics than of chemistry, that they resulted from excessive levels of temperature and humidity in the air. As scientists at the University of Pennsylvania wrote in 1895:

> The discomfort produced by crowded, ill-ventilated rooms in persons not accustomed to them is not due to the excess of carbonic acid [carbon dioxide], nor to bacteria, nor, in most cases to dusts of any kind. The two great causes of such discomfort, though not the only ones, are excessive temperatures and unpleasant odors. Such rooms as those referred to are generally overheated, the bodies of the occupants, and, at night, the usual means of illumination contributing to this result.[66]

This view was supported by extensive studies made in Germany controlling airflow, temperature, and humidity for human subjects closed in cabinets of little more than 100 cubic feet.

Through the nineteenth century there had been a constant increase in the amount of fresh air recommended. After Tredgold's figure of 4 cubic feet per minute per person came David Boswell Reid's 10 cubic feet (1835); by 1857 the English Barracks and Hospital Improvement Commission raised the figure to 20 cubic feet. By this time, requirements were often subtly varied according to the age, sex, and activity of rooms' occupants, and some authorities recommended as much as 30 cubic feet of fresh air per minute for each person. In the two decades that followed publication of the German experimental findings, fresh air requirements declined even more rapidly than they had climbed. By 1925 the consensus was that 10 cubic feet of outdoor air per person per minute should be supplied, the level forwarded by Reid almost a century before. Public health studies during the same period pointed out that no harm was done by recirculating air and that requirements of precise quantities of air supply were useless without consideration of the quality of that air. This resulted in the recommendation of percentages of air supply that might be provided through recirculation. Often as much as two-thirds of the required supply were allowed to be recirculated under what came to be know as the "effective air supply concept."[67]

Ventilation standards during the early part of the twentieth century accepted the germ theory of disease, and illness was no longer associated with atmospheres, vapors, gases, or miasmas. If contagion resulted from the transfer of microorganisms, that transfer was possible in almost any gaseous mixture. With disease prevention eliminated as a major objective of most room ventilation systems, the regulation of temperature and moisture remained. For this purpose the presence of a specific number of occupants required that a room be supplied with a flow of air that remained constant in quantity, but could be varied in temperature according to fluctuations of temperature inside and outside the building. Sturgis's *Dictionary of Architecture and Building* commented in 1901 that "the only method in vogue previous to about twenty years ago was to close the register when the room was sufficiently warm; which of course not only cut off the heat supply, but the air supply as well, interrupting ventilation."[68] In order to correct this condition, the "mixing valve" was developed, with provisions for varying the temperature of air entering a room by mixing fresh cold air, outside air that had been heated, and recirculated warm air.

Research done in the 1920s and 1930s showed that temperature control was related to the esthetic consideration of providing an odor-free environment. Thereafter, in spite of the difficulty of determining levels of odor and the extremely subjective factors involved, the provision of an odor-free atmosphere was accepted as a central goal in the design of ventilation systems. Instead of carbon dioxide being employed as an indirect indicator of healthful air, the smell of air was taken as a valid measure of the physical comfort and social conformity of conditions within a room.

In recent times there has been a tendency to return to some of the precautions that were embodied in earlier standards for ventilation. Carbon dioxide and hydrogen are no longer considered threatening in normal circumstances, but radon, the gases from urea-formaldehyde, and other substances appear to be extremely dangerous in relatively low concentrations. "Organic effluvia" may not be feared today, but modern ventilation systems have on occasion scattered dangerous microorganisms and gases have caused illnesses.

The specialized needs of some types of buildings assisted in the development of attitudes and devices that later found broader application. Because of their function, hospitals were a fertile area for experimentation. Louis XVI in 1786 instructed his architect to build new wards at Salpêtrière, Paris's major asylum for the insane, "so that the unfortunate women shall no longer be exposed to injury from the air." In the design that resulted, fresh air entered the door or window of each cell, rose through an opening in the ceiling, and was expelled through vented cupolas on the roof. This scheme was long used for hospital wards. By 1890 the type of ward design strongly recommended in France had beds arranged on both sides of a long, tall room with ventilators drawing out air at the top of an ogival ceiling (fig. 11.25).[69] Individual air intakes at floor level were assumed to surround each bed with an upward stream of ventilation, separated from the air moving around adjacent beds.

Florence Nightingale, adulated for her work in the Crimean War, favored wards of about 30 beds, a number that could be well cared for by a single head nurse. This led to a room 30 feet wide (it was believed that greater width between windows caused stagnation of air in the center of the room) and about 120 long, which avoided corners in which air did not move freely. In order that no air should linger overhead, the tops of windows

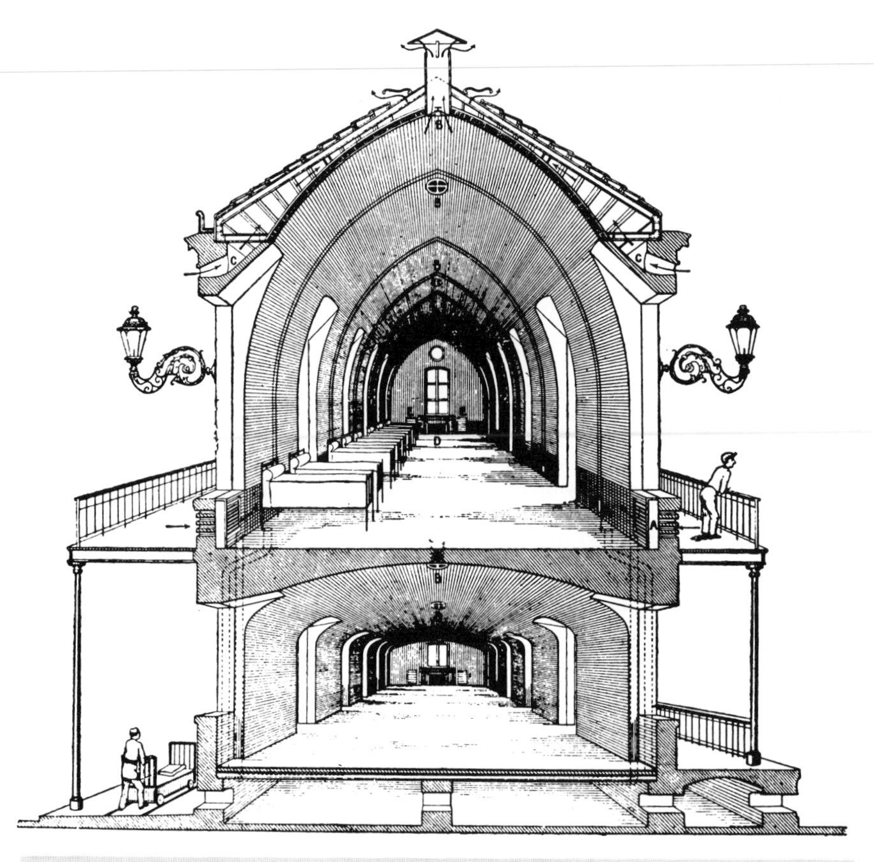

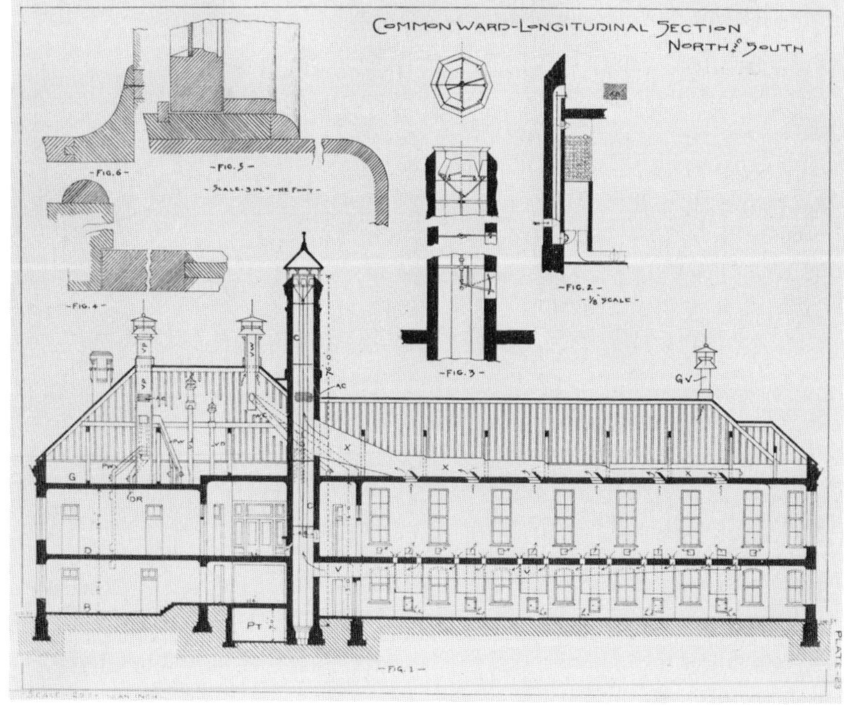

11.25 Before general acceptance of the germ theory of disease, hospital wards were designed to provide the best ventilation possible. In the Civil and Military Hospital, Montpellier, France (1884), the ground floor was a recreational area, and flues conducted air from beneath its floor to grilles beside the beds of the ward. Air entering at the eaves cooled the ceiling during summer weather. (C. Tollet, *Les édifices hospitaliers depuis leur origine jusqu'à nos jours*, 1892.)

11.26 In spite of the height of buildings at Johns Hopkins Hospital, many had only a single floor level for housing patients. The ground floor admitted air, which rose through grilles placed in the floor between the beds of the patients. Above the beds, air was led through the attic to a tall tower. (J. S. Billings, *Description of Johns Hopkins Hospital*, 1890.)

were to be near the ceiling, which was 16 to 17 feet high. Nightingale preferred heating and ventilating the ward by a fireplace in the center of the room. Her recommendations insisted on a plentiful supply and constant movement of air; however, she did not envision any method by which the circulation of air could be increased beyond that of an airy room.

In the United States, hospitals built for the Civil War had shown the value of a ward built rather like barracks, following most of the principles propounded by Nightingale, although often having twice as many beds. When the Baltimore hospital was founded by Johns Hopkins in the 1870s, the trustees asked for the recommendations of five experts. Of these, Dr. John S. Billings, assistant surgeon in the U.S. Army, was chosen to advise in construction of the hospital, for which J. R. Niernsee was architect.[70] The plan for Johns Hopkins Hospital consisted of a series of pavilions. The ground level of each pavilion was entirely devoted to ventilating and heating, the latter employing hot water (fig. 11.26). Air was delivered to the ward above through floor-level registers in each space between beds. In winter the air was removed through grilles in the ceiling. All foul air was led through ducts to an "aspirating chimney" that rose 76 feet above the fire that maintained its draft. In general, the effectiveness of such systems may well have been exaggerated. In the 1860s there were 220 tests made in the Munich Lying-in Hospital, where large chimneys were intended to exhaust air, and it was found that 58 percent of the time air moved in the desired direction, 25 percent of the time it was not in motion, and 17 percent of the time it went the opposite way.[71]

The New York Hospital, designed by George B. Post in the same period, was a five-story building as required by its urban location (figs. 11.27, 11.28). Its air movement was almost completely controlled by fans. Fresh air was admitted by grilles in the window sills, and air was exhausted through grilles high on the walls. Restricted sites required taller hospitals, as new institutions were built and old ones expanded. No longer could the ventilation of hospitals rooms be planned as simply as in one-story hospitals, where fresh air entered through the floor and rose through the roof. Channels, chases, and ducts were required, and the powerful force of fans was needed to draw air through them.

Like multistory hospitals, office buildings demanded special considerations. In early skyscrapers there were efforts to supply warm air through channels in the walls, but this method of heating proved impractical. The air had to travel so far that it was almost impossible to provide the required temperatures in every office; sound and fire could travel from floor to floor through the ducts; and the floor area occupied by such a heating system was unacceptable in buildings that demanded the most exacting utilization of their sites.[72] In many cases fireplaces were built in offices solely as outlets for air, their flues inadequate for actual fires. Hot water heating was the choice of most engineers, but steam was also used, either from independent boilers or as exhaust steam from boilers in the basement that operated engines, elevators, electric generators, and pumps. In New York the heating pipes of skyscrapers were usually concealed, but Chicago architects tended to expose them in offices and hide them only in the lower floors of very tall buildings,

11.27 The extension of a street required that the New York Hospital be rebuilt, providing 200 beds on a very limited site. More typically, hospital construction had involved large sites at the fringe of cities, which permitted their having only one or two stories of nursing rooms. George B. Post's design (1875) established a pattern for the ventilation of hospitals that were a part of the increasing urban density. (*American Architect and Architecture*, 17 March 1877.)

11.28 Ventilation problems of multistory hospitals in urban locales were solved through the introduction of mechanical ventilation. Fans in the New York Hospital brought in fresh air (*A*) and exhausted air from the attic (*N*). (C. Tollet, *Les hôpitaux modernes au XIXe siècle*, 1894.)

where supply mains were too large to be left visible and appearance was of greater concern.

The workings of a skyscraper's heating system were complex because of its dimensions, even if the basic design was relatively simple. The Equitable Life Building in New York, built just before World War I, contained 5,000 radiators. Its return riser was a mammoth tube 34 inches in diameter, which lengthened 8 inches when steam started flowing and the metal expanded.[73] Because the entire volume of a heated skyscraper served as a chimney with strong updrafts, air leaking into lower floors was a major heating problem. The installation of revolving doors lessened this problem and kept litter from being drawn into lobbies whenever doors were opened.

Sealed, well-insulated office buildings are today filled with equipment that provides high levels of illumination, computers generating heat, an assortment of business machines, and a host of employees. Heating has consequently become a different problem, always closely related to cooling.

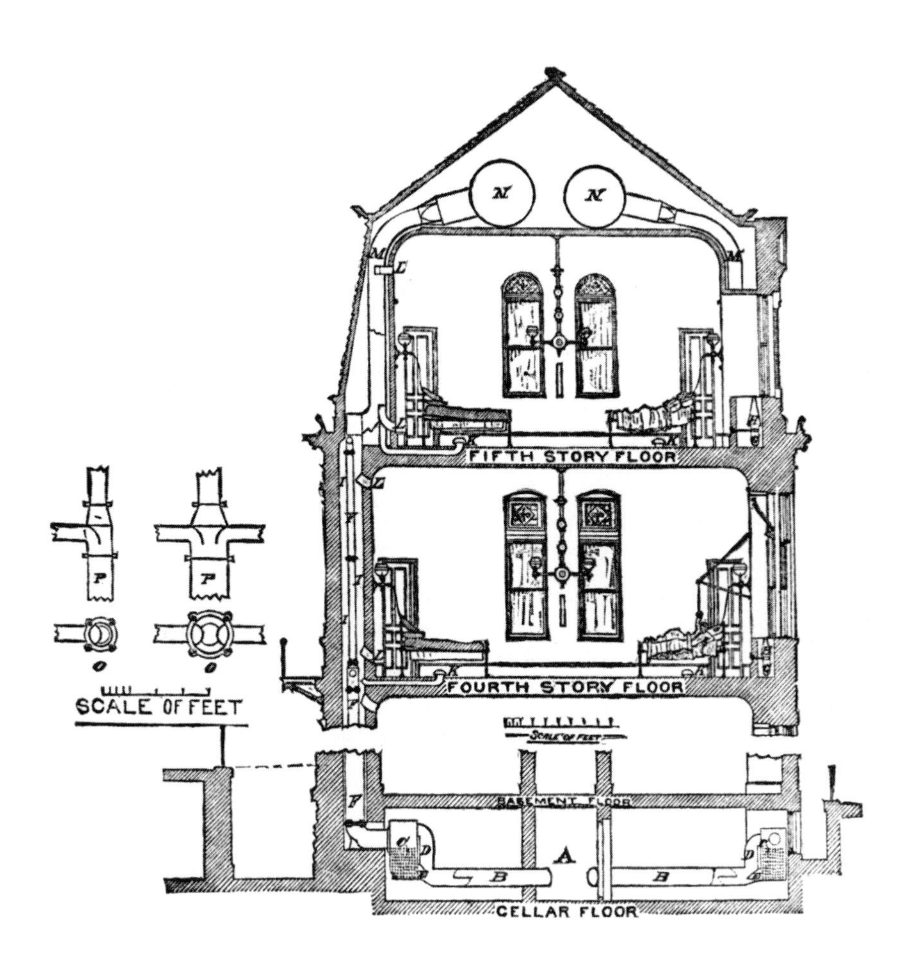
N
N
FIFTH STORY FLOOR
FOURTH STORY FLOOR
SCALE OF FEET
SCALE OF FEET
BASEMENT FLOOR
A
B
B
CELLAR FLOOR
P
P
O
O

12 Air Conditioning

1834 Ice-making machine patented (Britain) by Jacob Perkins

1851 Dr. John Gorrie patents (U.S.) an air cycle compression ice-making machine

1889 Carnegie Hall, New York, provided with an air-cooling system using blocks of ice

1904 New York Stock Exchange installs a cooling system using ammonia as refrigerant

1906 Willis H. Carrier patents (U.S.) "Apparatus for Treating Air"

1928 Milam Building, San Antonio, Texas, a milestone in air-conditioning office buildings

Introduction of the room air conditioner

12.1 When the ice tested thick enough for harvesting, ice drills were used to make holes in which saws could be inserted. Once cut, the blocks of ice were floated along channels to the shore. Teams of horses drew them up ramps and into the ice house, where they were laid in place with thick layers of sawdust. (*Scribner's Monthly*, August 1875.)

12.2 By the end of the nineteenth century, ice harvesting had become industrialized, far from the simple methods by which countryfolk had preserved ice from their farm ponds. The vagaries of winter weather and increasing pollution of lakes were powerful restraints on industrial development. (*Appleton's Cyclopedia*, 1880, 2:130.)

Throughout the fourteenth century the northern hemisphere cooled appreciably, so much that icebergs appear to have interfered with the Viking explorations of America. Some scholars find insufficient evidence to justify calling a long period between 1579 and 1880 the "Little Ice Age." However, cold periods occurred in about a quarter of the seventeenth century, and in eastern North America and western Europe 1816 was known as a year without a summer. In America there was a greater annual fluctuation of temperature than in most parts of Europe. Winters were bitterly cold in much of the area that was first settled, and summers could be blistering hot. Food was difficult to preserve. Smoking, corning, and drying were common methods of keeping food, and in hilly regions there might be a cool spring over which a shelter could be built as a place to chill milk and other perishables. In northern parts of the United States, icehouses were erected to store the winter's bounty for the summer days ahead. Only 12 designs for icehouses were granted patents in the United States from 1790 to 1873, and all those were awarded in the last 25 years of that period. Apparently European interest may have grown during that time, as witnessed by a German publication of 1864 in which the icehouse has a triple wall with the space between two of the walls large enough to hold barrels of beer at an intermediate temperature.[1]

In 1806 the brig *Favorite* sailed from Boston Harbor with a cargo of 130 tons of ice that Frederic Tudor had cut from a pond on his father's farm near Saugus, Massachusetts. Its destination was the island of Martinique, where ice was needed to cool victims of a fever epidemic. Unfortunately, the scorching sun of the Caribbean melted much of the cargo and Tudor's venture proved a financial failure. Two years later another shipment—this time to Havana—was equally unprofitable. Tudor began again after the War of 1812, and competitors joined him. In 1816 only five cargoes of ice were exported from Boston harbor, which was destined to dominate the American ice trade, and these shipments totaled 1,200 tons. Thirty years later the cargoes numbered 175 and the tonnage was 65,000. The range of the ice trade had by then extended from the West Indies around the Horn to Asia and Australia.[2] Most of this ice was intended to tinkle in the glasses of the wealthy, some was used in treating fevers, and much was purchased to refrigerate meat or maintain the temperature required to brew lager.

In the northern areas of the United States, local enterprises quarried and sawed blocks of ice from ponds and floated them along channels to points where horses could draw the blocks ashore. There they were loaded for transportation to a pit or an icehouse (figs. 12.1, 12.2). Often the harvesting of ice was an off-season activity of logging companies, which already had sleds and draft animals and produced the sawdust that was needed to insulate blocks of ice as they were placed in storage. Lumber mills, which had long contended with dangerously flammable piles of sawdust and shavings, now found buyers for those materials. In 1857 almost 100 wagons and about 150 horses were used to distribute 60,000 tons of ice to users in Boston and its environs.[3] Fishing boats began to set sail with their holds half-filled with ice to refrigerate their catch and keep it fresh for market. Passenger ships began to carry refrigerated meat, instead of slaughtering live animals during a voyage. Typically, double-walled icehouses, as developed by a Boston ice merchant, were filled by first spreading 6 to 8 inches of sawdust over the floor. About 4 inches of sawdust was placed in the spaces between blocks of ice and the icehouse walls and, when the building's capacity had been reached, 1 or 2 additional feet of sawdust were scattered over the top of the ice.[4] The profitable relationship between logging companies and the ice trade was equaled by that of urban concerns that used wagons and horses to supply coal in winter and ice in summer.

Before the Civil War, the city of New Orleans annually received some 50,000 tons of ice from Boston alone. In satisfying this demand and pricing the ice, the vagaries of winter weather always played a prominent role. Relatively warm weather in the northern states could reduce the winter's harvest and greatly increase the prices paid for ice during summer months.

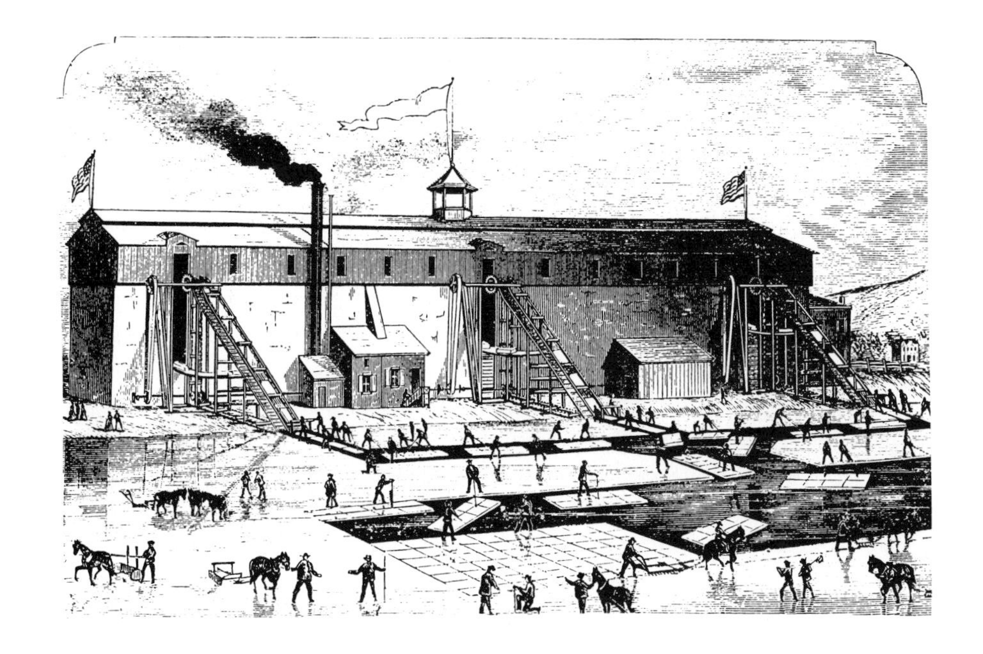

According to a grocer in Tennessee, shipments from the Northern Lakes Ice Company in La Porte, Indiana, lost as much as 40 percent of their weight while traveling over 600 miles in railroad boxcars.[5] Waste in storage and shipping often made natural ice costly, but it was a necessity for certain developing industries. By slaughtering animals in the West and shipping dressed meat in refrigerator cars, meatpackers could reduce freight charges, eliminate the cost of feeding in transit, and avoid the loss of animals' weight during shipping. Brewers, abundant after the German immigration of the 1840s, required low temperatures in order to regulate their preparation of lager. In some areas near metropolitan centers, natural ice continued to be cut and sold as late as 1919.[6]

Joseph Black, a young Scottish chemist, in 1760 evolved the theory of latent heat, which states that heat is absorbed as a liquid evaporates and heat is released by condensation of a vapor. This discovery was vital in the development of the steam engine, but it found other applications. More than 20 years later a Scottish physician, William Cullen, devised a machine in which air was pumped out of a closed vessel that contained water, the reduction of pressure increasing evaporation and thereby causing the water to freeze. It is these two phenomena, the latent heat of vaporization and the effects of pressure, on which refrigeration systems are based. In compression refrigeration, a liquid with a low boiling point is mechanically compressed before being released through a valve; in absorption systems two fluids (often ammonia and water) are heated until one evaporates and pressure rises to cause its condensation, then that fluid is released through a valve. Until the development of Freon in 1930, much attention was given to the characteristics of different refrigerants and the dangers of toxic and flammable materials escaping from the system.

By the nineteenth century, there was no lack of scientists and dabblers prepared to speculate about the methods by which cold might be produced mechanically. An American, Oliver Evans, proposed one in his book *The Abortion of a Young Steam Engineer's Guide;* a Scot described an absorption system that used sulphuric acid; and an Englishman gave demonstrations of an ice-making device that also employed sulphuric acid.[7] None of these forerunners were to have so much influence as an invention patented by Jacob Perkins, an American engraver and inventor who had settled in London to promote his method of printing currency in a manner that discouraged counterfeiting. Perkins's British patent of 1834 for ice making involved a chamber in which ether was compressed and evaporated according to the discovery of Cullen, and through that chamber snaked a system of pipes in which brine circulated and was cooled to about 5° F. From the evaporator, the brine flowed to another chamber containing boxes filled with water. As the water was

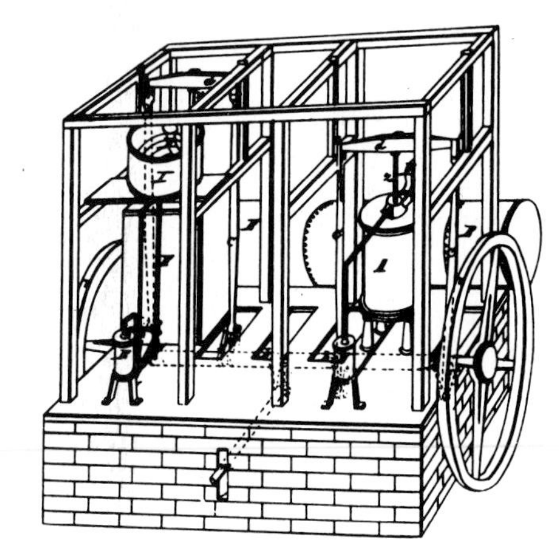

12.3 In his 1851 patent, John Gorrie declared that his ice machine would "accomplish my object with the least possible expenditure of mechanical force, and produce artificial refrigeration in greater quantity from atmospheric air than can be done by any known means." Gorrie died only four years later, frustrated in his efforts to obtain sufficient financing to develop his invention. (U.S. Patent no. 8080.)

12.4 James Harrison, an Australian journalist, succeeded around 1856 in developing an ether compression machine that was probably the first mechanical refrigeration system used in brewing. However, he did not succeed in his goal of finding a means of shipping the plentiful mutton of the Australian plains across the Indian Ocean to England. (*ASHRAE Journal*, July 1969.)

chilled to become ice, the brine recirculated to the evaporator where the ether once more lowered its temperature.[8] Perkins's invention did not gain wide commercial application, but an engineer in East London is said to have operated a Perkins compressor for a time and sold the ice it produced.

Years later John Gorrie, a physician in the busy cotton port of Apalachicola, Florida, became so absorbed in his theory that malarial fever could be cured or prevented by cooling the air of sickrooms that he abandoned his medical practice, which had included a government contract to care for sick sailors in a local hospital.[9] Written under a pseudonym, Gorrie's articles in the local newspaper show that he was fully aware of the nonmedical potential of his invention for producing cold:

> There are advantages to be derived from the generation of cool air within any building and this is equally applicable to ships as well. It might enable the hardy mariner to better serve mankind, he who contributes so much to our wealth and pleasure by transporting for us, from shore to shore, the rich production of the tropics—as animals when divested of life, and fruits which may be preserved entirely with all of their juices in a low temperature. This principle of producing and maintaining cold might be made instrumental in preserving organic matter for an indefinite time and thus become an accessory to the extension of commerce.[10]

His invention, patented in 1851 with the assistance of a Boston investor, was a compression design, releasing air at high pressure into a cooling chamber that was surrounded by water (fig. 12.3). Gorrie demonstrated his ice machine publicly, but he could not find sufficient financial backing to establish a commercial application.

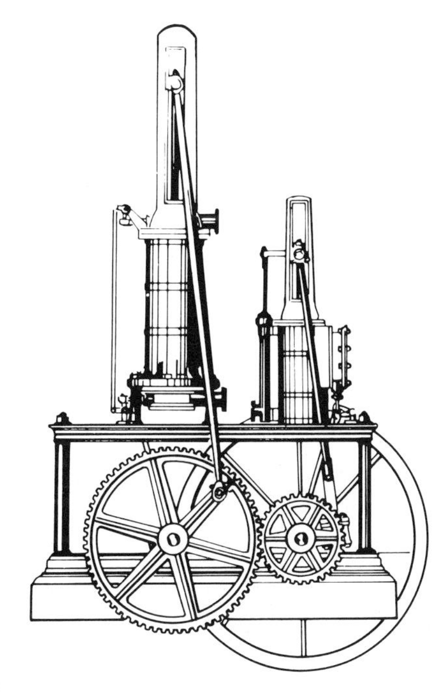

The air cycle compression system was clearly less efficient than others that used chemical agents, but air had the obvious advantage of being odorless, nonflammable, and readily available everywhere. It was natural that hospitals should prefer a cooling system that employed a harmless substance. Lloyd's, the London international market for marine insurance, and its underwriters gave their approval to no other system of refrigeration.

With the metropolitan population of Europe growing, the cattle and sheep that grazed on broad plains in Australia, Argentina, and the United States awaited a ready market. Shipping meat across stretches of equatorial water demanded a mass of natural ice that occupied much of a ship's hold, usually making the venture uncertain and unprofitable (fig. 12.4). In 1878 the steamship *Strathclyde*,

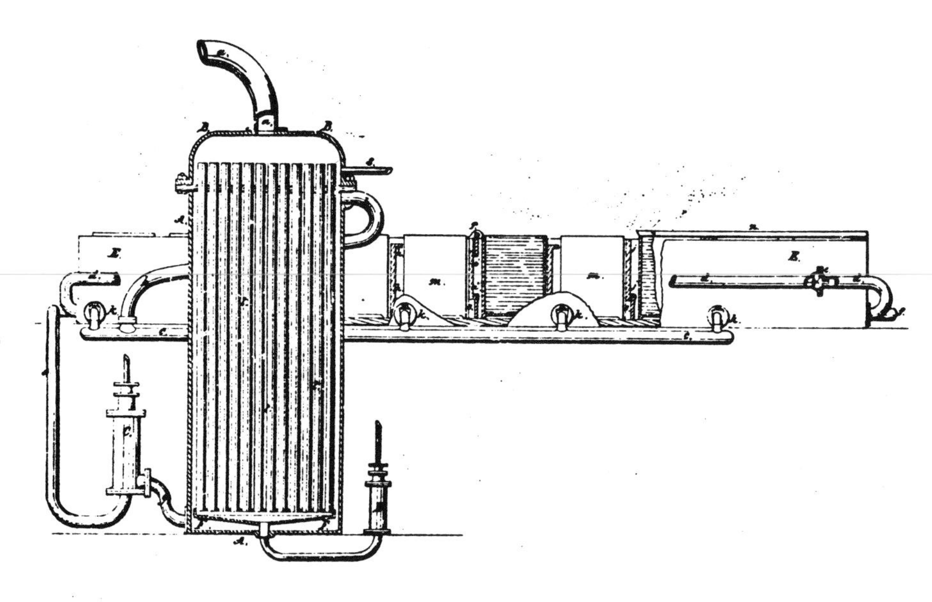

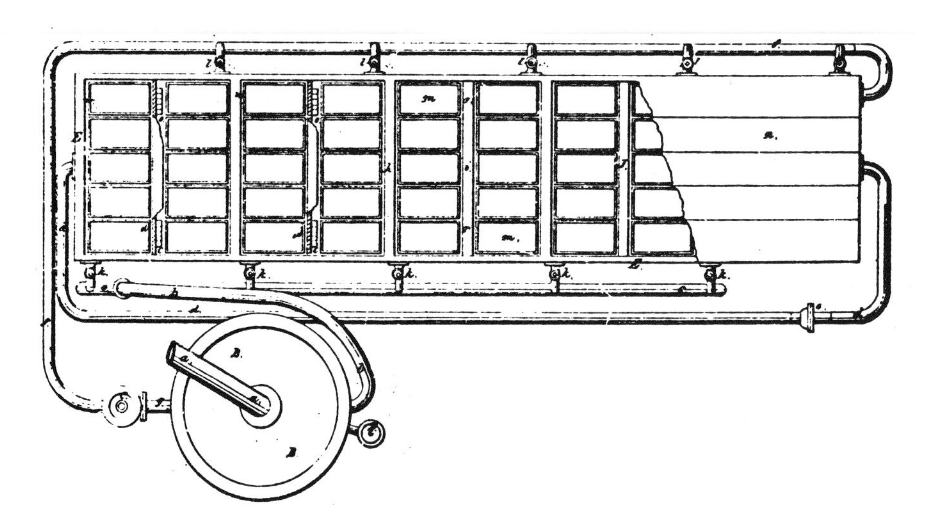

12.5 Alexander C. Twining's ice machine, patented in England (1850) and the U.S. (1850, 1862), was put in commercial operation in Cleveland, Ohio. This last patent uses a single tank of chilled brine to freeze 50 blocks of ice. Although the price of its product was almost competitive with natural ice, Twining's ice plant did not survive. (U.S. Patent no. 34,993.)

which had been outfitted with an air cycle system, sailed from Europe to Australia and returned with a load of beef and mutton in her cargo. Thereafter the air cycle system dominated in the marine transportation of foods, as well as in hospital installations. Air cycle equipment reigned until the 1890s, with many improvements being made upon the original models.

In 1853 Alexander C. Twining, an engineer living in New Haven, Connecticut, invented and patented a liquid vapor compression machine to make ice (fig. 12.5). Twining had studied engineering at West Point, and his early work laying out the northerly routes of the Hartford and New Haven Railroad led to a lucrative position as consulting engineer for other railroad companies that were engaged in westward expansion. After Twining resigned the chair of mathematics at Middlebury (Vermont) College in 1849, he devoted much of his time to development of his ice machine. The machine employed a system much like that used some 20 years before by Jacob Perkins. In the ice plant that Twining built in Cleveland, Ohio, a logical point to originate shipments of ice down the Ohio and Mississippi rivers, volatile sulphuric ether was employed.[11] The plant's output was said to have been 1,600 pounds of ice per day. For a time this type of machine was the most prevalent form of refrigeration, for it involved lightweight machinery, less expensive and more easily maintained than that of other systems.

The aqua ammonia system of refrigeration used as its refrigerant a mixture of water and ammonia, the latter usually procured from plants that manufactured illuminating gas.[12] A Frenchman, Ferdinand Carré, patented his machine before the Civil War began in the United States. New Orleans businessmen had a Carré machine shipped from France through the Union blockade, and it was set up at the military hospital in Augusta, Georgia, for the benefit of soldiers who suffered from fevers. This machine was small, producing only about 500 pounds of ice per day, and the process could be regulated only by stoking or banking the fire. Three more Carré machines were imported by these New Orleans investors, and they were shipped through Mexico in order to circumvent the Yankee blockade. One of the machines was purchased by a firm in San Antonio, Texas, where it was fitted with a steam coil to better regulate the process.[13] Because it could not provide clear ice from the city's water, the firm froze distilled water, producing transparent blocks that were considered superior to the natural ice shipped from Boston. In spite of such pioneer activity, in 1867 there were only five ice plants operating in the United States. It was the "great ice famine," caused by a very warm winter in 1890, that provided a major impetus for the American ice trade to convert to mechanical refrigeration.[14]

By the 1860s, refrigeration with natural ice had been adopted in most American slaughterhouses. In the next two decades, refrigerated railroad cars came into general use for shipping meat from the stockyards in Chicago. Shipments across oceans, the key that would open European markets, were attempted in the same period. Natural ice and a manually operated fan were used in 1875 to chill meat on the trip from New York to England. It was more than four times that distance to England from Australia and New Zealand, where wool had been the most marketable part of the sheep that grazed on rich grasslands. In 1879 two shipments of frozen meat reached England from Australia, and three years later a successful shipment was made from New Zealand, winning a prize that had been offered by the colonial government. By 1890 the number of sheep in Australia had greatly increased, and land values in both colonies had risen appreciably.

Early mechanical refrigeration found little application to the provision of human comfort, except for the cooling of feverish hospital patients. To some extent this may have been due to the fact that air cooling with ice had already been provided in some spaces intended for human occupancy, particularly places of assembly. The ventures into air conditioning were at first dominated by demands of industries in which there was a need for humidity control.

The French scientist Jouglet in the 1870s reported several attempts that others had made to bring cooler air into rooms that were intolerable for gatherings during the heat of summer. He described the frustrations that had been acknowledged in assemblies of the Académie des Sciences:

On that day the heat was suffocating, and the rays of the afternoon sun were glowing through the windows. Unable to control himself, Velpeau rose with protestations against the orb of day, and called upon the physical [science] section of the assembly to find some method of putting an end to such a state of things. General Morin, moved by the burning eloquence of his illustrious associate, rose, and stated that hitherto it had been found impossible to abate the nuisance . . . in fact, the brave general confessed himself vanquished by the rays of the sun itself.[15]

One design was tried in the very hall that had occasioned Velpeau's outburst. A vertical cylinder was placed on the roof and within it were set 104 pipes of 1½-inch diameter through which chilled water flowed. Air, cooled by contact with the pipes, descended into the room. This invention was found to be inadequate, and another experimenter made a futile attempt to improve it by wetting the external surfaces of the pipes.[16] A few years later a similar device was described to a meeting of the Institution of Civil Engineers in England. In this design, cool water from a nearby brook streamed over the surface of a perforated panel through which a fan blew air. It was reported that the temperature of the air going through this contrivance was lowered from 70° F to 56° F.

Another method of obtaining cool air was the use of underground sources. At the Necker Hospital in Paris a large masonry tunnel over 180 feet long ran 13 feet under the building's cellars, and it was reported to cool the air more than 8° F when outside air was 80° in the shade. The Conservatoire des Arts et Métiers simply drew air from the building's deep cellars; however, suggestions to use the air in the catacombs beneath Paris (which was said to be about 58° F) were rejected, because air from sewers and cemeteries was not considered healthful in a world that still believed in the miasmic principle of disease. As late as 1900 a more extensive system was used to cool the Court Theater in Vienna. An underground channel was divided by a center wall running its entire length. On one side water sprayed over 276 metal trays; on the other side were the cool moist walls of the tunnel. At the end of these two channels the air was directed to those spaces in the building that were at the time most urgently in need of cooling.[17]

Once machine-made ice was available, another option was considered, that of cooling air by passing it over ice or some surface chilled by ice. A French invention drew air through a round metal duct surrounded by a container filled with ice and salt (fig. 12.6). Metal vanes inside the duct insured that the air traveled a route long enough for it to be sufficiently cooled. An estimate indicated that cooling the wards of a Paris hospital by this method would cost fully as much as heating them through the winter, and therefore the proposal was rejected.[18] It is reported, however, that in 1880 an ice cream maker who had built a hotel in Staten Island, New York, began cooling his hotel dining room through ducts set in ice and salt. Three years later a restaurant at the Hygiene Exhibition in Berlin was cooled by air blown over ice, but the flimsy construction of this temporary building prevented the cooling system's being fully effective.

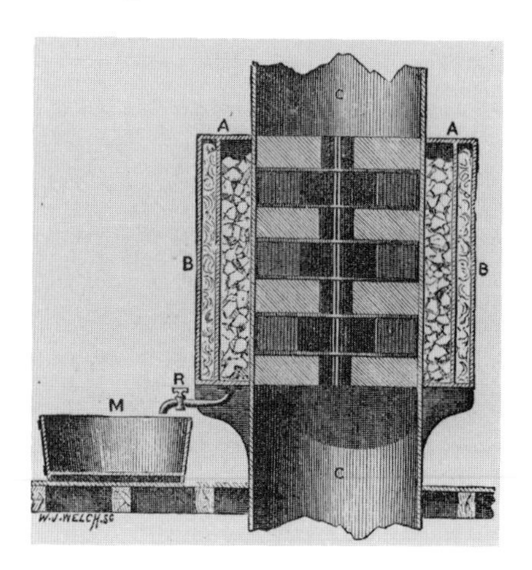

12.6 This double-jacketed flue, developed in France, was typical of many late nineteenth-century cooling experiments. The outermost cylinder was packed with an insulation material; the next held ice and salt. Metal fins within the flue increased the amount of chilled surface over which air passed. In the view of one writer, the high cost of natural ice limited the system to use as an "instrument of medical practice." (*Practical Magazine*, no. 6, 1873.)

12.7 Alfred Wolff, an American pioneer of air conditioning, designed this system for the New York Stock Exchange. Air entered at the right; after being filtered, it passed over coils that heated or cooled it. At the left an ammonia absorption machine chilled brine to be stored in a subterranean tank (center), allowing nighttime operation in anticipation of the daytime hours of frenzied trading in the Board Room. (*Metal Worker, Plumber, and Steam Fitter*, 5 August 1905.)

12.8 In spite of the windows and skylight, it was calculated that almost two-thirds of the heat in the Board Room of the New York Stock Exchange would rise from about 1,000 workers on the trading floor. (*Century Magazine*, November 1903.)

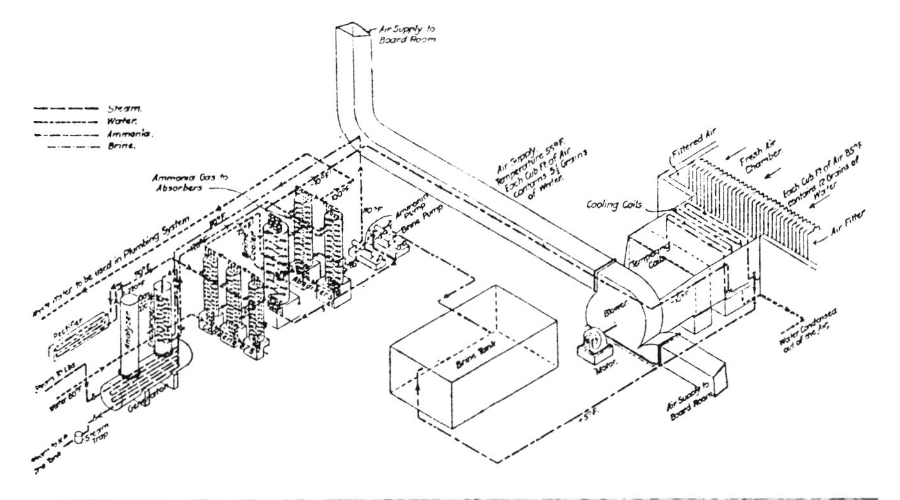
Air Supply to Board Room
Steam.
Water.
Ammonia.
Brine.
Ammonia Gas to Absorbers
Air Supply. Temperature 55°F. Each Cub Ft of Air Contains 5½ Grains of Water.
Filtered Air
Fresh Air Chamber
Each Cub Ft of Air 85°F contains 12 Grains of Water.
Cooling Coils
Air Filter
Brine Pump
Brine Tank
Blower
Motor.
Water Condensed out of the Air.
Air Supply to Board Room
Rectifier
Analyzer
Generator
Steam Trap

When Carnegie Hall in New York was built in 1889, the architect, William B. Tuthill, and his young consulting engineer, Alfred R. Wolff, provided in its basement a large rack on which blocks of ice were placed. Air was brought from an intake above the building's roof and drawn across racks of ice by four steam-powered blowers that stood 12 feet high. The blowers were necessarily powerful, for they forced the air again to the roof structure of the building, where it entered the concert hall through grilles in the ceiling.[19] Less complicated installations were used for some theaters and auditoriums. In fact, the 1901 high school graduation exercises in Scranton, Pennsylvania, were cooled by about 8 tons of ice, mounted on a wooden rack. Though temporary, this makeshift solution was reported to have lowered air temperature in the auditorium from the 90° F outside temperature to a moist 76° F.[20]

Systems remained much the same when mechanical refrigeration began to replace ice in cooling installations. An Indian rajah's palace was to have been cooled around 1890 by a mechanical refrigeration system planned by a Canadian company. Later a residence in Frankfurt, Germany, and a private library in St. Louis, Missouri, had mechanical cooling apparatuses installed. Even when a more complete treatment of air was provided for certain industrial purposes, gatherings of people were at best forced to rely on simple air cooling systems in which ventilating fans blew over refrigerated coils.

When it was announced that the New York Stock Exchange would be air-cooled, it was recalled that a frustrated proponent of cooling hospitals had bitterly commented years before, "If they can cool dead hogs in Chicago, why not live bulls and bear in the New York Stock Exchange?"[21] In New York's hot and muggy summer weather, the expanse of windows that opened into the trading room of George B. Post's design was responsible for about 25 percent of the heat in the room (figs. 12.7, 12.8). But this was a small matter when compared with the heat generated by the bodies of around 1,000 anxious traders in the market, which contributed, at maximum condition, at least 66 percent of the design load considered by Alfred R. Wolff, the mechanical engineer who had designed the ice-cooling system for Carnegie Hall.[22] The Stock Exchange's system, which used ammonia as its refrigerant, was the same that had been successfully applied by Wolff a short time before in the Hanover National Bank. In the Stock Exchange, as it was completed in 1904, air was taken in at roof level, drawn down a shaft to the basement, and went through a filter of 5,400 square feet of cheesecloth hanging in a zigzag pattern.[23] The refrigeration equipment was powered by exhaust steam from the electric generators that supplied the building's lighting system. Brine from a basement tank was chilled and then passed through coils

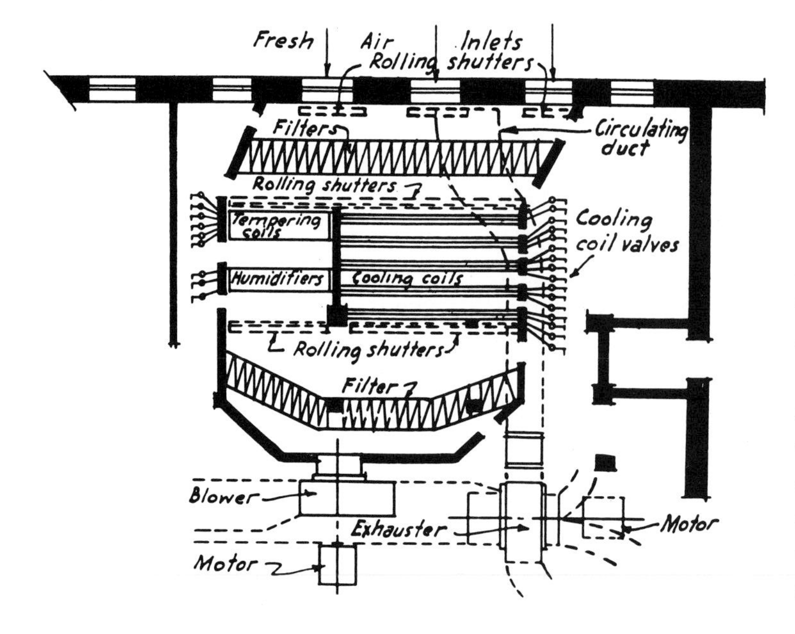

12.9 The system Alfred Wolff designed in 1907 for the Metropolitan Museum of Art in New York provided for warming, moistening, or cooling air as needed for the preservation of works of art. Fresh or recirculated air passed through cheesecloth filters before and after treatment, the route being determined by shutters that controlled the openings. The cooling equipment was used only a few years, because museum management felt that winter control of temperature and humidity was sufficient without cooling and dehumidification in the summer. (*Heating, Piping and Air Conditioning*, June 1936.)

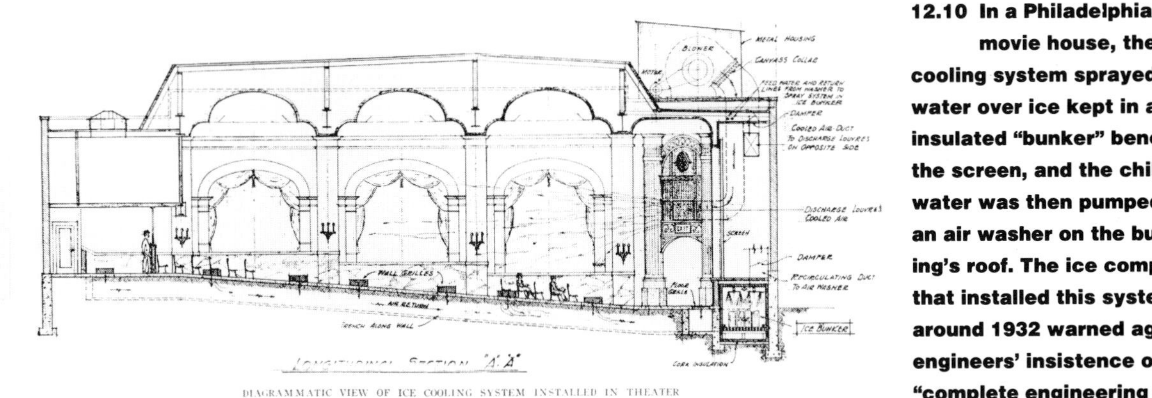

12.10 In a Philadelphia movie house, the air-cooling system sprayed water over ice kept in an insulated "bunker" beneath the screen, and the chilled water was then pumped to an air washer on the building's roof. The ice company that installed this system around 1932 warned against engineers' insistence on "complete engineering systems" and stoutly declared that "the progress of air cooling will take place not by the attainment of perfection and complete systems but by makeshifts and partial jobs." (*Refrigerating Engineering*, November 1932.)

located between the cheesecloth filters and the blower. One of the first large installations of air cooling outside industry, the Stock Exchange led to the application of cooling in theaters and department stores, where other crowds contributed body heat.

At about the same time, a theater in Cologne, Germany, installed a refrigeration machine to cool its audiences. During daytime hours the theater was cooled in preparation for the evening's performance by circulating cool well water through the coils of the system, and at the same time the refrigeration plant chilled brine, which was retained in a large storage tank. Once an audience filled the theater, cold brine circulated through the coils so that its lower temperature could compensate for the heat given off by bodies and lights.

A flourish of activity in cooling theaters began in the 1910s (fig. 12.10). The rapid development of cooled theaters was the result of a highly competitive business structure that came with the introduction of motion pictures. In 1917 the Empire theater in Montgomery, Alabama, installed a year-round cooling system. In the years that followed World War I, two Chicago movie houses led the way by installing air-cooling systems using mechanical refrigeration, and Grauman's Metropolitan Theater in Los Angeles, California, provided full air conditioning (fig. 12.11). It has been estimated that by 1931 about 400 movie houses, vaudeville palaces, and legitimate theaters were cooled.

Air-cooled interiors still had the problem of high humidity, and this was particularly true of systems that blew the air over melting blocks of ice. At the turn of the century it was not uncommon that beyond the ice racks would be placed shallow trays filled with calcium chloride, a by-product of the manufacture of soda ash that could absorb roughly three times its weight in moisture before dissolving. This was the same chemical used by Carrier in his first attempts to solve the problems of excessive humidity.

12.11 **Although early application of air conditioning centered in the United States, major European movie houses soon installed systems. By the late 1920s the "Paramount" in Paris had a Carrier machine in the basement. Cool air entered the auditorium through its ceiling and the undersides of balconies, and it returned to the basement through "mushroom" fixtures beneath the seats. (*Science et Vie*, February 1928.)**

In 1901 Willis Haviland Carrier graduated in engineering from Cornell University and took a position with the Buffalo Forge Company, a manufacturer of coils and fans for heating installations.[24] After only a year with the company, Carrier was assigned to solve the problems of the Sackett-Wilhelms Lithographing and Publishing Company of Brooklyn, New York, printers for *Judge*, the country's leading humor magazine, and other publications. Color lithography, which had only a short time before advanced to incorporate the half-tone process, demanded precise register of the several individual printings that might be required for a single colored picture. Fluctuations of humidity in the hot pressroom of the printer could cause paper to expand or contract and result in faulty register of the different colors of ink, and excessive moisture in the air impeded drying of inks and delayed the entire process. Sackett-Wilhelms had a dire need for humidity control, especially after the unusually hot and steamy summers of 1900 and 1901.

Carrier's first move was an experiment of drawing air through burlap soaked with a saturated solution of calcium chloride. This removed moisture from the air, but air coming from it seriously corroded the printing presses. In a second attempt Carrier carefully selected the dewpoint temperature at which the air retained the correct amount of water vapor for printing. For this purpose he used psychrometric tables provided by the Weather Bureau, charts that indicated the relationship of air temperature and the maximum amount of water vapor that it could retain. With this data, he determined the amount of coil surface, the required temperature of the coils, and the velocity of air movement that would be needed to accomplish his task. Money was saved the first summer by using cool water from an artesian well, rather than from city mains, but the second summer saw the addition of a refrigerating system. For winter months, when the heated air tended to be too dry, a spray of low-pressure steam was added. Having established the relationship of temperature and water vapor as the key to successful air conditioning, Carrier came upon the principle that was the basis of his work:

Here is air approximately 100% saturated with moisture. The temperature is low so, even though saturated, there is not much actual moisture. There could not be at so low a temperature. Now, if I can saturate

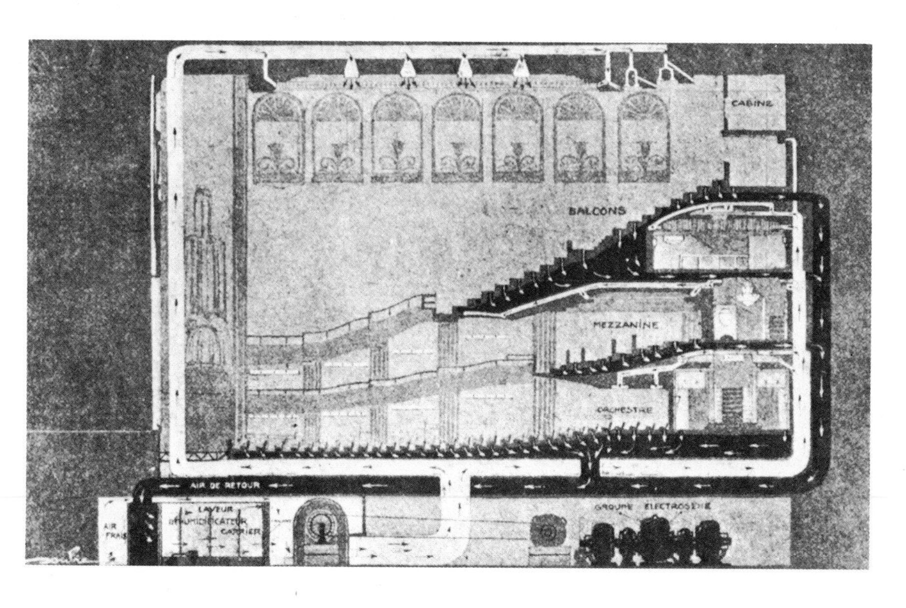

air and control its temperature at saturation, I can get air with any amount of moisture I want in it. I can do it, too, by drawing the air through a fine spray of water to create actual fog. By controlling the water temperature I can control the temperature at saturation. When very moist air is desired, I'll heat the water. When very dry air is desired, that is, air with a small amount of moisture, I'll use cold water to get low temperature saturation. The cold spray water will actually be the condensing surface. I certainly will get rid of the rusting difficulties that occur when using steel coils for condensing vapor in air. Water won't rust.[25]

One problem was that of finding a spray that would send out particles of water as fine as mist, and the solution lay in nozzles that had been designed to spread a fog of insecticide on plants. A U.S. patent for Carrier's "Apparatus for Treating Air" was issued in 1906. The first installation was made at the LaCrosse National Bank in Wisconsin, but there the system was used only to wash air in the ventilation system.

Carrier's work was first directed toward factory problems, just as refrigeration had been first used in meat packing, brewing, and other industries. Mills for cotton, rayon, and silk; factories making gelatin capsules for pharmaceutical companies and celluloid for movie film; and workshops where tobacco leaves were hand-rolled into cigars—there was a seemingly endless list of manufacturers who had for years tolerated waste, inefficiency, or imperfections that might now be avoided through control of temperature, humidity, and airborne particles of the materials used.

Factories and theaters had specific needs for cooling, and most of their problems of cooling centered on one or perhaps two large undivided spaces. Buildings that were subdivided into many separate spaces presented new questions. The Travis Investment Company of San Antonio, Texas, in January 1928 opened its new 21-story office structure, the Milam Building (figs. 12.12, 12.13). Although several earlier office buildings had been air-conditioned, the Milam Building was heralded as the tallest of these and the "most completely equipped." Eleven air conditioning units were distributed through the building. Each two typical floors of office space, arranged in a C-shaped plan that was common at the time, were served by a single air conditioning plant. In the basement, a refrigerating plant chilled water, which was stored in a tank beneath the basement floor. This tank provided the advantage of being able to operate the refrigerating plant, whose maximum daily load was the equivalent of 375 tons of ice, during the hours when electric rates were lowest. Cold water was pumped to each of the conditioning units, where fans pulled air through fine sprays of water and sent it down ducts above the corridor ceilings.

It was estimated that savings effected by the Milam Building's air conditioning system would compensate for about 40 percent of the cost of the system. Transoms over office doors could be eliminated, along with the ceiling fans that were customary in southern states. Other advantages that were put forth at the time included the office workers' escape from street noises, janitorial costs saved by preventing dust blowing into offices, and the tenants' willingness to pay higher rents for cool and comfortable offices. In addition, it was recommended that the occupants be allowed to control the volume of air entering their spaces, both because an excess of automatic controls would demand too much attention and the conviction

12.12 In addition to being the tallest reinforced concrete building of its time, the Milam Building, San Antonio, Texas, was also the most completely air-conditioned. In the city's sunny summers the highest cooling needs occurred around four o'clock in the afternoon, but that requirement was adjusted on the assumption that Texans were comfortable at temperatures slightly higher than those preferred by natives of other areas. (*Heating, Piping and Air Conditioning*, July 1929.)

12.13 From the mechanical rooms behind the elevators of the Milam Building (top of drawing), flat ducts ran above corridors, bringing warm or cool air to each office. Air returned through grilles in office doors, traveling through the corridors to reach the mechanical rooms. (*Heating, Piping and Air Conditioning*, July 1929.)

12.14 When *Fortune* magazine published this diagram explaining the air conditioning of office buildings in 1938, few skyscrapers were cooled. World War II delayed the construction of office buildings, but once the war ended air-conditioned office space rapidly became the standard in the United States. (*Fortune*, April 1938. Copyright 1938 Time, Inc. All rights reserved.)

that the office tenant "is considerably relieved merely to feel that he has taken some action, more or less regardless of the results." The journal *Heating, Piping and Air Conditioning* viewed the building's air conditioning system with enthusiasm:

> The windowless skyscraper, already envisioned by others and made possible by air-conditioning plus artificial illumination, will surely become a reality since, for one thing, it can be erected on relatively inexpensive property. Let those who cry for "fresh" air through open windows from the out-of-doors be reminded that it doesn't exist in the congested city. . . . So air-conditioning has come to make available every day the best in atmospheric comfort that nature offers so spasmodically.[26]

The second skyscraper to be so fully air-conditioned was probably the Philadelphia Savings Fund Society, a landmark of modern architecture built in the midst of the Great Depression, four years after the Milam Building.

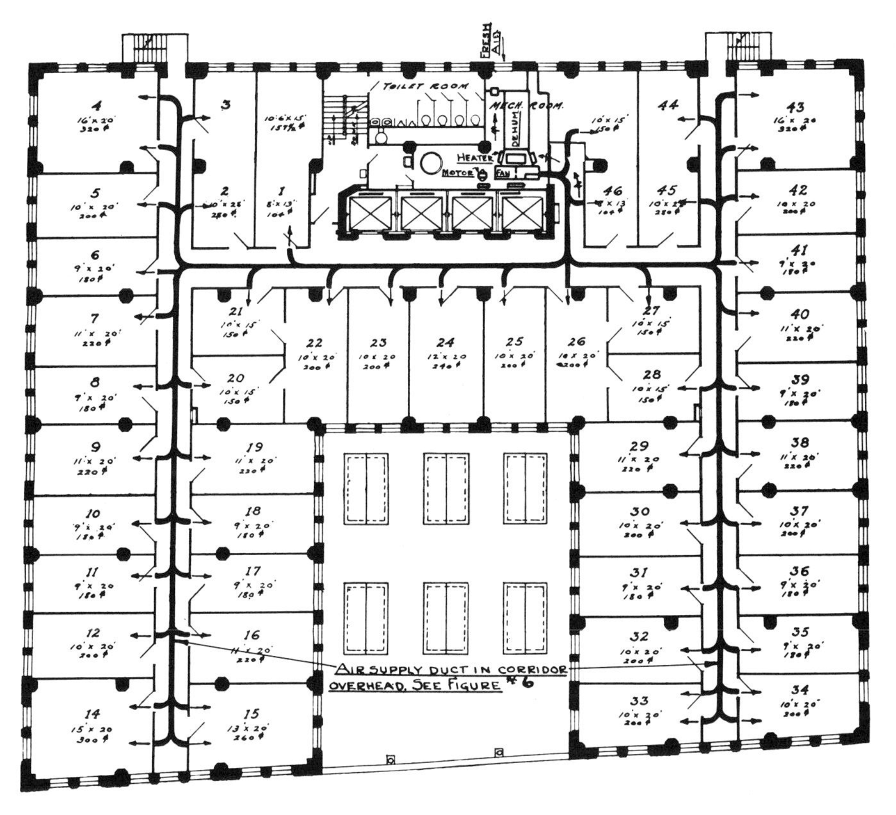
FRESH AIR
TOILET ROOM
MECH. ROOM
DEHUM
HEATER
MOTOR
FAN
AIR SUPPLY DUCT IN CORRIDOR
OVERHEAD. SEE FIGURE #6

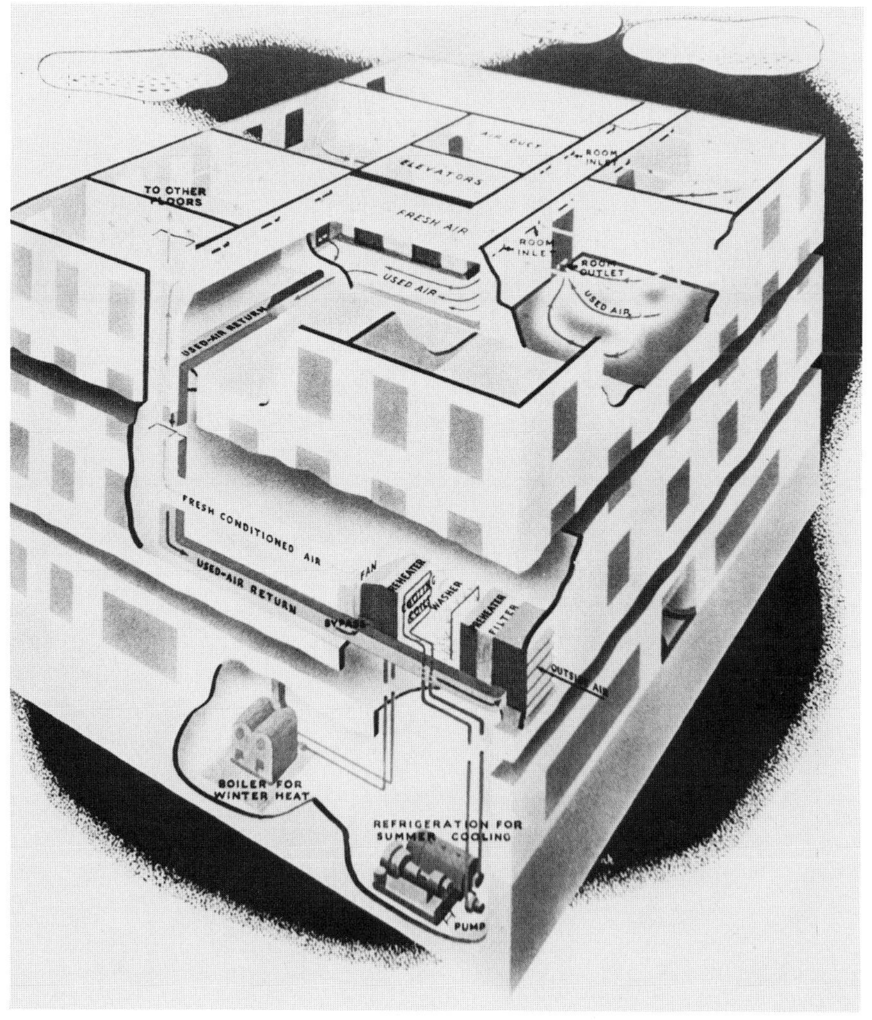
TO OTHER FLOORS
AIR DUCT
ELEVATORS
ROOM INLET
FRESH AIR
ROOM INLET
ROOM OUTLET
USED AIR
USED AIR
USED-AIR RETURN
FRESH CONDITIONED AIR
USED-AIR RETURN
FAN
REHEATER
WASHER
PREHEATER
FILTER
BYPASS
OUTSIDE AIR
BOILER FOR WINTER HEAT
REFRIGERATION FOR SUMMER COOLING
PUMP

12.15 By the late 1920s, major companies were manufacturing unit air conditioners, such as this one produced by Carrier. When several were spaced about a large room, they governed temperature and humidity without the expense of installing a system of ductwork. (*Mémoires, Sociétés des Ingénieurs Civiles*, 1929.)

Originally it had been planned to air-condition only the four lowest floors and a penthouse, the levels in which public spaces and executive offices were located. However, the desire to compete with other office buildings—and a sweltering summer during construction—caused bank officials to reconsider. With the steel frame completed to the twentieth floor, over half the building's height, it was decided that the entire PSFS building should be air-conditioned (fig. 12.14).

A Philadelphia newspaper in 1859 published an advertisement for a chair with an ice chest beneath the seat that offered a solution to the discomforts of hot summer days:

KAHNWEILER'S PATENT VENTILATING ROCKING CHAIR

This is a novelty that commends itself to all who value health and comfort. The invalid will recognize it as an invention especially conducive to his well being, inasmuch as with the addition of, say two cents worth of ice per day, the luxury of pure air may be fully enjoyed within doors, and the heat of the summer, or the vitiated atmosphere of a closed apartment, defied. Sold at the Manufacturer's Depot, Masonic Hall, No.. 715, Chestnut Street, by James Wilcox.[27]

Even with cooling later provided in some movie houses, restaurants, and department stores, long hot summers made people long for some means of cooling the offices and shops in which they worked and the rooms in which they endured steamy nights. In the first decades of this century most office buildings were heated by steam radiators and most central-heated houses had gravity hot-water or warm-air systems. There was little possibility of adapting those systems to cooling. Ice companies in the early 1930s saw their markets rapidly vanishing. Most of their industrial customers were installing mechanical systems to make their own ice and the household refrigerator was rapidly replacing the icebox. A faint and desperate hope of the ice industry centered on cooling buildings with ice. Small portable coolers, a fan and a chest that had to be filled with ice once a day, were introduced without great success.[28] Larger systems chilled water in an iced space and pumped it to fan units distributed about the building. More hope seemed to lie in systems still larger, such as those for restaurants and theaters, for which the ice industry valiantly claimed three advantages: low installation cost, flexibility to match varying loads, and the remarkable virtue of being "unautomatic."

Leadership in the development of small cooling units was provided by the railroad industry. Summer travelers had been forced to endure hours inside sun-baked cars, the open windows admitting blasts of hot air bearing granules of soot from coal-burning locomotives. In 1907 a company undertook the manufacture of passenger cars in which air passed over 600 feet of ice-covered tubes before entering the car, but it was the 1930s before successful cooling was provided in the Baltimore and Ohio Railroad's dining car, the *Martha Washington*. The following year an entire air-conditioned train was regularly scheduled between New York and Washington. Systems were quickly converted from ice-cooling to mechanical refrigeration, and by 1934 only slightly more than a third of the 2,494 air-conditioned railroad cars in the United States were still using ice.[29] Railroads that provided cooled cars had a distinct advantage in a competitive industry, and the president of the Baltimore and Ohio stated that air conditioning had done more than any other factor to increase that line's business.

By the 1920s, the design of small vapor compressors had advanced to the point that room air conditioners could be proposed. The introduction of household refrigerators had found an enthusiastic market, anxious to rid itself of the iceman's daily visit. By 1925, 75,000 electric refrigerators had been placed in homes in the United States, and that year they were first available in Britain.[30] A portable "cooling device" was patented in 1926 but never went into commercial production (fig. 12.16). Its cylindrical base housed a motor, compressor, and condenser, and at the top a fan blew across the cooling coil.[31] In the same year, application was made for a patent on an air conditioning console with intake and exhaust ducts that could extend to the outside through a window sash.

When the breakthrough in residential cooling came in the 1930s, many levels of complexity were included, but all depended on discoveries that had been made in the development of household refrigerators.[32] That decade saw unusually hot and dry summer weather in much of the United States. In some areas attic fans were effective at day's end to draw cool air into the house, thereby cooling the occupants and lowering the heat that radiated from walls and ceilings. Evaporative coolers, the first method studied by Willis H. Carrier early in the century, were constructed in their crudest form with water dripping over a layer of excelsior (curls of wood used for crating fragile objects) and a low-speed fan drawing air through this evaporative screen. In the drought years of the Great Depression those methods were effective, and their technological crudity was offset by their extremely low cost.

A more sophisticated product, though large and ugly, was the console water-cooled air conditioners introduced in the United States by the Frigidaire Corporation in 1929. Soon several other companies brought out competing models. Air-cooled models with humidity and temperature controls were on the market during the 1930s and found their way into many executive offices (fig. 12.17). In 1932 the Thorne company advertised a window-mounted room air conditioner, but it is doubtful that this design was ever actually put in production. It was four more years before an acceptable model was brought on the market by Philco Corporation. It did not provide a fresh-air intake, but a single motor drove the compressor, condenser fan, and evaporator fan to produce a capacity of 3,675 BTUs per hour. The Westinghouse Electric Corporation's design five years later had a capacity more than 50 percent greater. It offered a reverse cycle for heating and, as an additional advantage, was sold with a five-year warranty.

Little progress in the development of unit air conditioners was made during World War II, but immediately afterward window-mounted air conditioners were placed on the market by almost all American manufacturers in the field. Sizes were somewhat smaller than before, and the greater part of the chassis projected outside the window, thus eliminating the supporting legs and brackets that had been required by some earlier models. By this time, the same motor power obtained almost three times as great a capacity as in the first unit sold by Philco.

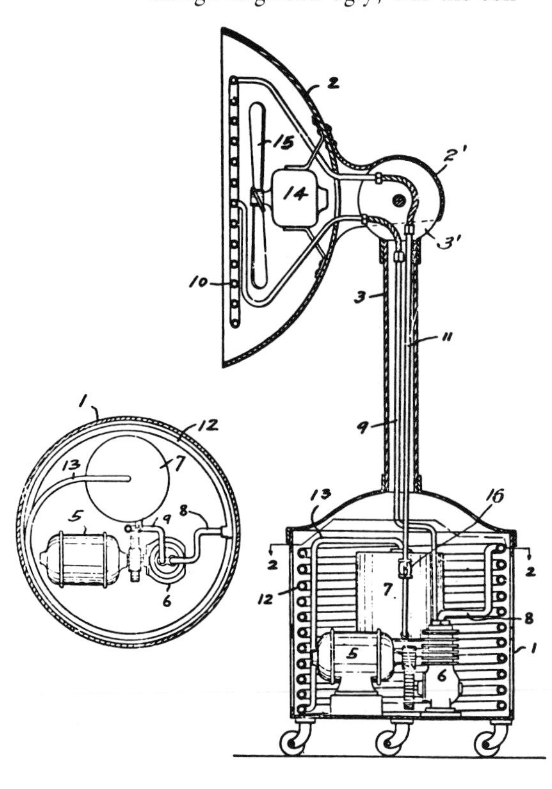

12.16 A 1926 patent for a room air conditioner was a harbinger of unitized cooling. With a water-cooled condenser the unit could have been easily moved about (had it ever been manufactured in quantity); however, no provision was made for the condensation that would form on the cooling coils. (U.S. Patent no. 1,831,825.)

12.17 Room air conditioners that were available after World War II often assumed the appearance of furniture. This ¾-horsepower console, manufactured by Fedders, disguised the machine with a jacket imitating mahogany. (*Refrigerating Engineering*, January 1950.)

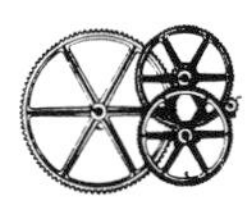

After World War II, applications of air conditioning were extended to include housing, schools, and nearly all other building types. Because it was costly to install systems of ducts and fans exclusively for cooling, warm-air heating methods were customarily installed in buildings that were to be air-conditioned. The seminal installations of Wolff and Carrier using mechanical refrigeration in air conditioning followed patterns that had long before been established for warm-air heating—air blown over coils filled with the temperature-altering agent. Therefore the combination of cooling and heating was easily accomplished.

Installations of air conditioning from the first suggested the option of completely sealing buildings and eliminating the operable sash that had been necessary to admit summer breezes. Stuart W. Cramer, a U.S. engineer, originated the expression "air conditioning" as a term for the improvement of the atmosphere within textile mills. The objectives he presented in 1909 were: "(a) Heating. (b) Air Moistening. (c) Ventilating. (d) Air Cleansing. (e) Air Cooling. (f) Automatic Regulation of humidity and temperature to a predetermined standard." Systems of the past, with sprays of moisture, filters of cheesecloth, and carefully placed intakes for new air, sufficed for several decades. At present it is the fourth of Cramer's criteria that is most questioned. Sealed buildings in a polluted setting and subject to health-endangering emissions from their own interiors give rise to new definitions of "air cleansing."[33]

13

Elevators and Escalators

1830s Hoists in mill buildings operated by steam power

1854 Elisha Graves Otis demonstrates his safety device at the New York World's Fair

1859 "Vertical screw railway" is installed in Fifth Avenue Hotel, New York

1870 The water balance elevator introduced by Cyrus W. Baldwin

1878 Vertical-cylinder hydraulic elevators introduced by Baldwin in the Boreel Building, New York; horizontal-cylinder elevators soon follow

1887 Electric elevator installed in Baltimore, Maryland

1892 Jesse W. Reno builds an experimental escalator in Brooklyn, New York

1900 Escalators of several makes operated at the Paris Exposition

1900–20 Popularity of air cushion safety systems

1903 Traction elevators introduced by Otis Elevator Company

1931 Two elevators per shaft installed in the Westinghouse office building, East Pittsburgh, Pennsylvania

1970 Double-deck arrangement of elevators revived in Time-Life Building, Chicago

Hoists were used very early in history for mining, loading ships, and other activities in which heavy loads were to be lifted. With power supplied by workmen, animals, or water wheels, hawsers and chains were drawn over pulleys to lift weights. Hoists usually operated by manpower in the Roman period, since there was little incentive to develop costly mechanisms to replace the labor of the Roman hordes on the dole.[1] During medieval times, with the decline of massive projects, hoists advanced little beyond the mechanical simplicity of winches turned by hand or by a treadmill. Often counterweights were added to compensate for the weight of containers in which loads were lifted. With the addition of the ratchet and pawl, a system of restraint that had been known in ancient times, the essentials of all hoists were present, and only the improvement of motive power was needed.

In English mills a hoist system, the teagle, was present early in the nineteenth century, and by 1835 power from the steam engines that ran spinning machines was used to operate teagles (fig. 13.1). The platform of the teagle was suspended by two ropes that went around a grooved pulley and supported counterweights at their opposite ends. Both the platform and the counterweights moved between vertical guide rails. The movement of the teagle platform was stopped by the operator's tugging a rope that shifted the belt onto a free-turning wheel.[2]

When the New York World's Fair opened in 1853, its displays were but a pallid imitation of the London event two years earlier. Visitors could leave the exhibition grounds and retire across Forty-second Street to Latting's Observatory and Ice Cream Parlor (fig. 13.2). Besides refreshments, this commercial appendage to the fair offered rides to the top of a slender conical structure from which one could view the Fair and the city, including the wooded tangle that would become Central Park.[3] Ascent was accomplished by a creaking elevator directly powered by a steam

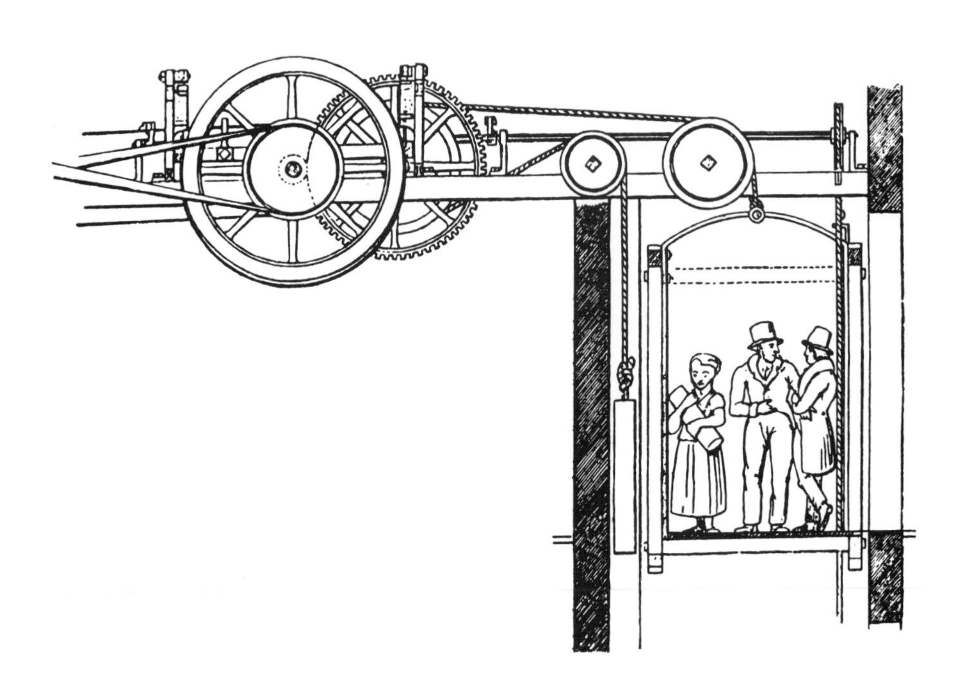

13.1 The teagle elevators employed in English textile mills in the 1840s were powered by belts from the main shaft of a factory, the shaft that drove all the machinery. (*Pictorial Gallery of Arts*, circa 1845.)

13.2 A short distance from the World's Fair, New York (1853), visitors could repair to the Latting Observatory and Ice Cream Parlor, in which an elevator would take them for a jerky trip to the top of the conical tower for a view of the city. (Courtesy Otis Elevator Company. Archive, United Technologies Corporation, Hartford, Connecticut.)

engine. During the second season of the fair, passengers traveling to the top of Latting's Observatory could hardly have been reassured by Elisha Graves Otis's prominent display in the Exhibition Hall, of his safety

device for hoists and elevators (fig. 13.3). For each demonstration of his device, Otis mounted a hoist platform and was drawn upward. Once at the top of the hoist framework, the rope by which the platform hung was cut, frightening the onlookers and activating the safety device Otis had invented.

The Otis safety mechanism used toothed metal runners at the sides of the shaft and pawls attached to the car. When the car was suspended, tension in the supporting rope or cable held back the pawls, allowing the car to move freely up and down the runners. When the rope was cut, the pawls were unrestrained and springs drove them forward to catch the teeth of the side rails and quickly halt the car's fall. Forced by illness to abandon work in building construction and carriage making, Otis in 1948 had established a small machine shop in Albany. When the city expanded its water system and took over the creek that powered his machinery, Otis moved to New Jersey and became superintendent of a bedstead factory, with his 16-year-old son Charles as the engineer-in-charge of the factory steam engine. When his employers began construction of a new factory in Yonkers, New York, Otis devised a means of preventing the plant's hoist from falling, a frequent industrial accident of that period. Orders from other factory owners and the financial collapse of the Yonkers Bedstead Manufacturing Company led Otis to establish his own business of building industrial hoists and to demonstrate his design at the World's Fair. The first order came from a New York furniture manufacturer whose hoists had recently fallen and killed two workmen.[4] Within two years Otis's workshop had produced over 40 elevators, all for installation in industrial buildings.

In a steam-powered factory, the engine drove an overhead shaft from which belts conveyed rotational force

13.3 In his World's Fair exhibition, Elisha Graves Otis had himself lifted on the platform of a factory hoist. When an assistant overhead cut the rope, the Otis safety device held the platform securely in place. (Courtesy Otis Elevator Company. Archive, United Technologies Corporation, Hartford, Connecticut.)

to individual machines. Adding a hoist to the array of equipment was a relatively simple matter, and shifting its belt to a neutral position would stop the hoist in the same way other machines were stopped. Provision of a counterweight was, of course, customary for factories' hoisting mechanisms. All of the elements of a passenger elevator—power, control, and balance—were present at a primitive level in factory hoists early in the nineteenth century. Only the improvement of comfort, distance, speed, safety, and economy remained as challenges.

For his industrial hoists Otis introduced a two-cylinder steam engine that could be powered by steam piped from a factory's boilers, and this compact unit proved to have many advantages over belt-driven machines connected to the factory's drive shaft.[5] Once the factory hoist became relatively independent in location and power, its adaptation to use for passengers was an obvious step. In 1857 the Otis Steam Elevator Company installed a passenger elevator, probably the first to be made, in the building of E. V. Haughwout and

Company, at the corner of Broadway and Broome Street in New York. The Haughwout Company's five-story building was crowded with displays of silver plate, gas-lit chandeliers, and the imported French china and clocks for which the store was famous. A store with a valuable location generously served by the city's transit systems, the Haughwout Company needed a means of attracting buyers to every corner of every floor of its building to see the full range of its merchandise. Hence, an elevator.

Elisha Graves Otis died in 1861, the same year in which he obtained the patent that would be the basis for his company's later growth. Reorganized as Otis Brothers and Company, the firm in 1871 secured a reissue of the fundamental patent that it had used for ten years. Under the direction of Charles R. Otis, who had worked alongside his father from the age of 13, the firm began to transfer its attention from steam-powered elevators to those driven by hydraulic systems, but recurrent depressions and economic panics made expansion difficult.

In 1859 the Fifth Avenue Hotel had been built facing New York's Madison Square. It was a splendid six-story structure of white marble in the classical style, a center of political intrigue, and it offered guests the convenience of the "vertical screw railway" installed by Otis Tufts, a Boston engineer. The *New York Times* admired "the car, or little parlor," which differed greatly from industrial hoists that were suspended on chains or cables, and described its equipment in detail:

> An open vertical space, some 10 feet square, extends through the house from ground floor to top floor, with openings to the intermediate floors. The car, a covered room, nearly filling this space, or "well," forms the nut of a screw which extends from top to bottom; so that as the screw revolves to the right or left, the car ascends or descends. A guide-way at one corner of the well prevents the car from turning around with the screw. The screw is 90 feet long, 12 inches in diameter across the (hollow) stem, and some 18 inches diameter across the thread. The threads are two inches thick or deep. It is made of the best gun iron, and is in several sections or lengths—so long a casting could not be well made in one piece. The different lengths of screw are joined together by pins of wrought iron. . . . On the bottom of the screw, in the cellar, is a large gear which is moved by a smaller one on a horizontal shaft, which shaft is revolved by a belt directly from the steam engine. This belt may be shifted by a wire rope passing through the car from the driving pulley on the said shaft, to a loose pulley on the same shaft. When so shifted, the car begins to descend by its own weight—*i.e.*, it begins to turn the screw, the gears, and the horizontal shaft in an opposite direction. By pressing a brake on a third pulley on this horizontal shaft, by means of a second wire rope in the car, the whole mass is stopped at any required point.[6]

The car's descent was regulated by pistons in two water-filled cylinders. As the elevator car reached the top or bottom floor levels, the mechanism was automatically switched between the two systems, the engine-driven screw and the piston-controlled gravity descent.

The installation was expensive to operate, both in the provision of steam power and the frequent repairs required, and the large wood-sheathed shaft in the center of the car restricted the space available to passengers. According to most descriptions, movement—whether upward or downward—was extremely slow. This stately travel was punctuated by jolt-

13.4 Vertical, two-cylinder steam engines for elevators briefly followed the use of belts from the engines that powered the other systems of a factory. At the top and bottom of the shaft a tug on a rope could activate the controls (*L*) that reversed the direction of the engine. (*Appleton's Cyclopedia*, 1880, 1:594, 595.)

ing stops for the floor levels at which passengers wished to leave. In spite of the safety devices provided by Tufts, a trip in the "vertical screw railway" was an unsettling experience for many. The children of a wealthy banker, who lived in one of the mansions across from the Fifth Avenue Hotel, were firmly instructed never to enter the elevator apparatus.[7] Only minor improvements were made a few years later when a Tufts elevator was installed in Philadelphia's Continental Hotel. In spite of their shortcomings, the two hotel elevators continued to be used for almost 20 years.

In an outburst of prophecy, a reporter for the *New York Times* enumerated the advantages that might be expected from a profusion of elevators such as that in the Fifth Avenue Hotel. First, he wrote, space in the top floors of buildings would become more desirable than in lower floors, altering the "compromise between high prices, dust and noise, on the one hand, and excessive weariness in stair-climbing on the other hand," and this would make possible a more profitable utilization of the ground level of building sites. Second, the convenience of office building elevators would, in effect, place "all offices on a level with the street, as far as physical exertion is concerned," thereby improving their quality and accessibility. Third, efficient hoists would also permit the use of upper stories of buildings for retail establishments and manufacturing. Fourth, greater comfort could be brought to private houses. Scorning all notions of healthful exercise provided by the chores of housekeeping, the writer proposed that residential elevators be used to free time for outdoor forays by the "ladies of New York."[8]

Few vertical screw elevators were built, and it was about 20 years before the William Miller Company of Cincinnati introduced a vertical screw elevator that eliminated the column in the center of the car. Up two sides of the elevator shaft were secured ridged strips of iron, each shaped like half a threaded hollow tube. At each side under the car floor, short threaded shafts turned within the guides, moving the elevator up or down. A mechanical system beneath the car was powered by a belt that ran the full height of the elevator shaft and was driven by a steam engine. This system was undoubtedly ingenious and safe, though the driving belt required frequent adjustments to keep it taut.[9] Miller's screw elevator proved impractical for tall buildings, but for almost 40 years it maintained a limited popularity for lifting heavy loads for short distances.

Early in the development of steam-driven passenger elevators, there were systems in which rope or cable was wound on a drum that was turned by a belt or gearing from the engine (fig. 13.4). Contemporary evaluations suggest that this system was not economically efficient, but its major drawbacks were of a practical nature.[10] Taller buildings required larger drums at the bottom of the elevator shaft, for the length of cable was necessarily equal to the maximum distance an elevator traveled. By increasing either the length of the drum or its diameter, more cable could be accommodated, but the size of the machine was often excessive and control became increasingly difficult. A drum elevator could seldom be used in a building higher than 10 stories. Direct use of a steam engine also meant that there was no simple method of limiting the movement of an elevator car at the top or bottom of its run. Safety devices could be placed at the ends of an elevator shaft, but if they failed and the elevator operator was not alert, a car might hurtle into the overhead structure or drop onto the bottom of the shaft.[11]

For warehouse and factory hoists, drum and direct-powered systems continued into the twentieth century, but they were seldom used for passengers after the introduction of hydraulic systems. The steam-powered elevator did not provide a ride that was either smooth or reassuring. In spite of many advances in the development of the machine, many of the chugs and lurches of the steam engine itself were transmitted to the elevator car and its passengers. Stops and starts were abrupt, and the public's fears were not greatly soothed by safety devices that were not visible to them.

An elevator system operated by water was patented in 1870 by Cyrus W. Baldwin, a Boston engineer (fig. 13.6). The Hale Elevator Company in Chicago, which undertook the manufacturing of the design, advertised it as the Hydro-Atmospheric Elevator, but it was more generally known as the water balance elevator. The principle was one that had been used in funicular railways in the Alps, where cars were drawn upward by filling a counterbalancing container with water.[12] In Baldwin's elevator the car was suspended by cables that ran over pulleys to a large iron bucket that weighed slightly less than the car. The bucket, traveling in a shaft beside that of the elevator, would be filled with water, and as the bucket descended the car and its passengers were drawn upward. As the bucket was emptied and rose from the bottom of its shaft, the car would descend. The only driving force required was a steam-powered pump to transfer water from the building's basement to a rooftop tank. Stops were made by applying pressure to the guide rails, and in most installations the springs that braked the car were so strong that the car probably moved more by release of the braking systems than by control of the water balance system.

The water balance elevator attained speeds as fast as 1,800 feet per minute (about 20 miles per hour).[13] Travel was smooth and operation of the system was economical. In practice, however, its high speed proved to be a disadvantage, even a danger. At the same time the ground area required for the water bucket was thought to be excessive. Although shafts for the several water buckets in a building's elevator system were much smaller than those required for the elevator cars themselves, when extended

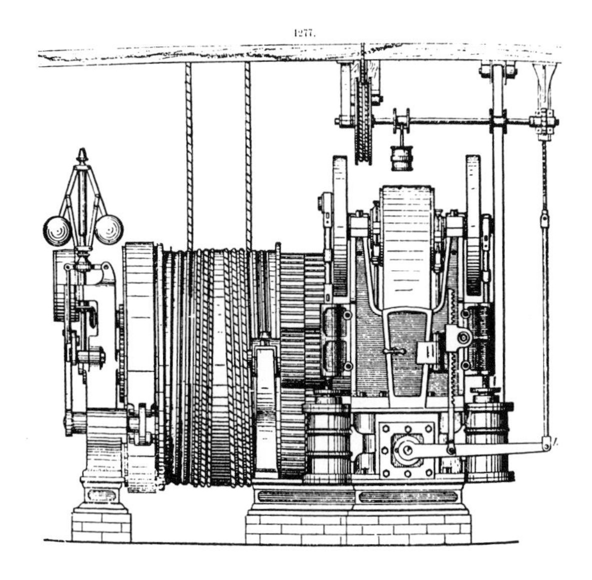

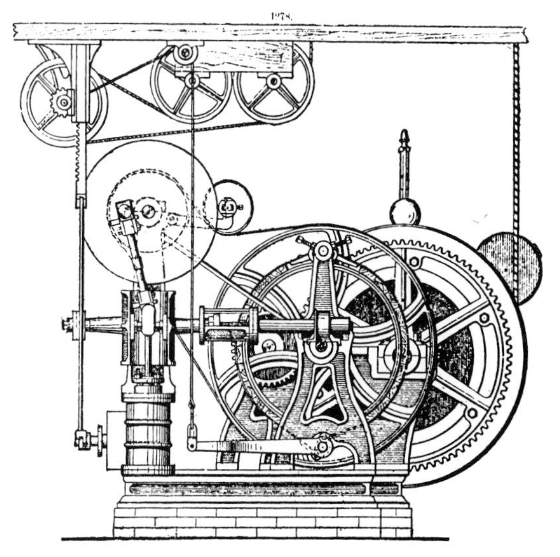

13.5 This elevator in Lord and Taylor's New York department store provided upholstered benches on which passengers rested, while an attendant tugged the rope that started and stopped the elevator. (Courtesy Otis Elevator Company. Archive, United Technologies Corporation, Hartford, Connecticut.)

13.6 The passenger car of the water balance elevator was linked to a sheet-iron bucket weighing slightly less. By adding or withdrawing water, speeds as high as 1,800 feet per minute could be attained, but it was necessary to provide for the metal tube a space as tall as the elevator shaft. (*U.S. Railroad and Mining Register*, 12 April 1873.)

13.7 Investors in Weehawken, New Jersey, built an elevator in 1891 to receive passengers off the ferry from New York and lift them to the elevation of the Palisades, where trains took them to the Eldorado amusement park and Gutenberg racetrack. Each of the three hydraulic elevators operating in this tower could carry 150 passengers 150 feet in 40 seconds. Within a few years the resort was bankrupt and the elevator tower was dismantled. (*Engineering Magazine*, June 1893.)

through all the stories of a building they occupied a considerable amount of potentially rentable floor area.

Though its popularity was short-lived, the water balance elevator found a place in many new skyscrapers. The Western Union Building in New York, second highest in the city and only 30 feet below the Tribune Building, had its direct-action steam engine elevators replaced with a water balance system shortly after the building was completed in 1875. This installation remained in service until the building burned in 1891.[14] Advances in elevator design were so rapid that by the time of the fire, water balance systems had long been replaced by hydraulic equipment and the electric elevator had been introduced.

A hydraulic crane invented in 1846 by a British engineer employed the force of water pressure to move a pis-

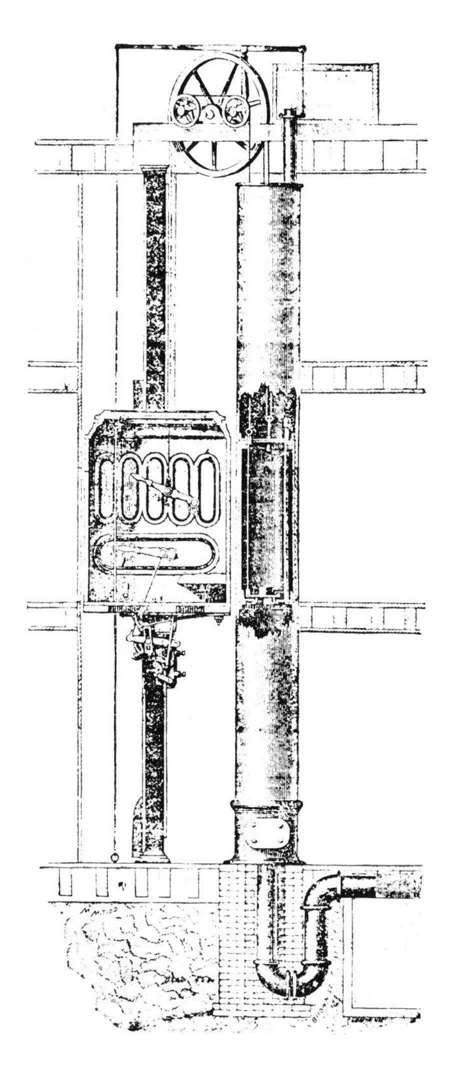

ton within an iron cylinder and thereby lift heavy objects. Unlike water wheels and steam engines, which required conversion of rotary movement to linear movement, the hydraulic cylinder produced linear movement.[15] An additional advantage of hydraulic systems was the ease with which they could be regulated with simple valves.

Cyrus W. Baldwin, inventor of the water balance elevator, was still associated with Chicago's Hale Elevator Company when in 1878 he introduced the vertical-cylinder hydraulic system in the Boreel Building in New York.[16] The vertical cylinder stood in the elevator shaft, directly behind the car, and to its piston was attached a frame bearing the traveling sheaves of the pulley system (fig. 13.8). Stationary sheaves were secured at the top of the elevator shaft. Water filled the vertical cylinder, entering under pressure near the top, and pushed the piston downward. As the piston moved, water was expelled from that portion of the cylinder beneath it, and the traveling sheaves of the pulley were drawn downward. Once the elevator had reached the top of the shaft, valves adjusted to stop water entering the cylinder and to open the way for water to move from the top of the cylinder to the bottom. The weight of the elevator slowly pulled the piston upward until the car reached the bottom of the shaft. With a counterweight somewhat lighter than the car itself, movement in either direction could be accomplished with relatively low water pressure. One distinct advantage of the vertical-cylinder hydraulic elevator was the ability to attach counterweights to the traveling sheaves of the pulley system. These weights had to be determined as the inverse of the gear ratio, so a pulley system producing three feet of car travel for each foot of piston movement would require a weight three times as great as that of an independent counterweight. Nevertheless, such counterweights had the distinct advantage of not requiring the installation of additional cables and pulleys.

On each floor, the vertical-cylinder hydraulic elevator required floor area above the cylinder. While the area needed was less than that for the water balance elevator, buildings were higher than they had formerly been, and astute landlords could not but covetously eye that small area reserved behind each elevator. The

cables extended upward to the pulleys at the top of the elevator shaft. Usually water was present only on the pressure side of the piston in a horizontal cylinder, and this permitted easy determination of the tightness of the piston's seal. In vertical cylinders, leakage around the piston could only be detected by comparing the readings of gauges that measured the pressure within the cylinder above and below the piston (fig. 13.10). The horizontal cylinder also avoided the differences of pressure that were present between the top and bottom of a vertical cylinder, which held a column of water that might be as tall as 40 feet.

Placed horizontally, the hydraulic mechanism occupied a great amount of floor area, but only in the building's basement. This location allowed walls to be constructed around the equipment, preventing the eerie sounds of creaking machinery from echoing throughout the elevator shaft. For installation of large numbers of elevators in a building, iron racks were constructed to stack the cylinders one above another, freeing basement area.

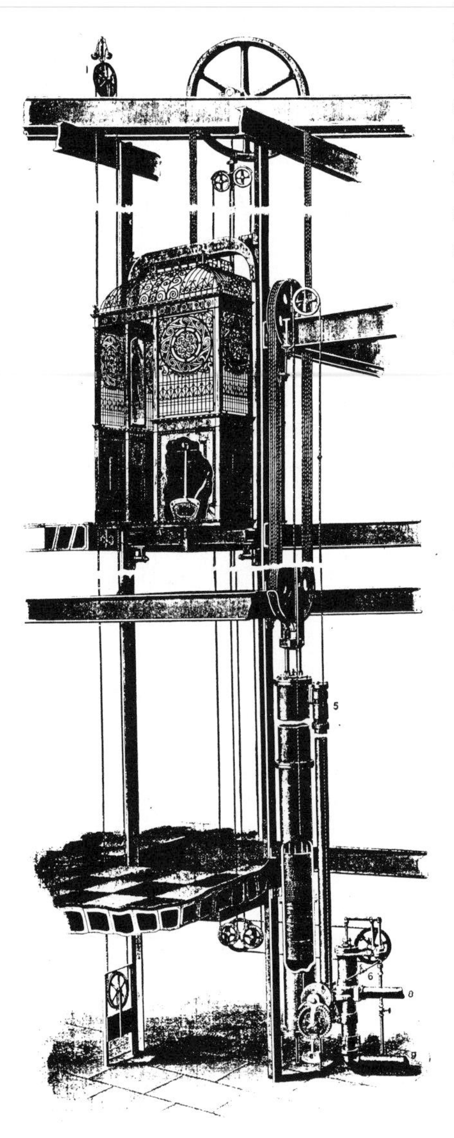

13.8 For hydraulic elevators with vertical cylinders, the traveling sheaves of the cable system attached directly to the cylinder's piston. Safety provisions included the devices shown beneath the car and the governor at the upper left of this drawing. (*Scientific American Supplement*, 12 August 1899.)

13.9 The horizontal hydraulic cylinder permitted a reduction of the area in elevator shafts. To save space in buildings' basements, often crowded with furnaces and generators of electricity for lighting, the cylinders could be stacked on steel frameworks. (Courtesy Otis Elevator Company. Archive, United Technologies Corporation, Hartford, Connecticut.)

horizontal-cylinder hydraulic elevator, introduced at about the same time as the vertical-cylinder system, was essentially the same but occupied basement floor area instead of rentable space in the upper stories (fig. 13.9). A piston rod moved horizontally with the traveling sheaves of the pulley system rolling on iron rails. Bending around a set of pulleys attached to the frame of the hydraulic engine, the

Both sorts of hydraulic elevator systems benefited from the use of the pulley. Pulleys could have been used in drum elevator systems, but they would have resulted in an awkward mechanism that magnified the unevenness of the movement. In the simplest form of hydraulic elevator, one end of the cable was secured to a crossbeam at the top of the elevator shaft. The cable extended down the shaft to turn through the traveling sheave fastened to the piston, then went back up the shaft to the top where it turned around a fixed sheave before turning down to be fastened to the elevator car. This allowed the car to travel two feet with only one foot of movement by the traveling sheave and the piston. Additional pulleys and more cable could increase this gearing ratio. Because of their locations, vertical cylinders tended to be made tall with small diameters and their gearing ratio seldom exceeded 4:1, four feet of elevator travel for each foot of piston movement. Horizontal cylinders were shorter and large in diameter, and consequently their ratio might be as high as 12:1. Longer piston travel offered advantages in smooth starting and stopping, but the choice between the two types of cylinder was more often made on the basis of the available water pressure and space allocations. A major improvement in the smooth operation of hydraulic equipment resulted from the use of valves that had been invented for the control of machinery in the manufacture of sugar.[17] By employing a series of openings, graduated in size, these valves produced a smooth acceleration and slowing of the elevator car's movement.

The simplest method of supplying power to a hydraulic elevator was the use of water from city mains. With relatively low pressures, ranging from 20 to 40 pounds per square inch, a horizontal cylinder was the necessary choice because of its larger piston head and the pressure variation that resulted from the height of the column of water contained in a vertical cylinder.[18] A more common procedure during the period of hydraulic elevators was the installation of a water tank on the top of the building. By pumping water from the basement into the tank, a high pressure could be obtained, although it was, of course, limited by the height of the building. Repeated use of the same water provided significant savings when several elevators were to be served. When the building's height did not afford sufficient water pressure, closed pressure tanks provided the pressure that was needed.[19] Most standard elevator

13.10 This drawing of a vertical cylinder hydraulic elevator shows a typical turn-of-the-century installation. Buttons in the car regulate valves that alter the admission of water and the flow of water from top to bottom in the cylinder. At the left, a governor and its cable control safety mechanisms beneath the car. (W. Baxter, Jr., *Hydraulic Elevators*, 1905.)

equipment was designed for a pressure of 150 pounds per square inch. As higher pressures were attained, smaller and less expensive tanks, cylinders, and pipes could be employed, but higher pressures demanded special and more expensive equipment.[20] Air trapped in the top 40 percent of a closed tank exerted pressure on the water below, and an automatically controlled pump maintained this pressure at a constant level.[21]

The plunger elevator was a hydraulic system in which the cylinder was placed in the ground directly beneath the car and the piston was directly attached beneath the car. The plunger had been employed in English factory hoists in the 1830s and had been admired for its safety and economy. Throughout much of the nineteenth century, plunger elevators were the dominant form in Europe for passenger service. With relatively few tall buildings being built there, the height limitations of the plunger elevator were of little significance, and consequently the European distrust of cars suspended by cables prevailed. With the assistance of a counterweight system, a plunger elevator could often operate on the water pressure available from city mains.

The increased use of plunger elevators in the United States late in the nineteenth century was influenced by the improvement of methods for drilling the holes into which hydraulic cylinders were inserted, a result of techniques developed in the oil industry. Advances in the manufacture of steel pipe brought about further improvements. A telescoping plunger was introduced, but it had little success. Although it required a much shallower cylinder, the difficulty of maintaining watertight connections between the telescoping sections made the system impractical.[22]

Plunger elevators at the end of the nineteenth century were installed with runs as high as 30 stories. Tests made by George I. Alden at Worcester (Massachusetts) Polytechnic Institute encouraged increased use of plunger installations. Although early plunger elevators had been slow, improved valves allowed speeds as high as 600 feet per minute. In 1904 a total of 110 plunger elevators were ordered for installation in the John Wanamaker stores in New York and Philadelphia.[23] In the end, electric elevators replaced plunger installations for all but relatively low runs and speeds not exceeding 200 feet per minute.

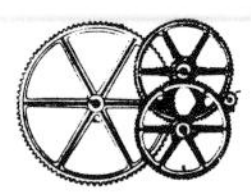

Once begun, the use of electrical power in passenger elevators advanced rapidly, greatly assisted by the development of electric motors for the streetcar. By 1873 both the Gramme dynamo and another manufactured by Siemens and Halske had been improved to the point of providing reliable and constant current. Many textile mills of the northeastern United States in the early 1880s installed hoists for which electric motors provided power to belt-driven winding drums, substituting electric power for steam power. Hoists of this sort were used for freight until the end of the nineteenth century.

Werner Siemens, the electrical wizard of Germany, exhibited an electric elevator in 1880 at the Industrial Exhibition in Mannheim. The system in many ways resembled the vertical screw elevator that had been developed by Otis Tufts about 20 years before. In the center of the shaft and the car, a notched steel column extended the full height of the elevator's run, providing the direct support usually demanded by the European public. Beneath the floor of the car,

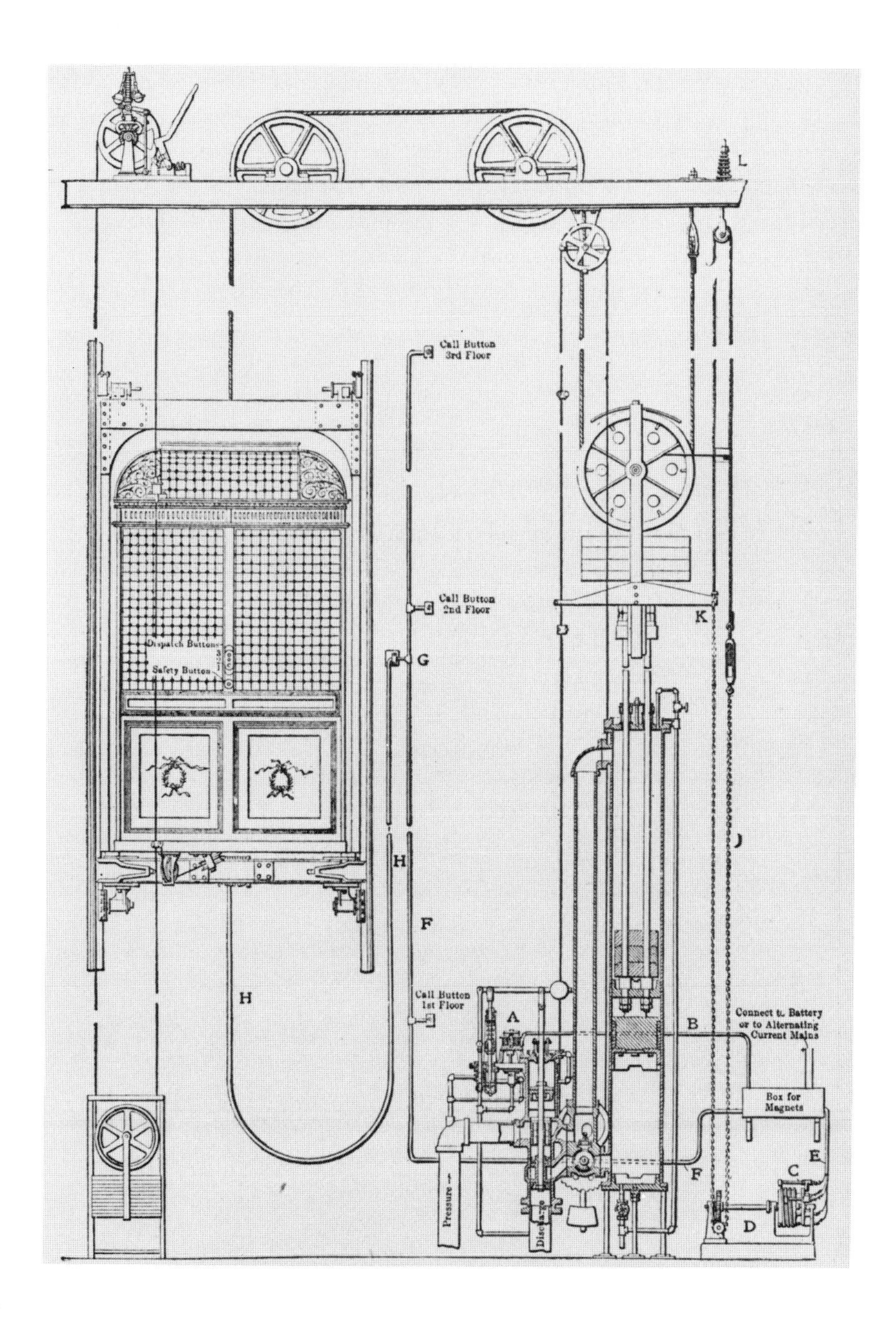

an electric motor turned cogwheels that raised or lowered it. Over 8,000 visitors at the Mannheim fair rode in Siemens's elevator at a speed of about 100 feet per minute, a jerky, thumping ride because the motion of the gears was transmitted directly to passengers.[24]

The first permanently installed electric elevator was designed by William Baxter, Jr., in 1887 for a Baltimore building. Power was supplied by a generator designed for arc lighting systems, and the electric motor was geared to a drum about which the cables were wound.[25] The whirling of a centrifugal governor controlled speed and, when the car descended with a full load, the governor could reverse the motor so it would act as a brake. Hand control was limited to a switch in the car that reversed the direction of the car's movement at the top and bottom of its run. Switches connected to the doors at every floor level stopped and started the motor, a

13.11 Open elevator cages gave an opportunity to explore the artistry of metalwork in the execution of grilles and gates. A French folio of 1898 shows an ornate design, set within a stairwell of more classical style. (T. Lambert, *Escaliers et Ascenseurs*, 1898.)

13.12 An electric elevator was developed by 1887, and nine years later the Otis Elevator Company reported that it had installed more than 4,000 of them. Acceptance of the electric elevator depended largely on determining its cost of operation. (*Electric Review*, 21 February 1896.)

13.13 The Sprague-Pratt elevator system substituted electric motors for hydraulic cylinders, threaded shafts taking the place of pistons. Like their hydraulic counterparts, these machines could be stacked in order to save basement floor area. (*Transactions, American Institute of Electrical Engineers*, 22 January 1896.)

system that resulted in relatively slow speeds of travel. Because electric motors produced rotary motion, the use of a drum was the most practical solution, but drums large enough to hold great lengths of cable were difficult to control and required extremely large motors (fig. 13.12). For this reason, hydraulic elevators continued to dominate large installations until the end of the century.

The problems of acceleration and deceleration troubled early electric elevator systems as much as they had limited hydraulic systems. In the latter, a series of valves, graduated in size and opened sequentially, had served to avoid lurching starts and jolting stops. In the former, a series of electrical resistances were engaged or disengaged, one by one, to achieve the same objective. By the end of the century the Otis Elevator Company had introduced a "magnetic dash-pot" that provided an alternative system for cushioning the electric motor's action. An additional advantage of this device was the fact that it reduced the drain on a building's total electrical system when an elevator motor pulled quickly against the inertia of a car and its counterweight. Before controlled acceleration was introduced, if a building operated its own steam-powered generators, a rapid start of one elevator car could suddenly dim lighting throughout the building.[26]

Charles R. Pratt, the young chief engineer of the Sprague Electric Elevator Company, in 1888 invented a new form of electric elevator that was extremely eclectic in its design. Pratt had worked for two other elevator manufacturers before joining Frank J. Sprague, who had played a leading role in the development of the first electric streetcar systems. The Sprague-Pratt elevator, as it was called, used a long horizontal screw driven by an electric motor (fig. 13.13).[27] Threaded on the screw was a block to which was attached the traveling sheaves of a pulley system, much the same as those used for horizontal-cylinder hydraulic elevators. Because they occupied about as much basement area as horizontal hydraulic cylinders, Sprague-Pratt machines were also often stacked on metal frames. In use they were plagued with mechanical difficulties and, because smoothness of motion was their only significant contribution, the Sprague-Pratt elevators were not popular.

Toward the end of the nineteenth century, the comparative advantages of hydraulic and electric passenger elevators were carefully studied. Few of the experts writing on the relative efficiencies of available systems were without a bias toward their own inventions, their company's systems, or a recent installation that they had designed. Hydraulic elevators, whether using vertical or horizontal cylinders, profited from the long experience that made them simple to operate and low in initial cost. Using water under high pressure offered a degree of economy, especially in English cities such as London and Liverpool, where mains in the central portions of the city furnished water at a pressure of 700 pounds per square inch.[28] However, the small valves that were required in high-pressure systems were difficult to maintain, and when they leaked the elevator controls were inexact in their action. Some high-pressure installations were

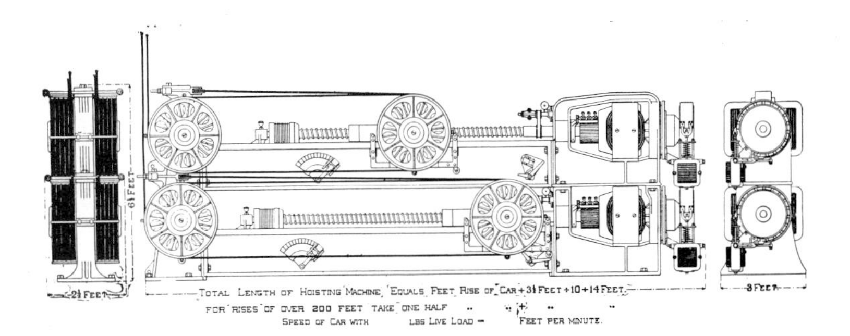

13.14 The Otis traction elevator of 1909 had its motor and drum at the top of the shaft with a braking system (*B, B′*) provided. A governor (*N*) and an undercar system (*I*) provided additional safety. The counterweight (*H*) and compensating system (*F*) controlled the load. (*American Review of Reviews*, December 1909.)

replaced only a few years after being first put in service. A system that provided water at both low and high pressure, employing high pressure only when the elevator load required it, offered definite advantages, but it had only a short period of application.

The electric drum elevator was so expensive to install that it was used only where the best of service was demanded. Such installations were unimpressive until improvements in the gearing almost tripled their efficiency.[29] The Sprague-Pratt electric machine had a reasonable efficiency, but the power used in starting it was costly. The power per mile traveled in the Marshall Field Building in Chicago when there were infrequent stops proved to be about two-thirds of that required when the cars stopped at every floor.[30] Weighing all the factors of maintenance, cost, and smooth control, it was extremely difficult at the turn of the century to make a choice among the available types of elevators.

Experiments at the Otis Elevator Company in 1903 led to the company's introducing the traction elevator system, which found a fresh combination of the customary elements in electric-powered elevators of that time (fig. 13.14). In all previous elevators, even the plunger elevator with its hydraulic cylinder beneath the car, counterweights served only the purpose of lightening the load to be moved by the driving mechanism. Theoretically, if sufficiently large motors were provided, these systems could have lifted and lowered their elevator cars without counterweights. The traction elevator abandoned the task of pulling up the elevator car and focused its attention on the balance between the car and its counterweight. Thus, there was no longer a need for the elaborate system of pulleys that had been used for both hydraulic and electric elevators, systems in which the pulleys, cables, and frames often weighed much more than the load to be moved. In its simplest terms, the traction elevator had cables attached from the car to the counterweight, and those cables passed over a drum that could be turned in either direction by an electric motor. The cables and the elevator car were moved by friction between the several cables employed and grooves in the drum, just as friction between wheels and steel rails moved locomotives and streetcars. One safety advantage of the traction elevator lay in the fact that if safety devices failed and the car or counterweights overran their limits at the bottom of the elevator shaft, the tension on cables would slacken, and the cables would no longer move as the drum turned.

Half-wrap traction systems had a single cylindrical drum at the top of the shaft with a series of V-shaped grooves cut into the drum to increase its friction with the cables (fig. 13.15). These grooves sometimes wore away after years of use, reducing the pinching action that grasped the cables, but the surfaces of drums were made in metal segments that allowed replacement. Full-wrap traction systems provided a second and smaller drum around which the cables ran before a second half-turn around the powered drum. The cables rested in grooves that were usually semicircular in shape, and they had twice the length of frictional surface provided by half-wrap traction designs.

The design of traction elevators permitted mounting the electric motor at the top of an elevator shaft, eliminating the congestion of basement areas that had resulted from the construction of towering office buildings needing many elevators on parcels of land that were extremely limited in area. By the turn of the century an increasing number of buildings in the

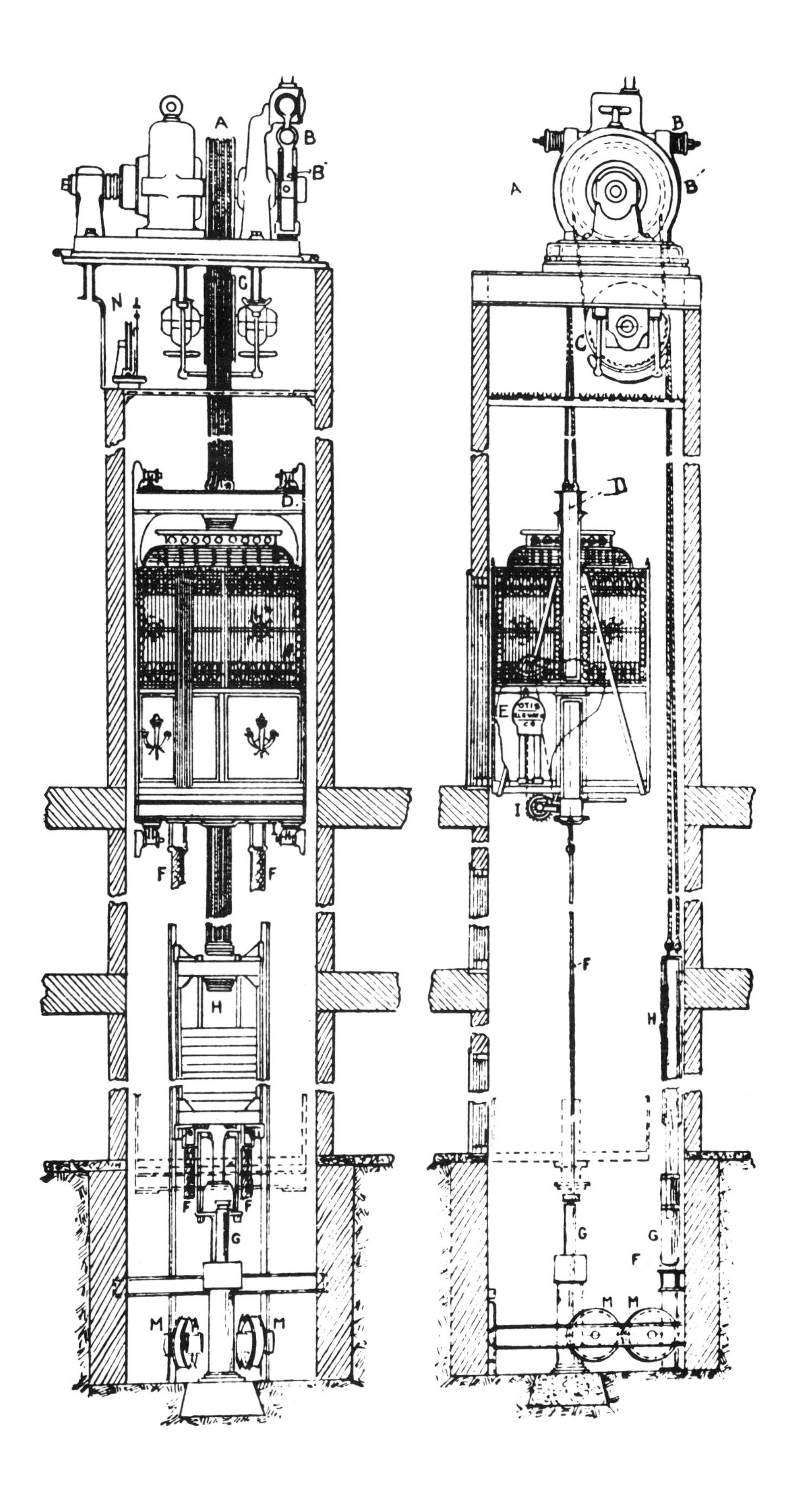
A
B
B'
N
C
D
F
F
H
F
F
G
M
M
B
A
B'
C
D
E
OTIS
I
F
H
G
G
F
M
M

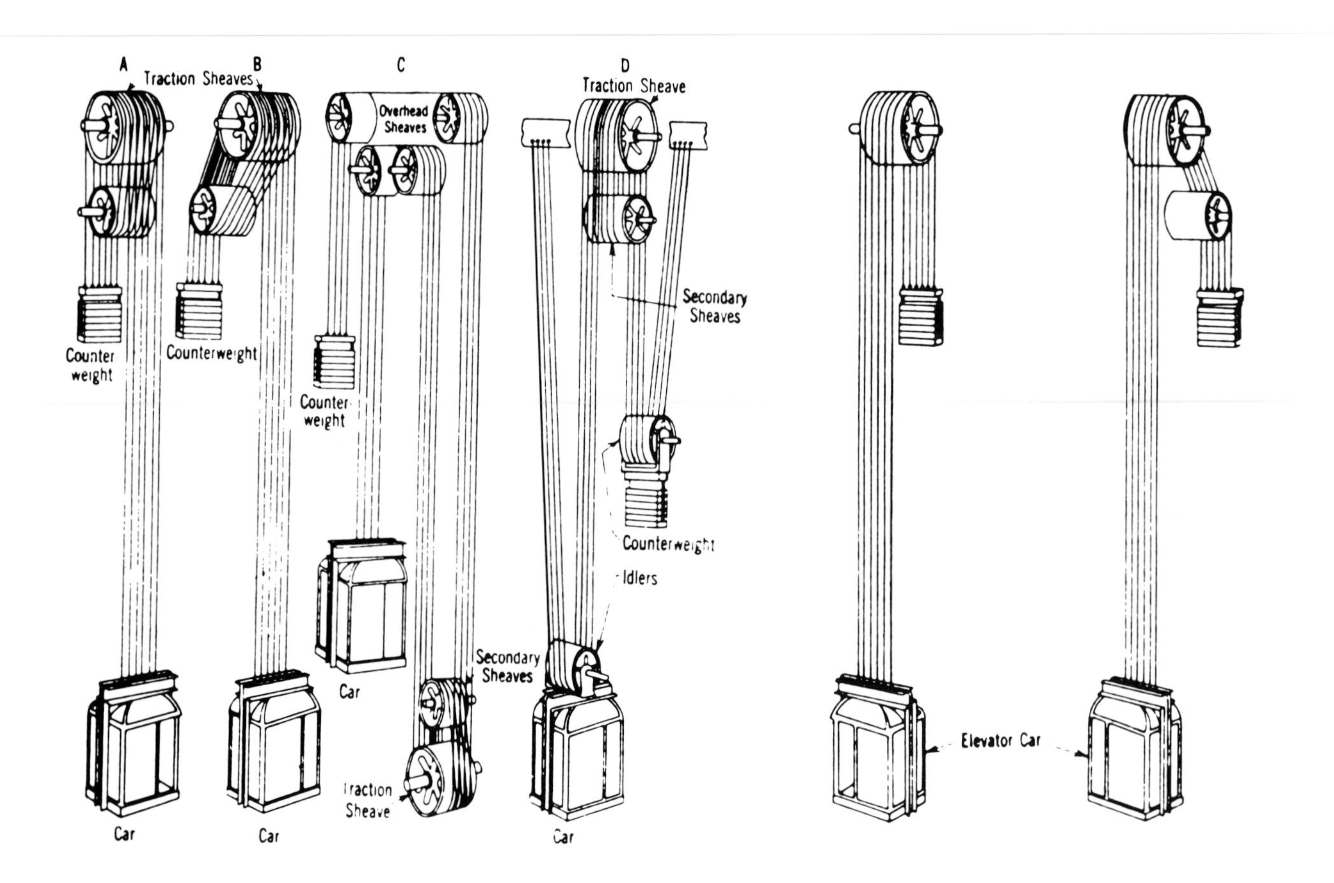

centers of cities had abandoned their individual steam-powered generating plants in favor of electrical supply from central plants, and basements were freed for other uses. By 1906 electric elevators comprised almost 90 percent of the new installations, and it was reported in 1922 that about 98 percent of new elevator installations in the United States were operated by overhead electric motors.[31]

Facilities for vertical traffic are the direct result of increased means of rapid transit on the level. While businessmen were restricted to the slow-moving horse-car or cab, business interests were scattered; the ship-owner had his office at the dock, the lawyer was at the courts, and the manufacturer at his factory; as a consequence, the building of five stories fulfilled all requirements, and the lift was a luxury. The advent of comfortable rapid transit and the telephone changes these conditions; the separate trades draw close to each other by means of business offices in the city, and the ensuing contraction causes the high building to spring up and the lift to become a necessary tool in the "manufacture of transportation."[32]

This comment, made in 1897, provides only one of the many reasons that have been given for nineteenth-century growth in the central portion of cities. There had long been systems of urban transportation, but none rapid or cheap enough to allow many to live far from their working place. Although many voices warned of the dangers of "high-building mania," it was inevitable once business had begun to concentrate at the cities' centers. A British journalist commented of New Yorkers, "They move almost as much on the perpendicular as on the horizontal plane. When they find themselves a little crowded, they simply tilt a street on end and call it a skyscraper."[33] This was in fact the image that excited many. The *Scientific American* reported with evident pride that "in the modern city the streets are often vertical. In a modern community like the Park Row Building in New York there are over six thousand inhabitants, with a vertical thoroughfare having twenty-five cross streets."[34]

No matter how much disapproval was expressed, the network of ground transportation spread farther and office buildings—and the elevators in them—climbed higher. At one time there had seemed a natural end to this quest for height, as the thick masonry walls required to support tall buildings began to fill the ground space that had become so valuable. Steel frame construction swept away this limitation, and the competitive craze for taller buildings was under way once more. By 1897 it could be said that in New York there were over 5,000 elevators of various kinds and that their daily travel was "as much, if not more, than the distance travelled horizontally by the various local cars and trains."[35]

If elevators were vertical streets, they also suffered from many of the psychological factors that could be observed in ground traffic. When the man in charge of the elevators of New York's Equitable Building was interviewed in 1924, he identified his principal problem: "Almost all of them [passengers] think they are in a hurry. Everyone rushes. . . . Men rush round a corner . . . so fast that they fairly pile into the cars. Once in, everyone feels cheated if the car does not start *at once*."[36] Most of this confusion occurred in the elevator lobby at the street level, where the ground floor of the building had become a frantic interchange between horizontal and vertical transportation systems.

Before World War I it could be said with accuracy that "one set of architects is wedded to the semicircular arrangement [of elevators]; another to their disposition in line; and another to divisions facing each other."[37] As long as a building did not require more than a dozen elevators, the semicircular array provided excellent visibility of all elevators arriving at the lobby level, but the open space in front of the arc of elevators was necessarily repeated on each floor, often at great cost of rentable office area. When a building site had one side abutting another structure, making that side unavailable for office windows, the curved bank of elevators could be placed there without much

13.15 In the full-wrap traction elevator (the four drawings at left) the bearings of the driving sheaves were heavily loaded, and this caused the system to be less efficient. The load on the bearings of the half-wrap machine (the two drawings at right) was half as great. Cables lasted longer on half-wrap machines, but the V-grooves on the sheaves, used to gain maximum friction, had to be replaced more often than the semicircular grooves of full-wrap machines. (*Journal, American Institute of Electrical Engineers*, April 1922.)

13.16 In the Metropolitan Life Building, Minneapolis, Minnesota, elevators were caged in corners of a central sky-lighted space. The effect was heightened by thick glass slabs that were the flooring of the galleries around the light well. (Courtesy of the Minnesota State Historical Society.)

13.17 The Havemeyer Building in New York (1897) had a ring of six elevators opposite the entrance and jutting into a narrow open space against an adjacent building. Two were express elevators, particularly useful for access to a rooftop restaurant and promenade, "where superb views of the bay and surrounding country can be had from this height" (15 stories). (W. Birkmire, *Skeleton Construction*, 1894.)

13.18 Arrangements of cables and sheaves could allow efficient location of the hydraulic machinery in the basement, as shown in this partial plan of the Empire Building, New York (1897). Below the third floor, the individual shafts were separated by masonry walls that were part of the air-cushion safety system. (*Scientific American Supplement*, 12 August 1899.)

floor area being wasted (fig. 13.17). For plans in which offices ran along both sides of corridors, whether C-shaped, E-shaped, or wrapped around a central light well, a linear arrangement was more desirable, since on upper floors little lobby area was required beyond that provided by the office corridors. Lines of elevators were limited to about seven cars, because a waiting passenger could not easily get to all of the cars in a longer line and the elevators farthest from the lobby entrance would not receive sufficient use (fig. 13.18). Later skyscraper towers often had parallel rows of elevators, a plan that was particularly appropriate when cars serving the same zones of height in the building could be placed facing each other. Also, when regulations required setbacks that reduced the area of upper floors of a skyscraper, rows of elevators could be eliminated as they reached the upper limit of their zone of service (fig. 13.19).

If saving floor area in an office building was a strong factor in elevator design, it was also vital to determine accurately the number of elevators that would be needed to serve a building. A British engineer stated the case in the 1920s: "I have said that, given the necessary particulars of a building, the lift engineer will be able to calculate the probable traffic. It must be admitted, however, that the lift engineer himself does not

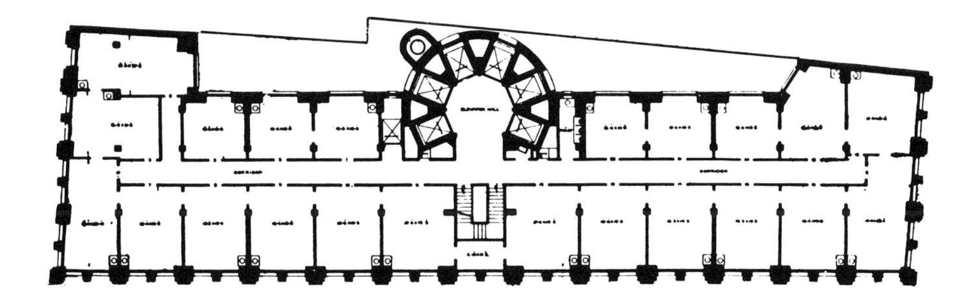

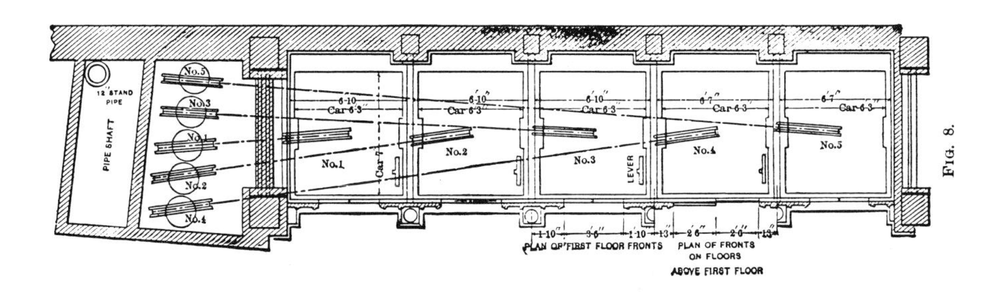

Fig. 8.

employ any scientific method in arriving at the number of passengers per minute which will require lift service on each particular floor during the busy part of the day. Each will draw upon his own experience and home-made formulae."[38] Like any form of transportation, elevators required a balancing of average and peak demands, with economy an inescapable consideration. Every type of building had a different relationship between floor area and the number of occupants to be accommodated, and there was much debate about the validity of comparing office building populations in London, New York, and other cities. Rapid elevator service was a marketable convenience, and the number of elevators might be greatly influenced by the level of rents to be charged in a building. All of these factors, subjective as they were, entered into the calculations by which elevators were selected. Unpredictable variations were so frequently encountered that at a Chicago convention one engineer warned against putting too much trust in estimated average conditions and concluded that "the psychology of it is more important than anything else."[39]

As early as 1898 the New York *Evening Post* reported that in "the dwelling houses of the rich" elevators were being provided for "men with money and a desire to be 'up to date'." In fact, elevators were so popular that "in cases where the elevator is not to be put in immediately provision is made for it either by space left for the shaft or by the actual construction of the shaft."[40] It was the electric elevator and push-button controls that made residential elevators practical; only five years after Baxter installed the first electric elevator in the United States, one company had installed 90

13.19 **Zoning of the elevators in the New York Telephone Building (1926) is indicated in this drawing. The locations of elevator groups are exaggerated in the diagram and can be more clearly seen in the plans of the elevator core at different levels. (F. Mujica, *History of the Skyscraper*, 1930. Courtesy of Da Capo Press.)**

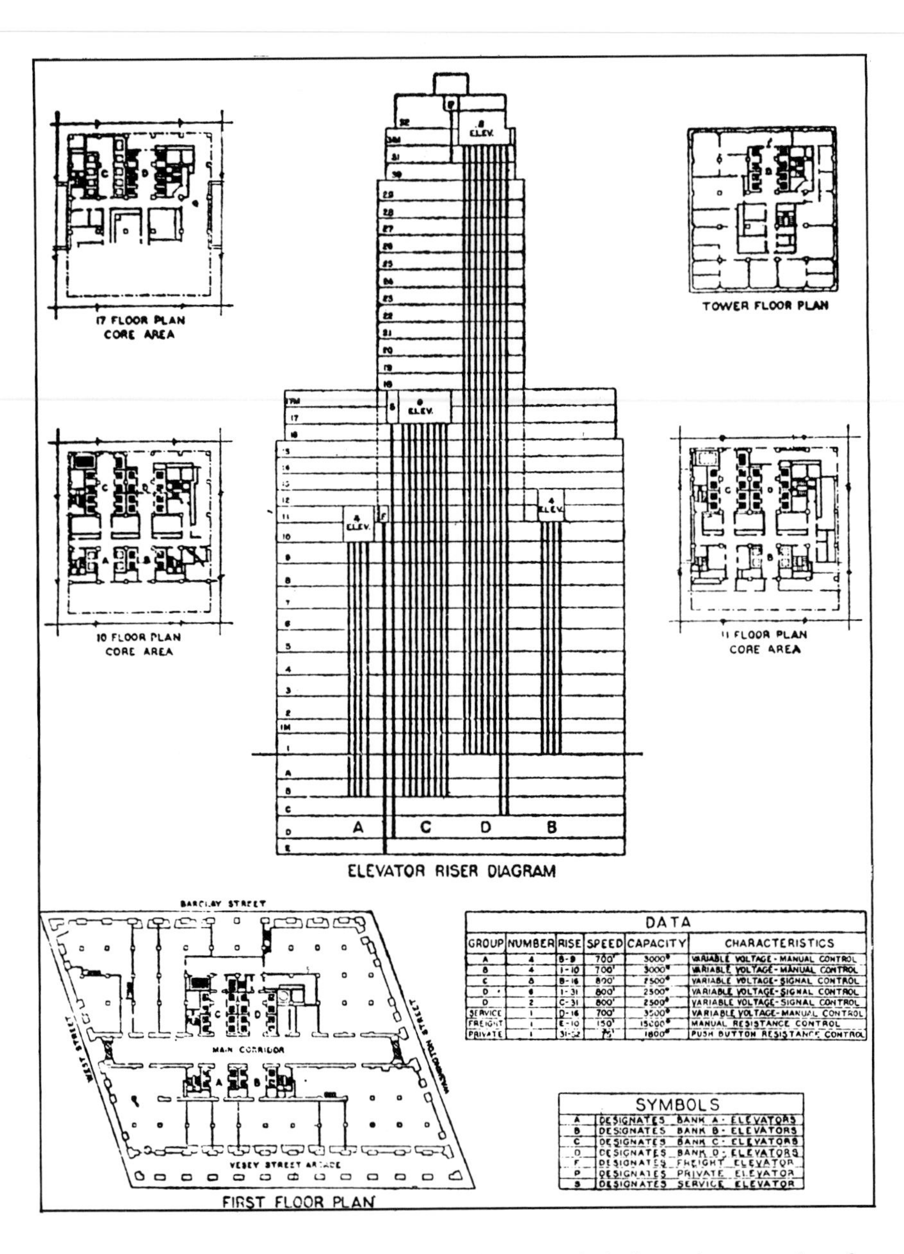

DATA					
GROUP	NUMBER	RISE	SPEED	CAPACITY	CHARACTERISTICS
A	4	B-9	700'	3000#	VARIABLE VOLTAGE-MANUAL CONTROL
B	4	1-10	700'	3000#	VARIABLE VOLTAGE-MANUAL CONTROL
C	8	B-16	800'	2500#	VARIABLE VOLTAGE-SIGNAL CONTROL
D	6	1-31	800'	2500#	VARIABLE VOLTAGE-SIGNAL CONTROL
D	2	C-31	800'	2500#	VARIABLE VOLTAGE-SIGNAL CONTROL
SERVICE	1	D-16	700'	3500#	VARIABLE VOLTAGE-MANUAL CONTROL
FREIGHT	1	E-10	150'	15000#	MANUAL RESISTANCE CONTROL
PRIVATE	1	31-32	75'	1800#	PUSH BUTTON RESISTANCE CONTROL

SYMBOLS	
A	DESIGNATES BANK A - ELEVATORS
B	DESIGNATES BANK B - ELEVATORS
C	DESIGNATES BANK C - ELEVATORS
D	DESIGNATES BANK D - ELEVATORS
F	DESIGNATES FREIGHT ELEVATOR
P	DESIGNATES PRIVATE ELEVATOR
S	DESIGNATES SERVICE ELEVATOR

residential elevators in New York. Push-button controls and safety door locks made it possible for the residential elevator to be operated safely by the "hands of children and servants."[41]

Residential elevators, although large enough to carry only three to eight persons, were customarily decorated lavishly:

The entire designing and construction of the elevator is done under the supervision of the elevator company, but the ornamentation is confided to the decorator and furniture makers. The elevator is almost never built of iron or steel, but of some handsome hard wood. Much carving sometimes goes into it, and the upholstery and other decoration are often very sumptuous.

Elevators are found very useful at large balls and receptions, and the time is rapidly approaching when it will be considered "bad form" if guests are asked to walk up a pair of stairs.[42]

Settees, gilded mirrors, and thick carpets were natural decorations for elevators in the houses of the wealthy, but there was a temptation to add ornament on other occasions. In spite of the protests of safety engineers, architects tended to view the elevator doors of office buildings as something to be made esthetically acceptable by "expressing their architectural ambition in iron and bronze."[43]

Around 1880, elevator cars for hotels and some department stores in the United States contained upholstered seats and benches. "Some of these cars are gorgeously inlaid and decorated; mirrors are inserted in the small amount of surface available for that purpose, and rich chandeliers hang in the middle."[44] Gas was brought to the light fixtures by a flexible tube that coiled on the top of the car as it rose. Later, faster service obviated the need for comfort in transit, and office building installations limited embellishments to the doors that were visible in the building's lobby. It was essential that the elevator car be kept as light as possible, and therefore it was long assumed that wood construction was highly desirable, if not necessary. But there were many devices that had to be attached to the elevator car: cables connecting the top of the car with counterweights and the driving mechanism; fittings that controlled the car's movement along the guide rails at the sides of the shaft; and brakes and safety devices that were usually attached to the bottom of the car. For this reason, whether operating in a closed shaft or the open cage that was popular in the United States, the elevator car came to be constructed within a sling made of iron and steel. Across the top a sturdy beam held the cables and was connected to two vertical members that ran beside the guide rails. Across the bottom a beam bore the weight of the car and its passengers and the safety devices that were provided.

The typical early elevator shaft was undeniably a fire risk. In 1880 it was pointed out: "Until recently freight elevators were generally open and passenger elevators were enclosed. Now many of the former are enclosed and some of the latter open."[45] When enclosed, the shaft became a chimney by which flames could spread quickly from floor to floor, and it was at the same time a potential source of fire. "Even in so-called fireproof buildings [in the United State] the posts are almost always of pitch-pine, soaked with grease which is used to lubricate the guide strips." At each story the elevator shaft would be closed off from the hallway with thin barriers of varnished wood, glass, or simply metal grilles. All these materials were ready to spread a fire that might easily be started in "oily rags or shavings, or by hot ashes from the boiler" which were often near the bottom of the shaft.[46]

In Europe it was different, and American inattention to fire precautions might have been "considered criminal abroad." According to one observer, "On the continent, open hatchways are forbidden within the walls of buildings, and such elevator-wells as are placed there are always enclosed from bottom to top by stout brick or stone masonry, with doors at each story."[47] In fact, whether as a fire precaution or an uncomplicated

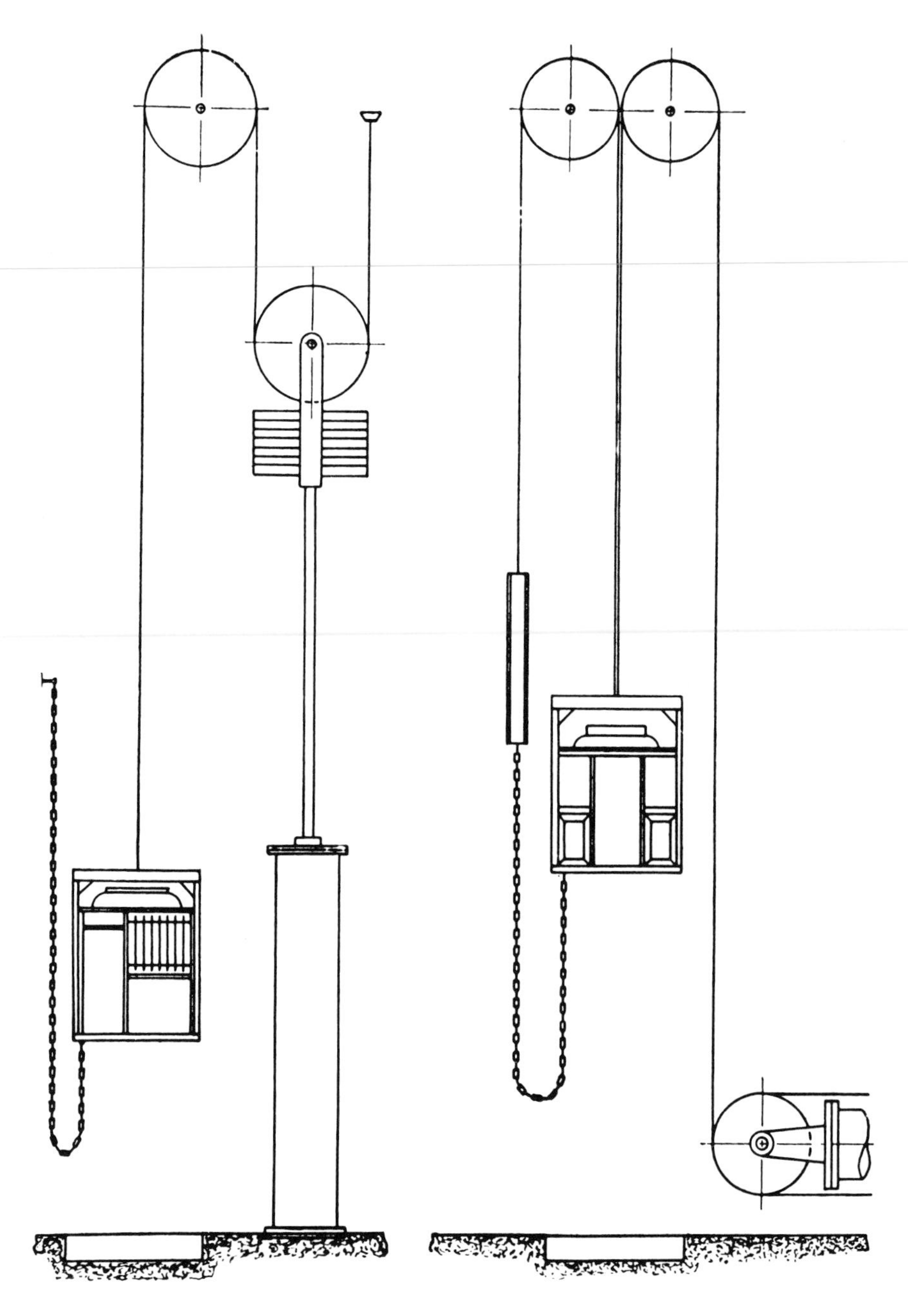

13.20 As buildings grew taller, it was necessary to compensate for the weight of cable in order to maintain a balance between the weight of the car and its counterweight. Compensator chains were attached to the car and to either the walls of the elevator shaft or the counterweight. The clank of chains could be reduced by weaving soft rope through the chain's links. (J. J. Jallings, *Elevators*, 1916.)

way of adding elevator service to existing buildings, elevator shafts in Europe were sometimes supported by an open framework of iron situated beside the building in a courtyard or areaway. Europeans were also inclined to construct their elevator cars of sheet metal, less flammable than their American counterparts.

A reason for open elevators may have been the early lack of lighted signals to warn passengers that a car was approaching. Many insisted it was essential that the waiting passenger see the elevator coming and that the operator in the car have a full view of every floor as he passed it. Other experts recommended that shafts should be built of solid masonry materials, extending above the building's roof and having a ventilator at the top. However, when Boston set about revising its building regulations, which had been enacted after the disastrous fire of 1872, the only requirement given serious consideration was one demanding that shafts should be separated from the remainder of the building by doors at each story.[48]

A smoother ride and more economical operation was possible through the use of counterweights. These could be calculated to have the desired weight in relation to the aggregate weight of the elevator car, parts of the pulley system, and pistons or other elements of the driving mechanism, but in tall buildings the weight of varying lengths of cables in the shaft with the elevator or on the side of the counterweight could become a significant factor in balancing the system to minimize the power required. An early method of compensating for the weight of cables consisted of heavy chains, one end fastened to the bottom of the elevator car and the other end to the wall of the elevator shaft (fig. 13.20). As the weight of the cables shifted, the proportion of the chain's weight hanging from the elevator car changed to equalize the system. Later a narrow upright pipe of water was placed near the vertical cylinder of hydraulic elevators and was connected to the water in the cylinder of the hydraulic system. As the piston rose or fell, the height of this column of water changed to provide a variable pressure that compensated for the shift of the cable's weight.

One factor that encouraged the abandonment of compensator chains was the clanking sounds they made, certainly not reassuring to nervous passengers. In their 1910 installation of electric traction elevators in the 44-story Metropolitan Life Building in New York, the Otis company substituted flat wire rope almost 4 inches wide in order to avoid the fearsome noises of chains.[49]

The earliest controls for steam-driven elevators were nothing more than loops of light cable that extended from the top to the bottom of the elevator shaft with one side of the loop running within the elevator car. By grasping the cable, an elevator operator caused the loop of cable to move with the car, and that movement activated a device that shifted the belt powering the elevator off the driving wheel and onto a freely spinning wheel. Another tug on the cable moved the belt back to its original position, and the elevator resumed its travel. More exact controls were needed. After all, if the elevator moved at 500 feet per minute, the maximum speed allowed in New York in 1913, a misjudgment of one-tenth of a second could result in the car's stopping almost one foot above or below the floor level intended. Obviously, many jolts and starts might be required for an inexpert operator to stop the car reasonably close to the desired floor level, and the warning "Watch your step, please!" was an essential part of the elevator operator's duties.

As speeds became greater and the hydraulic elevator came to dominate the field, the problems of control increased. At first, controls within the elevator car opened and closed a valve that governed the pressure of water against the piston, but the full force of water pressure made starting and stopping the elevator too abrupt. Similarly, electric elevator controls were relatively crude until around 1915, when an automatic leveling device was introduced that not only brought the floor of the car to the desired level, but also compensated for the cables' stretching when a large number of passengers caused a significant increase in the load. This advance allowed the night operation of office building elevators by late-working employees in the offices, sped the development of higher apartment buildings, and encouraged the installation of elevators in private residences (fig. 13.21).

A major factor in the development of elevators was the public's fear of elevator accidents. In 1878 the Paris newspaper *Figaro* printed a description of an accident at the Grand Hotel that soon found its way into the press of other countries:

The conductor touched the button, but instead of descending, the car began to mount with alarming rapidity. The casting which united the piston to the platform on which the car rested had broken. While the force of water was beneath the piston and the car was ascending, this was not felt, as the piston ran up and lifted the load as usual; but immediately the escape-valve was opened at the foot of the supply-pipe, the piston darted downward with fearful speed to the bottom, while counter-weights, now much heavier than the car and its load, began to run down, and pulled the car up at a dizzy rate. Arriving at the top floor, the car was violently rammed against the top beam. The shock was so great that it broke the chains which held the counterweights, and the car went flying down to the basement. The weights fell with a report almost equal to a cannon-shot, attracting nearly every person in the building. The three occupants of the cage were dead, all bleeding from the mouth and ears, showing cerebral congestion. At the inquest it was ascertained that a "blister" in the casting was the cause of the disaster.[50]

13.21 Push-button controls, leveling devices, and safety systems at elevator doors encouraged the use of elevators in apartment buildings. (*American Architect and Architecture*, 6 August 1919.)

The elevator was of the direct plunger type favored by Europeans for its apparent safety, and it had been installed only 18 months before the accident. The hydraulic cylinder and piston were located directly beneath the car, acting as a column of changing height, and a chain extended upward over a pulley to balance the car with a counterweight and thus reduce the work of the plunger system.[51] It had all seemed so safe that no safety devices were provided.

In 1857 a Boston freight elevator loaded with boxes of sugar fell to the bottom of its shaft, and the boxes were undamaged. Investigation showed that the car had been protected by relatively airtight construction at the bottom of the elevator shaft, the air entrapped there serving as a brake for the falling car. Several decades later the air cushion safety system became a common provision in elevator installations. The *Engineering News* in 1899 referred to the air cushion as having been "in considerable use some years ago" and explained that "a car falling freely down a 100-foot shaft, for example, would acquire a velocity at the bottom of about 80 feet per second, or 55 miles per hour, and unless a very deep shaft were made for the air cushion, the stop would be highly disastrous to the passengers."[52]

The year before these comments were made, there had been a demonstration of one of the most extensive and elaborate applications of the elevator air cushion in the Empire Building at Broadway and Rector Street, New York. As in most such installations, the walls of the air cushion were sloped in order to ease the escape of trapped air, and air escape valves were installed at the bottom of the shaft. In the Empire Building test "a car weighing 2,000 pounds was dropped from the twentieth story. The efficiency of the cushion was shown by the fact that the eggs and incandescent lamps carried upon the floor of the car were uninjured."[53]

The air cushions for elevators in the Philadelphia City Hall were similarly tested in 1902. Although the building stood 500 feet tall from the shoes of William Penn's statue to the sidewalk, the test allowed a car of about 1,500 pounds to fall free for 290 feet (a greater height than any previous test) before entering an air cushion 75 feet deep. In the car were a lighted lantern, six rats, fifty incandescent light bulbs, and several dozen eggs. "A New York reporter begged for the privilege of accompanying the rats, but was refused." A few seconds after the car was cut loose there was a "cannon-like report." When the car was entered it was found that "the lantern was upset, and a few of the eggs were broken, the rats were unscathed . . . and the incandescent light bulbs were unbroken."[54] F. T. Ellithorpe, president of the Ellithorpe Safety Air Cushion Company and principal promoter of the safety device, calmly reported to the press that he had himself many times dropped six stories without injury.

The last major installation of air cushions seems to have been that in the Woolworth Building, for which the architect Cass Gilbert had the assistance of Thomas E. Brown, Jr., consulting engineer for the Otis Elevator Company. For this installation the lower part of each shaft was enclosed with walls of steel and concrete, airtight elevator entrance doors were installed at every floor level, special outlet valves at the bottom of the shaft released air during impact, and inlet valves reduced suction as the car rose again. In November 1912 an article titled "Parachuting Down an Elevator Shaft" announced: "A safety expert, F. T. Ellithorpe, is to drop in an elevator seven hundred feet. . . . Mr. Ellithorpe is going to call into use this air to check the downward course only toward the end and not at the beginning of the elevator's fall. This would seem to add to the peril because it gives the plunging car just so much more time in which to acquire dangerous momentum."[55] Perhaps Ellithorpe had the same misgivings, for his name is not mentioned in a later report: "On the night of October 16, 1913, one of the forty-six story cars, loaded to a weight of 7,500 lb., corresponding to a carful of passengers, was dropped freely from the forty-fifth floor and stopped safely by the action of the air cushion. It fell 470 feet to the top of the air cushion in 5½ seconds and came to rest safely at the bottom of the air cushion in about 1½ seconds more."[56]

By the 1920s the air cushion had been abandoned as an elevator safety device. Smoother starting and stopping and the more facile control provided by new power systems had calmed the public's fears, and 50 years of experience with elevators had brought new generations, accustomed to travel in automobiles and inclined to trust mechanical contrivances. Nevertheless, the air cushion safety system had always offered distinct advantages. It was costly to construct but was not plagued by deterioration

13.22 The Chicago Time-Life Building (1971) was not large enough or tall enough to satisfy the usual criteria for the use of double-deck elevators. The prompt arrival of 2,800 clerical workers was the determining factor. (*Progressive Architecture*, December 1971.)

of springs, maintenance costs, or the tampering of building engineers. Furthermore, it lent itself to dramatic demonstrations of its capabilities, and that had been a significant factor in elevator development since the days of Elisha Graves Otis.

Almost 50 years after Otis had demonstrated his safety device at the New York World's Fair in 1854, elevator engineers were so confident as to refer to his invention as "a very pretty exhibition safety," because its action was triggered by a broken rope or chain, "the rarest cause of a car falling."[57] Once higher elevator speeds had been reached, the sudden shock that resulted from pawls engaging the ridges of the side rail was too great for the safety of passengers. Before steel rails replaced wood, this objection was somewhat alleviated by the fact that toothed pawls would gradually embed themselves in pine side rails, cutting deep gouges in the wood before bringing the car to a halt. Once the height of elevator shafts increased, the device was effective only when the cable broke near the car, for the weight of chain or cable in the shaft could maintain tension on the releasing mechanism.

Other means of stopping a falling car were triggered by a car's acceleration during a fall. In one system a cog rotated in a metal case, turned by the teeth of the metal side rail, and when the speed became too great the cog would be driven upward to wedge against the rail. Pneumatic devices included a pair of hinged boards on the bottom of the car that would be forced outward against the rails by the air pressure beneath the falling car, and a pneumatic piston applying restraint through its own cable attached to the car. At the end of the nineteenth century it was reported that "neither of these pneumatic devices is considered to be reliable."[58]

Many safeties were operated by centrifugal governors located on the top of the car or at the top of the shaft. A light cable attached to the car made a loop up and down the shaft, turning the governor. With excessive speed the spinning balls of the governor rose, and a safety device was activated. Multiple systems were common and there was always a buffer at the bottom of the shaft. For low speeds and shorter buildings the buffer was usually a simple, massive spring; for faster and higher installations the spring might be placed on an oil hydraulic piston and cylinder.

The fear of falling may have been a natural concern, but statistics indicate that other factors were a more common cause of accidents. Even a falling car could involve injuries from things other than its impact on the bottom of the shaft. Early counterweights were not always restrained to prevent their falling on a crashed car and, when the shaft was many stories in height, falling cables could themselves prove extremely dangerous.

There were repeated protestations of the safety of elevator transportation. In 1898 a Boston elevator inspector compared elevators to railroads: "Statistics of this country show that while one passenger among 166,000 is injured on railroads, only one out of 1,210,000 passengers, as nearly as can be determined, is injured and only one killed out of 2,000,000 passengers by reason of elevator operation."[59] Cables breaking or unwinding were responsible for only about 6 percent of elevator accidents. About 10 years later an 11-year study of fatal elevator accidents in Chicago and Manhattan showed that 85 percent took place at the elevator door and 15 percent in the shaft, whether from falling cars or workmen's activities.[60] Of the fatal doorway accidents, about half were a matter of the victims being crushed in

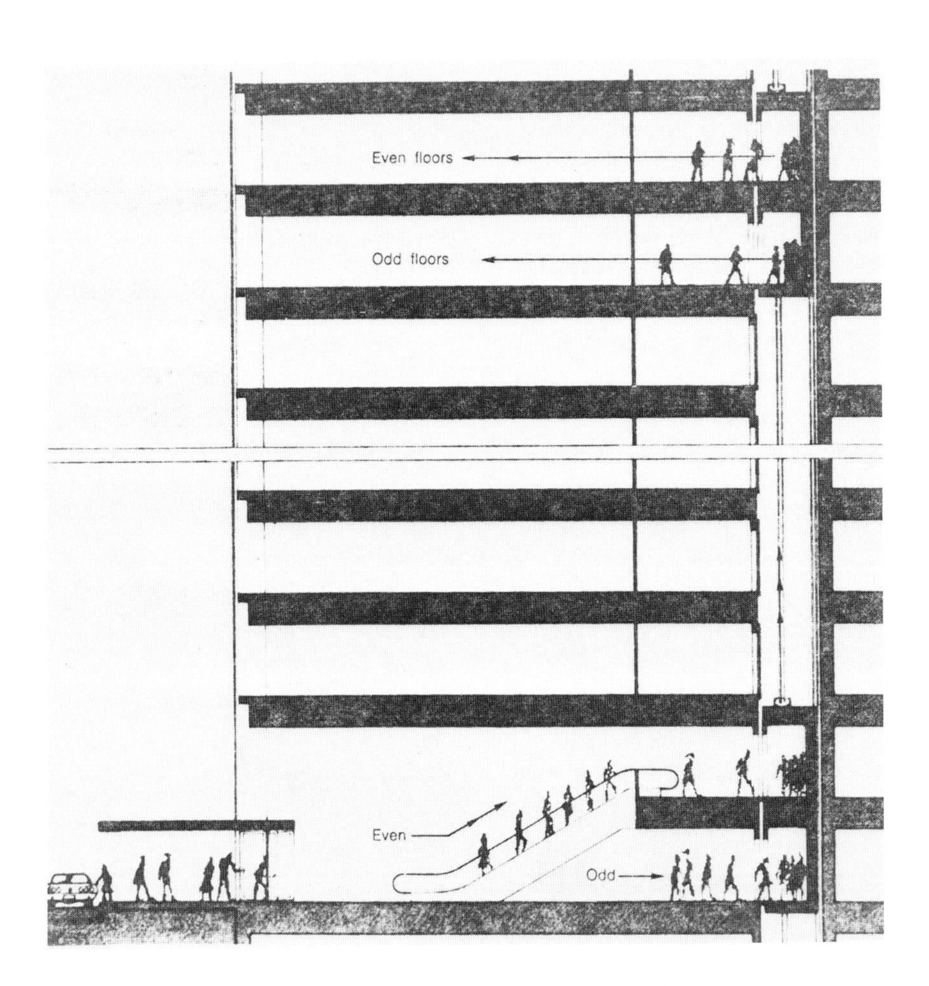

the doorway, and the other half were cases in which the victim fell through an open elevator door. Most of the accidents at the elevator door were the result of inadequate, broken, or unused safety devices that were intended to insure closure of the landing door until the car was level and prepared to be boarded.

Mrs. Elizabeth Insman last night walked into the elevator shaft at St. James Hospital where she was a patient, sustaining injuries which caused her death. She mistook the door for that which led to the bathroom.

Donald Meade, a clerk, couldn't understand why the elevator was so slow. He opened the safety gate and looked up the shaft. The descending car pushed down the gate, pinned him to the floor and crushed him to death.[61]

In most cases these accidents seem to have been related to human impatience and curiosity as much as they were to mechanical problems. As late as 1923, when mechanical and electrically controlled mechanical devices to prevent doorway accidents were well developed, it was reported that a check on the 56 elevator doors of the Film Building in New York showed

13.23 In the 1892 Reno patent, cast-iron strips moved along I-beam runners and turned on sprocket wheels at the top and bottom of the escalator. Reno's most significant feature was the comb plate (14 in the upper drawing), which extended into the grooves of the cast-iron strips (2 and 5 in the lower drawing). (U.S. Patent no. 470,918.)

that not one of the electro-mechanical safeties was in operating condition, and that in the second largest hotel in Richmond, Virginia, 17 of the 41 mechanical safeties were broken or dismantled.

Gradually municipalities enacted more stringent requirements for elevator safety at the door, but not without strong argument and the inclusion of exemptions for elevators already in operation. One safety engineer recalled a hearing in New York in which the official representative of building owners shouted to the aldermen, "Five million dollars, gentlemen—$5,000,000 for a lot of useless junk!"

A major economy in elevator systems was achieved in 1931 when the Westinghouse office building in East Pittsburgh, Pennsylvania, was designed to have two elevators operating in each shaft. The building's height required two elevator zones, the lower one running from the lobby to the eleventh floor and the upper zone starting at another lobby a half-story above street level, going directly to the twelfth floor, and thereafter serving the upper half of the building. The controls, in the design of which Frank J. Sprague took part, stopped either car at the top of its designated run until the other had also reached its top point. During the descent that was thus coordinated, safety devices prevented the cars from getting nearer to each other than a designated distance. The Westinghouse Electric Company estimated that this arrangement saved floor space that could have been rented at that time for a sum between $35,000 and $85,000 per year.[62] The electrical controls gave a degree of security, even in a period when elevators were run by operators at hand controls, but the fact that the lower elevator made more frequent stops on its downward trip, frequently bringing the express elevator to a complete stop, proved an irremediable disadvantage of the system.

The following year a New York office building of 60 stories used stacked elevator cars. Passengers wishing to reach even-numbered floors entered the upper part of the car from the building's street-level lobby; those wanting to go to odd-numbered floors went to the basement, where they entered the lower part of the car. The lower level was meant principally to serve workers arriving through a proposed subway entrance, which was unfortunately never built.[63] Consequently, the system was soon converted to single-car operation.

In 1970 the idea of double-deck elevators was revived in the design of the Chicago office of the Time-Life publishing house (fig. 13.22). These offices were occupied by the magazine's subscription service, and management required that all workers observe the same office hours. Therefore, although the building had only 25 stories, it was necessary in a five-minute period to move almost twice the number of occupants as usual in office buildings.[64] Shortly afterward three other buildings were designed with double-deck elevators.

Elevators satisfied the needs of hotels and offices, accommodating a moderate number of people who desired to reach their destinations within moderate intervals of time. For railroad terminals and busy department stores the larger numbers of people led to inevitable comparisons with the methods of movement used in mining and manufacturing, conveyor belts that delivered ores and raw materials to the carts and bins from which they would be used. Assembly line principles, the "iron foreman" that set the pace for factory workers, could also speed and regulate the flow of the throngs that crowded into the booming centers of metropolitan areas. Furthermore, the advantages of eliminating queuing through continuous movement could be realized by "inclined elevators," "moving stairways," and "electric stair lifts," as escalators were called before the turn of the century.

A U.S. patent for "revolving stairs" was issued as early as 1859, but the first steps toward development of the escalator were taken in 1892 by Jesse W. Reno, who at the same time proposed an underground rapid transit system for New York. Reno, the son of the Civil War general for whom the Nevada city was named, had worked in western mining activities for seven years after graduating from Lehigh University in 1883. Coming east to join the staff of the Thomson-Houston Company, he worked alongside C. J. Van Depoele, the Belgian engineer who contributed to the development of streetcar systems and electric motors. Steam-driven conveyor belts used to load ore, interest in the movement problems of metropolitan centers, and the potential offered by heavy-duty electrical motors were combined in Reno's mind.

In 1892 the *Engineering News* reported that an experimental "continuous passenger elevator" had been built by Reno in Brooklyn, and it suggested that the device would be useful "in places where the traffic is quite continuous, as at the downtown stations of the elevated railways and on the Brooklyn Bridge," giving people access to the streetcars from levels

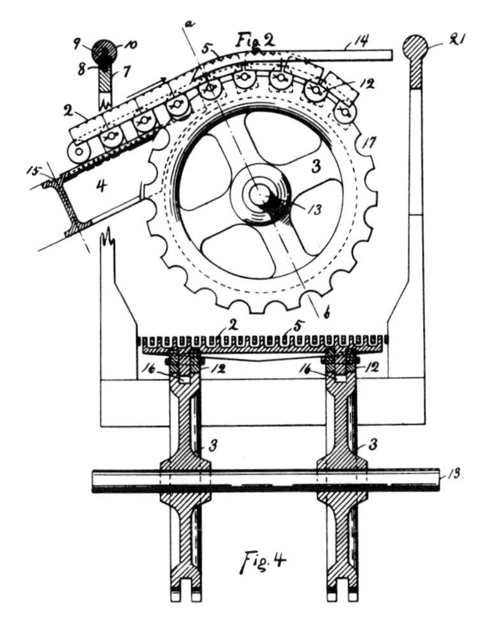

lower than the bridge's approaches.[65] This design was simply an inclined plane that traveled at the speed of 70 feet per minute (fig. 13.23). The moving surface was made up of cast iron strips, 3½ inches wide, that rolled along tracks at each side of the moving belt. These segments were grooved in the direction of travel and bands of rubber assured that users' feet would not slip on the slope of about 1:2. At landings passengers stepped onto an iron plate with projections like the teeth of a comb, fitting between the ridges of the cast-iron strips. Reno machines were installed at approaches to the Brooklyn Bridge and at Coney Island. Their principal shortcoming was the steep angle of the moving surface on which passengers stood. For safety and reassurance a moving handrail was provided on one sided, although it was found necessary to mold large white circles in the rubber covering in order

13.24 The Reno inclined elevator required that passengers stand on a steep slope, as shown in this early installation in a department store. Note the white circles that made the movement of the handrail evident to boarding passengers. (*Electrical Engineering*, 7 July 1898.)

13.25 In the 1892 Wheeler patent, steps appeared out of the floor and disappeared into the floor. At these points it was proposed to place barriers (*P* at left and right), which were to consist of belts rotating in the direction that would push away incautious toes. (U.S. Patent no. 479,864.)

to make it apparent that the handrail moved (fig. 13.24).[66]

In the same year that Reno patented his "endless conveyor or elevator," George H. Wheeler obtained a patent for an "elevator" that was similar. The Wheeler design did not have the early popularity of Reno's product, but it proved to be a basis for sound development. Instead of a steeply sloped plane, the passenger was confronted with familiar steps moving up the incline (fig. 13.25). Some were startled: "The sight of all those stairs gravely walking upstairs for ever and ever is calculated to seriously shock a man of nervous temperament, and is a thing to be avoided by one of uncertain or unsteady vision."[67] Each step was in itself a small cart carried on four wheels, the wheels at the front of the step riding on one pair of tracks and the wheels at the back riding on separate tracks. By altering the relationship of these tracks the steps could be positioned as a customary stairway or could transform the treads of the steps into a continuous plane. By this means the boarding passenger could step onto a flat moving surface, which would gradually become a stair and then return to a flat surface. The principal flaw of the

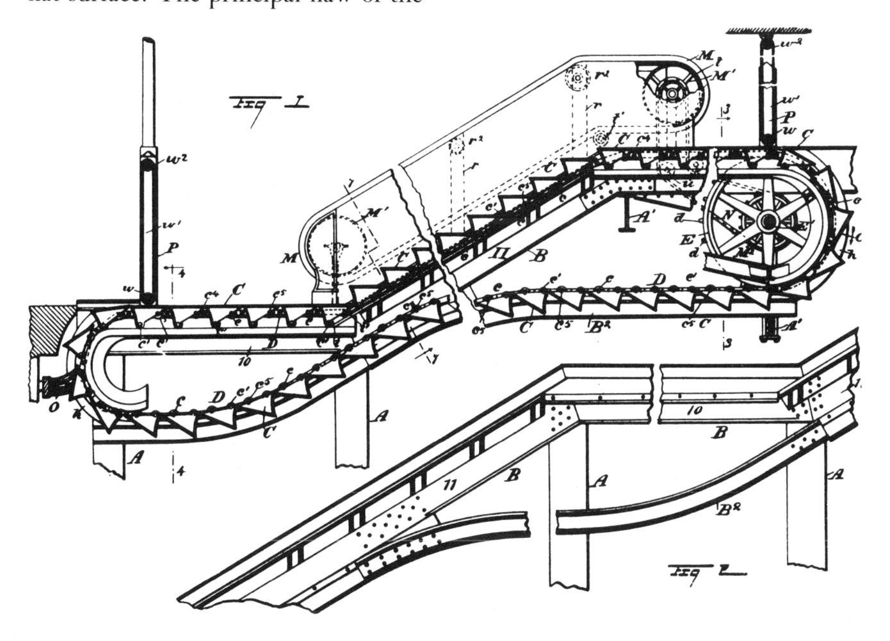

Wheeler design lay at the beginning and end of the escalator trip, where the steps were flattened to be even with the floor surface and then rolled down into or rose out of the floor. A large gap was necessary between the floor and the steps, a gap that had to be filled to "prevent the toes of the party from becoming wedged fast at the floor line," as Wheeler wrote in his patent. A barrier (or "shunt," as it was often called) covered the gap and, by its angled or curved shape, assured that passengers left the escalator by stepping off to the side and approached it from the side. The Wheeler patent was sold in 1898 to Charles D. Seeberger, an experimenter in escalators, who the following year joined forces with the Otis Elevator Company.

It was the Paris Exposition of 1900 that most dramatically brought attention to the advantages of the moving ramps, just then becoming known as escalators. A feature of the Exposition was an elevated moving platform that carried visitors along a circuit of about two miles, paralleling and crossing avenues. At two of the system's nine stations, escalators were used to lift passengers from the pavement to the level of the moving platform. In addition, planners of the Exposition, fearful that visitors would not bother to climb stairs to upper levels of some exhibition halls, installed 28 escalators within buildings, all required to be from French manufacturers. Over half were constructed after the Hallé system, a stiff belt supported on rollers every two feet. The LeBlanc system, which also provided escalators for the Exposition, employed bars between two sprocket chains that were supported on rollers. There was little fundamental difference between these two escalator systems and that of

Reno, five of whose machines were supplied to the Exposition through the company's French agent.[68]

Within the U.S. display in the Palace of Thread, Fabrics, and Clothing, Otis installed an escalator in addition to those selected by the authorities of the Exposition. A British engineering magazine made clear the way that it differed from others: "This device . . . is really a staircase, and is not a traveling band, such as was employed in places to carry passengers to the higher levels. It has treads and risers, and when at rest can be ascended and descended just as any other stair."[69]

The principal problem that remained in the stepped escalator was that of leaving it. At the Exposition of 1900 the single Otis escalator had been provided with the company's usual solution to this problem. At the top of the run, treads of the steps continued even with the floor for about five feet beyond the end of the incline. In this area a low V-shaped barricade was placed to shunt passengers to either side and off the moving surface (fig. 13.26). This awkward maneuver was eliminated shortly after the Exposition, when the treads of escalator steps began to be made of strips of wood or metal, which meshed with a comb plate such as that used in the early Reno machines. With this improvement shunts were no longer needed.

13.26 Escalators developed from the Wheeler patent required a barrier that covered the dangerous gap into which the steps disappeared. (*Cassier's Magazine*, March 1904.)

The logic of escalators demanded a heavy, even flow of passengers. From Reno's first installations at the stations of New York's elevated railways to present-day terminals, the escalator has always provided a means of moving crowds without delay. Early installations usually provided only upward transportation, an adjacent stair being provided for downward travel or for those who were reluctant to use the escalator. Eight escalators placed in a six-story Massachusetts textile mill in 1905 to speed the movement of its 6,000 workers were the first to be reversible. When workers arrived in the morning or returned from lunch, the escalators lifted them to the level on which they worked; going to lunch or leaving at the end of the day, the escalators were switched to the descending direction. Although such industrial uses were rare, it was reported to have been the careful decision of the mill's owners, based on such economic factors as the conservation of employees' energy and attracting the pick of available workers.

At stations of the New York elevated railway, escalators became extremely popular. When it was necessary to shut down the escalator at 23rd Street and 6th Avenue for a period, the station lost almost 65,000 fares. Early in this century, major railroad terminals began to install escalators to take passengers from the level at which trains arrived to the main concourse and waiting rooms. At the Gare du Quai d'Orsay in Paris in 1908, the decision to use a stepped escalator, rather than an inclined belt, was greatly influenced by the feeling that it was necessary that "the traveller's foot can rest on a plane that is absolutely horizontal" and that there be a place to rest hand luggage as one moved upward.[70] A few years later a similar escalator was installed in New York's Pennsylvania Station, but by that time stepped escalators had almost completely supplanted belt escalators, and the Reno patents had been sold to the Otis Elevator Company.

With improved city transportation, residents of metropolitan centers could take advantage of large department stores that sold a wide variety of articles at predetermined and clearly marked prices that were low in order to insure rapid sale. Any observer of the "restless activity and untiring energy of modern shoppers" realized that speeding them on their way about the sales areas increased sales, and escalators became an attractive solution to the problem. In the 1890s Reno escalators were installed in two New York department stores, Greenhut-Siegel-Cooper and Bloomingdale's. Similar escalators were also added at Harrod's in London and the Magasin du Louvre in Paris.

After the Paris Exposition of 1900 the enthusiasm of department stores for escalators blossomed. The escalator Otis had displayed at the Exposition was returned and installed in Gimbel's department store in Philadelphia. A few years later Gimbel's archcompetitor in New York, R. H. Macy and Company, connected all of its store's five floors with escalators. A measure of the escalators' value was found shortly before Christmas 1906, when the number of customers coming up to each floor by stair, escalator, or elevator was counted. At the third floor arrivals by escalator were double those by elevator; at the fourth the count was almost equal; and at the fifth arrivals by elevator were half again the number by escalator.[71] Store owners discovered that systems combining elevators and escalators drew buyers to upper floors, and as they traveled they saw the merchandise displayed on each floor. Escalators and elevators were also technological symbols of luxury. Because Berlin housewives were somewhat embarrassed to seek bargains, a major department store there was decked out with marble walls, glittering chandeliers, fountains, three escalators, and 83 elevators.[72]

Used in combination, elevators and escalators—like sidewalks and streets—made possible the increase of urban density and brought to realization the century-old characterization of vertical transportation within buildings as upended streets. Heavy-duty electric motors, the latest method of powering vertical transportation, had been developed for streetcars, and the controls needed for safely stopping and starting streetcars were not greatly different from those needed for elevators. Like horizontal transportation systems, elevators systems in skyscrapers were designed for peak hours of coming and going, and by zoning the heights to which different elevators provided service the system worked rather like a routing of streetcar lines. In fact, many of the tallest buildings have established mid-height lobbies in which travelers must transfer from one elevator to another that is routed to serve the uppermost floors, just as one waits on a street corner to transfer to another bus.

UNION
OFFICE

14

Fire Protection

1680 Nicholas Barbon establishes the earliest company to offer insurance against fire loss

1793 London committee of architects compares materials to control the spread of fire

1844 Chemical fire extinguishers introduced

1871–72 Chicago and Boston fires test the durability of buildings

1875 Automatic sprinklers offered for sale by Henry S. Parmalee

1880s Introduction of methods for protection of the metal frameworks of skyscrapers

"Slow-burning" construction encouraged by Edward Atkinson

14.1 An engraving by William Hogarth, intended as a satire of the British government's war policies in the middle of the eighteenth century, shows the ways of fighting a fire at that time. While hand-held squirts are aimed from windows of a building, an insurance company's fire engine is pumped and men carry leather buckets of water to keep it filled. (William Hogarth, *The Times* 1762.)

Ancient cities were always in risk of fire, whether started by carelessness within the city walls or by invading armies. Conquerors customarily set fire to the shrines and temples of a vanquished city, a practice so common that at the end of the Persian invasion in 479 B.C. the citizens of Attica pledged to eschew rebuilding their burned sanctuaries and leave the ruins "as memorials of the impiety of the barbarians."[1] In cities largely built of masonry walls and tile roofs it seems surprising that fire could quickly spread, but the explanation lies in the density of buildings within the cities. Around 480 A.D., radical regulations in Constantinople demanded that streets be at least 12 feet wide and balconies be no closer than 10 feet from the wall opposite.[2] Since the walls of most buildings were set at the edge of their owners' property and most rooms opened on small courtyards in the center of the buildings, it is reasonable to picture such a city as consisting of large districts of continuous building, with fire delayed by streets little more than it would be slowed by a hallway.

In medieval London, regulations enacted by subdivisions of the city included requirements that fireplaces be built with space left between them and wooden members of construction, that large houses have ladders present for fighting fires, and that during the months of June and July barrels of water be kept beside their doors.[3] Such controls were common in the densely populated and highly combustible warrens of medieval cities. Curfew (from *covrefeu*, or literally, "cover your fire") was usually imposed, and at night city patrols were alert for sparks rising from chimneys. Fire prevention regulations were based on both humane and economic factors, for James I in his Proclamation of 1620 observed:

In the time of King Richard the first, Henry Fitz-Ailwyn, then Mayor, for the prevention of casualties of fire, caused provision to be made, that Buildings in the sayd Citie [London] should be of Stone, which for many yeeres after was observed; But the neglect thereof in succeeding times, especially in the present, the great confluence of Our sayd Citie, pestering of the Streets with Jutties, Stalles, and other annoyances, scarcitie of Timber, and many other occurrents, have turned the policie of those ancient times from conveniencie to necessitie.[4]

Efforts to curtail fire risks in London centered on controlling the use of flammable building materials, and it appears that this was also true in provincial towns. From 1650 to 1850 the major fires in four counties of southern England diminished to a startling degree, the first three decades of that period having 36 major fires and the last three decades only one. The improvement of fire-fighting equipment and methods was the principal factor in this reduction, though improvements made while rebuilding also made a significant contribution. The hand-pumped fire engine was introduced into England from Germany during the seventeenth century, and from the Dutch later came the use of flexible hoses by which water could be directed toward the heart of a fire (fig. 14.1). More common were hand-held "squirts," syringes by which a small jet of water could be forced a short distance.

Winchester, a major market town, in 1656 required new kinds of roofing: "For ye avoyding of the greate inconveniences w[ch] are found by experience to grow by thatcht Houses within that citty, both in respect of the danger it occasions by Fire and the unseemliness thereof in soe auncient and famous a citty . . . all such Houses that are already thatched, [are] to be covered with Tyle or Slatt within One yeare next ensuing, upon the like payne of Tenn Pounds."[5] In general, the process of replacing timber and thatch with brick and slate reduced the flammability of each building that was rebuilt and deterred the spread of flames from the thatched roofs that remained.

In New England, the governor of Massachusetts Bay Colony in 1630 ordered that in Boston "noe man shall build his chimney with wood, nor cover his house with thatch," but this regulation was not enforced.[6] A series of fires that followed brought further restrictions, each fire frightening the townspeople into short-lived concern. After the 1653 fire, Boston householders were required to have ladders on their premises and the city government purchased "good strong Iron crooks" for pulling down burning buildings. The fire of 1676 led to restrictions on the spacing of houses, and three years later another and larger blaze caused the Boston General Court to demand rebuilding in brick and tile, a requirement that was soon set aside because of the expense it entailed.

The Great Fire of London started early on Sunday morning, 1 September 1666, and continued until it had laid waste to 395 acres of the city and demolished 13,000 houses. Unlike previous attempts to legislate fire-retardant construction, the Rebuilding Act of the following year had two vital characteristics: detailed requirements and a sound system of enforcement. New brick plants were soon set up in the outskirts of London to feed the feverish activity of reconstruction under the new regulations. The continued use of timbers for floor construction was somewhat offset by an increased popularity of plaster ceilings. In this Act and subsequent legislation, attention was repeatedly given to the thickness of brick party walls and separations between chimneys and the wooden members supporting floors, the first a restraint on the spread of fires and the second serving to lessen the risk of fires starting.

New York adopted its first building regulations in 1625; those simple requirements were succeeded by more detailed limitations meant to discourage the start of fires. Early in the eighteenth century, New Yorkers

were forbidden to distill rum or burn oyster shells for lime within a half mile of City Hall. Later the outdoor stacking of hay was prohibited, and regulations were imposed on the storage of pitch, tar, and turpentine. Expansion and construction continued apace, and around 1840 the New York *Mirror* longed for "the day when some portion of New York may be considered finished for a few years."[7] Such activity demanded further protection from fire, and soon New York regulations designated boundaries south of which wooden construction was outlawed. Because of the linear nature of New York's growth, the limits were simple to define: in 1860 at 52nd Street, in 1866 at 86th Street, and at the end of the century 149th Street on the east side and 190th Street on the west side of Manhattan. The flaw in this early determination of a fire zone was the law's being limited to the *construction* of wood buildings. Existing frame structures could still be shifted from lot to lot within the southern part of the city, and buildings erected in the northern fringes of settlement could be moved southward across the established line.[8]

After the Great Fire of London a young doctor, Nicholas Barbon, established the first system of fire insurance. Barbon was the son of Praise-God Barebone, leather merchant, member of Parliament supporting Cromwell, and lay preacher of the Anabaptist faith. Barbon was trained in Holland for the practice of medicine, but before the fire he had begun speculating in the construction of houses in the London area. Not the first to undertake the residential development of sizable tracts of land, Barbon advanced the system by laying out streets and squares, pricing lots by the lineal foot of frontage, and constructing houses that were—though narrow in plan and meager in form—attractive to naive buyers. Known for his showy clothing and love of food, Barbon considered building to be "the most proper and visible Distinction of Riches and Greatness, because the Expences are too Great for Mean Persons to follow."[9]

A year after the disaster, while the city was busily rebuilding, Barbon began offering fire insurance on real estate. In the beginning Barbon conducted his insurance business alone, but in 1680 he combined with other investors to establish a company called the Fire Office. In the six years following 1686, the company provided coverage for 5,650 houses with premiums set twice as high for timber construction as for brick. The company's monopoly was ended with the founding of the Friendly Society for Insuring Houses from Fire, and the later competition of The Amicable Contributorship for the Assurance of Houses and Goods from Fire. (The latter company, with good reason, soon changed its name to Hand-in-Hand Fire and Life Insurance Society.) In 1710 the Sun Fire Office opened, and it came to dominate the English fire insurance activity along with the Phoenix Assurance Company and the Royal Exchange Assurance.[10]

Nicholas Barbon hired a corps of men to fight fires at properties on which his company had written policies. Other companies formed their own fire brigades, and soon each insurer installed cast lead insignia to mark the buildings on which it held insurance. Being privately funded, a fire brigade only fought fires on buildings insured by its company, and it withdrew if the unfortunate householder's insignia proved to be that of another underwriter. When there was no insignia, all fire brigades that might be present watched as flames consumed the structure, thereby clearly demonstrating the wisdom of a householder's contracting for insurance.[11]

In the United States a fire in Philadelphia during the spring of 1730 razed a number of shops and some residences. After this fire, city officials purchased additional fire-fighting equipment and instituted new regulations, including one that forbade smoking on the streets.[12] In addition, citizens organized the Union Fire Company, a volunteer group of fire fighters, self-equipped and limited to 30 in number, which included Benjamin Franklin. This organization was the springboard from which came the first stable fire insurance company in the United States. (A previous company had been established in Charleston, South Carolina, but it had failed after the 1741 fire in that city.) Begun as a simple compact among men who each contributed to the group's funds for investment, within two years the Philadelphia company adopted the name Philadelphia Contributionship for the Insurance of Houses from Loss by Fire, and it began selling policies to nonmembers. For the next 32 years the Contributionship was the sole fire insurance company in the United States. Unlike English insurance firms of the time, the Contributionship inspected each building before issuing a policy on it, and rates were determined individually for each building rather than being set for broad categories of function and construction. The rate determinations of the Philadelphia Contributionship sometimes included requirements that alterations or improvements be made to the property considered for insurance. One of the most controversial of these measures was the company's decision in 1781 that it would not insure buildings with trees nearby, a declaration that led to the formation of a competing insurance company more generous toward the Philadelphia landscape.

As the actuarial determination of fire insurance rates developed, competition quickened. In 1810 there were few companies in the United States, but their number gradually increased. New York had a rate-fixing agreement among all of its companies in 1821, but within a few years 17 new companies were formed and open competition was restored. In 1826 another agreement on rates was reached, but the New York fire of 1835 ruined many companies and rates soared, to be followed once more by a flurry of new companies and the concomitant rate cutting. This cycle was repeated before and after the New York fire of 1845. Between 1849 and 1865 there were at least 70 new fire insurance companies started in New York. Throughout the United States competition and rate fixing alternated until conditions were stabilized by the formation in 1866 of the National Board of Fire Underwriters.

Attempts were made to limit the likelihood of fires starting or spreading. In the 1770s at the edge of London on Putney Heath the Lord Mayor attended a demonstration arranged by David Hartley, a member of Parliament and diplomat who later signed the treaty ending the American Revolution. A fire roared in the lower room of a house, while the inventor and trusting friends waited confidently in the upper room. Hartley's fireproofing method, which was patented in 1773, consisted of thin plates of iron fastened beneath, above or on both surfaces of wooden floor construction. Spaces between the plates and wood were filled with sand or other noncombustible granules. The results of the test on Putney Heath were so impressive that the Lord Mayor marked the event by constructing a commemorative obelisk nearby, and in 1777 Parliament extended Hartley's patent for 31 years.[13]

14.2 The Chicago fire burned more than 2,024 acres and 17,450 buildings, consuming property valued at $200,000,000. Many insurance companies collapsed, but those surviving poured over $40,000,000 into Chicago's reconstruction. (*The Conquest of Fire*, 1914.)

14.3 A cartoon after the Chicago fire showed insurance agents bemoaning their situation. Insurance adjusters were viewed with suspicion, a sentiment encouraged by the fact that almost two-thirds of the city's losses were held by insurers outside Illinois and almost all companies within that state were forced to suspend operations. (*Fighting Fire*, 1873).

14.4 The 1872 Boston fire razed 776 buildings in an area of 65 acres, and these were largely masonry buildings of size and pretension. Rebuilding was completed in several years, despite the world-wide depression that began about six months after the fire. (*The Conquest of Fire*, 1914.)

The Royal Society of Arts in 1778 heard a paper presenting the method of fireproofing floors that had been devised by the young Lord Mahon, later the third Earl Stanhope, who long combined scientific investigations with his political career. Mahon's fireproofing relied on the fire resistance of plaster covered with a grout made by boiling together pine tar, chalk, and sand.[14] Wood laths were nailed about 1½ inches below the top edge of floor beams, and the stucco was applied to both upper and lower sides of the laths, filling the space almost to the top level of the wooden beams. Wood flooring was added above this and a ceiling of plaster was fastened to the bottoms of beams. Lord Mahon provided another dramatic display, sitting with friends in the upper room of a test structure, while a fire blazed beneath with enough heat to melt the glass in the windows.

In 1792, an organization of the most prominent London architects, originally intended only for social purposes, launched a search for means of preventing the spread of fire within buildings. Henry Holland, a foremost architect, son of a builder, and known for the sturdy construction of his buildings, led the investigation, perhaps because of his outspoken conviction that the current building regulations in London were "insufficient, unintelligible and the source of perpetual contention."[15] Two houses in Hans Place were obtained for tests much like the dramatic demonstrations that had been staged more than a decade before. Wood shavings and tar barrels were set afire in rooms protected by various fireproofing measures, principally those of Hartley and Stanhope. The fires blazed for one or two hours and both methods proved successful, limiting the spread of fire although wood was charred and plasterwork damaged. Another test proved the effectiveness of "Wood's liquid," a paint whose formula we do not know today, but which probably paralleled the alum solution that had been used on wooden beams by ancient Greeks and had served to fireproof the fabric of Montgolfier's balloon.

Holland and his committee reached conclusions that were published in 1793 as *Resolutions of the Associated Architects with the Report of a Committee by them appointed to consider the Causes of the frequent Fires and the best Means of preventing the like in the Future.* The primary causes of fires were declared to be the unwise location of fireplaces and furnaces; flames rising through stairways and behind paneling; and the liberal payments of fire insurance companies. Only in England did fire underwriters pay the amount of a policy or replace a building, even though the burned structure might have been old and ramshackle. Only in England did a fire underwriter accept the owner's claims on a building's burned contents without requiring corroboration or a prior listing. Apparently the sharp competition among insurance companies had led to terms that encouraged and even rewarded arson.

A remarkable effort to develop fireproof construction for dwellings was accomplished around 1800 when Sir Robert Peel, cotton manufacturer and father of a prime minister, built fireproof row houses for workers in his Staffordshire mill. One must assume that Peel had sound and humane reasons for taking such extraordinary precautions in the construction. Decreasing the possible damage from fire protected a landlord's investment, and the quality of housing often affected the quality of workers available. The system employed in the houses at Peel's Fazeley mill used brick vaults with concrete filled between the vaults and the roof or floor.[16] Over the lower rooms the transverse semielliptical vaults were little more than customary basement construction of the period. Above the upper story there was a longitudinal semicircular vault with a wrought-iron tie rod holding the exterior walls against the vault's outward thrust. Roof tiles rested directly on the concrete above this vault.

In November 1872, thirteen months after Chicago burned to the ground, Boston suffered a disastrous fire (figs. 14.2, 14.3, 14.4). The Chicago fire had been spread by strong winds, but the night of the Boston fire was calm. Panicked crowds in Chicago had handicapped all efforts to halt the flames, but Bostonians remained remarkably calm. Men had to drag fire-fighting equipment through the streets of Boston because the fire department's horses still suffered from "horse distemper," an epidemic from which most of Boston's draft animals had recovered.

A full year before, a letter published in a Boston newspaper had predicted catastrophe:

When that dozen lumber-yards on the roof is once well on fire, it will be taken, not by little sparks only, but by cords, into and upon every building within half a mile! Every window on the line of the gale will be broken into by the fiery brands, every place where there is wood for fire to catch upon, and fires will soon be rushing from fifty of those windows or roaring from the exposed wood. . . . Then would come the story, so lately told of Chicago: "Awful conflagration! Boston in ruins! Thousands of houses and the business portion of the city in ashes!"[17]

This warning was not heeded, but it proved to be horribly accurate. In a meeting held in Faneuil Hall three days after the Boston fire, a resolution condemning mansard roofs was passed

by a large majority. One calmer participant pointed out that roof shapes were not at fault, that instead blame lay in building them of wood even when a building's walls were of masonry. Fire insurance companies then declared war on mansard roofs, forcing construction of noncombustible materials if not the abandonment of the style.

Blame was also placed on open elevator shafts and the height of some Boston buildings, beyond the reach of water from volunteer firemen's hoses. This did not cause limitations to be placed immediately on the height of buildings in Boston, but it strongly encouraged later efforts in that direction. Combined, the tragedies of Chicago and Boston led to the inception of municipal fire departments in the United States. Volunteer organizations, often more attentive to social and political interests than fire-fighting skills, were soon supplanted by municipal employees, and their equipment was modernized. In 1866 only 15 cities in the United States had steam fire engines, but ten years later the number had grown to 275.[18]

On the day following the Boston fire, Henry Ward Beecher, one of the most popular speakers in an age that admired oratory, addressed a hall of mournful Bostonians. In his opening remarks he declared:

> There is no other city that could have offered such buildings to destruction. Granite—it is a child of fire, and would seem to be able to defy the flame; but it sparkled and cracked and was destroyed as if it were but chalk. . . .
>
> Was it wise to lay the foundations of [buildings] solid, to carry up the first story fire-proof, the second story fire-proof—the third, the fourth, the fifth story all fire-proof—and then put a Mansard roof on the top of all, to take fire and scatter sparks around the neighborhood? Those great buildings were admirable for business purposes, and now, as it proves, although not intended by the architect, admirable for fire.[19]

Not only granite, the hardest stone commonly used in building construction, but other sorts of stone were found seldom to survive the concentrated heat of a building fire, especially when suddenly drenched by cold water from firemen's hoses. (It had long been a practice of New England farmers to crumble granite boulders in their fields by building a fire around a stone and then throwing water on it.) From sandstone to granite, the durability of building stones under normal conditions did not necessarily relate to their resistance to the intense heat to which they were subjected when entire business districts burned. However, it was observed that brick had performed well in both the Chicago and Boston fires.

At that time a topic of discussion among architects and engineers was the problems that arose from the expansion of material under the fierce temperatures of a conflagration. When buildings had been built of masonry and wood, the wood trusses and joists burned and fell, while heat merely caused overall expansion of the masonry. With iron beams supporting the floors, there was a real danger of the metal's expanding and thrusting outward against the surrounding walls of masonry. For a complete iron framework, with columns adjacent to or embedded in the exterior walls, expansion became even more threatening.

Cast-iron storefronts, which had been introduced at the middle of the nineteenth century, presented particular dangers. In the heat of fire, their metal cracked or melted; long before that, fire was often transmitted from building to building through the voids left between the cast-iron panels and masonry walls behind them. Those spaces should have been filled with masonry and mortar, but in practice they rarely were.

The characteristics of wrought iron were quite different. It would not crack in a fire and seldom melted, but instead it expanded, warped, bent, and twisted.

Most of Boston's mansard roofs had been built of quick-burning light wood framing, and once fire reached the roof of a building it leaped to the next. Many of the buildings that perished in Boston were reported to have burned from the top down. At all levels of the blazing buildings, window glass broke in the glare of heat and firebrands entered the openings, often setting the structure afire at all levels from sidewalk to mansard.

On an April day in 1723 the Lord Chancellor, Sir Hans Sloane, physician and scientist, and several other gentlemen of note gathered in Belsize Park, north of London. There a three-story wood house, filled with oil and kindling, was set afire. Once the flames ran high, Ambrose Godfrey's fire extinguisher was rolled into the fire, and with a loud explosion the fire was suppressed. Another demonstration was conducted about two months later and it is recorded that this fire was extinguished in three minutes. Godfrey's fire extinguisher consisted of a small wooden keg filled with water, in the center of which was set a pewter sphere containing gunpowder.[20] From the sphere to the top of the barrel a metal tube held a fuse by which the powder could be ignited before the barrel was rolled to the center of a fire.

There is no evidence that Godfrey's extinguisher was often used in the 20 years following those tests, but interest was revived by the inventor's son in the 1760s, when he persuaded the Royal Society of Arts to test the device. A three-room, three-story structure of brick was built in a field outside London. With the Duke of York, Prince William, Prince Henry, and distinguished scientists attending, the lower two rooms were set afire and the flames were quickly suppressed when three extinguisher kegs were rolled into the building. The experiment was repeated with fire in the second story, and again in the topmost room. All these tests were successful, but there is no evidence that such extreme measures were particularly popular.

In 1844 a fire extinguisher better suited to stopping flames before they became a conflagration and without broad damage to the surroundings was patented by an Englishman, H. Phillips (fig. 14.5). He patented an improvement on the device five years later, and soon the chemical extinguisher was accepted in its home country and the United States. A "portable machine for domestic use," Phillip's design consisted of a metal cylinder whose diameter was about half its length. Its action was activated by pushing a spike that broke open a small vial of sulphuric acid.[21] This acid flowed into a bottle of chlorate of potassium and sugar, which caused combustion, the products of which fell on a surrounding block formed of a mixture of charcoal, potassium, and a bit of gypsum. This in turn produced intense heat and a violent expansion of the gases in the vessel, which was filled with water. The steam produced by these chemical reactions issued from a hose attached to the extinguisher.

Carbon dioxide extinguishers were brought into use during the 1860s. One of the first was patented in France by Carlier and Vignon, but all were similar. In them a container of sulphuric acid was poured into a solution of water and bicarbonate of soda, either by inverting the entire extinguisher or by pulling a plunger that opened the bottle of acid. The soda and acid combined to produce quantities of carbon dioxide gas about 300 times the original volume of water, and the formation of this gas produced sufficient pressure to shoot gas and water out of the extinguisher with considerable force. A large version of this device was built for the fire department of New York City, and in its trials the stream of gas and water was driven as far as 250 feet through the hose and 50 feet beyond the nozzle of the hose.[22]

For two sound reasons the owners of textile mills often led in investigating ways of preventing and extinguishing fire: the mills were veritable tinderboxes, and capital investments in buildings and machines were sufficiently large to justify expenditures for preventive measures. An atmosphere filled with floating particles of lint, lighted by oil lanterns, gas burners, or electric arc lamps, made fire a constant threat. Heating was customarily provided by pipes containing exhaust steam from the factory's engines, and valves in that system could be opened to release hissing jets of steam into any portion of the mill in which fire was detected. Where floor areas were relatively small, this precaution was sufficient, but in larger mill buildings it appears to have been of little value.[23] An obvious improvement, encouraged by underwriters, was the English manner of installing parallel runs of perforated water pipe at the ceiling level. By turning a valve, water from a tank atop the mill's hoist tower could be sprayed over a given area of the mill. By 1859 perforated pipe sprinklers were required in Lowell, Massachusetts, for any part of a mill having particular susceptibility to fire, but in other locales they were usually installed only in the highly flammable picker buildings, where bales of cotton were broken open and debris was blown out of the cotton with air jets.

A system of automatic sprinklers was offered for sale in 1875 by Henry S. Parmalee, a piano manufacturer in New Haven, Connecticut. Overhead piping was laid out in the same manner used for perforated pipe sprinklers: a water pipe crossing the factory space in the center of each structural bay, fed from a main along one side wall of the building. The automatic sprinkler, as it developed through several years, consisted of an outlet from the pipe, capped with a small grooved button that whirled as water passed, showering a fine spray over an area of about 100 square feet (fig. 14.7). Over

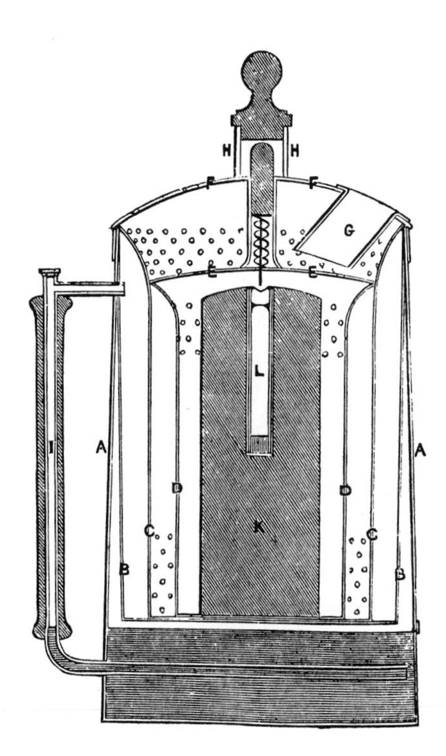

14.5 Phillip's Fire Annihilator had a handle on the side, from which issued the hot gases generated by chemicals and the steam from a water-filled jacket. (*Illustrated London News*, 8 September 1849.)

14.6 The first automatic sprinkler installed in the Parmalee piano factory was activated by heat melting a metal ring (*A*). Once the lever at the top was no longer held by the spring, water pressure in the valve raised the lever and filled the perforated sprinkler. (G. Dana, *Automatic Sprinkler Protection*, 1914.)

14.7 In one of the most popular sprinklers around 1880 a brass cap was soldered over the slotted spinner. A disadvantage of this design was the tendency of the water inside the cap to conduct heat away and thus delay the solder's melting. (G. Dana, *Automatic Sprinkler Protection*, 1914.)

14.8 A sprinkler introduced early in the 1880s replaced the perforated head with a jet of water that was directed against a deflector above it. The valve was kept shut by two levers held together by a clip of fusible material. (G. Dana, *Automatic Sprinkler Protection*, 1914.)

this was soldered a brass cover. When the temperature reached about 160° F the solder melted and water pressure blew the cap away, letting the spray soak the fire area. It took some time to heat the brass cover to the temperature required, but there was no danger that a forgetful or frightened worker might fail to open a valve.

Within five years of Parmalee's introduction of his automatic sprinkler there were 18 other types on the market (fig. 14.8). Fire insurance authorities soon discovered that of almost 1,000 mill fires, about 80 percent occurred in mills without automatic sprinklers, and the average loss in those instances was about 16 times the average for mills equipped with automatic sprinklers.

The convenience of illumination has always been accompanied by a threat of fire. Whale oil lamps might overturn, and later illuminants, such as camphene, coal oils, and kerosene, were so volatile that explosions were frequent. Gas lighting too had obvious dangers. By the turn of the century, Buffalo, New York, required at least 3 feet between a ceiling and any gas burner, 18 inches when a protective shield was provided. The New York Charter required globes or other glass surrounds for all gas burners in "theaters and other places of public amusement, manufactories, stores, hotels, lodging-houses, and in show-windows."[24]

The introduction of electric lighting did not entirely eliminate the danger of fire being caused by an illuminating system. When textile mills began to install them in the 1870s, arc lamps led to 23 fires within half a year. Fluctuations in current supplied to arc lamps could cause bits of the lamps' carbons to fly off, and imperfections in the quality and consistency of the carbons could cause

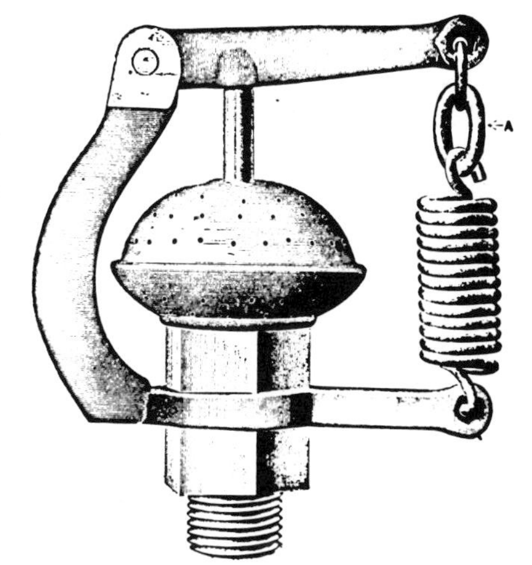

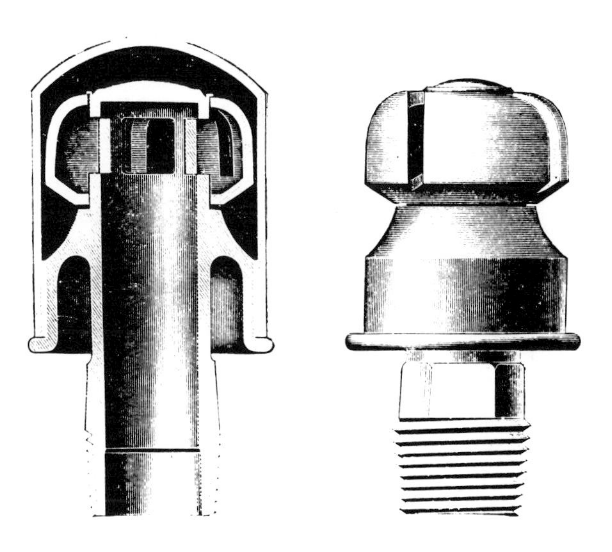

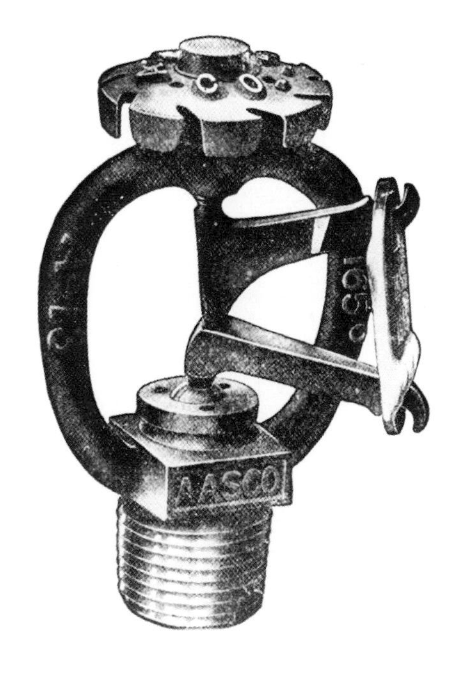

them to break or scatter sparks. The Brush Electric Company placed asbestos trays at the bottom of the glass globes around their arc lamps, and other manufacturers employed an assortment of guards to prevent sparks and flaming shards of carbon coming near flammable materials.

Less risk came from incandescent lighting, and it must be remembered that electricity, unlike gas and kerosene, does not add fuel to the flames it might cause. At first, caution was the rule. According to the regulations published in 1882 by the Phoenix Assurance Company, an English fire insurance firm, wires for lighting were required to be insulated and installed with no less than 2½ inches between the two wires serving a fixture. In "non-hazardous" buildings, wires could be covered by a wood molding, but only if a band of wood 1 inch wide separated the two wires.[25]

In the United States the National Board of Fire Underwriters in 1892 summoned its electrical inspectors to an emergency meeting from which arose the Underwriters International Electric Association and, eventually, the National Electric Code. One member of the Board declared:

> We cannot assume that the most reputable merchants have all at once become criminals. We find that our better class of risks is burning in a greater ratio than ever before, and that there are mysterious causes at work. . . . That mysterious element I believe to be electricity. . . . When we consider the appalling increase in fires during the last eighteen months we may well be startled. We are standing, I repeat, in the presence of a mysterious element which no one is at present able to fathom.[26]

Such anxiety was understandable, for, according to an insurance journal, between 1890 and 1900 the number of electrical fires per year grew six times as great, and the financial losses rose to almost 6½ million dollars. By 1900 U.S. fire underwriters had mounted a powerful campaign for adoption of the National Electrical Code. In 1893 a laboratory had been established in Chicago to advise on the dangers that might be involved with the electrical lighting installed for the Columbian Exposition, the first large public gathering to be lighted with electricity. This pioneer effort became the basis for the National Underwriters Laboratory.

New England owners of textile mills found early in the nineteenth century that their own efforts to make their mills less combustible did not bring corresponding reductions of insurance rates. Resentment led to the formation in 1835 of the first factory mutual fire insurance company in the United States. Under a mutual company, fire losses were recompensed by payment from funds contributed by all members or, in extreme cases, by assessments made on all member

14.9 This 1889 catalog shows a sprinkler system installed in a typical textile mill building. Although the work space of the mill maintains the flat-roofed form of slow-burning construction, a romantic tower holds a water tank that feeds the sprinklers. The spray of one sprinkler is shown at right. (Insurers' Automatic Fire Extinguisher Company, *Automatic Sprinklers for Extinguishing Fires*, 1889.)

companies; therefore, mutual companies were more strongly motivated than proprietary companies to discover methods of fire prevention and to provide favorable rates to their members who employed those methods. Edward Atkinson, who had been an official of several textile companies, became a leader in the development of factory architecture while president of the Boston Manufacturers' Mutual Fire Insurance Company, a position to which he was elected in 1878. A large man, characterized as "positive often to the point of obstinacy," Atkinson campaigned to improve the construction of factory buildings, although his manner of presenting his opinions alienated many architects and drew biting rejoinders from his opponents.

Basically, Atkinson attacked a method of building developed "after the pestilent invention of the buzz-saw," a method that consisted principally of "destructible granite, exposed iron and light wood."[27] Ample evidence was cited to show that the intense heat of a fire caused stone to crumble, melted or twisted iron, and fed upon the multitude of small wooden members. It was found in one period that the majority of mill fires were being caused by overheated oils on machinery, and research to determine standards for lubricants was thereupon funded at the new Massachusetts Institute of Technology. When kerosene began to replace other oils for lighting, Atkinson sent out a warning that all kerosene should be tested and a flash point of 124° F or higher should be required. Many factories had installed iron fire doors that twisted and sometimes melted during a fire. New metal-clad doors of heavy wood were strongly recommended by Atkinson, who invented a device by which they closed automatically when heat melted a soldered connection. He also strongly favored the use of automatic sprinkler systems. With the authority of the insurance rate schedule, Atkinson was able to press for conformity among the owners of factory buildings that his organization insured; others he lectured, chided, and reprimanded with unrelenting passion.

Though economical and strong, the American system of wood frame construction had faults, as John Wellborn Root pointed out: "The bonfire thus carefully prepared has the peculiar merit of having each stick of wood provided with its own flue and draft."[28] Edward Atkinson's principal prescription for industrial buildings was "slow-burning" construction, now more commonly known as mill construction. For this system of building, cast-iron columns and masonry piers were replaced by heavy timber columns; iron beams and light wood joists were supplanted by large wood beams; and floors were made of thick wood planking. In the construction of mills, stylish roofs of turrets, mansards, and gables gave way to flat roofs, also built of heavy wood planks (figs. 14.10, 14.11). The advantages of this system of factory and warehouse design lay in the fact that heavy timber construction burned slowly and remained capable of supporting its load even after a considerable amount of the wood had been charred by a fire. Even with mill construction, subsequent decisions could negate those advantages, and in 1891 Atkinson wrote:

Architects who are not conversant with all the rules of safety prescribed by the mill engineers have adopted the heavy timbers set wide apart and the thick plank floors covered in by a solid and suitable roof; but they have then converted these primary elements of slow combustion into very quickly combustible buildings, first, by connecting floor with floor by open stair-

14.10 With the introduction of slow-burning mill construction, the complex silhouettes of mansard roofs and towers (top) were succeeded by a more severe outline and roofs that were nearly flat (bottom). (E. V. French, *1860—Fifty Years—1910*, 1910.)

14.11 A drawing of typical slow-burning construction shows heavy rafters (10 by 12 inches) and floor beams (10 by 14 inches). Echoing the principles of skyscraper construction, Atkinson insisted that "the mill should hold up the walls, rather than that the walls should support the mill." (*Engineering Magazine*, November 1891).

14.12 Twenty years of tests on the relationship of heat and the strength of iron and steel were summed up in a graph published by the British Institution of Civil Engineers in 1890. (*Minutes of Proceedings, Institution of Civil Engineers*, 1890–1891.)

ways or stairways sheathed with wood; second, by sheathing the walls of the lower stories and sometimes the whole building with wood; third, by setting up partitions containing many cords of light wood; and lastly, making the most fatal mistake of all in finishing the woodwork with ordinary varnish, over which a fire runs with the rapidity of a race-horse.[29]

Steel construction too had dangerous reactions to fire. William Fairbairn, a Scottish engineer, was one of the first to study the behavior of iron and steel when subjected to changes of temperature, although his tests were narrow in scope. For instance, plate iron had been found to suffer little change in strength until it was raised to a red heat, whereupon its strength rapidly decreased. Studies made by the Franklin Institute in the 1880s supported previous findings that upon reaching a temperature around 400° F iron bars began to lose strength quickly, only a third of the strength remaining once the temperature had reached 1,000° F.[30] With the advent of iron construction it soon became evident that, although the material was incombustible, iron structures were not to be relied on at high temperatures (fig. 14.12). Since the advantages of iron and steel construction were obvious, it was necessary to find ways by which structural members could be protected.

In England the prevalent method of protecting columns was to form concrete around them, using a wire mesh to reinforce it. After its introduction around 1889, expanded metal lath was employed as reinforcement for concrete fireproofing and as a base for plastering on the underside of floor construction. In the United States the protection of structural frameworks was long dominated by the use of hollow terra-cotta tiles, shaped to fit around the steel members (fig. 14.13).

Expert opinion on fireproof construction at the end of the nineteenth century was summarized in the pamphlet *How to Build 'Fireproof'* by Francis C. Moore, an insurance official in the United States. Published in 1898 by the British Fire Prevention Committee, Moore's paper set about informing his British readers of practice in the United States, because "the general introduction into the Metropolis [London] of what are termed 'frame buildings' as used in America for warehouses and offices, cannot be far distant."[31] Moore was insistent about the need for covering structural metals: "All ironwork, columns and pillars, beams and girders, should be 'fire-proofed,' *i.e.*, covered with at least 4 ins. of incombustible material, terra-cotta or brick. . . . Brick-work is a good covering but porous terra-cotta, or even wire lath and plaster, may prove effective. Where wire lath and plaster is used, the column should first be wrapped with quarter-inch

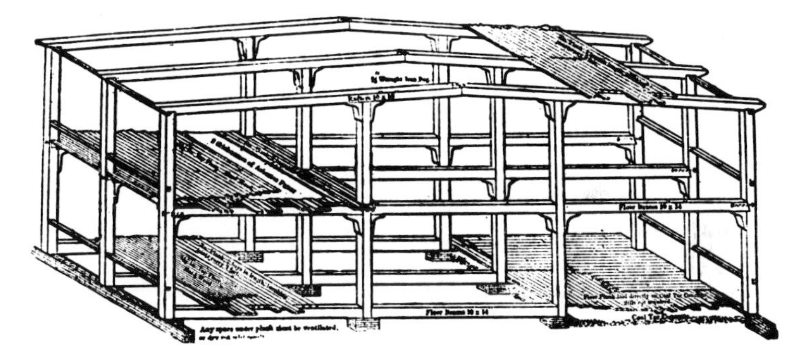

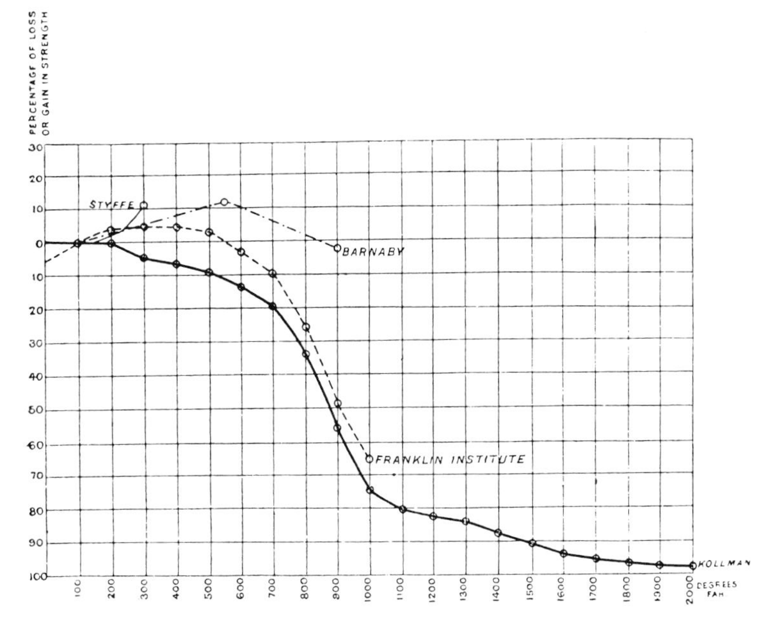

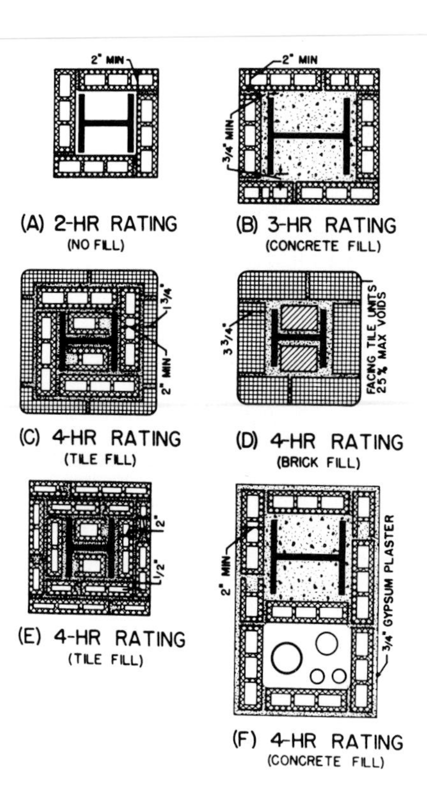

14.13 A handbook of the 1960s illustrates typical column fireproofing methods and their fire-resistance ratings. The rating corresponds to the protection provided around metal columns in terms of the time during which adequate protection might be expected. (H. C. Plummer, *Brick and Tile Engineering*, 1962.)

14.14 This stack of typical terra-cotta fireproofing shapes includes pieces meant to protect round columns, several with recesses to fit around the flanges of I-beams, and skewed shapes for use in flat-arch floor systems. (*Insurance Engineering*, September 1907.)

asbestos, bound with wire. This would prove reliable and inexpensive."[32] Porous terra-cotta was made by mixing such materials as sawdust or hay with the clay from which hollow tiles were to be made (fig. 14.14). The high temperatures of the kiln burned away those materials, leaving a terra-cotta that was light and could be sawed and nailed. Hence, the product was sometimes referred to as "terra-cotta lumber." Terra-cotta's light weight led many to question its adequacy as fireproofing. Moore pointed out that when the ten-story Leonard Building in Detroit, Michigan, had burned in 1897, "as fast as the columns or wall girders were warped by the heat the tiling dropped out like loose bricks."[33] A similar report came from the ruins of the Pittsburgh fire in the same year, where expansion of the heated building frames had caused the bottom faces of terra-cotta floor arches to scale off. Although popular because of its low cost, wire lath and plaster was always considered questionable protection for structural metal. In 1896 a New York building in which wire lath and plaster had been used to reduce the area occupied by columns, thereby gaining the last bit of rentable floor area, was quoted an insurance rate 15 percent greater than that for buildings with columns surrounded by brick or terra-cotta.[34]

More costly than office building fires were those in department stores and multistory warehouse buildings, where both the structure and the merchandise were lost. A New York architect, F. H. Kimball, recommended that in buildings where merchandise was stored the first-floor columns should have a double jacket of fireproofing with a 2-inch air space between them. However, most experts opposed any hidden spaces—no matter how small—within which fire could rapidly extend itself. Warnings were given to provide blocking in the vertical shafts left for piping or wiring. Investigation of fires showed that protective walls and floors meant little if they were penetrated by even small openings.

Often the heat from a blazing building would crack windows across the street, but in the 1889 Thanksgiving fire in Boston the collapse of a nearby building blew in the large glass show windows in the front of the Ames Building, and the structure was quickly enveloped in fire. Flames in the lower floors of a building could leap out windows and enter the windows above. Around the turn of the century, New York required that any building over two stories high and not

more than 30 feet from another building should have metal shutters on all windows above the ground floor and that the shutters be closed each night.[35] Similar regulations were proposed for London, but they were not made part of the Building Act.

City regulations were often strengthened in response to fires in which large numbers of people were threatened. In 1847 the German theater in Budapest, one of the city's largest buildings, was razed by flames that started in its new heating system. This conflagration, luckily occurring when the theater was empty, came shortly after theater fires in Karlsruhe and Stuttgart, and German architects quickly responded by proposing legislation that would dictate the number, location, and size of exits from theater seating.

Lighting was essential to the theatricality of performances, and the introduction of gas illumination did little to make a stage area less dangerous. In England it was reported that the 69 theater fires of the 1850s were followed by 99 in the 1860s and 181 in the 1870s.[36] By the 1880s deaths in theater fires averaged 187 per year. Many of the larger British cities required fireproof curtains between the stage and the audience before a

14.15 Although terra-cotta blocks that fit around I-beams' flanges were widely used (top), the floor construction indicated in the lower drawing was often recommended. Deeper blocks reduced the concrete required, which reduced both weight and cost, and small reductions in floor thickness were advantageous when accumulated through 20 or more floors of office building construction. (*Engineering Magazine*, October 1892.)

theater was licensed to operate, although their building regulations contained no such requirement. With gas lighting, glass chimneys and protective gratings provided a degree of safety, but accidents were almost as inevitable on stages as they were in textile mills. The French world of ballet was shocked in 1862 when Emma Livry, the rising star of the Paris Opéra *corps de ballet*, was enveloped in flames as her tutu touched a backstage gas flame. Her lingering death only served to emphasize the dangers that lay behind the proscenium. Most theater fires could be attributed to faulty lighting. A backstage light started a fire in a theater in Vienna's Ring area during a performance of Offenbach's *Tales of Hoffman* and, although the proscenium of the theater was outfitted with an iron curtain, it was not lowered soon enough to prevent flames trapping those who were seated in the balconies.

Six years later the Opéra-Comique in Paris burned during the first act of *Mignon*. The official reports listed only 77 fatalities, but an additional number died after weeks of suffering. A public prosecutor claimed that the fire had been started when a poorly placed stage light with a damaged protective grating came in contact with cluttered pieces of scenery. The building engineer failed to lower the iron curtain and firemen panicked in the face of the flames. Exit in one direction was seriously limited by a revolving door the theater management had installed in response to patrons' complaints about drafts of cold night air.[37] More than a year later Paris authorities enacted new regulations for theaters requiring the use of electric lighting, incombustible materials, sprinkler systems, and improved emergency stairs, including external stairs. Several years before the fire in the Opéra-Comique, comparative studies had been made for the Paris Opéra, Charles Garnier's building of the 1860s. As a result of this investigation, 7,455 gaslights were eliminated in that building and were replaced by 18 arc lamps on the building's facades and 6,131 Edison incandescent lamps inside.[38] Nevertheless, a British committee reported the Opéra's stage to be one of the most dangerous in Europe, a height of 13 stories in which hoses were provided with water pressure of only 70 pounds.

The accounts of theater fires form a somber litany of accidents, neglect, and indifference. Disaster was usually required before stronger regulations were enacted and enforced. An investigation of London theaters by the Royal Society of Arts in 1883 showed that almost all of the city's theaters used scenery treated with some chemical preparation that might lessen the danger of fire, usually having a borate or silicate as its basis. Electric lighting was adopted rapidly by theaters, for it offered advantages in staging as well as safety.

One of the most influential fires of the nineteenth century took place in the Iroquois Theater in Chicago during a matinee performance of *Mr. Bluebeard* during the Christmas season of 1903. The theater was new and soundly built, but sprinklers above the stage were not yet operable. At the end of the second act, when lights had been adjusted for the song "In the Pale Moonlight," sparks issued from an uncovered arc light. As the scenery caught fire, stagehands unsuccessfully used tubes of Kilfire, a powder extinguisher, which were hung about the stage area. The fireproof curtain was lowered but jammed before reaching the stage floor, and, as fleeing actors left by the only stage exit, the draft of outside air blew flames under the curtain into the auditorium. Ventilators above the stage were inoperable. While most of the audience seated at ground level escaped, those in the balconies and boxes were trapped. Over 570 died, and many others were trampled on stairs and in hallways.

The horror of the Iroquois Theater fire moved officials in Chicago and elsewhere to look more closely at their regulations and inspection procedures. As in most theater fires, no single factor could be blamed. It was necessary to view all precautions as parts of an interrelated system, the failure of any one part capable of proving the others inadequate.

Fortunately, most theater fires occurred before and after performances, when audiences were not present but fewer actors and stagehands were present to notice flames in time to extinguish them. Electric lighting, sprinkler systems, and advancing regulations regarding exitways led to some amelioration of conditions, but theater fires in the 1870s were almost three times as many as in the 1850s.[39] It became evident that the principal precaution that could be taken was frequent inspection and stricter enforcement of the laws that were being enacted to control the design of theaters.

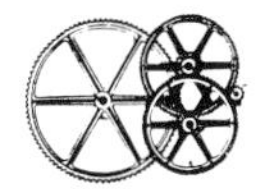

During the nineteenth century the professional press principally con-

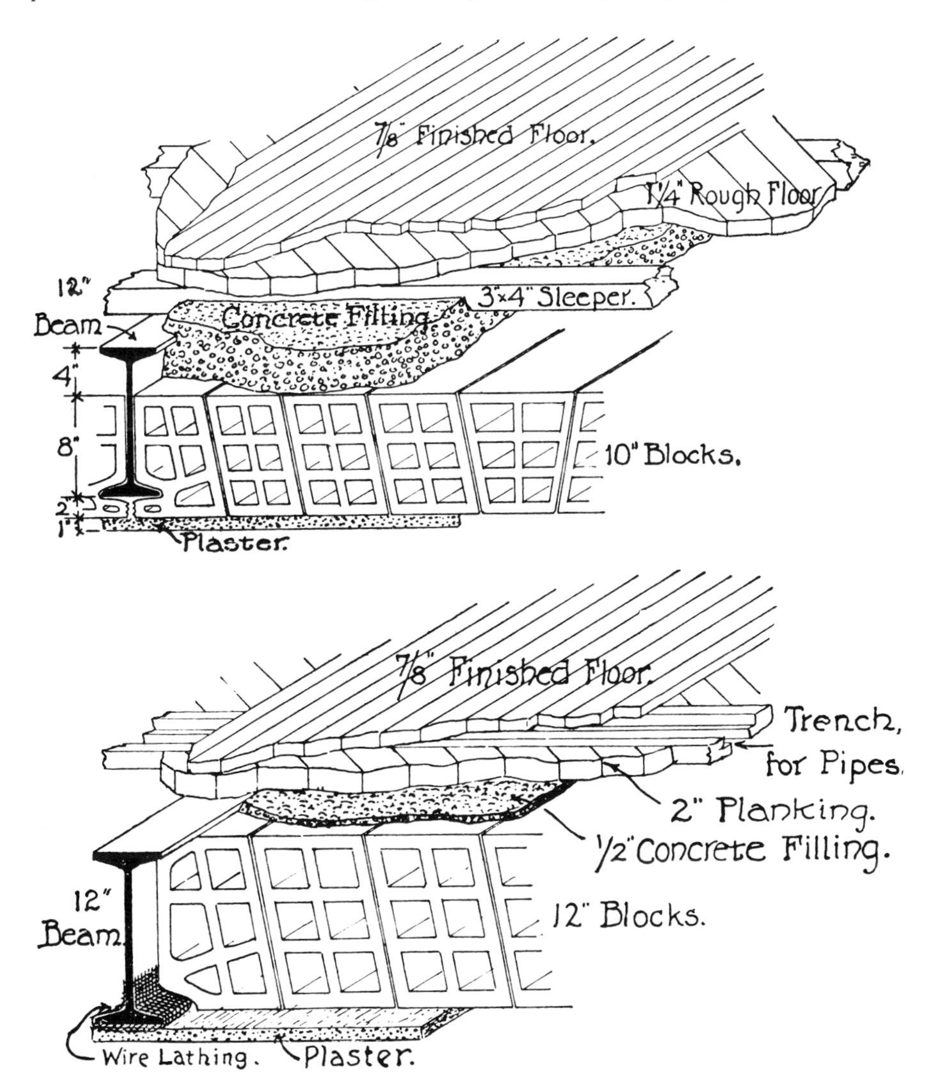

14.16 Hotels, apartment buildings, and theaters were often touted as being "fireproof" because their principal materials were incombustible. Frequent fires in "fireproof" structures made such advertising another symbol of the greed and deceit attributed to real estate investors. (Bettmann Archive.)

cerned itself with fireproofing buildings, in terms of reducing the way in which they might catch fire or be damaged by fire. Little attention was given the problem of insuring means of safe exit for the occupants of buildings. In this regard a speaker before the Royal Institute of British Architects pointed out early in the twentieth century that "experience in all countries had indicated that the remedial measures must in the first instance be indicated by the public authorities. Private enterprise in this matter has not been satisfactory."[40]

It was reported in 1898 that the Berlin Building Act required that any tenement or commercial building of four or more stories have two enclosed stairwells and that no part of the building be more than 100 feet from a stair.[41] In 1905, amendments to the London Building Acts allowed city authorities to require satisfactory means of escape in both old and new buildings, but there were so many offending structures that the requirements were not enforced.[42] Cities in the United States, according to an English insurer reporting on his tour of the eastern states in 1902, had found no consensus and presented a variety of regulations.[43] For purposes of fleeing a fire, buildings at that time often contained such provisions as ropes, canvas chutes, and wire ladders. The exterior iron fire escape, peculiarly American, included types having ladders from level to level, which merited the criticism that "only athletic and cool-headed men could be expected to descend such a contrivance."[44]

Installation of fire escapes with iron stairs was a common method of improving precautions in existing tenements and factory buildings. In some cities' codes the provision of fire escapes could do much to reduce the requirements for stairways within a building. Fire fighters considered external fire escapes to be a good form of exit and also a useful method for their gaining access to fires. At one time the Chicago building ordinances required that fire escapes be provided at each level with standpipes into which firemen could fit hoses. New York authorities in 1902 demanded that all new stores, factories, and hotels of 2,500 to 5,000 square feet of floor area should have two stairways. It was not required that these stairs be enclosed, but if they were enclosed exterior fire escapes might not be insisted on. No such requirement was imposed for office buildings, though they were on the whole taller. Chicago and other cities had ordinances that were similar. In searching through published plans of the major skyscrapers of the period, one finds that until World War I it was common for buildings as high as 25 stories to have only a single unenclosed central stairway. It was also common to have only ornamental grillwork closing elevator shafts from corridors on every floor.

At the beginning of skyscraper construction, enthusiasts spoke of such buildings as cities within themselves and elevators as thoroughfares turned on end. The simile unfortunately also applied to the dangers of fire. Crowded as a walled medieval town and almost as combustible, skyscrapers had few gates through which their inhabitants could escape, and the vertical thoroughfares could act as chimneys filled with flames and smoke. The assumption that buildings could be made fireproof was repeatedly disproved by events. But tall buildings were born from the urge to gain maximum profit from a restricted area of land, and confidence in fireproofing may well have been encouraged by economic considerations. Probably the most realistic attitude is

that of a civil engineer writing in 1890 on the subject of "Fire-proof Construction": "It is intended that the word 'fire-proof' shall apply to any materials or combination of materials which have been designed to resist the action of fire; whether successfully or not is immaterial. A comparison between various structures or materials can thus be instituted, and one can be described as being more, or less fireproof than another."[45]

15 Structural Engineering

1586 Fundamentals of graphic statics published by Simon Stevinus

1638 Galileo Galilei presents his explanation of beam action

1678 Robert Hooke publishes theory of elasticity

1757 Publication of the first of Leonhard Euler's papers on the buckling of columns

1775 The engineering program of the Ecole des Ponts et Chausées is reorganized

1776 Charles Augustin Coulomb states the conditions governing strength of beams

1857 Theorem for continuous beams published by B. P. E. Clapeyron

1865 David Kirkaldy's testing establishment opens

1882 Grillage foundations constructed for the Montauk Building, Chicago

1893 Pneumatic caissons used in constructing foundations for the Manhattan Life Building, New York

1932 Moment distribution introduced for the design of building frameworks

Structural design must balance the realities of construction practices and the discipline of structural engineering. The former is largely empirical, based on experience gained in building and the skills of the building crafts. The latter, usually expressed in mathematical terms, is founded on theoretical knowledge, experience, and the profession's responsibility for public safety. To clarify the nature of structural design, let us consider a simple formula that is used to determine the size of an evenly loaded beam needed to support a floor area:

$$\frac{(L + D)\,l}{8} = fS$$

where:
L = live load of the supported area, the estimated maximum weight of fixtures and occupants, as set in regulations;
D = dead load of the supported area, the calculated weight of building materials;
l = distance spanned by the beam;
f = allowed stress for the material according to regulations, as determined by tests and adjusted by a factor of safety;
S = section modulus, a function of the cross-sectional shape of the beam.

The left side of the equation describes the conditions imposed on the beam, and the right side determines the resistance of the beam. On both sides there are factors that can be determined with considerable accuracy (D, l, and S). On both sides there are factors arising from the theoretical understanding of the conditions imposed on the beam and the manner in which a beam responds (S and the number 8). In addition, on both sides there are factors that are adjusted by the wisdom of the engineer and decisions of official agencies in the interests of safety (L and f). This simple case demonstrates the manner in which structural engineering, as an "empirical technology," combines theoretical investigation, data, and social responsibility in its designs.

15.1 The skill of medieval architects was such that they could propose artful conceits, as shown in the handbook prepared by Villard de Honnecourt in the thirteenth century. In the pair of arches, joints of the stones in both are aligned toward a point on the center support. Hence, the stones act as a single arch and would remain stable after removal of the wood column. (Drawn from the *Album of Villard de Honnecourt*, circa 1235.)

No such complexity is to be found in ancient times, when the empirical knowledge of building crafts, taught by master to apprentices, provided the tradition and theory on which structural design was based. The Roman architect Vitruvius in his *Ten Books of Architecture* compared the qualities of stone taken from various quarries and the wood of different trees. Referring to the four elements as they were understood at that time, Vitruvius characterized fir as having "a great deal of air and fire with very little [of] moisture and the earthy," because of which it is "naturally stiff, it does not easily bend under the load, and keeps its straightness when used in the framework." In contrast he viewed oak as "having enough and to spare of the earthy among its elements, and containing but little moisture, air and fire," which caused it to warp.[1] Although Vitruvius wrote at length about the traditional proportions of columns and the spaces between them, little was said of structural considerations, except for a declaration that in one style of temple (araeostyle) columns were spaced "farther apart than they ought to be" and, hence, wooden beams were required, rather than stone.[2]

Medieval masons during their apprenticeships were introduced to the involved geometric techniques required to lay out plans and prepare the templates and models from which stonework would be cut. The traditional methods and "rules of thumb" of their craft were imparted to them as part of the mysteries of the mason's lodge, and all guild members were sworn to secrecy. The Ordinance of the German Lodge Masons (1459) required that "no craftsman, whether master or warden or fellow shall teach by means of any extract how to set up from the ground-plan to anyone whomsoever who is not of our handicraft, who has not employed his days on stone masonry."[3]

The transformation of the massive stonework of Romanesque architecture into the delicate tracery of the Gothic presents clear evidence of the powerful logic of the trial and error methods employed by medieval master builders. Without mathematical theories or predictive methods, but with great geometrical skill, Gothic builders learned to fashion stone—a material limited to receiving compressive forces and requiring careful fitting—into ribs and vaults that display a profound understanding of the action of forces within the structure (fig. 15.1). Little has been discovered to indicate medieval knowledge of elementary engineering as it is known today, and it is evident that builders of that period did not employ any form of structural analysis (if analysis is taken to mean assigning numerical values to the events taking place within a structure and the assumption of basic physical characteristics). But through experience with the actual performance of a relatively small group of structural elements (arches, vaults, ribs, buttresses, and counterweights being foremost) an impressive empirical wisdom was achieved. Still, a degree of prediction was furnished by the building process. When famine, plague, or warfare halted construction of a cathedral, the tops of walls and vaults were protected with thatch, and the stonework settled through the years it waited for work to be resumed. The appearance of cracks or the spalling of stone indicated incipient dangers, and corrections were made when work resumed.[4]

While craft traditions had sufficed for the remarkable traceries of Gothic construction, theoretical explanations were sought in the Renaissance, and at the heart of the earliest discoveries was the study of mechanics and statics. Mechanics describes the ways in which forces, gravitational and others, act on objects; and statics—as opposed to dynamics—is that branch of mechanics that considers those actions that result in equilibrium. Archimedes, who was inclined to rely more on experimental evidence than the many-layered theoretical explanations of other Greek writers on science, had arrived at the principle of the lever, although the nature of nongravitational forces is not clearly stated in his works. Despite suggestions of the subject that are to be found in writings of the scholastic period, the fundamentals of statics were not known until the Renaissance.

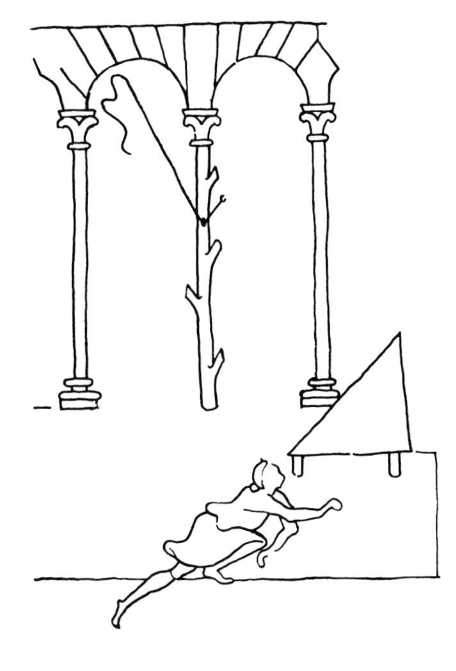

By the fifteenth century, knowledge of construction had escaped from the secrecy of masons' lodges and came to be published in books on architecture. Most often the controlling proportions for construction were described as the extreme conditions at which failure threatened—probably with the numbers altered to provide additional precaution (fig. 15.2). For instance, Leon Battista Alberti stipulated the dimensions of arches for a bridge: "And there should not be a single Stone in the Arch but what is in Thickness at least one tenth Part of the Chord of that Arch; nor should the Chord itself be longer than six times the Thickness of the Pier, nor shorter than four times."[5] Renaissance writers on structure did not know exactly what path the pressure within a vault took as it made its way down to a support. Indeed, engineers since that time have pursued this problem with differing assumptions leading them to a variety of conclusions. In Leonardo da Vinci's notebooks is found a drawing of a pointed arch that is loaded only at its top point, a problem simpler than the usual loading. Leonardo contended that a straight line drawn between the top and bottom ends of the outer curve of the stones of the arch should not meet the inner curve.[6] Almost two hundred years later, François Blondel proposed a method in which three equal chords were inscribed in the curve of an arch and one chord was extended an equal length to determine the required thickness of the pier that would adequately support the arch (fig. 15.2). For semicircular arches Blondel's method results in a pier whose width is one-fourth of the span of the arch, the narrowest pier that had been permitted by Alberti.

A lengthy study of structure commenced in the Renaissance, much of it focused on the construction of domes. Since they were the central and dominant architectural feature in many of the most important buildings of the sixteenth, seventeenth, and eighteenth centuries, the design of domes was an attractive and rewarding challenge to experts. Since ancient times, tie rings, superimposed masses, and buttressing to resist outward thrusts had been used to stabilize domes. Tie rings of bronze chains or linked oak timbers were the most popular method, probably because they were least likely to obscure or encumber the dramatic silhouette that architects sought when using a dome. In experiments, chains were draped to represent the curves that might be the inverted lines of thrusts, and intricate graphic solutions attempted to follow forces from stone to stone. The objective was, as with Leonardo's drawing of an arch, to make sure that the thrusts followed a path that remained within the thickness of the dome's shell. In a hemispherical dome of uniform thickness, outward thrusts begin at an angle 52° down from the vertical, but, with cupolas at the top and colonnades around the lower portion, there was significant variation of the level at which tension might be first encountered.

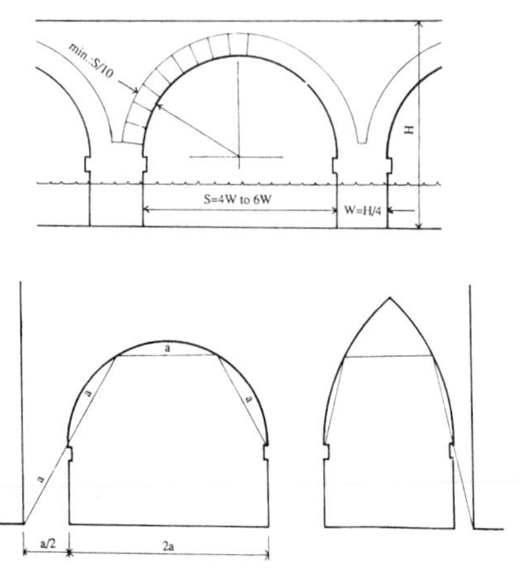

15.2 Top: Alberti's recommendations on bridge building required that the footings be half and the pier width one-quarter of the vertical distance from the roadbed to the streambed. The arches' spans could range from a distance equal to that height to one-half again as great. Bottom: Blondel's system advised drawing three lines of equal length within an arch and extending one for the same distance to determine the required thickness for a pier. (Drawing by author.)

The notebooks of Leonardo include sketches that demonstrate his understanding of forces in terms of direction and magnitude, as they are described in modern mechanics. In his drawing of a weight suspended by two cords of unequal length, a superimposed rectangle is marked off in units that indicate the distribution of the load between the cords. In writing a prospective employer, Leonardo presented himself as an architect who knew "what is the nature of weight and of energy in force, and in what manner they should be combined and related to one another and what effect they will produce when combined."[7] However, Leonardo's studies of statics, mingled among his many ideas, are limited to speculative consideration of the topics, and apparently were neither tested nor communicated to others.

Long after Leonardo, in 1586, a Dutch army quartermaster published a book on statics; its translation into Latin in 1608 as *Mathematicorum Hypomnemata de Statica* made his knowledge of the subject accessible to scientists and mathematicians throughout Europe. Simon Stevinus acted as a tax collector before studying mathematics at the University of Leyden. His army responsibilities included superintending dikes and waterways, critical factors in Holland. Recognized as a founder of the science of hydrostatics, he also introduced the use of decimal fractions and devised methods of navigation. Stevinus was clearly aware of the principle of moments, which describes the action of levers, and of the force triangle, a fundamental means by which the resultant action of two or more forces may be determined in both direction and magnitude. By representing forces as lines drawn at slopes related to their directions and lengths proportional to their magnitudes, Stevinus provided the basis for nineteenth-century advances in graphic statics, which enables the solution of structural problems through drawings.

While the science of statics defines the nature and power of forces acting on parts of a structure, there remains the investigation of the means by which the parts or the entire structure resist external forces and maintain stability. This study, known as the strength of materials, focuses on the characteristics of the material of which the structure is built, the shape in which it is present, and the manner in which internal forces respond to forces that are imposed on the structural element.

As the dominant architectural material of early times, masonry had limited capabilities and so was relatively simple to understand. Although stone could to a degree sustain tensile stresses (those that pulled on it), there was always a lurking possibility that any stone had within it some undetected flaw, some weakening stratum created when the stone was first formed. Mortar caused one stone to adhere to another, but weathering and uneven settlement could cause hairline cracks to appear. Because of these vulnerabilities of masonry, it has usually been assumed that only compressive (pushing) forces can be effectively resisted by it, and since the time that they were developed by the Romans the spanning forms of arches, vaults, and domes have been the most common structural forms in which masonry has been used. (The outward thrusts could be counteracted by the addition of materials capable of withstanding tensile forces.) When stone was used as beams, the spans between the capitals of columns was small, in Greek temples ranging only from around 1½ to 2½ times the depth of the beam. As a precaution such stone beams were often made of three

15.3 In this drawing from *Discorsi e dimostrazioni matematiche* (1638), Galileo posed the classic problem of the cantilever beam. He assumed that the beam's fibers acted around the point *B* with no variation in the intensity of their action. This results in the beam's strength being evaluated as three times that assigned by present-day theory. (Galileo, *Discorsi e dimostrazioni matematiche,* 1638.)

15.4 Arrows in these diagrams represent the relative strain on parts of a beam. Mariotte at first assumed that tensile action pivoted around the bottom of the beam, but later he recognized the significance of Hooke's Law of elasticity and designated the neutral axis at midheight of the beam's cross section. (Redrawn from *Proceedings, Institution of Civil Engineers,* 1952.)

pieces, placed side by side, so that if one should split the others would continue to support the load.

Understanding of the internal action of beams was advanced when Galileo Galilei published his book *Two New Sciences* (1638) in Holland, after he had been condemned by the Roman Inquisition for his espousal of the Copernican theory of the planetary system. The first two parts of Galileo's book present his findings regarding the strength of materials, and his explanation of a cantilevered beam reveals his insights and errors (fig. 15.3): "Fracture will occur at the point B where the edge of the mortise acts as a fulcrum for the lever BC, to which the force is applied; the thickness of the solid BA is the other arm of the lever along which is located the resistance. This resistance opposes the separation of the part BD, lying outside the wall, from that portion lying inside."[8] In other words, Galileo believed the cantilever beam was held fast solely by tensile resistance of its fibers and that they were equally stressed. These assumptions may now seem naive, but if one were confronted with a failed cantilever, all the fibers torn except the bottom slivers from which the beam dangles, this conclusion would seem more plausible.[9]

It was around 50 years later that Edme Mariotte, a French priest, physicist, and one of the founders of the French academy, concluded that fibers in the upper half of a cantilever beam would be in tension, pulling to hold up the beam, and fibers in the lower half would be in compression, pushing to support the beam. Unlike Gali-

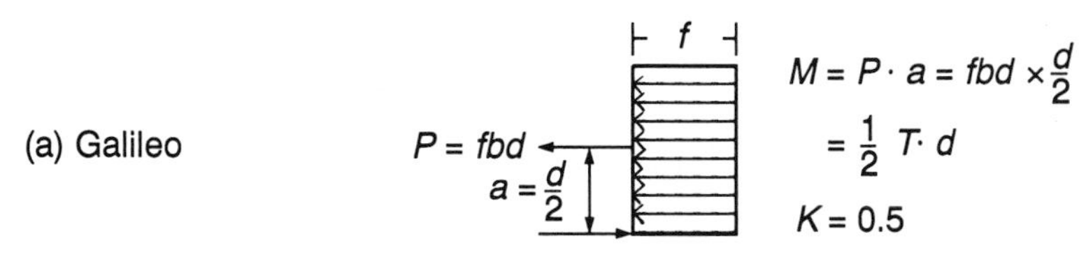

M denotes moment of resistance under extreme fiber stress f in a beam of rectangular section

P denotes resultant longitudinal force in tension or compression

T denotes "absolute resistance" or ultimate tensile strength of the section of area $b \times d$

K denotes the ratio of M to $T \cdot d$

leo, Mariotte surmised that these two forces, tensile and compressive, acted about a level located at mid-height of the beam. He also reasoned that the strain on individual fibers would increase from zero at this axis to a maximum at the top and bottom of the beam. Mariotte's explanation of the cantilever beam initiated recognition of the fact that within a beam a distribution of tensile and compressive responses of the fibers must equal the action instigated by any load carried by the beam (fig. 15.4).[10] As Mariotte put it: "You may conceive that for half the thickness the parts are pressed together, those near the outside more than those near the middle and that for the other half of the thickness the parts are extended."[11] Mariotte's structural theories were developed from his experiments with rods of wood and glass, undertaken while he was in charge of designing a system that would bring water to the palace at Versailles. The validity of his theory of stress distribution depended largely on Hooke's Law, which was developed by experiments with springs: "It is very evident that the Rule or Law of Nature in every springing body is, that the force of power thereof to restore itself to its natural position is always proportionate to the distance or space it is removed therefrom, whether it be by rarefaction, or the separation of its parts the one from the other, or by a Condensation, or crowding of those parts nearer together."[12] Therefore, for the loaded cantilever, which had become known as "Galileo's Problem," when the beam bowed under a load, its fibers were stretched (rarefaction, tension) on the convex side and were pressed (condensation, compression) on the concave side. And the same would be true for a simple beam spanning between two supports, where fibers would be in tension on the convex (lower) side of the beam and in compression on the concave (upper) side.

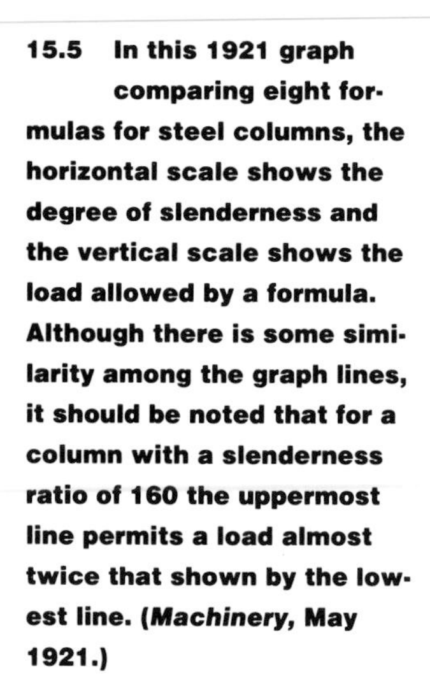

15.5 In this 1921 graph comparing eight formulas for steel columns, the horizontal scale shows the degree of slenderness and the vertical scale shows the load allowed by a formula. Although there is some similarity among the graph lines, it should be noted that for a column with a slenderness ratio of 160 the uppermost line permits a load almost twice that shown by the lowest line. (*Machinery*, May 1921.)

For almost a century, mathematicians and physicists debated the action of a beam's cross section. Some clung to Galileo's theory that the neutral axis (the horizontal about which forces within the beam rotated and where neither tension or compression are present) was located at the bottom of the beam; others were uncertain about the location of the neutral axis and furthermore attached no importance to its exact placement; and still other agreed with what became known as the Mariotte-Leibniz theory. Then Charles Augustin Coulomb, a French physicist and military engineer, in 1776 stipulated the basic conditions that govern the strength of beams: (1) The neutral axis of a rectangular cross section of a single material is at the middle of its height. (2) The total tensile force in beam fibers on one side of the neutral axis must equal the total compressive force on the other side of the neutral axis. (3) The resistance within the beam must equal the bending induced by the loads on the beam.

Among a multitude of other structural problems that required solution was the design of columns, a problem of greater complexity than that of beams. For centuries, columnar supports had taken the form of massive piers or cylinders of sufficient size to counteract the thrusts of arches and domes and to support the weight of massive stone construction. With the introduction of iron columns it became necessary to ascertain the characteristics of columns that were more slender. It was not the vertical pressure that was the problem, for, according to calculations using today's standards, a column of sandstone, if it could be braced in a vertical position, could be over a mile high before it would be crushed by its own weight.[13] The primary hazard for iron columns was buckling, bowing out to one side or another. Once a column begins to buckle, any load it bears has an increasingly strong effect. Therefore, theorists sought a means of describing the stiffness of columns and a way to determine the degree of stiffness required when a given load was imposed on a column.

The first influential theory of columns came from Leonhard Euler, a Swiss mathematician who spent much of his life at the Prussian and Russian academies under Frederick the Great and Catherine the Great. In 1757 Euler published a paper in which he presented his formula for determining the maximum load that could be applied to a given column before it would buckle. This formula combined the elasticity of the material from which the column is made, the shape and size of the column's cross section, and the height of the column. Euler's formula was corroborated by the results of tests and other mathematical studies of columns, but the limits generally accepted for its validity restricted it to very slender columns. As builders shifted from cast iron to wrought iron and steel, tests were made of each material using different methods of fastening the columns at top and bottom. In the century and a half that followed Euler's publication of his column formula, many additional formulas were developed, either based on the results of tests or on other theoretical studies.

In 1913 an engineer of the United States Bureau of Standards compared an array of 27 formulas for steel columns. Fifteen of these were used by American railway engineers; two were noted as being popular among English engineers; and two were favorites of German engineers.[14] A few years earlier the writer of a similar article comparing column formulas had said that "there is no better device for 'whipping the devil round the stump' than the invention or promotion of a new column formula. In times gone by it was a rather popular amusement; today it is out of fashion. Not that the utmost pitch of perfection has been attained; rather, a conviction of the fruitlessness of the diversion has impressed itself quite deeply into the current tradition of engineering."[15] In 1926 another article compared some 50 requirements for columns found in different United States municipal building codes. An editorial note observed: "There seems to be an almost unlimited array of column formulas which all seem to reach about the same result. If that is the case why not throw them all together and strike an average?"[16]

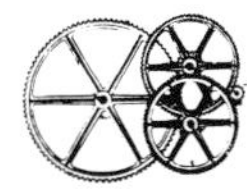

The complexity of most column formulas demonstrated that many engineers at the end of the nineteenth century had sound theoretical grounding in physics and mathematics. It was France that had set the pattern for training engineers. In 1671 the Académie Royale d'Architecture was founded, the first French school of higher education dealing with technical subjects, and its curriculum stressed engineering subjects fully as much as instruction in the fine arts. Five years later France's army corps

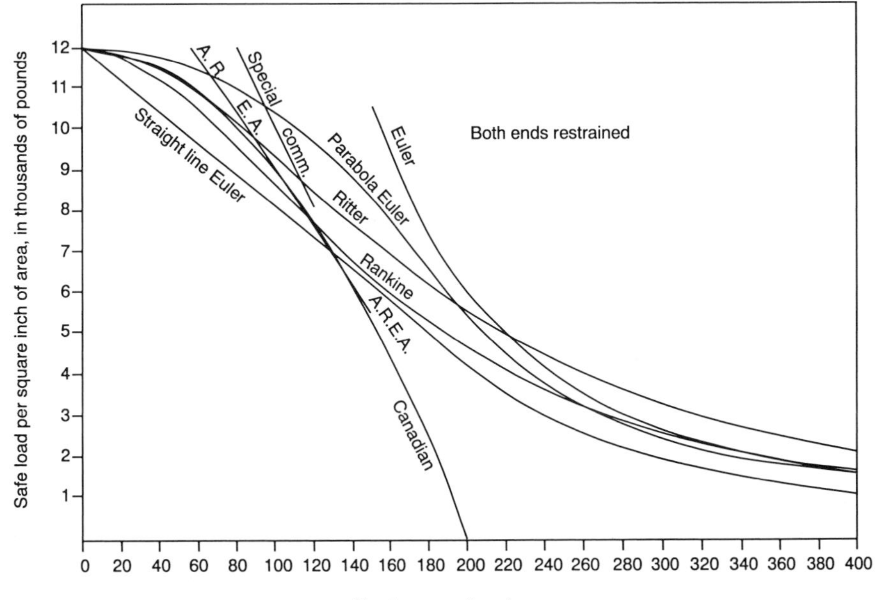

of military engineers was formed. In 1775 the Ecole des Ponts et Chaussées, where most French military engineers were trained, reorganized its program of instruction, which had been in operation for several decades. After the Revolution the Ecole des Ponts et Chaussées became one of the schools designated for advanced instruction, and before entering upon the advanced studies its students were expected to have completed study at the Ecole Polytechnique, which had been set up in 1794.[17] This system of technical education provided France with a succession of brilliant engineers, versed in actual construction as well as theory and mathematics.

15.6 This machine was used by Eaton Hodgkinson around 1840 to test cast-iron beams of different shapes. By simple leverage, the weight at the left applies a greatly magnified force on the beam, causing the failure shown. (E. Hodgkinson, *Experimental Researches on the Strength and Other Properties of Cast Iron*, 1846.)

Early in the nineteenth century, German polytechnical schools were established; unlike the French schools, they accorded as much attention to training engineers for industry as for the military.[18] Britain, the leader in mechanization, lagged behind the Continent in the education of engineers, and it was 1840 before its first engineering professorship was established at the University of Glasgow. Professional opinion in Britain was divided on the virtues of classroom training or apprenticeship. Thomas Telford, England's most accomplished civil engineer at that time, favored the latter means of preparing for the profession, but at the same time he believed a knowledge of the French language was essential in order to have access to the information in the manuals and handbooks used in the French engineering schools.[19] Technical education in the United States was not formalized until the last half of the nineteenth century. In 1862, during the Civil War, Congress appropriated land that states could sell in order to establish colleges whose curriculums should include instruction in the "mechanic arts." Before the land-grant colleges' engineering curriculums were well established, much of the construction in the United States depended on many immigrant engineers and on others who had received brief training during military service in the Civil War.

Each nation seems to have developed its own attitude toward structural design during the years that followed. Around 1920 the contrasts that had developed were spelled out in letters published in the *Engineering News-Record* and other journals.[20] It was contended that British engineers did not use books in their practices to the extent that they were used by engineers in the United States and that the British relied much more on experience. One party to the brief crossfire questioned why British writers on the design of structural steel preferred long and complex mathematical methods to simple graphic analysis. A British writer deplored American "profound faith in 'figures' and 'figureability,' " and American engineers were depicted as being "influenced by the German thirst for unnecessary detail."[21] Similarly and more recently, others have written of "German science and French daring."[22] However accurate such stereotyping may have been, these comparisons demonstrate that within the scope of structural engineering there can be sharp and clear distinctions between the operative points of view of individual engineers and engineering specializations (such as bridges and buildings), as well as among national groups.

The validity of formulas could be judged by testing the actual structural members in place, loading parts of the building with bags of sand or bars of metal far beyond the load for which it had been designed; or members could be tested separately. Since all formulas included factors based on the nature of the material to be used, a

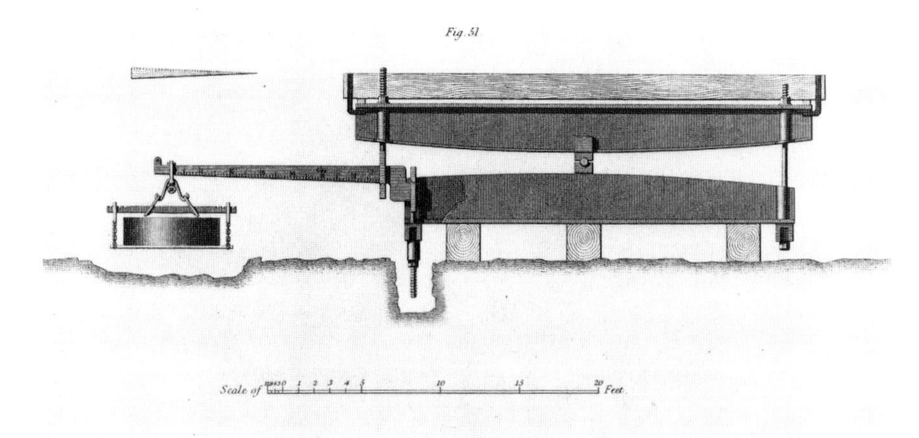

sample of the material could be tested to determine its characteristics. All of these possibilities, except the first, called for the services of testing machinery. Many formulas in use were developed from test data, and others were verified by tests that supported their theoretical predictions. By the middle of the eighteenth century several experimenters had constructed machines in which small samples of wood, iron, or stone could be tested to determine their strength in tension, compression, or bending. A series of tests made during construction of the French Panthéon (Ste.-Geneviève, Paris) attracted much attention and, as discussion continued, improvements made in the testing machines made their findings more accurate. Improvements continued through the next century. Efforts were made to lessen friction in the machines; loads were placed on the test samples with greater accuracy, and experiments were conducted to ascertain the sizes and shapes of samples that should be tested for the most accurate results.

Because several of the early English cast-iron bridges had failed by the 1840s, a Royal Commission on the Application of Iron to Railway Structures recommended extreme caution in the design of metal bridges. Isambard Kingdom Brunel, the most prominent of railway engineers, called it the "Commission for Stopping Further Improvement in Bridge Building."[23] Nevertheless, railway construction thrived and much of the engineering work in England and the United States related to the use of metal construction in railway work. In England, William Fairbairn and Eaton Hodgkinson conducted exhaustive tests of the material to be used in a railway bridge over the Menai Straits, a bridge in the form of a rectangular tube across an unprecedented span of 459 feet (fig. 15.6).

During the 1860s the testing laboratory of David Kirkaldy was set up in London, and many foreign companies sent samples there, often in an effort to check the accuracy of their own testing machines.[24] It was in Kirkaldy's laboratory that the tests were

15.7 In the construction of the Crystal Palace (1850), a hydraulic press was used to test the 214 cast-iron girders that were major elements of the structural system. The girders spanned 24 feet, and those supporting galleries were tested at 22 and 15 tons. This procedure was less newsworthy than the troops of running cadets that simulated the loads that might be induced by the public's footsteps (fig. 4.28). (*Illustrated London News*, 30 November 1850.)

conducted through which Thaddeus Hyatt discovered the principles of reinforced concrete construction. Kirkaldy's testing laboratory was one of the first established in Great Britain, where such work was usually carried out by private concerns. On the Continent most laboratories operated under government auspices. In Germany universities were the usual locale, permitting use of the equipment for research as well as for the investigation of samples presented by manufacturers.[25] In America, an organization of engineers petitioned the government for construction of a machine suitable for gaining data about American iron and steel. It took about four years to construct the "United Testing Machine," which was installed at a Massachusetts arsenal in 1879.

All of these machines were based on eighteenth-century models. Lever-fulcrum machines applied test loads by placing weights on a lever far from the fulcrum's knife-edge. Other designs applied loads by hydraulic pressure. Such simple principles required augmentation by a host of refinements that assured even application of the loads and accurate methods of measuring the stresses that caused materials to fail.

The testing of a completed building or its parts was the most definitive manner in which strength could be determined. However, loading a floor with sand or lead, as was sometimes required by city officials, merely indicated that the tested portion satisfied the minimum requirements of regulations. Testing every member before it was put in place—a process more suitable for wood, iron, or steel than for reinforced concrete—or testing randomly selected samples of members provided a limited reassurance for, as editors of the *Engineering News* wrote in 1906: "Few structures have such simple service; most structures are designed for (and actually exposed to) distributed loads and concentrated loads, quiescent and moving loads, shock, wear, corrosion, freezing, and other forces of attack, each of which may be destructive independently of the others."[26]

The variation probable in a material, the narrow set of circumstances considered in many design formulas, and the unpredictable loads inflicted on the structure all encouraged caution. For that purpose, engineers employed a factor of safety, which has also been realistically defined as a "factor of uncertainty" or even a "factor of ignorance."[27] Probably there was always a custom of exaggerating the magnitude of threatening conditions and underestimating the strength of structure, for only in the Gothic world does there seem to have been a sense of daring that made precautions relatively rare. A clear-cut determination of a factor of safety appears in the middle of the eighteenth century when three mathematicians, investigating cracking in the dome of St. Peter's basilica, doubled their calculation of the amount of steel that should be provided, or in other words halved the stress that was to be imposed on the steel.[28] Through the two centuries that followed, the problem of safety occupied the attention of many organizations and their committees on safety.

Engineers on the Continent tended in the early nineteenth century to relate the stress allowed for materials to the maximum lengthening of the material from which it would elastically return to its original length. English engineers, on the other hand, related allowable stress to the maximum stress, the loading conditions at which the material would fail. In 1849 a German engineer visiting the United States expressed his surprise that American engineers based their calculations on lower loads and higher allowable stresses for their materials than was European practice.[29] Late in the nineteenth century, W. J. M. Rankine in a standard reference book presented a more complex procedure for determining a factor of safety, which he defined as the ratio of the load causing failure to the load that might be safely utilized in calculations. Rankine recommended factors of safety that varied with the type of material and the engineer's judgment of whether the structure would have "perfect materials and workmanship" or "good ordinary metals and workmanship." The same sort of distinctions were made in the United States with different standards for buildings, bridges, and machines.[30]

In other approaches, factors of safety were applied to the assumed loading of members as well as to the strength of the materials of which they were fashioned. By 1962 the extenuating conditions were so carefully considered by an English committee that it recommended estimated loads be adjusted by a factor calculated by a formula having six factors: seriousness of the member's failure, quality of workmanship, loading conditions, importance of the member, estimated warning of incipient failure, and the manner in which members might collapse. From these subjective evaluations, a number was to be derived, and this would be used as the factor of safety for the particular member under consideration.[31] As with the design of columns, a choice might be made among available formulas for designing a structural member, but the variation of cost

15.8 Different cities' building codes in the United States required that engineers' calculations assume 40 to 150 pounds per square foot as live load for office buildings. In 1905 a professor of engineering displayed photographs of his students standing in a box, 6 feet square, to demonstrate the conditions that produced loadings of 41.8, 100.0, and 154.2 pounds per square foot. (*Transactions, American Society of Civil Engineers*, June 1905.)

associated with the choice of formula was usually much less than the cost imposed by the stipulation of a factor of safety or the assumption of a specific loading condition.

Rankine, a professor at the University of Glasgow, also introduced the distinction between dead loads, the actual weight of the fabric of the building, and live loads, the estimated maximum weight of furnishings and occupants. Dead loads were easily calculated with reasonable accuracy, but live loads were largely guesswork. In 1905 an American engineer compared the regulations of American cities and found that the prescribed live loads for apartment buildings and hotels ranged from 40 to 75 pounds per square foot; for office buildings from 60 to 150; for theaters from 80 to 150; and for schools from 75 to 150 (fig. 15.8).[32] Through most of the nineteenth century columns had been calculated as carrying all the required live load above them, but with the advent of the skyscraper some cities allowed reductions in live loads. Following the lead of New York, many requirements assumed that all floors would never be fully loaded at the same time, and the top floor was considered to have the entire prescribed live load, the next lower floor 95 percent, the next 90 percent, and so to a 50 percent load, which would be used for all remaining floors.

These arbitrarily reduced loads were critical in designing foundations. Even with live loads reduced, live loads and dead loads in tall buildings caused pressures on foundations much greater than had been encountered previously. As an engineer wrote in 1908:

> In New York, there are two buildings over forty stories high. These [tall buildings] develop concentrated loads of 3,000,000 or 4,000,000 lb. in a single column and their heights, of from 200 to 500 ft., are so great that the slightest irregularity of settlement in the foundation produces a greatly magnified variation from the vertical in the superstructure and causes great injury or destruction to the delicate and accurate machinery installed, besides settlement, and cracks and displacement of the beautiful cut-stone walls and lintels which would be great blemishes and would inspire public distrust sufficient to seriously impair the value of the building. It is therefore necessary that the foundations should be more than safe; they must be absolutely immovable within the smallest fraction of an inch; or that, if displacement occurs, it must be perfectly regular and uniform.[33]

Traditionally, most walls and piers had been supported by extending stone to a level below the frost line, each course of stone widened to enlarge the area at the bottom. There were, of course, marshy sites that demanded other treatment, but in earlier times such places very seldom were selected as sites for buildings of significance. It was more likely that ancient buildings would be located in a healthful, well-drained site, perhaps with solid rock beneath them.

In Chicago, where construction of tall commercial buildings began, foundations were first made of rubble or trimmed dimension stone, a choice made on the basis of cost. However, Chicago's water level, roughly that of Lake Michigan, was about 15 feet below ground level and only 1 or 2 feet above a crust consisting of 3 or 4 feet of stiff clay. Beneath that crust, soft clays extended for about 45 feet, and firm stone was first encountered around 80 feet below water level.[34] In order to provide basement areas and rest foundations on the layer of stiff clay, it was often necessary that the stepped portion of foundations extend upward into the basement area, which was sorely needed for the building's equipment. Indeed, calculations of spread footings for the Montauk Building, Chicago (1882) showed that the slopes of footings would extend into the ground floor and leave insufficient space in the basement for placing dynamos. Therefore, the architect, John Wellborn Root, employed old steel railway rails, placed in layers at opposite directions.[35] The steel rested on a broad slab of concrete and was itself surrounded by concrete. This rail-grillage system saved weight, basement space, and time in its construction, and the same principle, using I-beams instead of rails, offered even greater advantages. This form of foundation could spread a load as much as 10 feet beyond the sides of a column, and required no more than 3 or 4 feet of depth. Steel grillage foundations were commonly used in many cities in the United States (fig. 15.9). As buildings grew to be taller, the grillages were often extended to cover the entire base of the building, but this method did not sufficiently recognize the large variations in the loads on different columns. Soon areas of grillage were separated, each being used for a single column or a group of three or four columns. In New York special care had to be exercised when pouring the concrete layer over quicksand.[36]

Simple spread footings were usually sized according to rules of thumb. In several parts of the United States the continuous footings beneath bearing walls were given a width equal to one foot for each story of the building.[37] An 1873 pamphlet by Chicago architect Frederick Baumann advised that each footing be of a size proportional to the amount of dead load resting on it and that it be made certain that the center of each footing coincide with the center of the loads resting on it. Furthermore, he recommended the use of allowable

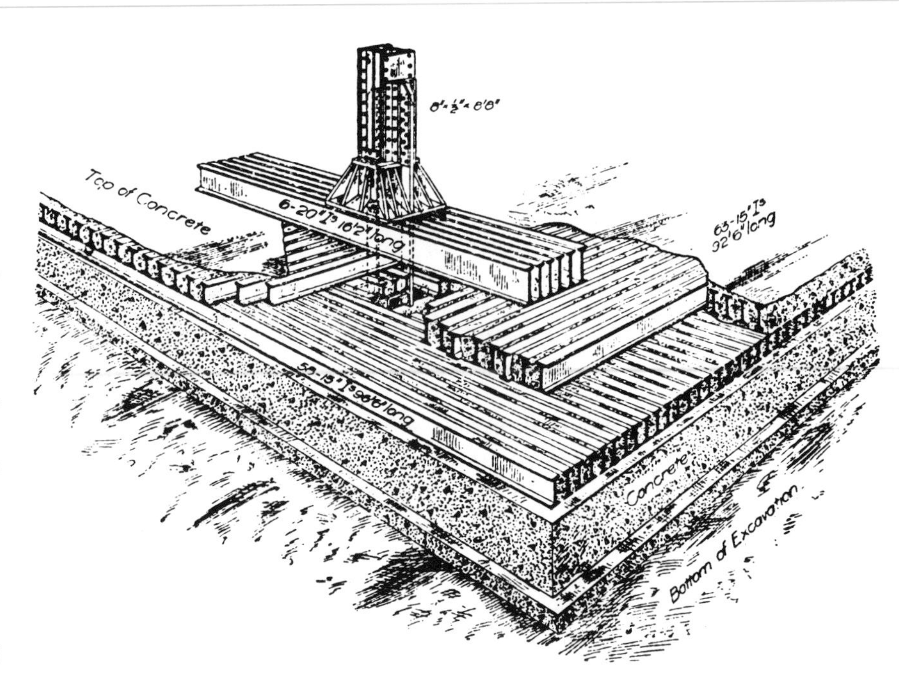

15.9 The Spreckels Building in San Francisco was 75 feet square in plan and 15 stories high. It rose from a grillage foundation, resting on a concrete slab 2 feet thick and 96 by 100 feet in area. Each column bore on a weight-distributing pad of six I-beams. The structure was sturdy enough to survive the San Francisco earthquake of 1906 and support the additional six stories of a 1938 remodeling. (*Engineering Record*, 4 April 1908.)

15.10 For this turn-of-the-century caisson, pressure was provided by a steam-powered air pump and regulated according to a gauge atop the pier. Earth was either hoisted out the airlock tube or, as shown here, blown out by the air pressure. As excavation progressed, masons continued building the pier, its weight forcing the caisson down. (*Engineering Magazine*, April 1897.)

bearing capacities that ran as low as 20 percent of those that had been commonly used in Chicago.[38] Baumann's principles came to be standard practice among many architects and engineers in the United States during the period in which spread footings and grillages were most often employed. Nevertheless, settlement continued to be a major building problem in Chicago. At the end of the nineteenth century it was accepted that a new building would settle, and they were built at a level that anticipated a settlement of 2 to 9 inches. The Masonic Temple building (1892), a skeleton construction and the first building to reach 21 stories, settled over 11 inches in the first five years it was occupied, and 4 inches more in the following ten years. By 1913 some portions of the building had sunk at least 6 inches more than other parts. Since a greater control of settlement would have required the considerable expense of going much deeper, owners considered settlements of this order acceptable. In New York few tall buildings settled more than a fraction of an inch, solid rock or "hardpan" usually providing a firm base of construction.[39]

With buildings taller and consequently heavier, piles were introduced, an ancient method of building in wet places such as for bridges and harbor works. In principle, piles may be viewed as columns, taking the weight of the building to a supporting stratum of rock and braced by the surrounding soil; and as supported by friction between their surfaces and the soil; and they may be used to compact the building area, converting loose soil into a denser material. For early skyscrapers the piles, if wooden, were cut off a short distance below the water level in order to avoid their rotting. Around and over the tops of wooden piles a concrete slab was poured, and

this was capped with spread footings or steel grillages.

Piles to compact the soil for the construction of the Park Row Building, New York, 30 stories high and the tallest building in the world when it was completed in 1898, were spruce trunks about 12 inches in diameter, and they were driven so close together that their cross sections filled over a fourth of the area on which the foundation bore.[40] For some New York projects for which piles were driven into dense, fine sand, they could not be forced down more than 10 or 12 feet. In Chicago, piles were driven much deeper, and piles for the Chicago Post Office were so closely spaced that driving the last ones caused others to rise several feet.[41] In cases where a constant level of ground water could not be safely assumed, steel tubes were driven into the soil, then removed, and the holes filled with concrete.

Excavation and pile driving for buildings could often require precautions to avoid disturbing the soil and foundations beneath existing adjacent buildings that rested on shallower foundations. At the end of the nineteenth century the building regulations of New York made the owner of a new building financially responsible for correcting damage done to any adjacent building with foundations 10 feet or farther beneath street level.[42] When the Manhattan Life Building in New York, 17 stories high, was being designed in the 1890s, the architects were forced to turn to a new system of support. There was not sufficient space on the site to drive the number of piles needed; grillages would have risked the soft soil being squeezed out sideways, and an open excavation 45

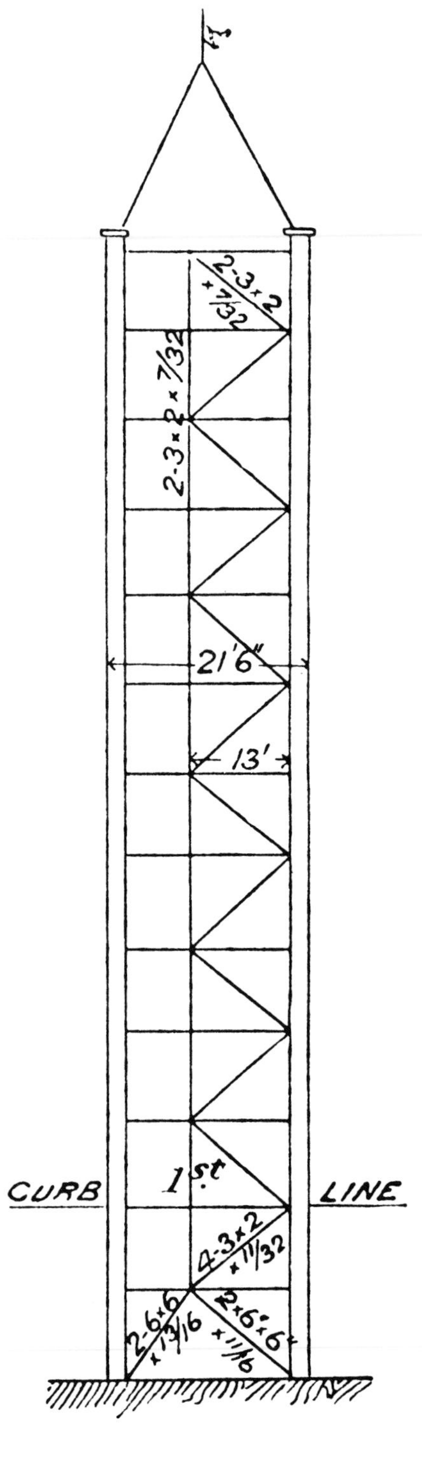

15.11 The Tower Building (New York, 1888) included an 11-story wing on a single lot 21½ feet wide, which provided an enviable Broadway address. To neutralize wind pressure on the side walls, diagonals were introduced at five points in the wings' depth of 108 feet, acting as upended trusses from the basement to the roof. (*Transactions, American Society of Civil Engineers*, September 1892.)

15.12 The Venetian Building (Chicago, 1891), a 12-story office building, employed portals (shop drawings shown here) as wind bracing for two ground-floor bays where shops were to be located. In the office floors above, diagonal rods (upper right and upper left) set in partitions provided bracing. (*Engineering News*, 26 December 1891.)

feet to rock would have endangered adjacent buildings.[43] The architects and engineers decided to use pneumatic caissons, a method that had been employed in the construction of many major bridges. A box of steel, the size of the required pier foundation, 7 or 8 feet high and its bottom open, was located on the site. From the center of this caisson a tube rose so that the soil removed by workers within the box could be taken away. On the top of the caisson, construction of a masonry pier was begun, and its increasing weight forced the box downward as workers dug. Air pressure within the caisson was increased as needed to keep water out. Using the pneumatic caisson, foundations of a typical large New York building at the end of the century took only two or three months to prepare, and engineers foresaw greater speed as more experience was gained (fig. 15.10).[44]

As skyscrapers grew to be much higher than the dimension of their bases, problems arising from wind pressure demanded attention. In 1895 a New York engineer pointed out that in the city "there are many instances of buildings varying from 23 to 25 feet in width and from 120 to 175 feet in height, with the uppermost 100 feet exposed to wind."[45] Although the depth of such buildings' sites might commonly be around 100 feet, it was the typical narrow lot in New York that often established the smaller dimension of a site, the dimension that would be needed to resist the overturning pressure of winds blowing against a side wall above any adjoining structures. Many failures occurred during construction, at a stage of incompletion when winds could blow into partially enclosed spaces but little of the building's weight was in place. It was assumed that the weight of walls and floors would counteract part of a wind's overturning force, a resistance estimated to be as much as 30 percent of the needed force in the case of one Chicago building.[46] Though economical, cast-iron columns were used less often than previously because that brittle metal could not satisfactorily withstand the bending forces produced by winds. Some architects believed that the partitions within a building could effectively resist wind forces, but most partitions were interrupted by door openings and were made of flimsy masonry materials.[47] The most effective methods of combating wind pressures were diagonal bracing and portals. Diagonals had been used in wind bracing for the towers of suspension bridges, but in most buildings the diagonals between the ends of columns, whether rods in two directions or stiffer members in only one, interfered with the placement of windows (fig. 15.11). Therefore, it became customary to arrange the diagonals as knee braces, members slanting from midheight of a column to a point near the middle of the beam above. When knee braces were provided for at least two corners of a bay, these corners were stiffened without disturbing the customary window locations. The portal was an archlike construction of sheet metal, squared on the outside to match the rectangle of the bay and arched inside the top to provide rigidity at the upper corners of the bay (fig. 15.12). By using either of these methods in vertical bands up a tall slender building, the framework was made to act as a truss, anchored to the ground at one end and cantilevered upward to resist the wind.

The general practice in wind bracing in 1915, as reflected in city building regulations in the United States, exempted buildings less than 100 feet in height "in which the height does not exceed four times the average width of the base."[48] New York regulations, which many other cities followed, permitted inclusion of the action of partitions and a 50 percent increase in the stress allowed in the framework. Both of these provisions were considered unwise by some prominent engineers. The wind pressure to be assumed in computations was at that time usually required to be around 30 pounds per square foot, although engineers often believed it wise to increase that number.

During a flurry of skyscraper construction in the 1930s, tests were made in smooth-flow aeronautical wind tunnels to determine wind con-

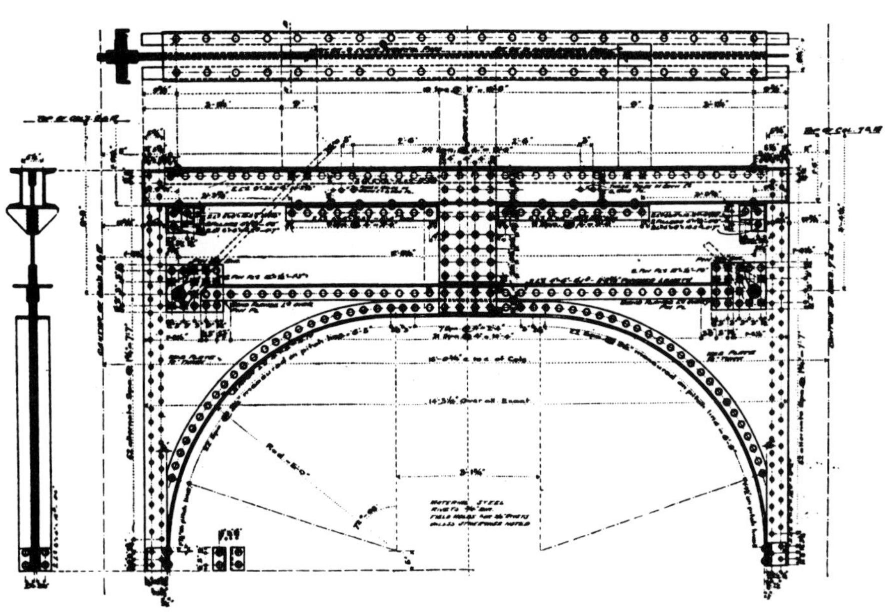

ditions more accurately. It was perhaps those findings that prompted adjusting the New York City code to ignore conditions below a height of 200 feet, to assume a pressure of 20 pounds per square foot, and to increase design stresses allowed for the steel by one-third. Since the 1930s the changes in the materials and methods of constructing tall buildings have introduced new considerations in problems of wind bracing. The strength of the materials used in construction has increased and the weight of buildings has decreased—from about 20 pounds per cubic foot to 10 pounds.[49] Taller buildings with lighter walls have made wind pressure a much more important factor in structural design.

Continuous beams and statically indeterminate systems were paramount problems in structural theory during the last half of the nineteenth century. When a beam rests on two supports, the condition is *statically determinate*, because the span can be considered in isolation, the action of external forces being entirely counteracted by the action of forces within the span; when a beam extends over three or more supports it is classified as *statically indeterminate*, in which case no span can be isolated because each is influenced by the action of external and internal forces of adjacent spans. This case is much more complex theoretically and mathematically. Attention was drawn to the subject of continuous spans at midcentury by the desire of railway engineers to construct metal framework bridges across several supports. Toward the end of the nineteenth century the metal frameworks of tall buildings came into consideration, for the rigid connections between beams and columns brought about complex interactions that transferred stresses among beams and columns throughout the entire framework. This structural problem is much more difficult to solve in theory and execute in application than are simple beams acting in relative isolation from the remainder of a structure.

Around 1850 B. P. E. Clapeyron, a French engineer working principally for railway companies, began using a new method of solving the problem of continuous beams, the theorem of three moments, in the design of bridges. In the next two decades the theorem was extended and study of the three-dimensional frame was advanced. A wide variety of approaches appeared, many of them related to the extensive study of railroad bridges. Much of this highly detailed work was done by Germans, who felt that the availability of iron in Britain made British engineers less concerned with theoretical principles than "the poor devils of the Continent."[50] In the 1870s, a period when the British were thought to be notably negligent in the mathematical education of their engineers, a German engineer wrote that "continuous beams are popular only in countries where engineers can calculate."[51]

Many engineers had received little mathematical training or had been trained by years of experience as draftsmen, and both sorts were enthusiasts of systems of graphical analysis that could remove the mathematical mystery from problems and utilize graphics, a form of investigation with which they were more familiar. The simple depiction of forces as lines drawn at angles that indicated their direction, with a length that showed their magnitude, had been introduced by Stevinus at the end of the sixteenth century. By 1860 these principles were sufficiently advanced to be used in the complete analysis of simple trusses, and during the three decades that followed, graphical methods for studying continuous beams were developed.[52] Until the 1930s, whenever new formulas for column or floor girders appeared, there could be heard an outcry questioning the need for such mathematical intricacies and lauding the simplicity of graphical methods.

A succession of events made the study of three-dimensional structural frameworks even more complex: the use of welding in steel construction gave the building frame a greater rigidity; reinforced concrete construction demanded the investigation of composite construction in which two or more materials acted in combination; and new methods of lighting, temperature control, and ventilation permitted the use of different plan forms for tall commercial buildings, allowing a greater depth for workspaces, now less dependent on being near windows. Many methods for the analysis of frameworks were introduced. Some were based on assumptions about the "energy" or "work" involved in resisting loads, and found mathematical expressions for these factors. Others were derived from the small deflections (sagging and bending) that would result from loading a structure, and endeavored to trace the movements of the framework in its efforts to distribute these movements. Outstanding was the "moment distribution" method of 1932, by which the individual elements of a framework were initially considered in isolation, as statically determinate, and a method of mathematical approximation was used to bring these numbers into balance.

Always there has been a startling contrast, increasing as buildings' complexity increased, between the sophistication of the formulas and theories employed by engineers and the generalizations found in establishing factors of safety, loads, and the characteristics of materials. This is, it must be remembered, the inevitable expression of the different interests and capabilities of physicists and mathematicians, engineers, the construction industry, and those who regulate construction standards in the interest of the public.

The failure of a building does not necessarily mean its falling down. Undue settlement and the appearance of large cracks in an office building would probably result in its failure as a profitable investment for the owners. At the same time, the provision of every conceivable structural precaution would make construction of the building so expensive that financial failure might be inevitable. Through the nineteenth century the factors of safety required for wrought iron and steel by some British governmental agencies was around 4, and in the United States current codes for concrete construction require a factor of safety of 2.2 for concrete and 2.5 for reinforcing steel.[53] We have previously noted the variation of live load requirements, an irregularity that has diminished through the years, and the arbitrary assumption of a method for reducing the loads considered for the design of columns and foundations for tall buildings. No likely change of structural theory or the formulas employed in structural design would affect the cost and stability of buildings nearly so greatly as do the justifiable precautions that are fundamentally measures of the probability of failure.

16

Acoustics

1820s John Blackburn constructs a paraboloid sounding board behind his pulpit

1820–36 Attempts to correct the acoustical faults of the House of Representatives, U.S. Capitol, Washington

1832 Benjamin Henry Latrobe's principles of acoustics published in the American edition of the *Edinburgh Encyclopedia*

1847 The "isacoustic curve" presented by John Scott Russell in a lecture before the Royal Institute of British Architects

1886–91 Construction of the Auditorium Building and Schiller (Garrick) Theater, Chicago

1895 Wallace C. Sabine commences his study of the acoustical problems of the Fogg Museum, Harvard University

1900 Sabine publishes his acoustical principles

1927 Construction of Salle Pleyel, Paris, according to acoustical design of Gustave Lyon

1964 Initiation of experiments that lead to the electronic adjustment of the acoustics of the Royal Festival Hall, London

16.1 Dots in the plan indicate the location of acoustical vases in the vaulting of side aisles in the church of St. Victor, Marseilles. While the necks of the vases result in openings only 2 or 3 inches in diameter, depths vary from 7 to almost 27 inches. The variety of shapes employed and ridges on the interior surface of one vase suggest a planned acoustical treatment rather than a casual use of potters' discards. (*Archeologia*, May–June 1971.)

16.2 Geometric concentration of sound reflections is illustrated by this seventeenth-century drawing showing a speaker and listener at the foci of a semiellipse. Reflection and the "whispering gallery" phenomenon found in many buildings' domes drew the attention of early scientists. (A. Kircher, *Phonurgia Nova*, 1673.)

Vitruvius described the acoustical details of Roman amphitheaters in the fifth book of his *Ten Books of Architecture*. Like Aristotle, Vitruvius was aware that sound was the result of air waves, and in writing on theaters he clearly described the phenomena of reflection and reverberation. He mentioned an interesting acoustical device:

> Let bronze vessels be made, proportionate to the size of the theater, and let them be so fashioned that, when touched, they may produce with one another the notes of the fourth, the fifth, and so on up to the double octave. Then, having constructed niches in between the seats of the theater, let the vessels be arranged in them, in accordance with musical laws in such a way that they nowhere touch the wall, but have a clear space all round them and room over their tops. They should be set upside down, and be supported on the side facing the stage by wedges not less than half a foot high. Opposite each niche, apertures should be left in the surface of the seat next below, two feet long and a half a foot deep.[1]

For a small amphitheater Vitruvius recommended that 13 niches be provided in a tier of seats halfway up the slope of the seating. For a larger amphitheater he advised niches in three tiers.

Although Vitruvius stated that such vessels were not to be found in Rome at that time, he claimed that amphitheaters in some Italian provinces and Greek states used bronze vessels, or ceramic urns where cost was a significant factor. There was long no concrete evidence of such vessels in ancient amphitheaters, but archaeologists have discovered niches of the kind described by Vitruvius, niches that were used as ovens by modern generations of peasants. More recently, ceramic urns about 3 feet in diameter and 5 feet high have been found in the stage structure of an ancient Sardinian theater.[2] Although Vitruvius's prescriptions were perhaps more advanced than common practice in the first century B.C., it seems likely that Romans employed resonant vessels located in both the audience seating and stage, as well as megaphones built into the actors' masks. Since modern tests show typical ancient theaters to have excellent acoustics, bronze vessels may seldom have been needed.[3]

Medieval builders on occasion embedded clay pots in ceiling vaults of churches and chapels, and sometimes in walls or beneath floors, for what were obviously acoustical purposes. As early as 1824 earthenware vessels were discovered in the vaulting of St. Blaise, Arles, and other examples have since been found throughout much of Europe.[4] In the case of the small medieval church at St. Victor in Marseilles, each of the 12 clay pots discovered in its vaulting had a decidedly different shape (fig. 16.1). Modern tests suggest that the pots did not significantly amplify sound at the floor level, but they effectively absorbed sound, smoothing the frequencies and decreasing contrast.[5]

By the middle of the seventeenth century the geometry of sound reflection was familiar to many scientists, principally because of its assumed similarity to the behavior of light. Using a newly developed system for pumping air from a container, Athanasius Kircher proved that sound was not conducted by a vacuum, and in *Phonurgia Nova* he presented descriptions of the manner in which sound is reflected, dispersed, and concentrated by the shapes of interior spaces (fig. 16.2).[6] By this time the requirements of the speaking voice, delivering a lecture or sermon, had become as important as the needs of actors and choruses. Speakers insisted that the

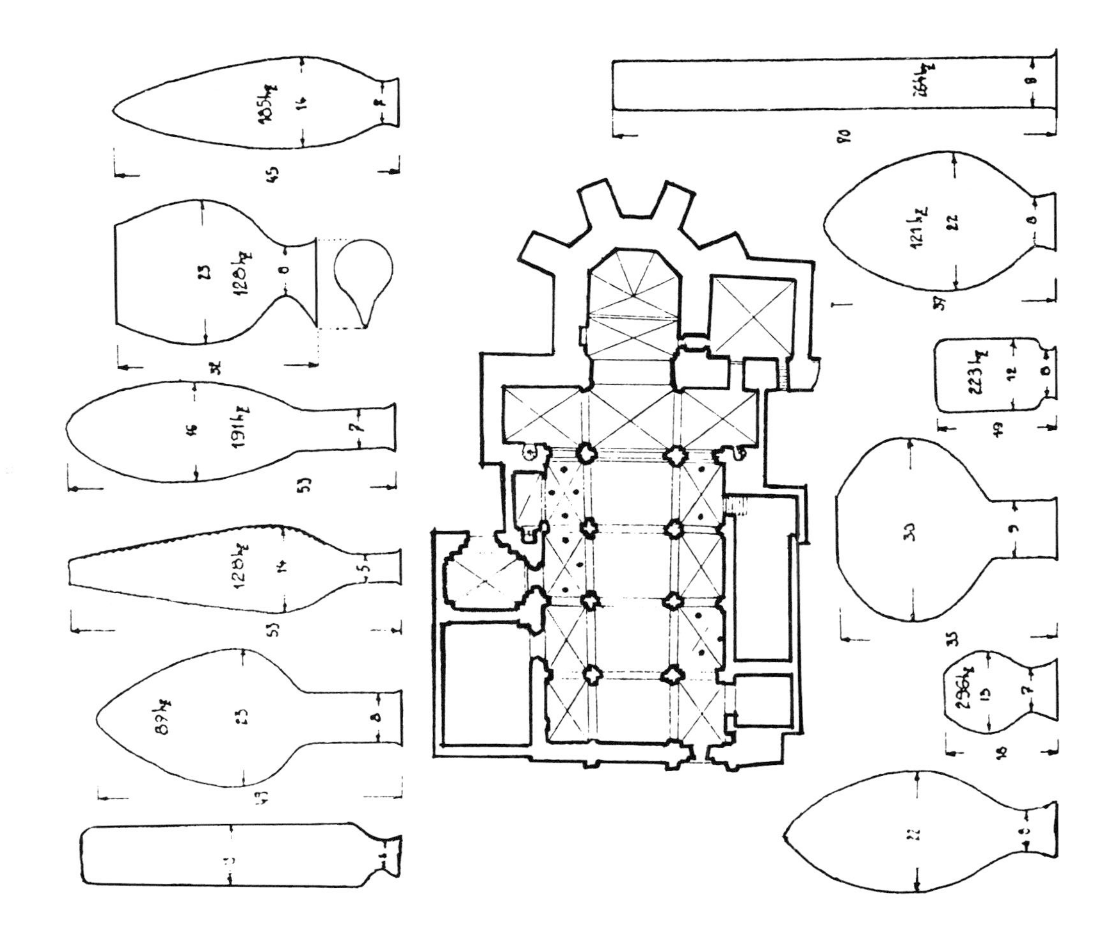
89 hz
23
49
8
128 hz
14
53
5
16
191 hz
53
7
32
23
128 hz
8
45
185 hz
16
8
22
296 hz
13
18
7
30
35
9
223 hz
49
12
8
121 hz
22
37
8
80
264 hz
8

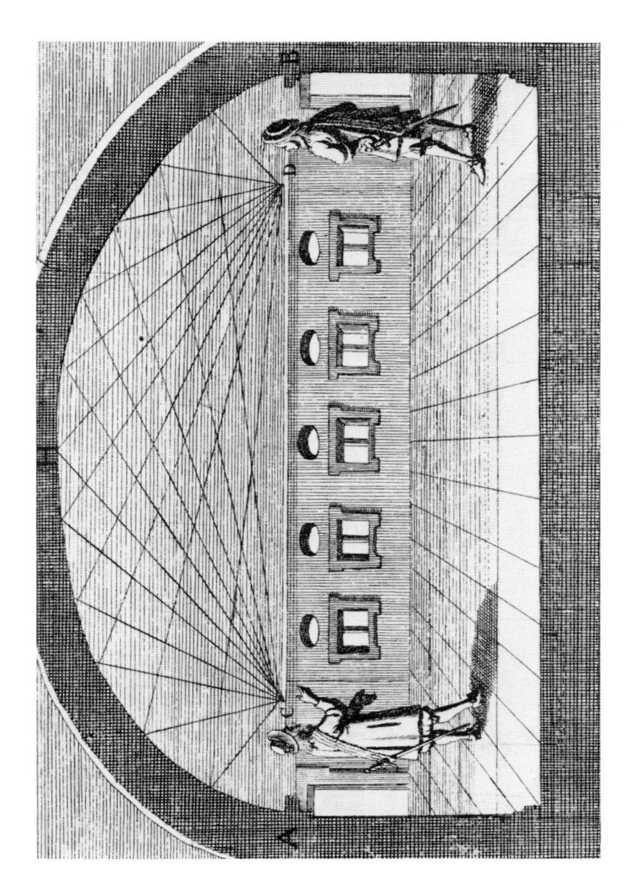
A
C
D
B

16.3 **The range of the speaking voice, as stated by Sir Christopher Wren, is here superimposed on the plan of St. Bride's, one of his London churches. The nave of the church had dimensions very near those Wren considered to be the maximum for good hearing. With the pulpit brought well forward, Wren's conservative estimate of vocal range meant that almost all of the pews at the lower level and more than half those at the upper level could hear a sermon clearly. (Drawing by author.)**

16.4 **The chamber of the House of Representatives in the U.S. Capitol had the centers of the arch and half-dome at about the same level as the heads of the congressmen. This engraving shows draperies hung between the columns of the gallery, but such corrections were not sufficient to overcome the acoustical problem. (G. Brown, *History of the U.S. Capitol,* 1899–1900.)**

form of buildings assist in their efforts to be heard in every cranny of the lecture halls and churches in which they spoke.

Sir Christopher Wren, scientist and geometer as well as architect, was perforce concerned with the placement of pulpits in order to assure that sermons might be heard clearly in all the 51 London churches that he designed to replace those destroyed in the catastrophic fire of 1666 (fig. 16.3). Wren insisted that a church's nave should not exceed 90 feet in length and 60 feet in width, and in his designs for places of worship the pulpits were customarily placed well forward of the altar, near parishioners in the front pews. Wren's conclusions on church acoustics appear to have been principally based on his observation that "a moderate Voice may be heard 50 feet distant before the Preacher, 30 Feet on each Side, and twenty behind."[7] More than a century later Benjamin Wyatt, architect of Drury Lane Theater, in *Observations on the Principles of a Design for a Theater,* estimated a much larger scope of the human voice, 92 feet in front, 75 on the sides, and 30 to the rear.[8] The listening area described by Wyatt is roughly four times that established by Wren, but this difference may have resulted from Wyatt's considering the trained voices of actors rather than the intonations of vicars.[9]

Common professional knowledge of architectural acoustics in the United States is indicated by the explanations provided in 1804 for the congregation of St. Michael's Church, Charleston, by the young architect Robert Mills. Asked to prepare plans for an addition to the church (a project never completed), the native Charlestonian provided his clients with a proposal accompanied by explanations that included his "Doctrine of Sounds." Mills recommended removing the church's flat ceiling and installing a vault the length of the nave. In justifying his design, Mills spelled out the basic principles of acoustics as they were accepted at that time:

1. Sound has some of the properties of light. It is radiant, that is, it proceeds when excited from one centre, in every possible direction in which it meets with no obstruction.

2. It is reflected, and follows the general laws of reflection, its angle of *incidence* and *reflection* being equal. Its reflection is called *Echo.*

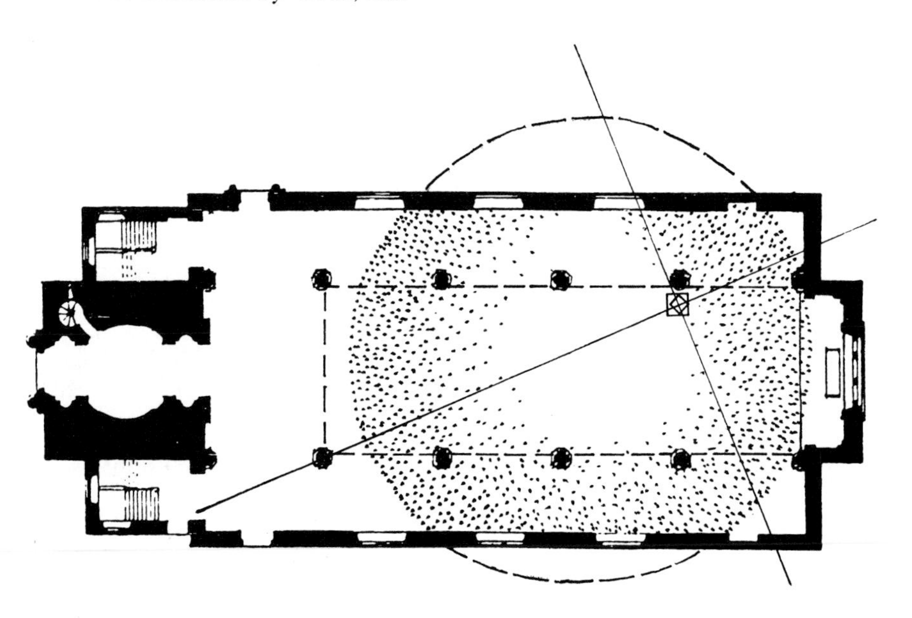

3. It is probably also refracted. Its peculiar properties are not so well understood.[10]

Most of Robert Mills's technical knowledge came from his early association with Benjamin Henry Latrobe, who had endeavored to solve some of the acoustical problems of the U.S. Capitol before it was burned by the British during the War of 1812. When Latrobe contributed the article on "Acoustics" for the 1832 American edition of *The Edinburgh Encyclopedia*, he declared acoustical considerations to be particularly important in the United States, "a government in which public debate precedes every public measure."[11] Latrobe's explanation of room acoustics is firmly based on the geometric reflection of sound.

The object then would be attained were the room so constructed that no secondary and subsequent echoes could reach the audience, or that they should be so weak as not to have any perceptible effect.

The most effectual means, which could be adopted, would probably be, to prevent all echo excepting from the ceiling, by hanging the walls with drapery, or otherwise covering them so that they should not reverberate sound. Rooms, the walls of which are broken into sunk pannels enriched by relievos, or which are decorated with fluted pilasters, or otherwise so varied in their surface as to offer to the *rays* of sound, which in this respect resemble those of light, no regular mirror from which they can be uniformly reflected, are better calculated to render the voice distinctly audible, than those, the walls of which are unvaried in their surface. . . .

I cannot help regretting that the abuse, attributed to the use of pictures and statuary in churches, has expelled them from most of the religious edifices of our country. Independently of the operation of sensible representations of the objects of our veneration or faith on our minds, pictures and statuary have a great effect in suppressing interfering echoes in churches.[12]

Latrobe provided no standards by which other architects might determine in advance the extent to which reflective and absorptive surfaces would be required. He discussed geometric principles at length, and the corrective functions of soft or diffusing treatments were emphasized, for these indefinite standards were the

16.5 **In 1826 Robert Mills's recommendations for correction of the acoustics of the House of Representatives contrasted the existing conditions (upper drawing) with his proposal of a curved back wall (lower drawing). Six years later Mills proposed raising the floor and reversing the seating in the chamber. ("Alteration of Hall, House of Representatives," Rep. No. 495, 22d Congress, 30 June 1832. Courtesy of the Architect of the Capitol.)**

period's response to failures of acoustical design. While the most common practice of the time was the addition of absorptive materials, when that failed efforts were made to reshape rooms.

When the U.S. Capitol was rebuilt after the War of 1812, the Hall of Representatives suffered even more acoustical problems than its predecessor, which had received Latrobe's attention (fig. 16.4). Dr. William Thornton, designer of the original building, recommended hanging heavy curtains to absorb the sound, the same advice he had offered 12 years before for the previous hall. After the room had been occupied only a year, the bothersome conditions were attributed principally to dampness within the new construction, but draperies were hung between the columns and the gallery floor was covered with carpet. The following year complaints were renewed, and a congressional committee sought advice from a variety of sources. Charles Bulfinch, Capitol architect, in 1822 recommended that a flat ceiling of fabric be stretched beneath the dome, a solution said to have been taken from Saunders's book on theaters. This was done and was successful in ending echoes, but it drastically reduced the available light and absorbed so much sound that speakers could hardly be heard. After only a few days the fabric was removed. Wooden walls were built between the columns, but these too were soon removed.[13]

For more than 15 years there were complaints from the congressmen, recommendations from all sides, and no successful solution. Then Robert Mills executed a solution that had been proposed long before. He raised the floor a distance of about 4 feet, to the top of the pedestals upon which the columns stood. The center of the domical ceiling's curvature had been at almost the same height as speakers' heads, and therefore elevating the floor served to reduce the focusing of sound as it was reflected from the dome (fig. 16.5).[14] Mills also recommended reversing the arrangement of the representatives' seats so that the speaker's chair would be in the center of the semicircular colonnade, but this scheme was used for only one session. Complaints continued throughout the years that the House of Representatives met there and ended only when the space was converted to use as Statuary Hall.

Churches were often plagued by acoustical problems. The elaborate carvings we see directly above the pulpits of Baroque churches could fulfill little more than a decorative function, for they reflected little sound toward the pews. Later great faith was placed in the use of sounding boards, sound-reflecting surfaces set at speakers' positions to direct the sound of a voice toward its audience. Usually of light wood construction, sounding boards in churches were intended to bolster the preacher's voice so that it could be clearly heard in the pews at the rear of the nave. Speaking before the Royal Institute of British Architects in 1860, T. Roger Smith stated the common practice:

> It will be readily understood that a slanting reflector overhead, to beat downward and forward rays of sound that would otherwise escape towards the ceiling and be lost, is likely always to do good, and can in no case be so injurious as one behind the speaker; and it need scarcely, I think, be added that the only reflectors that *can* be of advantage are those that throw the sound forwards in the same direction as that in which the speaker is speaking. An echo reflected down from a high ceiling, or worst of all, back from an opposite wall, will always be disagreeable.[15]

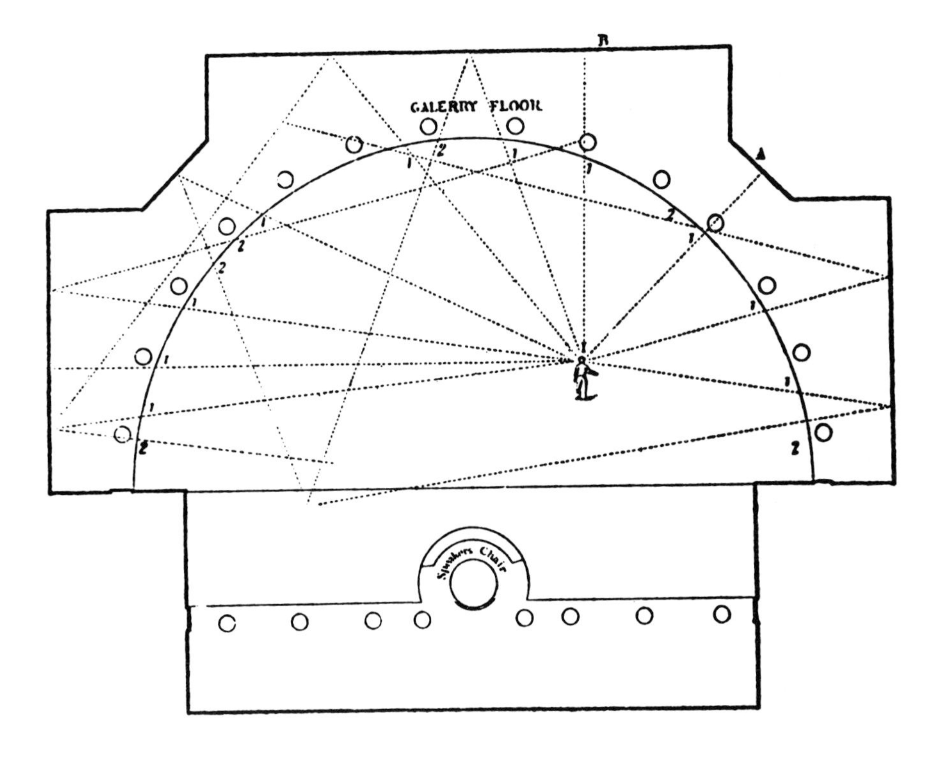
B
A
GALLERY FLOOR
Speakers Chair

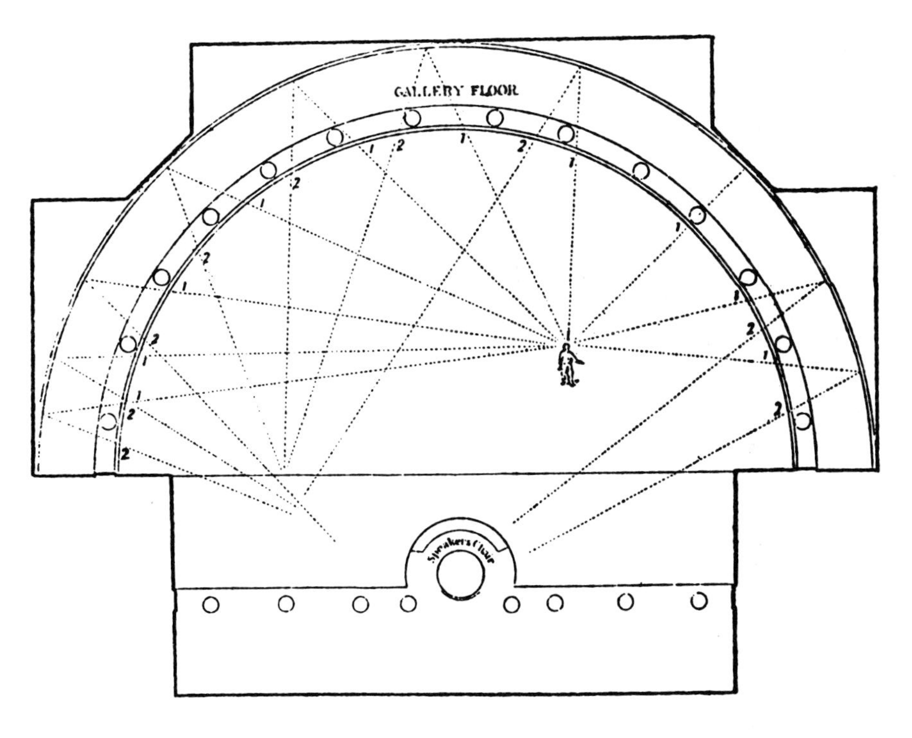
GALLERY FLOOR
Speakers Chair

16.6 In the Hall of the Senate House, University of Calcutta, where graduation ceremonies were conducted, a trial reflector was made in 1896 of tin sheets mounted on a framework of iron angles. The final version attempted to gain resonance through a 2-inch space between its wooden front surface and a tin back surface. The shape is a truncation of a paraboloid with its focus 5½ feet above the floor of the dais and its axis directed toward the floor at the opposite end of the hall. (*Indian Engineering*, 21 March 1896).

Sometime before 1829 the Reverend John Blackburn, finding it difficult to make himself heard in his church, had erected behind the pulpit "a sounding board like a hood—parabolic in section, with his head in about the focus of the parabola."[16] The parabolic form was chosen because rays radiating from the focus would be reflected as parallel rays extending the length of the church. Apparently Blackburn's scheme fulfilled its purpose, for we are told by Smith, "The congregation were now able to hear the preacher, and the remote seats of the church became some of the best." Blackburn published his design in a small pamphlet that was circulated to scientific groups, and a model was presented to the Royal Society of Arts. The carpenter who had built Blackburn's reflector went on to construct at least 29 others. However, these sounding boards were spoken of as being "villainously ugly," and most were dismantled soon after they were built. Indeed, many clergymen found the parabolic reflectors to be more bothersome than beneficial. Not only did the preacher's head need to be kept at a relatively constant position within the sounding board, but the parabolic form received sound in the reverse of the manner in which it projected sound. One visiting clergyman who preached in a Cambridge church having such a reflector reported that he was startled by the clarity with which he heard "the whisperings of the charity children in the remotest part of the west gallery."[17]

A similar device was constructed in 1895 to correct the acoustics of a long rectangular room used for the annual graduation convocation at the University of Calcutta (fig. 16.6). A large, shallow paraboloid dish was constructed on an iron framework. The diameter of the reflector was about 37 feet and its concavity had a depth of about 10 feet. After a first trial model of this giant reflector had been constructed and found helpful during one graduation, a permanent reflector was built, its front surface covered with thin boards of teak and the back covered with sheets of metal. It was believed that this double surface and the air space between would provide a useful resonance.[18] In practice it was discovered that, as with Blackburn's reflectors, should a speaker shift his position appreciably, acoustic conditions were little better than before construction of the reflector.

Not all acoustical problems were those of an inadequate strength of the sound. When Thomas U. Walter in 1832 began construction of Girard College in Philadelphia, an institution for training "poor white male orphans, between the ages of six and ten years," his design was severely restricted by the stipulations of the will left by Stephen Girard, the leading American banker of the period. In extraordinary detail the college's benefactor spelled out the plan of the buildings that were to be built and the materials and methods that were to be used in their construction.[19] The architect had little choice but to follow those instructions, although problems of reverberation in the classrooms were evident early during construction. In his final report to the building committee Walter commented wryly: "As Mr. Girard left no discretionary power in reference to this part of the design, we were compelled to take the letter of the will as our guide, let the results be what they might."[20] The result was eight classrooms, each 50 feet square, 25 feet high, and having vaulted ceilings. When J. B. Upham, a Boston physician who wrote on architectural acoustics, visited the college in 1847 he found that the reverberation of a

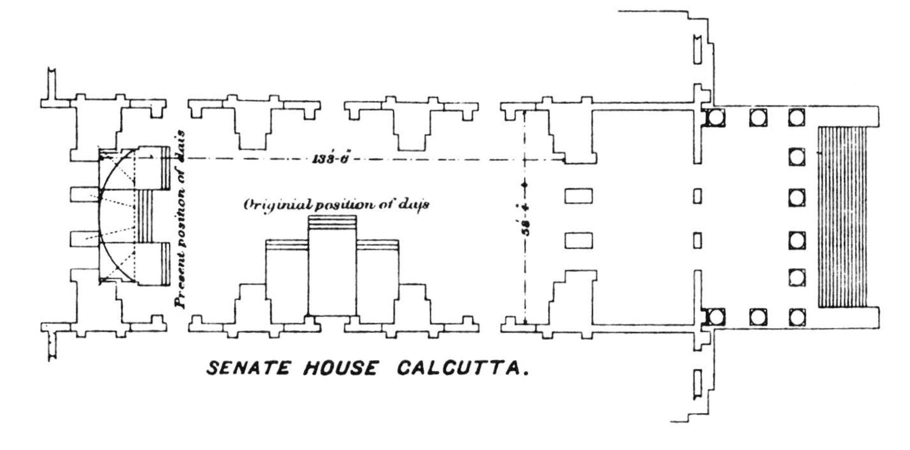

sound in these rooms lasted 6 seconds, a noisy condition that prevented the rooms being used for classes. The usual corrective procedure at that time would have been the introduction of carpets and curtains to absorb sound waves. Such furnishings were not, however, appropriate to the use of the rooms at the institution and other means were found. Upham described the results:

> In one room, which had been treated simply by papering upon the solid walls and extending festoons of cotton cloth from the apex of the dome to the corners and centre of the cornices on each side, the reverberation was reduced to four and a half seconds; and in others, in which a partition of cloth was stretched across the room horizontally, from the opposite cornices, thus completely shutting off the arched ceiling of stone, and substituting a level surface of yielding canvas, its duration was only half a second. By whose suggestion these simple contrivances were tried, we could not learn, but presume they originated with the skillful architect of the building.[21]

The solution was not unusual. It had been used about three decades before in one of the attempts to correct the foul acoustical conditions of the Hall of Representatives in the U.S. Capitol.

By the middle of the nineteenth century a general knowledge of architectural acoustics was established, partly theoretical and partly empirical, that permitted a designer to avoid disastrous acoustical conditions. Nevertheless, there was no method by which predictive study of conditions could be used to enhance the acoustical characteristics of a space. In fact, early in the nineteenth century it was said that when a friend told a prominent English jurist that he wished to build a room with splendid acoustics, the justice immediately warned, "Then be sure you don't tell your architect so!"[22]

Some believed that a semicircular hall, modeled after Greek and Roman amphitheaters, provided an ideal form, although this view may have resulted largely from an infatuation with classical examples or from the fact that an audience seated around a 180° sweep would be nearer to the speaker than in most rectangular arrangements. Others opposed circular or semicircular forms with equal fervor. Two major English architects at the start of the nineteenth century held contrary views:

> [Sir John] Soane.—Is it not probable that you will be more successful in having a good sound by having a circular end than by having square sides? Yes, you will find that in the Olympic theater at Vicenza there is a circular end; so had all

the ancient theaters, both Greek and Roman. . . . [I have] concluded that shape to be peculiarly favorable to hearing.

[Sir Robert] Smirke.—I think if the circular form were preserved . . . above the proposed seats, it would be a very inconvenient room for hearing; there would be in all probability such reverberation of the voice, that it would be scarcely possible to hear with distinctness.[23]

For rectangular rooms there was a strong belief that acoustical conditions were improved by having a simple ratio among the room's dimensions, a typical Renaissance recommendation for obtaining proportions that were visually harmonious. The ballroom at Buckingham Palace was considered excellent for musical performances, and this was attributed to the fact that its dimensions were about 110 feet by 60 feet by 45 feet, nearly a relationship of 8:4:3. Simpler ratios were commonly recommended , such as 2:3:5 or 2:3:4, just as they had once been preferred for esthetic reasons. In the discussion following Smith's paper at a 1860 meeting of the Royal Institute of British Architects, John Scott Russell made the proud announcement that a hall to be built at the University of Edinburgh—the dimensions determined by a professor of science—had been set at 96 feet in length, 48 in width, and 32 in height (6:3:2), "the three proportionate numbers of musical harmony."[24] Much later Wallace C. Sabine was to point out that in a great many halls it was impossible to determine if the measurements to be considered in calculating proportions should be taken at the front or rear of balconies and stage platforms.

There was even debate on the materials to be used in building a hall. Some, many musicians among them, swore that only wood could produce the desired resonance, and here the recurrent analogy of the violin is evidenced. Others accepted plaster and stone.

There was also an inexplicable belief that wire strung overhead across a hall would counteract undesirable acoustical conditions that were due to reverberation or would vibrate in sympathy with a speaker's voice, thus strengthening the sound. Wires were to be seen in many English churches at the turn of the century, and a troubled lecture room in H. H. Richardson's Sever Hall at Harvard University had a multitude of overhead wires. A British physicist later remembered that wires were one of the first remedies proposed by architects in the period before World War I.

Although the belief strangely persisted that wires could augment or clarify sounds in an auditorium, the most imaginative extension of this notion seems to have been proposed in England around 1880. Inspired by the examples of the sounding boards of stringed instruments and the vibrations of tuning forks, a man named Engert invented a complex acoustical contraption of wires and springs. A report on one demonstration described the hall: "One or more layers of steel wires were stretched along a [room] lengthwise, connected by cross wires and spiral springs, and properly tuned, so that the vibration may be absorbed and conveyed from one to another, and instantaneously spread over the whole building."[25] Another report described what is apparently a different system devised by Engert. In this case five boxes fronted with louvers were located at the back of the stage and two were placed in front of the orchestra. Each box contained five steel plates of different dimensions, placed about an inch apart and suspended by wire springs. These were claimed to resonate with different tones of the music, thereby reinforcing the sounds produced by orchestral instruments. Unfortunately observers attending Engert's demonstrations were unable to detect significant changes in the acoustical quality of the performances due to the devices provided by the inventor. In fact, the *Builder* complained that Engert's demonstrations were held in halls that had no apparent acoustical flaws. Soon mention of these intricate schemes, examples of both the wire theory and the musical instrument analogy, vanished from the architectural press.

Constant references were made to the resonance of musical instruments, particularly stringed instruments. A French consultant as late as 1933 compared concert halls and instruments, saying: "All enclosed space, of the size of a violin or that of a concert hall, in order to send out a pure sound wave or to resonate without echoes, must have the form of two cylinders."[26] He went on to compare that form with the violin, the Mormon Tabernacle in Salt Lake City, and the Salle Pleyel in Paris.

When Charles Garnier was designing the Paris Opera House in the 1860s, he grew concerned about the acoustical properties of the dome over the hall. His study of the problem was not easy:

> It is not my fault that acoustics and I can never come to an understanding. I gave myself great pains to master this bizarre science, but after fifteen years of labor, I found myself hardly in advance of where I stood on the first day. . . . I had read diligently in my books, and conferred industriously with philosophers—nowhere did I find a positive rule of action to guide me; on the contrary, nothing but contradictory statements. For long months, I studied, questioned everything, but after this travail, finally I made this discovery. A room to have good acoustics must be either long or broad, high or low, of wood or stone, round or square, and so forth.[27]

For advice he consulted Charles Joseph Sax, the famous Belgian instrument maker. Asked whether a metal dome would be less desirable than a traditional dome of stone, Sax stated categorically: "The nature of the material used for this dome should have no influence on the reverberation of the sound."[28] Indeed, in spite of Garnier's frustration with the mysteries of architectural acoustics, the dome appears to have contributed to the conditions that make the Paris Opéra one of the better halls of Europe.

There was great concern about the influence the movement and temperature of air might have on sound waves. Certainly, anyone who had shouted far across an open field knew that the sound carried farther with the wind than against it. As a result there were often proposals that air brought into auditoriums for purposes of heating or ventilation should originate in the vicinity of the stage and be withdrawn at the opposite end of the hall, carrying sound waves along as it traversed the length of the auditorium. A physicist at Johns Hopkins University in 1878 published results of an experiment that seemed to show that listeners, unaware of the change, detected distinct and unfavorable alterations in an orchestra's sound after the hall's fans had been reversed to no longer draw air from the stage.[29] At the end of the nineteenth century this logic was recommended by some for the new Boston Symphony Hall, designed by Charles McKim. Wallace C. Sabine, acoustical consultant to the architects, calculated the influence of air movement in such circumstances as being about the same as a forward tilt of the listener's head, and he curtly pointed out that "the problem of properly heating and ventilating a

16.7 The isacoustic curve, as presented in John Scott Russell's lecture in 1847, recommended that the floor of a hall should be shaped so that rays representing sound would pass 18 inches above a head in the row in front of a listener. (*Edinburgh Philosophical Journal*, 1838.)

room is sufficiently difficult in itself."[30]

In addition, there were observed difficulties in sound waves crossing zones of very warm air. This appears to have been a genuine and frequent problem at a time when public spaces were often heated by stoves placed in the center of the floor. A member of the House of Commons complained to David Boswell Reid in 1835 that when the large stove beneath the middle of that chamber sent up a current of hot air through the floor grill, a speaker on the opposite side of the room became almost inaudible, though he had been clearly heard before the heater started. Another instance was reported to Sabine by William LeBaron Jenney, the Chicago architect who built the first skyscraper. Consulted regarding a courtroom in which observers could not hear judges, lawyers, or witnesses, Jenney diagnosed the problem as arising from the placement of a stove between the major areas of the courtroom and also between two doors, one on each side of the courtroom. A band of air moving through the doors and toward the heavy upward draft of the stove appeared to be the muffling obstacle, and Jenney recommended "that the stove be removed and that the warm air should be let into the room from steam coils below."[31] This phenomenon, which Sabine explained as resulting from both reflection and refraction of the sound waves as they experienced a radical alteration of air temperature, became a less significant problem during the late nineteenth century. In his advice about the courtroom Jenney had forecast the solution, for the development of improved methods of heating and ventilating replaced the use of central stoves and so eliminated such problems.

From the time of Newton's development of a value for the velocity of sound in air (*Principia*, 1687), the attention of physicists and mathematicians was directed toward the phenomena of sound and hearing. By the early nineteenth century, fairly reliable measurements had been made for the speed with which sound traveled through water and metals, as well as air and other gases. The production of sound was investigated by study of the basic tone-producing elements that were present in traditional musical instruments, vibrating strings, open and closed pipes of air, bars of metal, and taut membranes like drum heads. Fundamentals and harmonics were identified, and the relationship between pitch and frequency was established. This patient investigation was part of that period's exhaustive scientific study of all phenomena; and it may also have been influenced by the musical enthusiasms of some gentleman of science. At the same time, economic interests of trading nations encouraged the study of the ocean wave's action against ships and the sounds of warning bells in heavy fog.

When John Scott Russell, a young professor of natural philosophy and geometry at the University of Edinburgh, was approached around 1835 by businessmen to study the possibility of using steam power to move barges on the Edinburgh and Glasgow canal, he was assigned an unused section of canal in which to study the wave phenomena that often impeded the movement of barges when they were pulled too rapidly by horses. Russell had been born in the Vale of Clyde near the shipbuilding fringes of Glasgow; although studying to follow his father's life as a minister, his summers had been spent working in engineering shops, and his keen interest in science was not neglected. When he was 27 years old, Russell delivered his first paper on waves, and he also completed his first ship as the new manager of the Greenock shipworks near Glasgow. Several ships later, including work in collaboration with the foremost British engineer, Isambard Kingdom Brunel, Russell moved to London where he continued his study of ships' hulls and became a leader in shipbuilding.[32]

In 1847 Russell delivered an address, "On the Arrangement of Buildings with Reference to Sound," to the assembled membership of the Royal Institute of British Architects. In his book *Waves of Translation in the Oceans of Water, Air and Ether,* Russell treats sound waves in the "air ocean" in the same terms as waves in the sea. Little direct application is provided for architectural conditions, and one can only ponder the manner in which Russell may have developed the extremely specific recommendations he presented to architects. The lecture was not published at the time it was delivered, and therefore it had little immediate influence, but 11 years later it appeared in three issues of the *Building News,* as reconstructed from an editor's notes made during Russell's lecture.[33] This publication seems to have reached many eyes and to have become part of architects' working knowledge of room acoustics.

Russell's first principle was the isacoustic (equal hearing) curve, which he had first presented in the *Edinburgh New Philosophical Journal* for 1839 (fig. 16.7). Based on the same logic as that employed to determine lines of sight in theaters, the isacoustic curve derived the slope of an auditorium floor through the requirement that a line from the speaker's mouth to a listener's ear should pass 12 to 18 inches above the ear of a listener in the next row forward. (To avoid excessively steep floor slopes in large halls, the isacoustic curve was often laid out with much less clearance above the head of the person in front.) In addition, Russell stated that each row of seats should be arranged as a circular arc having its center at the speaker's position. This system was claimed to have the advantage of "simply taking the great body of sound and dividing it equally between all the hearers," although this could not actually be the case, due to the reflection of sound

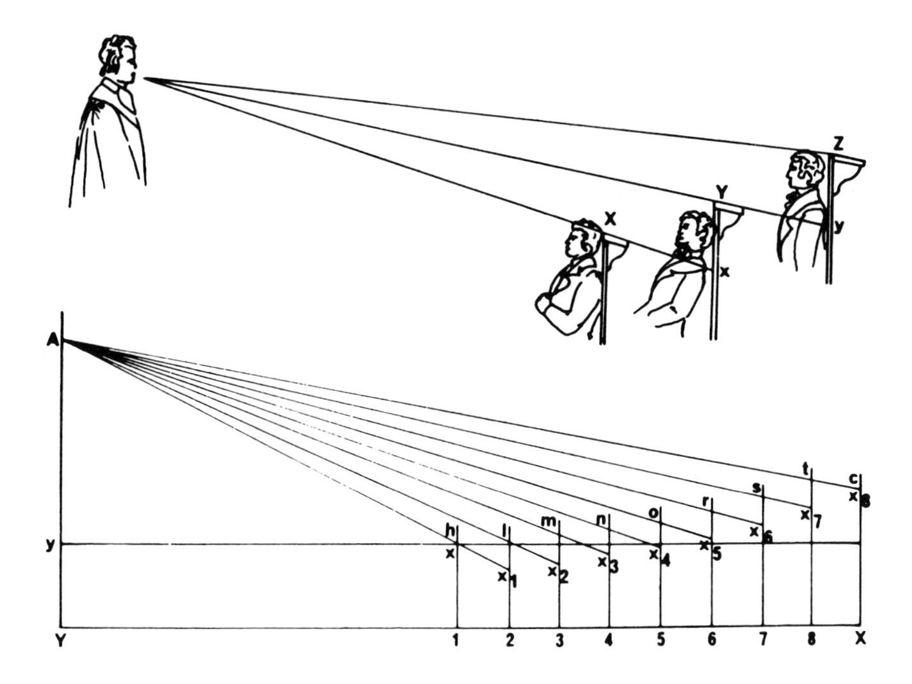

16.8 Theaters or opera houses, such as Covent Garden, had their floors and walls covered with seats, boxes, balconies, and audience, leaving only the ceiling surface bare. This arrangement provided ample absorption of sound and brought the listeners near to the stage. (*Builder*, 10 April 1847.)

16.9 In the Chicago Auditorium (Adler and Sullivan, 1889), the successive surfaces of the proscenium arch extend almost to the midpoint of the seating area, including the balcony levels. In addition to helpful reflection of sound waves and providing a relatively low ceiling, this shape fulfilled Adler's dictum on "formation of a stage picture." (P. E. Sabine, *Acoustics and Architecture*, 1932.)

and the fact that the strength of the sound would necessarily diminish as the distance between the speaker and the listener increased.

The second principle stated by Russell had to do with room dimensions. As was common in that day, the length and breadth of a room were associated with musical tones ("thirty feet long gives out C natural") and dimensions that were arithmetically related were believed to make certain that the sound could be clearly heard throughout ("forty-eight feet long, twenty-four feet wide, and sixteen feet high . . . the room will be easily voiced").[34]

As a third principle, Russell stated the well-established fact that sound is reflected at an angle equal to that at which it strikes a surface. Like Latrobe, "he thought it desirable that the voice should be but once heard, and that echoes should be entirely got rid of."[35]

For curved spaces, the "following" action found in domes and their whispering galleries was explained by Russell's having observed that ripples of water approaching a wall at a small angle tend to follow the wall instead of being reflected. For the horseshoe-shaped plans of opera houses Russell recommended that the boxes around the back wall should be separated by partitions and drapery should be hung to prevent the reflection of sound waves from their back walls.

A somewhat different view of reflected sound had been stated in 1835 by David Boswell Reid, a British scientist better known for his study of ventilation: "In the most perfect form of building for the communication of sound, any reflected sound must be prevented from continuing so as to interrupt any new tone by being thrown upon a non-reflecting floor. So long as the reflected sound comes up in time to strengthen the primary impulse before any new sound is heard, it is to be taken advantage of; beyond that it is injurious."[36] Having measured reverberation times as long as 7 seconds in hard-surfaced rooms, Reid recommended auditoriums be built with low walls, high pitched ceilings, and floors covered to absorb sound. This viewpoint was later supported by Joseph Henry, director of the Smithsonian Institution, in describing his preparation for the construction of the Institution's lecture hall in Washington.[37] Through experiments made by clapping his hands before hard smooth walls, Henry concluded that a distance greater than 30 feet from the wall resulted in an identifiable echo, making one-sixteenth of a second the limiting interval between two sounds. Clarity of hearing, he said, depended on four factors: room size, strength of the sound, placement of reflecting surfaces, and the materials of these surfaces. These are in essence the variables found in much modern acoustical calculation.

Early theaters with several balconies and ceilings nearly flat had few echoes, because the audience was brought close to the stage and most of the surfaces were made absorptive by the vertical tiers of audience. For eighteenth-century theaters and concert halls, domed ceilings became popular. By the nineteenth century, theaters had fewer balconies, sometimes only one. These changes in theater design brought increased acoustical problems, both in the amount of sound reflected and the geometry of the hall.[38]

A notable application of Russell's isacoustic curve was the Auditorium in Chicago, where music had become a part of civic ambition. Opera started in Chicago in 1850 with a performance of Bellini's *La Somnambula*, and in the period before the Chicago fire English, Italian, and German troupes vied for the public's ear. Chicago's Central Music Hall was designed in 1879 by Dankmar Adler, a young architect who had just established his own practice. Another music hall had

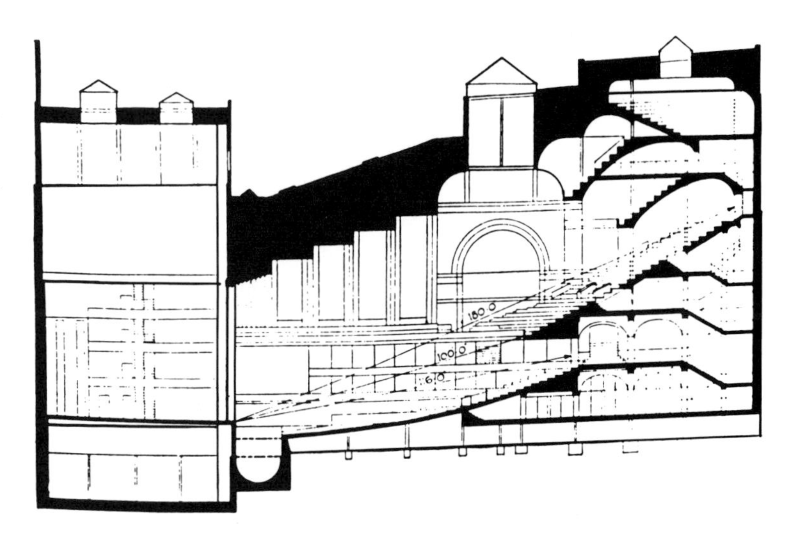

been designed by Adler about seven years before, and it seems to have been his first venture into a building type for which he was to become justly famous. The Central Music Hall's acoustic qualities were widely admired and helped establish a local reputation for Adler in that architectural specialization. When a board of public-spirited citizens in 1884 began planning the first Chicago Grand Opera Festival, there was no hall available large enough to seat the audiences that were required if the performances were to be financially successful. An ingenious solution was found when the board engaged the architectural firm of Adler and Sullivan—for young Louis Henry Sullivan had become Adler's partner—to construct a hall of 6,000 seats entirely within the Interstate Industrial Exposition Building, located where the Art Institute now stands. Only five weeks and about $60,000 were available for construction of this temporary opera house. The festival was a cultural and financial success, in spite of the moderate prices charged for tickets. The following year Adler and Sullivan built similar facilities inside the Milwaukee Exposition Building, seating 10,000 and having a stage large enough to hold the massed choruses of the North American Sänger-Bund. This auditorium too was admired for its acoustics.

When Chicagoans, encouraged by their success in presenting opera, organized to build a suitable permanent hall for musical performances, Adler and Sullivan were the architects to whom they turned (fig. 16.9). The Chicago Auditorium was planned with offices, a hotel, and retail establishments, in addition to the performance hall that had originally suggested the project.

In a paper read at the annual convention of the American Institute of Architects in the year that his firm received the Auditorium commission, Adler stated the acoustic principles on which he worked:

> It should be said, in a general way, that in the construction of the banks of the seats, Scott Russell's isacoustic curve should be adhered to as far as practicable. That wherever possible resonant materials should be used in the construction and facing; that large, hard, smooth surfaces should be avoided; that walls and ceilings should be well broken; that the width and height of the house should be least at the stage, and that these dimensions should be increased with the distance from the stage, and that all our measures should tend toward the reduction to a minimum of the volume of air to be set in motion by the voices of speakers and singers. . . .
>
> I will add in this connection that a comparatively low proscenium . . . is desirable as one of the first conditions of this system of construction for acoustic effect. . . .
>
> A modification of Scott Russell's isacoustic curve should be used in laying out the banking of the seats. This modification . . . consists in shifting the level of the focus to which the curves are drawn from the level of a speaker's mouth to the floor-line at the front of the stage, and in substituting for a single focus to middle of the stage, foci tending toward the sides of the curtain opening for the respective sides of the house. . . . [39]

In this and other writings on theater design, Adler stressed the logic behind the features that identified the theaters of Adler and Sullivan: the deeply splayed proscenium arch reflecting sound to the audience; a ceiling springing from the top of that low arch, limiting the volume of the hall and directing sound toward the rear; and a floor sloped steeply enough to allow ample visual and auditory connection between performers and audience.[40]

When completed in 1889, the Auditorium proved the wisdom of these measures, but the transverse arches of the ceiling concentrated sound toward the center seats and the hall's size (4,237 seats) demanded that the orchestra sacrifice subtleties of interpretation to the production of volume. The Schiller (Garrick) Theater, completed about two years later, was more successful. This record of accomplishment led to Adler's being called as acoustical consultant on Andrew Carnegie's Music Hall in New York, for which the architect was William B. Tuthill, himself well versed in acoustics.

In 1891 the widow of William Hayes Fogg, a merchant, died in New York, leaving Harvard University funds to build and maintain a small museum to house the university's meager art collection. Designed by Richard Morris Hunt, the original Fogg Museum building (later assigned to other functions and renamed Hunt Hall after its architect) included gallery rooms and a semicircular lecture hall, which was to some extent modeled after the classical theater. It was to be a modest building placed among a polyglot group of architectural examples. For the Fogg Museum, Hunt chose a rather pure and refined classicism, stricter in style than the mansions he had built for the Astors and Vanderbilts and more restrained than his Administration Building at the Columbian Exposition in Chicago. No one has suggested that the design has particular merit among Hunt's works.

The lecture hall of the Fogg Museum had been included in the building program principally to provide space for the lectures on art history that were delivered by Charles Eliot Norton, then approaching retirement from his professorship at Harvard University. These lectures, described by his son as "Lectures on Modern Morals as Illustrated by the Art of the Ancients," were presented with a heady mixture of charm and sarcasm that attracted large numbers of students. To Professor Norton's dismay the lecture hall of Fogg Museum when completed was ill suited for lecturing because its acoustics were dreadful. Norton, who had long ridiculed the jumble of architectural styles built at Harvard during his decades there, did not hide his dissatisfaction with the lecture hall: "Had it been intended as an example of what such a building should not be, it could hardly be better fitted for the purpose." Declaring the building to be an indication of the "lack of civilization" in the United States, Professor Norton refused to lecture in the room.[41]

The president of Harvard University, Norton's cousin Charles Eliot, tolerated the controversy about the Fogg Museum, even when student pranksters painted "Norton's Pride" in red letters on the walls of the museum. Eliot had studied and taught mathematics and science, and it was not surprising that he should call upon the scientific resources of the university to relieve his embarrassment. He assigned the task to Wallace C. Sabine, a 27-year-old instructor in physics. The investigation that followed would establish the quantitative science of architectural acoustics, replacing rules of thumb that had been used through most of the nineteenth century.[42]

The lecture hall of Fogg Museum was semicircular in plan and most of its seating fell under a half-dome (fig. 16.10). The floor was gently sloped, and a semicircle of columns set off an

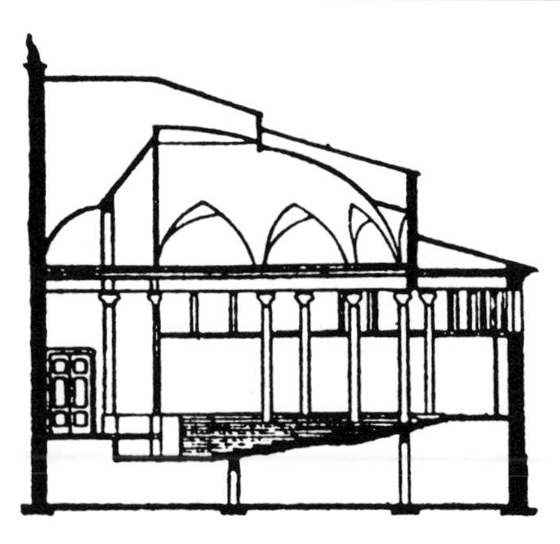

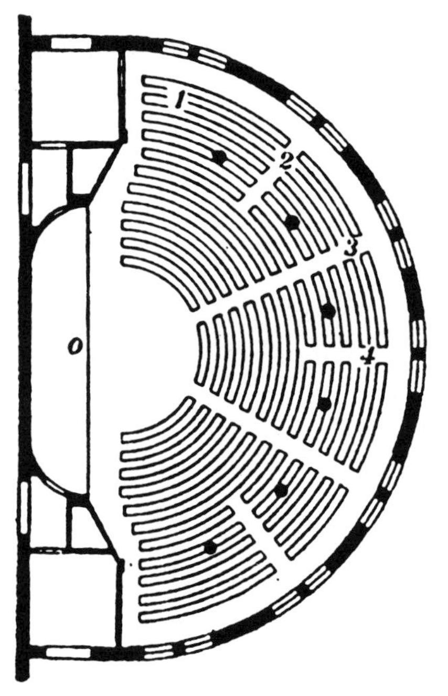

16.10 The acoustical shortcomings of the somewhat classical form of the lecture hall in the original Fogg Art Museum, Harvard University (Richard Morris Hunt, 1895), were the focus of investigations begun soon after the building was completed. Equipped with an organ pipe and almost 1,500 seat cushions from another hall, Wallace Sabine studied the hall's reflection and absorption of sound, working "every second or third night for two months." (*American Architect and Building News*, 21 April 1900.)

aisle around the back of the hall. Acoustical conditions within the lecture hall were such that when it was empty a word spoken in an average tone faded away only after a bit more than 5½ seconds, interfering with the intelligibility of several syllables that might follow. If a full audience were present, there was a slight improvement, but reverberation still made it almost impossible to endure a lecture.

Looking around for readily available absorptive materials, Sabine brought loose seat cushions, hair stuffing covered with fabric, from another Harvard auditorium and stored them in the vestibule of the lecture hall. In all there were about 1,500 cushions, enough to cover all the seats, aisles, speaker's platform, and the back wall of the room and allow some additional cushions to be hung from the ceiling in the final stages of experimentation. Working late at night so that the sounds of carriages in the streets would not interfere with his measurements, Sabine started by placing 27 lineal feet of cushions on the front row of seats and again timing the reverberation. Row by row, cushions were added to the lecture hall, and measurements of the reverberation were made as the room was gradually padded. The results were plotted as a graph, a hyperbolic curve that indicated the relationship between the number of cushions used and the reverberation of the museum lecture hall. To extend his data, Sabine removed the cushions and tested other materials. He draped chenille about the room, rolled out heavy Oriental rugs, hung pads of hair felt, and filled the seats with listeners, men and women in separate tests. For each change, measurements were obtained as they had been for the seat cushions. But since the characteristics of these materials could be mathematically related only to a number of cushions borrowed from another auditorium, a more appropriate unit of scientific measurement had to be found. This was accomplished by opening the windows of the lecture room. Assuming open windows to provide the ultimate level of absorption, the performance of cushions and other materials was mathematically related to them.

When measurements had begun on the Fogg Museum lecture hall, it was suggested that rough troweling or a sand-finish of plaster on the walls might solve its problems. Sabine showed that textural variation was useless when he scattered a layer of uniformly sized pebbles on the floor of a test room and found no appreciable change from the reverberation measured with the concrete floor bare. The young physicist experimented with other rooms and often repeated tests as his techniques and equipment improved. He was assigned a Harvard laboratory that had been originally constructed underground for experiments requiring a constant temperature, made possible by structurally isolating it from the remainder of the building. This made it possible for much of his testing to take place during the day, protected from outside noises.

By 1900 at least a dozen faulty spaces on the Harvard campus had been studied by Sabine, and his fund of data had greatly expanded. In less than five years he developed a formula for the prediction of reverberation time, analyzing the contributions of absorptive or reflective materials that might be present. By that time his tables of the absorptive qualities of materials had grown to include even houseplants and oil paintings.

The remedy that Sabine applied to the lecture room in the Fogg Museum was rectangular panels of hair felt under thin sheets of asbestos paper, which were placed over rectangular sections of the upper walls and in the semicircular recesses of the ceiling. Although Sabine pronounced the results "entirely satisfactory," other corrective measures were made periodically until the building was demolished in 1973.

H. H. Richardson's Sever Hall, a neighbor of the Fogg Museum on the Harvard campus but built more than a decade earlier, was another subject for Sabine's diagnostic studies. It too had been a target for Professor Norton's scorn: "Its interior arrangement was sacrificed to its exterior appearance."[43] Sever Hall contained a large lecture room that was long and low-ceilinged, a shape that led to acoustical difficulties. When Sabine investigated the room, he found a profusion of wires stretched across the ceiling with lengths of cloth hanging from some of them. His diagnosis was that the wires had "the merit of being harmless," while the yards of fabric were "like bleeding a patient suffering from a chill."[44]

There was no shortage of acoustical problems in the Boston area. The Boston Public Library, completed by McKim, Mead and White a few years before the Fogg Museum, had a lecture hall that was an acoustical tragedy. Its hard surfaces sustained sound for almost 9 seconds. Sabine was called upon to remedy this condition. Sabine's work on this project, as well as Charles Eliot's recommendation, led to his being chosen as consultant on the design of Boston Symphony Hall, one of the world's greatest spaces for music.

Charles McKim had been approached to design a hall for the Boston Symphony in the fall of 1892, when he was deeply involved in completing designs for the Chicago Columbian Exposition. During the following summer McKim submitted drawings of three designs to Major Henry Lee Higginson, founder and perennial benefactor of the Boston orchestra. One design, much preferred by McKim, was laid out as a classical theater, semicircular in plan. Another was elliptical in form, and the third was a rectangular hall. Drawings of the classical scheme and a model of its interior were displayed

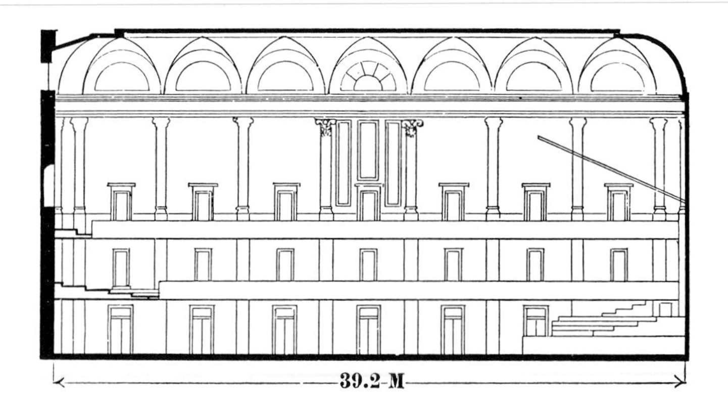

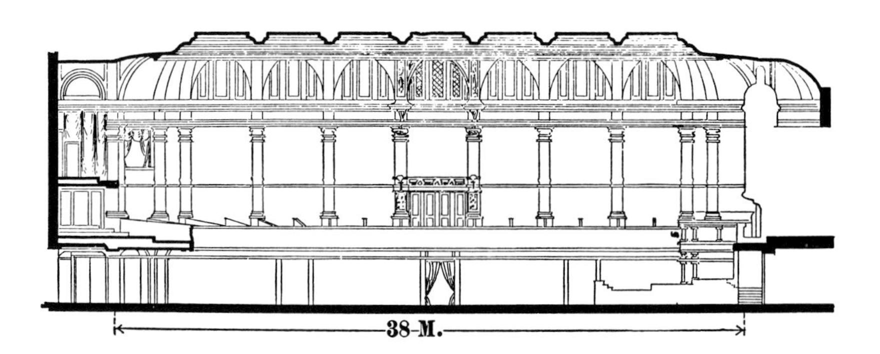

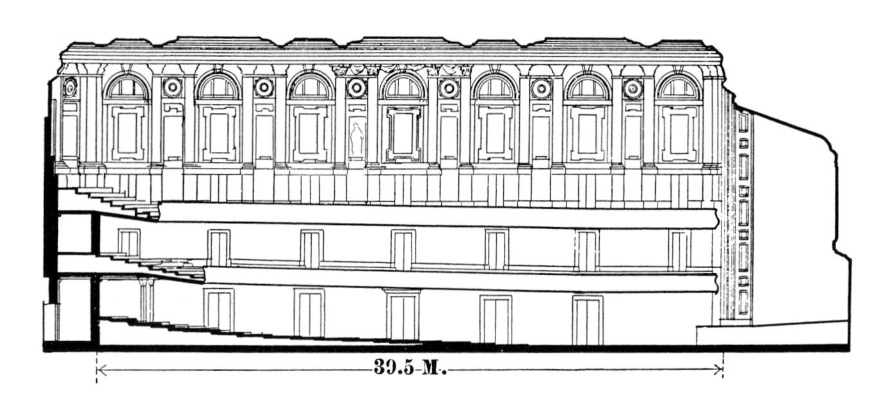

16.11 Top: old Music Hall, Boston (George Snell, 1863); center: Neue Gewandhaus, Leipzig (Martin Gropius and H. Schmieden, 1887); bottom: Symphony Hall, Boston (McKim, Mead and White, 1900). McKim's semicircular design for Symphony Hall, modeled after ancient amphitheaters, was abandoned; work on the project resumed after a delay of ten years. Although the committee wished to emulate Leipzig's new hall, some decisions were obviously influenced by the hall in which the Boston orchestra had performed for many years. (*American Architect and Building News*, 16 June 1900.)

in the Boston Public Library early in 1894, and it is reported that the design was favorably viewed by the public.

In the years before construction began, McKim visited Europe and had an opportunity to visit concert halls there. The celebrated Viennese conductor Hans Richter is said to have told him, "I don't know anything about acoustics, but my first violin tells me we always get the best results in a rectangular hall." Major Higginson and the directors of the Symphony resisted the semicircular scheme that McKim favored, perhaps because of their attachment to the rectangular shape of the old Music Hall in which the orchestra had been performing. Several years later a rectangular plan was chosen for Symphony Hall. That decision was strengthened by the acclaim given the Neues Gewandhaus in Leipzig (1887), said to be better for Bach and early Classical music than for Romantic works (fig. 16.11)

The Leipzig example was not closely followed in Boston. Symphony Hall seated an audience of 2,631, over a thousand more than were seated in the Neues Gewandhaus. The proportions of Symphony Hall were made much the same as those of the old Boston Music Hall, which was wider and lower-ceilinged than the Neues Gewandhaus. In describing the design of Symphony Hall, Sabine stressed the fact that the duplication of proportions cannot produce identical acoustical qualities. He pointed out: "Our increasing demands in regard to heat and ventilation, the restriction upon the dimensions enforced by location, the change in size imposed by the demands for seating capacity, have prevented, in different degree, copies from being copies, and models from successfully serving as models."[45] With all such differences of materials considered, particularly those required by new standards of fireproofing, Sabine adjusted the conditions in the new Symphony Hall to produce a reverberation time of 2.31 seconds according to his calculations, a figure almost the same as that for the Neues Gewandhaus.[46] (Sabine calculated the reverberation time of the Neues Gewandhaus from published plans; the reverberation time was, in fact, 1.9 seconds.)[47] Both models on which the Boston Symphony Hall was based had placed the orchestra platform within the hall, although the old Music Hall in Boston had a sloped reflective plane above the platform. In Symphony Hall the orchestra was placed in a shallow recess with splayed surfaces at the top and the sides reflecting sound toward the audience.

Sabine had begun in 1895 with the assigned task of searching for an escape from an embarrassing situation on the Harvard campus; only three years later he addressed the annual convention of the American Institute of Architects on his findings. From the many acoustical problems that were found on the campus, his work spread to include consultation with architects over a broad geographic area. More important, the articles he wrote for major American architectural journals (*American Architect and Building News*, 1900; *Brickbuilder*, 1914–1915) long served as texts for alert architects.

In the first half of the twentieth century few large concert halls were built and only one, Salle Pleyel, attracted attention. France's foremost manufacturer of pianos, the Pleyel company was directed by Gustave Lyon, an accomplished musician, graduate of

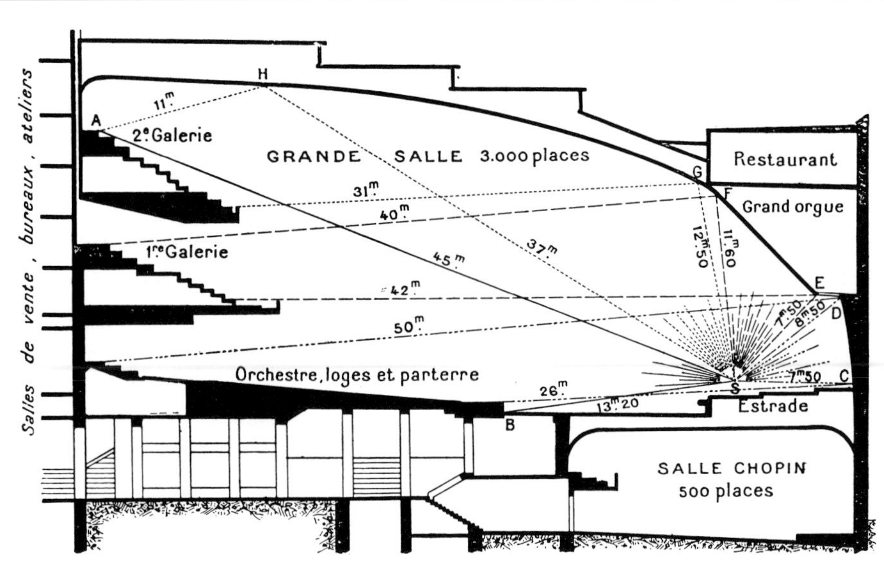

16.12 Although Gustave Lyon emphasized geometric acoustics, the Salle Pleyel, Paris (Auburtin, 1927), did not result in a hall of distinction. Similar forms were understandably fashionable among early functionalist architects. (*L'Illustration*, 10 September 1927.)

the Ecole Polytechnique, inventor of improvements in the piano, and ranking French expert on architectural acoustics.[48] Before his company undertook the construction of Salle Pleyel in 1927, Lyon had gained a reputation for imaginative solutions to acoustical problems. At the turn of the century he had been approached to organize annual series of symphonic concerts in the Trocadéro, a massive cylindrical building across the Seine from the Eiffel Tower. From the time it opened in 1878, the Trocadéro had been branded a failure as a hall for musical performances. Its acoustics were dreadful, there were not enough fire exits, and the toilets were totally inadequate.[49] Lyon wisely insisted that acoustical corrections be made before he assumed responsibility for scheduling concerts. Vaulting of stretched cloth had been installed above the stage organ, concave shapes that Lyon felt should be replaced by convex. After estimates revealed the prohibitive cost of merely erecting the scaffolding necessary for any change, Lyon found a startling alternative: "He had a cylindrical balloon made, almost ten feet long and thirty-two inches in diameter, along its length a rigid member made of aluminum. This balloon when inflated with hydrogen had an upward force strong enough to require a silk anchor cable. . . . The acoustics of the hall were found to be appreciably improved."[50] Lyon relied greatly on reinforcing sound by the reflective shaping of halls.

In Salle Pleyel the ceiling formed a giant parabola, curving upward from the rear of the stage platform to the ceiling of the second balcony (fig. 16.12). Side walls were canted to project additional sound toward the back seats. At the time, the French architect Le Corbusier was preparing his entry in the competition for the design of the Palace of Nations, the League of Nations' center to be built in Geneva. The plan and ceiling shape of the auditorium in his competition entry were those of Salle Pleyel, "conceived exactly after the principles of Mr. G. Lyon."[51] Incensed by the rejection of his design by a "dull-witted gang of architects affiliated with the Institute and some Academies," Le Corbusier wrote glowingly of Lyon's design for Salle Pleyel, citing it as an example of the new functionalist point of view: "Here today is the Salle Pleyel, it too is *the truth*, the truth of reality, the functioning truth, as opposed to the fake truth of the Institute; and the Institute is bent before this new irrefutable truth. There is the historical value of the Salle Pleyel: with neither ambiguity nor limitation, *it is true*, true as the aircraft that flies and the fish that swims."[52]

When the hall opened, the *New York Times* critic reported, "This immense auditorium possesses the most perfect acoustics that I have ever known. Never an echo, never an unpleasantly resonant sound."[53] But not all opinions have been so favorable. One American expert felt that the high ceiling caused a "slight interfering effect . . . during the very rapid movements."[54] Another said that sound waves returning from the back wall and balcony fronts produced echoes. It is interesting to note that drawings published by Le Corbusier in connection with the Palace of Nations competition showed a lower ceiling, while maintaining a generally parabolic form by using sections of four parabolas.[55] In analyzing the entries in the Palace of Nations competition, a Swiss expert wrote admiringly of the auditorium designed by Le Corbusier and Lyon, though he judged that the hall would have been too reverberant for its purpose since the ceiling and walls were proposed to be made of double layers of plate glass.[56]

After the ravages of World War II and the massive social and economic readjustments that immediately followed, the British proudly staged the Festival of Britain on the south bank of the Thames in 1951, the centenary of the Great Exhibition of Victoria's time. The focal point of this event and of the complex of cultural facilities that were built there later was the Royal Festival Hall, an auditorium seating around 3,400 for symphony concerts. From the first, external acoustical factors had to be considered along with internal conditions, for the site was near trains crossing Hungerford Bridge, and London's Underground ran directly below (figs. 16.13, 16.14). Hope Bagenal, the foremost British authority on architectural acoustics, advised the architects of the London County Council, and the hall was built as an independent shell, carefully insulated from the outer shell of the building. Interior shapes and materials were designed with the most advanced techniques, and prior to the first official performance four test concerts were held over a two-month period.[57] Among the usual audiences there were selected groups of listeners, including frequent concertgoers, professional music critics, and musicians. In response to these tests some minor changes were made in materials used in certain areas of the hall. Scientific measurements of the hall's acoustical characteristics were conducted before the test concerts, during the test period, and after the hall was formally opened. Similarly, subjective opinions continued to be sought after the opening, and the press was scanned for comments on the acoustics of the hall. When 18 acoustical experts, in London for an international conference, attended a concert, 13 evaluated the hall as "excellent to very good," but six wanted more "fullness," a view that was paralleled by many press comments during the first 18 months the hall was in use.[58]

After more than a decade, the reverberation time of the Royal Festival Hall was found to have slightly decreased, probably due to minor changes made in the building's ceiling. Subjective opinions of critics and musicians still tended to indicate that its sound was too "dry" and needed to be "warmer"—satisfactory, that is, for the quieter passages of Classical works, but not fully supporting the climaxes of Romantic symphonies. Officials of the Royal Festival Hall decided to test a new corrective procedure. Their acoustical consultants noted:

> Even if opinions had been unanimous that the Hall should be made acoustically warmer—which we assume means lengthening the r.t. [reverberation time]—there was nothing that could be done by conventional methods to get a substantial increase of the r.t., short of such impracticable measures as removing the ceiling, raising the roof, or reducing the audience capacity by, say, 25%.[59]

The essential device of the new method comprised a microphone, an amplifier, and a loudspeaker, all commonplace items in the 1960s. To test the procedure, 89 such units were installed, each serving to reproduce only one frequency. By putting the microphones toward the rear of the hall and locating loudspeakers toward the front, about 50 to 75 feet away, it was assured that the sound from the speakers (called "assisted resonance") would reach the listener after the sound heard directly from the orchestra.[60] The sound traveled farther before reaching the listener's ear, and the effect was much the same as if the roof and ceiling *had* been raised. Although the equipment was simple, the planning was understandably intricate.

The test of assisted resonance was begun quietly. Unannounced, the system was operated at a low level at first and was gradually advanced to its intended setting during a series of 29 concerts. Public announcement of the change was made only after eight con-

16.13 The Royal Festival Hall, London (Architects of London County Council, 1951), was built on a site where trains ran above and below ground level in the immediate vicinity. The problems of the surroundings were adequately solved, for the most frequent complaints after the hall's completion involved the interior, its lack of "fullness" or a "singing tone." (*Acustica*, 1953.)

16.14 A permanent system of "assisted resonance" was installed about 15 years after the first opening of the Royal Festival Hall. (*Acustica*, 1953.)

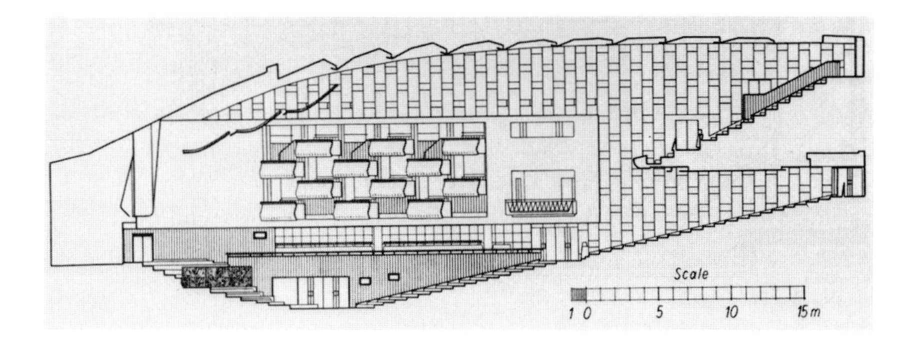

certs at full setting. Comments of critics, musicians, and acousticians, as well as concertgoers, were mostly favorable. Some heard faults in the orchestral sound, minor but troublesome to them, and some had misgivings about the propriety of what had been done. The music critic for the *New York Times* reflected on a concert:

> One was conscious of a smoothness, richness and instrumental fusion that definitely puts the Festival Hall into a superior class . . .
>
> The total result is an amazing improvement. Before last March the hall had a reverberation time of 1.35 seconds. Now it is 2.1, about ideal for a 3,000 seat hall.
>
> But in so doing, the sponsors of "assisted resonance" have raised many esthetic and even moral questions.[61]

By 1968 a full and permanent system had been installed in the Royal Festival Hall using 172 units, almost twice the number of microphones and loudspeakers used during the test period. Years later another music critic recalled: "It was too late for indignation: the musical merits of the system had been proved and approved. Festival Hall performers and audiences have long since stopped remembering that what they listen to is electronically 'fudged.' Is it a moral matter? I think not."[62]

Increasingly, systems of "electronic acoustics" have been used to attain the desired conditions for differing sorts of performances. Halls built for a broad range of presentations—concerts, plays, and popular music—have been equipped for instantaneous alteration to the acoustical characteristics most flattering to the evening's event. Technological systems have made it possible to "hear" a hall that has been electronically shaped, without the limitations imposed by the hard and heavy materials of which traditional architecture is made.

Notes

Publishers are listed for all monographs of the twentieth century.

I Materials

1 Wood

1. Spiro Kostoff, "The Practice of Architecture in the Ancient World: Egypt and Greece," in *The Architect*, ed. Spiro Kostoff (New York: Oxford University Press, 1977), 16.

2. A. Richards, "Roman Timber Building," in *Studies in Building History*, ed. E. M. Jope (London: Odham Press, 1961), 25.

3. L. F. Salzman, *Building in England down to 1540* (Oxford: Clarendon Press, 1952; Millwood, N.Y.: Kraus Reprint Company, 1979), 238.

4. John Harvey, *The Gothic World* (London: Batsford, 1950), 12–13.

5. Salzman, *Building in England*, 238.

6. Bryan Latham, *Timber: Its Development and Distribution: A Historical Survey* (London: George G. Harrap, 1957), 32.

7. Latham, *Timber*, 204.

8. Benno M. Forman, "Mill Sawing in Seventeenth-Century Massachusetts," *Old Time New England* 60 (April-June 1970): 111. Quoting Rhys Jenkins MS. folder, Science Museum Library, London.

9. Forman, "Mill Sawing," 112.

10. Latham, *Timber*, 214.

11. Nollie Hickman, *Mississippi Harvest: Lumbering in the Longleaf Pine Belt, 1840–1915* (University, Miss.: University of Mississippi, 1962), 22.

12. E. K. Spring, "Development of Materials for Woodcutting Tools," in *One Hundred Years of Engineer Progress with Wood*, Proceedings of Wood Symposium (Washington: Timber Engineering Company, c. 1952), 82.

13. Agnes M. Larson, *History of the White Pine Industry in Minnesota* (Minneapolis: University of Minnesota Press, 1949), 19.

14. Nathan Rosenberg, "America's Rise to Woodworking Leadership," in *America's Wooden Age: Aspects of Early Technology*, ed. Brooke Hindle (Tarrytown: Sleepy Hollow Restoration, 1975), 47.

15. Latham, *Timber*, 218–219.

16. Latham, *Timber*, 209.

17. William H. Seward, *Opinion of William H. Seward, on the Rights of the Patentee, during the Third Term of Woodworth's Patent* (Auburn: 1848). *Report of the Case of William W. Woodworth as Administrator, &c. vs. Rodolphus C. Edwards, et al., Tried at Boston (Mass.)* . . . (Auburn: 1848).

18. Latham, *Timber*, 213.

19. John Burroughs, "American versus English Woods," *Critic* 3 (13 January 1883): 15.

20. Theodore Roosevelt, *The Winning of the West* (New York: Review of Reviews, 1904), 1:107–109. Quoted in Richard A. Bartlett, *The New Country: A Social History of the American Frontier, 1776–1890* (London: Oxford University Press, 1974), 12.

21. Bartlett, *New Country*, 237–240.

W. G. Youngquist and H. O. Fleischer, *Wood in American Life, 1776–2076* (Madison: Forest Products Research Society, 1977), 20.

22. Mitchell Mannering, "The Wealth of American Forests," *National Magazine* 7 (March 1898): 491.

23. Carlile P. Winslow, "The Economic Aspects of Wood Preservation for Structural Purposes," *Engineering Magazine* 35 (August 1908): 700.

24. Bartlett, *New Country*, 240–241.

25. Donald McKay, *The Lumberjacks* (Toronto: McGraw Hill Ryerson, 1978), 80.

26. Richard G. Lillard, *The Great For-*

est (New York: Alfred A. Knopf, 1947), 225.

27. Quoted in McKay, *Lumberjacks,* 40.

28. Stewart H. Holbrook, *Holy Old Mackinaw* (New York: Macmillan, 1938), 190.

29. M. J. A. Stoeckhardt, "List of the Different Processes Adopted for the Preservation of Wood . . . ," *Journal of the Franklin Institute* 46 (July 1848): 56–58.

30. N. A. Richardson, "The Development of Coal Tar Creosote as a Wood Preservative," *Wood* 19 (December 1954): 484.

31. F. M. Potter, "Creosote Oil," *Wood* 19 (September 1954): 368.

32. Sherry H. Olsin, *The Depletion Myth* (Cambridge: Harvard University Press, 1971), 64.

33. Rosenberg, "America's Rise," 43.

34. Robert F. Fries, *Empire in Pine: The Story of Lumbering in Wisconsin, 1830–1900* (Madison: State Historical Society of Wisconsin, 1951), 135.

35. Larson, *White Pine Industry,* 391.

36. Richard L. Williams, *The Loggers* (New York: Time-Life Books, 1976), 216.

37. Andrew Dick Wood and Thomas Gray Linn, *Plywoods: Their Development, Manufacture and Application* (Brooklyn: Chemical Publishing Company, 1943), 38.

38. Thomas D. Perry, "Plywood Is Engineered Wood," in *One Hundred Years,* 58.

39. Wood and Linn, *Plywoods,* 188–189.

2 Masonry

1. L. F. Salzman, *Building in England down to 1540* (Oxford: Clarendon Press, 1952; Millwood, N. Y.: Kraus Reprint Company, 1979), 132–133.

2. Brunilde Sismondo Ridgway, "Stone Carving: Sculpture," in *The Muses at Work,* ed. Carl Roebuck (Cambridge: MIT Press, 1969), 101.

3. J. B. Ward-Perkins, "Quarrying in Antiquity: Technology, Tradition and Social Change," *Proceedings, British Academy* 57 (1971): 140.

4. William L. Saunders, "Notes on Quarrying," *Scientific American* 64 (7 March 1891): 149.

5. Peter Stanier, "The Granite Quarrying Industry in Devon and Cornwall. Part 1, 1800–1910," *Industrial Archaeology Review* 7 (Spring 1985): 180.

6. Saunders, "Notes on Quarrying," 149.

7. Ward-Perkins, "Quarrying in Antiquity," 143.

8. George A. Thiel and Carl E. Dutton, *The Architectural, Structural, and Monumental Stones of Minnesota* (Minneapolis: University of Minnesota Press, 1935), 29.

9. Stanier, "Granite Quarrying Industry," 172.

10. T. Donnelly, "Structure, Technology and Demand in the Aberdeen Granite Quarrying Industry, 1880–1914," *Construction History* 1 (1985): 41.

11. T. K. Derry and Trevor I. Williams, *A Short History of Technology* (London: Oxford University Press, 1970), 547.

12. Edward Owen, *Observations on the Earth, Rocks, Stones and Minerals . . . about Bristol* (London: 1754). Quoted in F. J. North, *Limestones, Their Origins, Distribution and Use* (London: Thomas Murby, 1930), 370–371.

13. William L. Saunders, "Dimension Stone Quarrying—The Blasting Process," *Transactions, American Society of Civil Engineers* 25 (November 1891): 502–503.

14. Saunders, "Dimension Stone Quarrying," 514. Stanier, "Granite Quarrying Industry," 179.

15. Oliver Bowles, *The Stone Industries,* 2d ed. (New York: McGraw-Hill, 1939), 258–259.

16. Bowles, *Stone Industries,* 50.

17. Salzman, *Building in England,* 123.

18. "Building Stones—The New Houses of Parliament," *The Penny Magazine* (1 July 1843): 254–255.

19. Joseph Gwilt, *The Encyclopedia of Architecture: Historical, Theoretical and Practical* (London: 1876; New York: Crown Publishers, 1982), 483.

20. Kenneth Hudson, *The Fashionable Stone* (Park Ridge, N.J.: Noyes Press, 1972), 4.

21. M. H. Port, ed., *The Houses of Parliament* (New Haven: Yale University Press, 1976), 98.

22. Donnelly, "Structure, Technology and Demand," 42.

23. Bowles, *Stone Industries,* 57, 161.

24. Henry Thompson Rowell, *Rome in the Augustan Age* (Norman: University of Oklahoma Press, 1962), 138.

25. Salzman, *Building in England,* 140–142.

26. Fernand Braudel, *Civilization and Capitalism, 15th–18th Century,* trans. Sian Reynolds (New York: Harper and Row, 1987), 1:267–268. John Summerson, *Architecture in Britain, 1530 to 1830* (Baltimore: Penguin Books, 1954), 62.

27. C. G. Powell, *An Economic History of the British Building Industry, 1815–1979* (London: Architectural Press, 1980), 38.

28. A. Marl, "Brickmaking," *Clay Record* 2 (30 August 1897): 16.

29. W. Foster Hidden, "The History of Brick Making in and around Vancouver," *Washington Historical Quarterly* 12 (April 1930): 131.

30. Harley H. McKee, "Brick and Stone: Handicraft to Machine," in *Building in America: Contributions toward the History of a Great Industry,* ed. Charles E. Peterson (Radnor, Pa.: Chilton, 1976), 82.

31. M. D. P. Hammond, "Brick Kilns: An Illustrated Survey," *Industrial Archaeology Review* 1 (1977): 171.

32. *Industrial Chicago* (Chicago: 1891), 1:383.

33. Humphrey Chamberlain, "The Manufacture of Bricks by Machinery," *Journal of the Royal Society of Arts* 4 (6 June 1856): 493.

34. Chamberlain, "Manufacture," 493.

35. "City Authorities Don't Want the Wire Cut Product," *Clay Record* 12 (22 April 1898): 21. Reprinted from *Minneapolis Journal.*

36. G. C. Mars, "Brick—Its Manufacture," *American Architect—Architectural Review* 123 (11 April 1923): 315.

37. Charles T. Davis, "Pressed and Ornamental Bricks," *American Architect and Building News* 17 (25 April 1885): 196.

38. *A History of Real Estate, Building and Architecture during the Last Quarter of a Century* (New York: 1898; New York: Arno Press, 1967), 403.

39. *Industrial Chicago,* 1:372.

40. *Industrial Chicago,* 1:386.

41. W. Johnson, "Brickmaking in America," *American Architect and Building News* 28 (7 June 1890): 147. Reprinted from *Architect.*

42. W. Noble, "The Heavy Clay Industry, 1900 to 1950," in *Ceramics: A Symposium,* ed. A. T. Green and Gerald H. Stewart (Stoke-on-Trent: British Ceramic Society, 1953), 763.

43. W. C. Lemert, "Brickmaking Machinery," in *World's Columbian Exposition, Chicago, Illinois, 1893. Special Reports upon Special Subjects or Groups,* 57th Cong., 1st sess, 1901, H. Doc. 510, 68.

44. C. E. L. Franklin, "Changes in the Theory and Practice of Drying," in *Ceramics: A Symposium,* 333–334.

45. Noble, "Heavy Clay Industry," 761.

46. E. Rowden, "Firing in the Heavy Clay and Refractories Industries, 1900–1950," in *Ceramics: A Symposium,* 776.

47. S. B. Hamilton, *A Short History of the Structural Fire Protection of Buildings* (London: HMSO, 1958), 7.

48. Letter, John Walker to William Strutt, 29 October 1792, in Turpin Bannister, "The First Iron-Framed Buildings," *Architectural Review* 107 (April 1950): 235.

49. Letter, Matthew Boulton to William Strutt, 8 May 1793, in H. R. Johnson and A. W. Skempton, "William Strutt's Cotton Mills, 1793–1812," *Transactions, Newcomen Society* 30 (1955–1956): 184.

50. Hamilton, *Structural Fire Protection,* 14.

51. Peter B. Wight, "Origin and History of Hollow Tile Fire-Proof Floor Construction," *Brickbuilder* 6 (March 1897): 54.

52. Wight, "Origin and History," second part (April 1897): 74.

53. S. B. Hamilton, "The History of Hollow Bricks," *Transactions, British Ceramic Society* 58 (February 1959): 52.

54. "History of the Manufacture of Hollow Bricks," *Building News* 4 (26 March 1858): 317. Benford Denton is given as the name in some descriptions.

55. "History of the Manufacture of Hollow Bricks," 317.

56. Hamilton, "Hollow Bricks," 48–49.

57. Hamilton, "Hollow Bricks," 49.

3 Terra-Cotta

1. K. A. Esdaile, "Coade Stone," *Architect and Building News* 161 (19 January 1940): 94.

2. Esdaile, "Coade Stone," 95.

3. S. B. Hamilton, "Coade Stone," *Architectural Review* 116 (November 1954): 295–296. "Mrs. Coade" directed the work, but, because "Mrs." (mistress) was used as a title of respect for unmarried women, it is uncertain whether the name Eleanor Coade in company records refers to the mother, who was sixty when the family came to London, or the daughter.

4. J. Travenor-Perry, "An Episode in the History of English Terra-Cotta," *Architectural Review* 33 (June 1913): 120.

5. Alison Kelly, "Mrs. Coade's Stone," *Connoisseur* 197 (January 1978): 16.

6. Alison Kelly, "Sir John Soane and Mrs. Eleanor Coade," *Apollo* 129 (April 1989): 253.

7. *Somerset House Gazette* (1824). Quoted in Gilbert R. Redgrave, "Terra Cotta, and Its Employment as a London Building Material," *Building News* 15 (7 February 1868): 92.

8. Redgrave, "Terra Cotta," 92.

9. William F. Jelke, "Terra Cotta," *Yale Scientific Monthly* 3 (December 1896): 79–87.

10. Robert C. Mack, "The Manufacture and Use of Architectural Terra Cotta in the United States," in *The Technology of Historic American Buildings,* ed. H. Ward Jandl (Washington: Foundation for Preservation Technology, 1983), 122.

11. James Taylor, "The History of Terra Cotta in New York," *Architectural Record* 2 (December 1892): 145.

12. Walter Geer, *The Story of Terra Cotta* (New York: Tobias A. Wright, 1920), 54–56.

13. "Discussion on Mr. C. Barry's Paper.—'Memorandum on the Works Executed in Terra Cotta at New Alleyn's College, Dulwich'," *Royal Institute of British Architects, Papers read at Session 1868–1869* (London: 1869), 27–28.

14. "Use of Terra Cotta in the Philadelphia Museum of Art," *Ceramic Age* 9 (May 1927): 140.

15. Cervin Robinson and Rosemarie Haag Bletter, *Skyscraper Style* (New York: Oxford University Press, 1975), 38.

16. "Machine Made Terra Cotta Faces McGraw-Hill Building," *Brick and Clay Record* 79 (14 July 1931): 30.

17. Michael Stratton, "Science and Art Closely Combined: The Organization of Training in the Terra-Cotta Industry, 1850–1939," *Construction History* 4 (1988): 39.

18. Geer, *Terra Cotta,* 35–37.

19. "The Manufacture of Terra-Cotta in Chicago," *American Architect and Building News* 1 (30 December 1876): 420.

20. Margaret Henderson Floyd, "A Terra-Cotta Cornerstone for Copley Square: Museum of Fine Arts, Boston, 1870–1876, by Sturgis and Brigham," *Journal of the Society of Architectural Historians* 32 (May 1973): 84–88.

21. John H. Sturgis, "Terra-Cotta and Its Uses," *American Institute of Architects. Proceedings of the 5th Annual Convention* 5 (1871): 39–41.

22. Geer, *Terra Cotta,* 69.

23. Geer, *Terra Cotta,* 85.

24. "In the Early Days of Terra Cotta," *Ceramic Age* 10 (November 1927): 184.

25. *A History of Real Estate, Building and Architecture during the Last Quarter of a Century* (New York: 1898; New York: Arno Press, 1967), 522–523, 527.

26. American Terra Cotta Company records as cited in Mack, "Architectural Terra Cotta," 138.

4 Iron and Steel

1. R. J. Forbes, "Metallurgy," in Charles Singer and others, eds., *History of Technology* (Oxford: Clarendon Press, 1956), 2:58.

2. S. B. Hamilton, "The Structural Use of Iron in Antiquity," *Transactions, Newcomen Society* 31 (1957–1959): 33.

3. Hamilton, "Antiquity," 42.

4. L. F. Salzman, *Building in England down to 1540* (Oxford: Clarendon Press, 1952: Millwood, N.Y.: Kraus Reprint Company, 1979), 289.

5. Salzman, *Building in England,* 294.

6. Christopher Wren, *Parentalia.* Quoted in Hamilton, "Antiquity," 43.

7. Thomas Southcliffe Ashton, *Iron and Steel in the Industrial Revolution,* 3d ed. (Manchester: Manchester University Press, 1963), 9.

8. Several studies have established the basic information about the history of iron construction in mill buildings: S. B. Hamilton, "The Use of Cast Iron in Building," *Tranasactions, Newcomen Society* 21 (1940–1941): 139–155; Turpin Bannister, "The First Iron-Framed Buildings," *Architectural Review* 107 (April 1950): 231–245; A. W. Skempton, "The Origin of Iron Beams," *Actes du VIIIe Congrès International d'Histoire des Sciences* (Vinci: Gruppo Italiano di Stori delle Scienze, 1956), 1029–1039; and A. W. Skempton, "Evolution of the Steel Frame Building," *Guilds Engineer* 10 (1959): 37–51.

9. Skempton, "Evolution," 41.

10. Bannister, "Iron-Framed Buildings," 237–238.

11. Hamilton, "Cast Iron in Building," 145.

12. "Charpente métallique de l'ancienne Halle aux Blés," *Construction Moderne* 4 (3 November 1888): 47. Taken from *Génie Civil,* 18 August 1898.

13. Bertrand Gille, "Devis pour la coupole de la Halle aux Blés de Paris," *Revue d'Histoire de la Siderurgie* 8, no. 2 (1967): 105–106.

14. Frances H. Steiner, "Building with Iron: A Napoleonic Controversy," *Technology and Culture* 22 (October 1981): 720.

15. Arthur Vierendeel, *La construction architecturale en fonte, fer et acier* (Louvain: A. Uystpruyst, 1903), 26.

16. E. Rivoalen, "La Bourse de Commerce à Paris," *Construction Moderne* 5 (21 December 1889): 122.

17. "Charpente métallique," 48.

18. Charles Fowler, "Metal Roof at Hungerford Market," *Transactions, Royal Institute of British Architects,* (London: 1835–1836), 44–46.

19. T. L. Donaldson, "Memoir of the Late Charles Fowler, Fellow," *Royal Institute of British Architects, Papers read at Session 1867–1868* (London: 1868), 10.

20. Vierendeel, *Construction architecturale,* 90.

21. James Bogardus, *Cast Iron Buildings: Construction and Advantages* (New York: 1858), 9.

22. "Men Who Have Assisted in the Development of Architectural Resources—John B. Cornell," *Architectural Record* 1 (October-December 1891). Quoted in Alan Burnham, "Last Look at a Structural Landmark," *Architectural Record* 120 (September 1956): 278.

23. Esmond Shaw, *Peter Cooper and the Wrought Iron Beam* (New York: Cooper Union School of Art and Architecture, 1960), 31.

24. "Our New York Letter," *American Builder and Journal of Art* 1 (February 1869): 49. Quoted in Jay E. Cantor, "A Monument of Trade," *Winterthur Portfolio* 10 (1975): 180.

25. Cervin Robinson, "Late Cast Iron in New York," *Journal of the Society of Architectural Historians* 30 (May 1971): 164.

26. "Iron Architecture in the United States," *Architect* 28 (30 December 1882): 407.

27. James J. Davis, *The Iron Puddler* (Indianapolis: Bobbs-Merrill, 1922), 106–107.

28. R. J. M. Sutherland, "The Introduction of Structural Wrought Iron," *Transactions, Newcomen Society* 36 (1963–1964): 67.

29. See "Corrugated Iron Roofs," paragraph 420 in J. C. Loudon, *An Encyclopedia of Cottage, Farm, and Villa Architecture* (London: 1863), 207.

30. H. W. Dickinson, "A Study of Galvanised and Corrugated Sheet Metal," *Transactions, Newcomen Society* 24 (1943), 30.

31. Vierendeel, *Construction architecturale,* 469.

32. Vierendeel, *Construction architecturale,* 70–71.

33. Vierendeel, *Construction architecturale,* 87.

34. Vierendeel, *Construction architecturale,* 104–105.

35. "The Development of Iron and Steel Roof Design," *Builder* 91 (22

September 1906), 339.

36. "The Development of Iron and Steel Roof Design," *Builder* 91 (29 September 1906), 365.

37. Obituaries: "Rowland Mason Ordish," *Engineering* (17 September 1886): 198. "Roland Mason Ordish," *Engineer* (17 September 1886): 232–233.

38. Vierendeel, *Construction architecturale*, 124.

39. Vierendeel, *Construction architecturale*, 151. Quoting César Daly, *Revue de l'Architecture* (1878).

40. Vierendeel, *Construction architecturale*, 220.

41. August Choisy, *Histoire de l'architecture* (Paris: 1899; Paris: Editions Vincent, Fréal & Cie., 1954), 2:593.

42. Cyril Stanley Smith, *The Science of Steel, 1532 to 1786* (Cambridge: MIT Press, 1968), 116–117.

43. Guyton de Morveau, *Encyclopédie méthodique*, 1:450. Quoted in Cyril Stanley Smith, "The Dictionary of Carbon in Steel," *Technology and Culture* 5 (Spring 1964): 166.

44. W. K. V. Gale, "The Bessemer Steelmaking Process," *Transactions, Newcomen Society* 46 (1973–1974): 17.

45. Gale, "Bessemer Steelmaking," 21.

46. Peter Temin, *Iron and Steel in Nineteenth-Century America* (Cambridge: MIT Press, 1964), 145.

47. Philip W. Bishop, *The Beginnings of Cheap Steel*, U.S. National Museum, Bulletin 218 (Washington: Smithsonian Institution, 1959), 43.

48. William Kelly, "William Kelly's Own Account of His Invention of the Pneumatic Process," *American Iron and Steel Association. Bulletin* 30 (1 June 1896): 1.

49. Kelly, "Own Account," 1. The tale of William Kelly is embellished with accounts of spying English workmen, who were said either to have included Henry Bessemer or to have quickly transmitted information to Bessemer, and stories of Kelly's Kentucky father-in-law who believed him to be mad.

50. William M. Sweet, "Iron and Steel Manufacture," in U.S. Census Office, *Manufacturing Industries*, 11th Census (1890), 3:392.

51. Temin, *Iron and Steel*, 143.

52. W. F. Durfee, "The Early Use of Rolls in the Manufacture of Metals," *Cassier's Magazine* 15 (April 1899): 481.

53. W. K. V. Gale, "The Rolling of Iron," *Transactions, Newcomen Society* 37 (1967): 37.

54. Robert A. Jewett, "Structural Antecedents of the I-Beam, 1800–1850," *Technology and Culture* 8 (July 1967): 349.

55. Robins Fleming, "Evolution of the Steel Skeleton Type of Building," *Proceedings, Engineers' Society of Western Pennsylvania* 43 (February 1927): 4.

56. Shaw, *Peter Cooper*, 23–24.

57. John Fritz, "The Development of Iron Manufacture in the United States in the Past Seventy-five Years," *Journal of the Franklin Institute* 148 (December 1899): 443.

58. Fritz, "Past Seventy-five Years," 453–454.

59. Temin, *Iron and Steel*, 165.

60. W. Basil Scott, "Constructional Steelwork: A Short History," *Architects' Journal* 68 (11 July 1928): 56.

61. J. C. Carr and W. Taplin, *History of the British Steel Industry* (Cambridge: Harvard University Press, 1962), 227.

62. T. Good, "The American Iron and Steel Industry," *Cassier's Magazine* 36 (May 1909): 87.

63. Good, "American Iron and Steel," 87.

64. Scott, "Constructional Steelwork" (4 July 1928), 21.

65. Scott, "Constructional Steelwork" (4 July 1928), 21.

66. "The Disaster at Charing Cross Railway Station," *Iron and Coal Trades Review* 72 (22 June 1906): 8. Summarizes and quotes report of Major Pringle to the Board of Trade.

67. Scott, "Constructional Steelwork" (11 July 1928), 55.

68. Carr and Taplin, *British Steel Industry*, 163. Quoting Sir James Kitson.

69. Vierendeel, *Construction architecturale*, 33.

70. Vierendeel, *Construction architecturale*, 35. Quoting M. Mary, inspector-general of Ponts et Chausées de France.

71. A. M. Robb, "Ship-Building," in Charles Singer and others, eds., *A History of Technology* (Oxford: Clarendon Press, 1958), 5:368.

72. C. G. Poore, "The Riveter's Panorama of New York," *New York Times*, 5 January 1930.

73. Gilbert D. Fish, "The Practicability of Welded Steel Construction," *Western Architect and Engineer* 97 (June 1929): 103.

74. M. S. G. Cullimore, "Joints of Metal Structures," in *Engineering Structures*, ed. P. S. Bulson, J. B. Caldwell, and R. T. Severn (Bristol: University of Bristol Press, 1983), 67.

75. *A History of Real Estate, Building and Architecture During the Last Quarter of a Century* (New York: 1898; New York: Arno Press, 1967), 466.

76. "Discussion of the Papers on Influence of Steel Construction," *American Architect and Building News* 54 (28 November 1896): 71–72.

77. Joseph Gwilt, *The Encyclopedia of Architecture; Historical, Theoretical and Practical* (London: 1867; New York: Crown Publishers, 1982), 395.

78. J. Lincoln Steffens, "The Modern Business Building," *Scribner's Magazine* 22 (July 1897): 48–49.

79. Corydon T. Purdy, "The Steel Skeleton Type of High Buildings," *Engineering News* 26 (5 December 1891): 535.

80. H. J. Burt, "Growth of Steel Frame Buildings," *Engineering News-Record* 92 (17 April 1924): 683.

81. Alan Burnham, "The Rise and Fall of the Phoenix Column,"

Architectural Record 125 (April 1959): 225.

82. Fleming, "Evolution," 12.

83. Purdy, "Steel Skeleton" (12 December 1891), 560. Burt, "Growth," 683.

84. Dankmar Adler, "Tall Office Buildings—Past and Future," *Engineering Magazine* 3 (September 1892): 765.

85. John Beverley Robinson, "The Tall Office Buildings of New York," *Engineering Magazine* 1 (May 1891): 201.

86. Homer Hoyt, *One Hundred Years of Land Values in Chicago* (Chicago: University of Chicago Press, 1933), 153.

87. Scott, "Constructional Steelwork" (11 July 1928), 56.

88. Fleming, "Evolution," 22.

89. "Limitations to the Production of Skyscrapers," *Atlantic Monthly* 2 (October 1902): 486. Quoted in Fleming, "Evolution," 23.

5 Glass

1. D. B. Harden, "Domestic Window Glass: Roman, Saxon and Medieval," in *Studies in Building History,* ed. E. M. Jope (London: Odham Press, 1961), 42–43.

2. L. F. Salzman, *Building in England down to 1540* (Oxford: Clarendon Press, 1952; Millwood, N.Y.: Kraus Reprint Company, 1979), 185.

3. Henry Chance, "On the Manufacture of Crown and Sheet Glass," *Journal of the Royal Society of Arts* 4 (15 February 1856): 232.

4. L. M. Angus-Butterworth, "Window-glass," in Charles Singer and others, eds., *A History of Technology* (Oxford: Clarendon Press, 1957), 3:237.

5. Joseph D. Weeks, "Report on the Manufacture of Glass," in U.S. Census Office, *Report on Manufacturing in the U.S.*, 10th Census (1880), 2:65.

6. Warren C. Scoville, "Labor and Labor Conditions in the French Glass Industry 1643–1789," *Journal of Modern History* 15 (December 1943): 293.

7. The Venerable Bede, quoted in R. W. Douglas and Susan Frank, *A History of Glassmaking* (Henley-on-Thames: G. T. Foulis, 1972), 134.

8. Harden, "Domestic Window Glass," 40.

9. Sheridan Muspratt, *Chemistry as Applied to Arts and Manufacture* (Glasgow: 1860). Quoted in Douglas and Frank, *History,* 139.

10. Chance, "On the Manufacture," 225.

11. Chance, "On the Manufacture," 225.

12. "A Day at a Glass-Factory," *Penny Magazine* 13 (29 June 1844): 255. C. Hanford Henderson, "Glass-making," *Journal of the Franklin Institute* 124 (September 1887): 208.

13. Henderson, "Glass-making," 209.

14. Arthur E. Fowle, "Flat Glass," *The Glass Industry* 5 (June 1924): 102.

15. Chance, "On the Manufacture," 226.

16. Warren W. Scoville, "Technology and the French Glass Industry, 1640–1740," *Journal of Economic History* 1 (November 1941): 156–157.

17. Daphne du Maurier, *The Glass-blowers* (1963; Harmondsworth, Middlesex: Penguin Books, 1966), 19.

18. Scoville, "Labor and Labor Conditions," 275, 293.

19. Reasons that have been given for the noble status granted some glassworkers have included: their participation in crusades; their association with the construction of stained glass windows in medieval cathedrals; and the relationship between the French love of wine and the glasshouses' production of bottles.

20. Weeks, "Report," 2:77.

21. "Manufacture of Glass in the United States," *Hunt's Merchant's Magazine* 28 (January 1853): 120.

22. Henry James Seymour, "Pittsburgh: Glass and Glassmakers," *Magazine of Western History* 3 (February 1886): 371–372.

23. Douglas and Frank, *History,* 104.

24. Douglas and Frank, *History,* 108.

25. Ronald L. Michael and Ronald C. Carlisle, "Monongahela and Pittsburgh District Glass: 19th Century," *APT: Bulletin of the Association for Preservation Technology* 7, no. 1 (1975): 58.

26. Data for the table derive from Weeks, "Report," 19; Julius Goldschmidt, "Austrian Sheet and Mirror Glass," *Manufacturer and Builder* 24 (September 1892): 193; W. E. S. Turner, "The Modern Production of Sheet Glass," *Journal of the Royal Society of Arts* 73 (24 July 1925): 824.

27. E. W. Tillotson, "Twenty-five Years Progress in the Glass Industry," *Journal of the American Ceramic Society* 6 (January 1923): 246.

28. Chance, "On the Manufacture," 224.

29. Robert J. Montgomery, "Present and Future Walls for Use with Molten Glass," *Journal of the American Ceramic Society* 8 (April 1925): 205–206.

30. W. E. S. Turner, "The British Glass Industry: Its Development and Its Outlook," *Journal of the Society of Glass Technology* 6 (August 1922): 125.

31. Douglas and Frank, *History,* 30–31.

32. T. C. Barker, *Pilkington Brothers and the Glass Industry* (London: George Allen and Unwin, 1960), 82.

33. Turner, "British Glass Industry," 129.

34. Pearse Davis, *The Development of the American Glass Industry* (Cambridge: Harvard University Press, 1949), 109.

35. Davis, *American Glass Industry,* 197, 198.

36. "The Knights of Labor encompassed all fields of work, only barring membership of 'gamblers, saloonkeepers, bankers, lawyers, and stock brokers.'" Allan Nevins and Henry Steele Commager, *A Pocket History of the United States*, 5th ed. (New York: Washington Square Press, 1970), 284.

37. Henry A. Gemery, "Productivity Growth, Process Change, and Technical Change in the U.S. Glass Industry" (Ph.D. diss., University of Pennsylvania, 1967).

38. "Collective Agreements as to Production as Restraint of Trade," *Monthly Labor Review* 18 (February 1924): 197.

39. Richard Guenther, "Belgian Glass Trust," *Monthly Consular Reports* (Department of Commerce and Labor) 74 (February 1904): 347.

40. "The First Machine for the Commercial Production of Window Glass by the Sheet Process," *Scientific American* 62 (1 December 1906): 400.

41. Edmond Deffernez, "Des souffliers de verre—Hygiène, maladies et accidents," *Mémoires, Académie Royale de Médecine de Belgique* 5 (1880): 12.

42. Deffernez, "Des souffliers," 11–12.

43. Weeks, "Report," 2:6.

44. The development of greenhouses is described in Ken Butti and John Perlin, *A Golden Thread* (New York: Van Nostrand Reinhold, 1980); John Hix, *The Glass House* (Cambridge: MIT Press, 1974).

45. Susan Lasdun, *Victorians at Home* (New York: Viking Press, 1981), 118.

46. Paul Hollister, "The Glazing of the Crystal Palace," *Journal of Glass Studies* 16 (1974): 95–110.

47. Charles Dickens, *Household Words.* Quoted in Raymond A. McGrath and A. C. Frost, *Glass in Architecture and Decoration,* 2d ed. (London: Architectural Press, 1961), 129.

48. In 1854, when the Crystal Palace was dismantled and rebuilt in the London suburb of Sydenham, 8 percent as much glass was supplied to replace that broken and to accommodate changes in the shape of the building.

49. Robert Linton, "Continuous Operation in the Manufacture of Window Glass," *Engineering Magazine* 16 (November 1898): 252–254.

50. Weeks, "Report," 2:38.

51. Alton D. Adams, "Gas Plants in Large Buildings," *American Architect and Building News* 68 (23 June 1900): 95.

52. Davis, *American Glass Industry,* 124.

53. Douglas and Frank, *History,* 130–131.

54. Robert Linton, "The Window Glass Machine," *Proceedings, Engineers Society of Western Pennsylvania* 33 (February 1917): 6.

55. Barker, *Pilkington Brothers,* 193.

56. Barker, *Pilkington Brothers,* 194.

57. Robert G. Skerrett, "Modern Development in Manufacture of Building Glass," *Compressed Air Magazine* 26 (July 1921): 10138.

58. E. Gentil, "La fabrication mécanique et automatique de la verrerie; ses récents progrès," *Technique Moderne* 18 (1 January 1926): 8.

59. "The Glass Industry in France," *Review of Reviews* 29 (January 1904): 111. Based on report in *Revue des Deux Mondes,* November 1903.

60. Trevor Bain, "The Impact of Technological Change on the Flat Glass Industry and the Union's Reaction to Change: Colonial Period to Present" (Ph.D. diss., University of California, Berkeley, 1964).

61. Raymond Chambon, *L'histoire de la verrerie du Belgique du IIme siècle à nos jours* (Brussels: Editions de la Librairie Encyclopédique, 1955), 287.

62. "The Drawing of Sheet Glass (Fourcault System)," *Glass Industry* 2 (August 1921): 191. Translated from *La Verre,* January 1921.

63. Davis, *American Glass Industry,* 192.

64. Turner, "Modern Production," 832.

65. "The Colburn Window-glass Machine," *Scientific American Supplement* 65 (16 May 1908): 312.

66. Douglas and Frank, *History,* 156.

67. Fred J. Huntley, "Making Glass for Ford Windshields," *Glass Industry* 4 (January 1923): 2.

68. J. Wilson Robinson, "River Rouge Automobile Plate Glass Factory," *Engineering News-Record* 93 (4 September 1924): 378.

69. "Continuous Production in Glass Trade," *Glass Worker* 42 (25 August 1923): 10.

70. Barker, *Pilkington Brothers,* 204.

71. Douglas and Frank, *History,* 160.

72. Alastair Pilkington, "A New Kind of Glass," *New Scientist* 5 (22 January 1959): 168.

73. James Ferguson, as quoted in Peter Collins, *Changing Ideals in Modern Architecture, 1750–1950* (London: Faber and Faber, 1965), 138.

6 Cements

1. Vitruvius, *The Ten Books of Architecture,* trans. Morris Hicky Morgan (1914; New York: Dover Publications, 1960), 2.5, 2.6.

2. James Elmes, *Sir Christopher Wren and His Times,* 1823. Quoted in George Godwin, Jr., "Prize Essay upon the Nature and Properties of Concrete," *Transactions, Royal British Institute of Architects* 1 (1835–1836): 8.

3. John Smeaton, *A Narrative of the Building . . . of the Eddystone Light,* 2d ed. (London: 1793), 181.

4. P. Gooding and P. E. Halstead, "The Early History of Cement in England," in *Proceedings, Third International Symposium on the Chemistry of Cement, 1952* (London: Cement and Concrete Association, 1954), 10.

5. Arthur Charles Davis, *A Hundred Years of Portland Cement, 1824–1924* (London: Concrete Publications, 1924), 24–25. A. J. Francis, *The Cement Industry 1796–1914: A History* (London: David and Charles, 1977), 24.

6. British Patent no. 2,120, James Parker, 28 June 1796. Quoted in Gilbert R. Redgrave and Charles Spackman, *Calcareous Cements: Their Nature, Manufacture and Uses* (London: Charles Griffin, 1905), 22.

7. Francis, *Cement Industry,* 36.

8. Francis, *Cement Industry*, 32.

9. ["J. A. A."], "Portland Cement," *Nature* 97 (15 June 1916): 329.

10. Redgrave and Spackman, *Calcareous Cements*, 41.

11. Redgrave and Spackman, *Calcareous Cements*, 42.

12. A. W. Skempton, "Portland Cements 1843–1887," *Transactions, Newcomen Society* 35 (1962–1963): 147.

13. R. H. Bogue, *A Digest of the Literature on the Constitution of Portland Cement Clinker*, National Bureau of Standards (Washington: 1927), 4.

14. Bogue, *Digest*, 5.

15. Robert W. Lesley, *History of the Portland Cement Industry in the United States* (Chicago: International Trade Press, 1924), 10.

16. H. F. Gonnerman, *Development of Cement Performance Tests and Requirements*, Research Bulletin, Portland Cement Association (Chicago: Portland Cement Association, 1958), 2–3.

17. C. W. Pasley, *Observations on Limes, Calcareous Cements* (London: 1847), 109.

18. Gonnerman, *Cement Performance Tests*, 7, 11.

19. Davis, *Hundred Years*, 151–154.

20. Davis, *Hundred Years*, 159–162.

21. Francis, *Cement Industry*, 233.

22. Gooding and Halstead, "Early History," 25–26.

23. Lesley, *Portland Cement Industry*, 111.

24. Lesley, *Portland Cement Industry*, 13.

25. Harley J. McKee, "Canvass White and Natural Cement, 1818–1834," *Journal of the Society of Architectural Historians* 20 (December 1961): 194–197.

26. Lesley, *Portland Cement Industry*, 33.

27. F. H. Doremus, "The Passing of Natural-rock Cement," *American Architect and Building News* 79 (March 1903): 99.

28. Lesley, *Portland Cement Industry*, 26.

29. Lesley, *Portland Cement Industry*, 115.

30. Robert W. Lesley, "History of the Portland Cement Industry in the United States," *Journal of the Franklin Institute* 146 (November 1898): 341, table opposite page 336.

31. Lesley, "History of Portland Cement," 68.

7 Reinforced Concrete

1. James E. Packer, "A Contribution to the Study of Roman Imperial Architecture," *Technology and Culture* 9 (July 1968): 359.

2. Packer, "Contribution," 363.

3. D. S. Robertson, *A Handbook of Greek and Roman Architecture* (Cambridge: Cambridge University Press, 1927), 234.

4. John Harvey, *The Gothic World, 1100–1600* (London: B. T. Batsford, 1950), 16.

5. Quoted in Thomas Potter, "Early Uses of Concrete," *American Architect* 89, no. 2 (1906): 209. Reprinted from *Builder's Journal*.

6. George Godwin, Jr., "Prize Essay upon the Nature and Properties of Concrete," *Transactions, Royal Institute of British Architects* 1 (1835–1836): 15.

7. J. Mordaunt Crook, "Sir Robert Smirke: A Pioneer of Concrete Construction," *Transactions, Newcomen Society* 38 (1965–1966): 10.

8. Crook, "Sir Robert Smirke," 17.

9. W. Fisher Cassie, "Lambot's Boats," *Concrete* 1 (November 1967): 380.

10. Joyce M. Brown, "W. B. Wilkinson (1819–1902) and His Place in the History of Reinforced Concrete," *Transactions, Newcomen Society* 39 (1966–1967): 135.

11. Brown, "W. B. Wilkinson," 129–130.

12. W. Fisher Cassie, "Early Reinforced Concrete in Newcastle-upon-Tyne," *Structural Engineer* 33 (April 1955): 135.

13. W. Fisher Cassie, "The First Structural Reinforced Concrete," *Structural Concrete* 2 (July-August 1965): 447.

14. Andrew Saint, *Richard Norman Shaw* (New Haven: Yale University Press, 1976), 187.

15. Alexander Payne, "Concrete as a Building Material," *Royal Institute of British Architects. Sessional Papers, 1875–76*, 187.

16. British Patent no. 2151, William Lascelles, 11 June 1875: "Improved Method of the Construction of Buildings." "Ninety-year-old Precast System: 1882 Croydon Houses in 'Patent Slab Construction,'" *Concrete* 6 (April 1972): 28.

17. From the Thaddeus Hyatt papers, Kansas State Historical Society, and information provided by Hyatt's descendents.

18. During his lifetime Hyatt took out 190 patents: 86 in the U.S., 91 in Britain, and 13 in France. About half of these related directly to the sidewalk gratings he manufactured or his experiments with concrete. Others ranged from "Fireproof Buildings" (British patent no. 3124) in 1871 to "Navigating the Atmosphere" (British patent no. 10816) in 1901, the year of his death.

19. W. K. Hatt, "Genesis of Reinforced Concrete Construction," *American Concrete Institute, Proceedings of the 12th Convention* 12 (1916): 28.

20. Thaddeus Hyatt, *An Account of Some Experiments with Portland-Cement-Concrete Combined with Iron, as a Building Material* (London: 1877), 15–16.

21. U. S. Patent no. 206,112, 16 July 1878, Thaddeus Hyatt, "Composition Floors, Roofs, Pavements, &c."

22. Ernest L. Ransome and Alexis Saurbrey, *Reinforced Concrete Buildings* (New York: McGraw-Hill, 1912), 33.

23. S. B. Hamilton, *A Note on the History of Reinforced Concrete in Buildings*, National Building Studies, Special Report no. 24 (London:

HMSO, 1956), 3.

24. Hamilton, *Note*, 3.

25. David Molitor, "Masonry Construction," *Journal of the Association of Engineering Societies* (January 1900). Quoted in E. Lee Heidenreich, "Monier Construction," *Journal of the Western Society of Engineers* 5 (June 1900): 209.

26. O. Kohlmorgen, "Evolution of Reinforced-Concrete in Germany," *American Architect and Building News* 90 (15 December 1906): 189. Reprinted from *Concrete and Constructional Engineering*.

27. Jasper O. Draffin, "A Brief History of Lime, Cement, Concrete and Reinforced Concrete," *Journal of the Western Society of Engineers* 48 (March 1943): 40–41.

28. Edouard Suenson, *Jaernbeton*, 4th ed. (Copenhagen: P. E. Bluhme, 1931), 5. Quoted in Ransome, *Reinforced Concrete*, 27.

29. In Ward's presentation to the American Society of Mechanical Engineers there is no mention of his architect Mook or of earlier experimenters in reinforced concrete. Yet the design, although extraordinarily clumsy in parts, indicates a professional hand, and Ward referred to concrete as *béton*, the term used by the French.

30. Ward's residence is described in W. E. Ward "Beton in Combination with Iron as a Building Material," *American Society of Mechanical Engineers* 4 (30 March 1883): 361–365; Peter B. Wight, "The Pioneer Concrete Residence of America," *Architectural Record* 25 (May 1909): 359–363; "Concrete as a Building Material—A Remarkable House at Portchester," *American Architect and Building News* 2 (18 August 1877): 266; Ellen W. Kramer and Aly A. Raafat, "The Ward House: A Pioneer Structure of Reinforced Concrete," *Journal of the Society of Architectural Historians* 20 (March 1961): 34–37.

31. Ward, "Beton in Combination," 391.

32. Ransome, *Reinforced Concrete*, 3.

33. Ransome, *Reinforced Concrete*, 4–5.

34. D. G. McBeth, "François Hennebique—The 'Ferro-concrete' Man," *Concrete* 10 (May 1976): 19.

35. Patricia Cusack, "Agents of Change: Hennebique, Mouchel, and Ferro-concrete in Britain, 1897–1908," *Construction History* 3 (1987): 63.

36. Peter Collins, *Concrete: The New Vision of Architecture* (New York: Horizon Press, 1959), 69.

37. "Origin and Development of Armoured Concrete," *Engineer* 96 (9 October 1903): 346–347.

38. Robert Conot, *A Streak of Luck* (New York: Seaview Books, 1979), 346–348.

39. E. S. Larned, "The Edison Concrete House," *Cement Age* 6 (March 1908): 269.

40. Larned, "Edison Concrete House," 272–273.

41. Larned, "Edison Concrete House," 277.

42. Paul Planat, "La théorie des ciments armés," *Construction Moderne* 9, 10 (30 December 1893 and following issues).

43. Hatt, "Genesis," 23.

44. "The Literature of Reinforced Concrete," *American Architect and Building News* 89 (12 May 1906): 159.

45. J. J. L. Bourdrer, "Concrete and Concrete-steel in Holland," *Cement Age* 2 (1 July 1905): 115.

46. "Relative Permanence of Steel and Masonry Construction," *American Architect and Building News* 78 (11 October 1902): 12. Reprinted from *Proceedings. American Society of Civil Engineers*.

47. Hamilton, *Note*, 17.

48. Jacob Feld, *Lessons from Failure of Concrete Structures* (Detroit: American Concrete Institute, 1964), 13–14. Quoting *Engineering News* (1903).

49. Feld, *Lessons*, 62–63.

50. "The Use of Concrete-Steel Construction in New York," *American Architect and Building News* 82 (17 October 1903): 21.

51. Kohlmorgen, "Evolution,: 189–190.

52. N. de Tedesco, "The Historical Evolution of Reinforced Concrete in France," *Concrete and Constructional Engineering* 1 (July 1906): 167, 170.

53. Frank Kerekes and Harold B. Reid, Jr., "Fifty Years of Development in Building Code Requirements for Reinforced Concrete," *Journal of the American Concrete Institute* 50 (February 1954): 442.

54. Kerekes and Reid, "Fifty Years," 462.

55. C. B. Porter, "Railway Concrete," *Journal of the American Concrete Institute* 23 (May 1952): 726.

56. H. W. Bryson, "Pouring Concrete by Gravity," *Architect and Engineer of California* 20 (April 1910): 70.

57. Duff A. Abrams, *Design of Concrete Mixtures*, Structural Materials Research Laboratory, Lewis Institute, Bulletin no. 1 (Chicago: 1918). Quoted in Howard Newlon, Jr., "From a Diverse Heritage . . . Teachers I Never Knew," *Journal of the American Concrete Institute* 72 (February 1975): 74.

58. Anatole de Baudot, *L'Architecture, le passé—le présent* (Paris: Librairie Renouard, 1916): 175–176.

59. Hamilton, *Note*, 17.

60. Carl W. Condit, "The First Reinforced-concrete Skyscraper," *Technology and Culture* 9 (January 1968): 20.

61. A. O. Elzner, "The First Concrete Skyscraper," *Architectural Record* 15 (June 1904): 544.

62. Auguste Perret, "Thoughts on Architecture," *Arkhitektura SSSR* 4 (January 1936): 12–13. Unpublished translation by R. Branson, Avery Architectural Library, Columbia University.

63. Joseph Abram, "An Unusual Organization of Production: The Building Firm of the Perret Brothers, 1897–1954," *Construction History* 3 (1987): 82.

64. Collins, *Concrete*, 245.

65. U. S. Patent no. 206,112, Thaddeus Hyatt, 16 July 1878, "Composition Floors, Roofs, Pavement, &c."

66. J. L. Peterson, "History and Development of Precast Concrete in the United States," *Journal of the American Concrete Institute* 50 (February 1954): 482.

67. "An Example of Modern Concrete Construction," *Scientific American Supplement* 72 (9 December 1911): 372.

68. Richard Moore, "An Early System of Large-Panel Building," *Journal of the Royal Institute of British Architects* 76 (September 1969): 385.

69. British Patent no. 6115, March 1901, John Alexander Brodie, "Improvements in, and in the Mode of Constructing and Erecting Buildings."

70. Peterson, "History," 484.

71. Daniel L. Schodek, "Precast Concrete Housing: The Youngstown Project," *Journal of the Prestressed Concrete Institute* 23 (November-December 1978): 64.

72. James O'Gorman, "O. W. Norcross: Richardson's 'Master Builder,' " *Journal of the Society of Architectural Historians* 32 (May 1973): 109.

73. Henry. T. Eddy and C. A. P. Turner, *Concrete-Steel Construction* (Minneapolis: Heywood Manufacturing Company, 1914), 158–159.

74. K. A. Faulkes, *The Design of Flat Slab Structures—An Historial Survey*, UNICIV Report no. R-129 (Kensington, N.S.W.: University of New South Wales, 1974), 3.

75. Faulkes, *Flat Slab Structures*, 5–6.

76. The Bercy station is discussed in R. Vallette, "Considérations sur les voutes minces autoportantes," *Génie Civil* 104 (27 January 1934), 85–88. Freysinnet's hangars at Orly and the earlier hangars are discussed in Eugène Freysinnet "Hangars à Dirigeables en Ciment Armé," *Génie Civil* 83 (22 September 1923), 266–268. A list of early thin-shell structures is provided in Charles S. Whitney, "Reinforced Concrete Thin Shell Structures," *Journal of the American Concrete Institute* 24 (February 1953): 524–527.

77. The collaboration of the Zeiss corporation and Dyckerhoff and Widmann is detailed in *Weit spannt sich der Bogen 1865–1965. Die Geschichte der Bauunternehmung* (Wiesbaden: Dyckerhoff and Widmann, c. 1965); Walther Bauersfeld, *Projection Planetarium and Shell Construction*, James Clayton Lecture (London: Institution of Mechanical Engineers, c. 1957); Günter Huberti, *Die erneuerte Bauweise*, Volume B of *Von Caementum zum Spannbeton: Beiträge zur Geschichte des Betons* (Wiesbaden and Berlin: Bauverlag GmbH, 1964).

78. Walther Bauersfeld, "Development of the Zeiss-Dywidag Process," in Jürgen Joedicke, *Shell Architecture* (New York: Reinhold, 1963), 281.

79. David P. Billington, *Thin Shell Concrete Structures*, 2d ed. (New York: McGraw-Hill, 1982), 10.

80. Wayne M. Faunce, *Problems of Construction* (New York: American Museum of Natural History, 1935), 209.

81. Henry J. Cowan, *Design of Reinforced Concrete Structures*, 2d ed. (Englewood Cliffs: Prentice Hall, 1989), 4.

II Systems

1. Quoted in Peter Collins, *Changing Ideals in Modern Architecture 1750–1950* (Montreal: McGill University Press, 1967), 99.

8 Lightning Protection

1. H. Prinz, "Lightning in History," in *Lightning*, ed. R. H. Golde (London: Academic Press, 1977), 1:6.

2. James George Frazer, *The Golden Bough* (New York: Macmillan, 1922), 10:183.

3. Frazer, *Golden Bough*, 1:82, 10:248.

4. B. Dibner, "Benjamin Franklin," in Golde, *Lightning*, 1:44.

5. Letter, Benjamin Franklin to Cadwaller Colden. Carl Van Doren, *Benjamin Franklin* (New York: Viking Press, 1937), 158.

6. Edwin J. Houston, "Franklin as a Man of Science and an Inventor," *Journal of the Franklin Institute* 166 (April 1906): 281.

7. *Poor Richard's (Improved) Almanac*, 1753. Quoted in Abbot Lawrence Rotch, "Did Benjamin Franklin Fly his Electrical Kite before He Invented the Lightning Rod?" *American Antiquarian Society* 18 (October 1906): 122.

8. Eleanor M. Tilton, "Lightning-rods and the Earthquake of 1755," *New England Quarterly* 13 (March 1940): 96. Zoltán Haraszti, "Young John Adams on Franklin's Iron Points," *Isis* 41 (March 1950): 12.

9. Information about Prokop Divis can be found in I. Bernard Cohen and Robert Schofield, "Did Divis Erect the First European Protective Lightning Rod, and Was His Invention Independent?" *Isis* 43 (December 1952): 358–364; Joseph J. Kral, "The Inventor of the Lightning-rod," *Popular Science Monthly* 42 (January 1893): 356–361.

10. *Tableau de Paris* (Amsterdam: 1782–1783). Quoted in H. C. Browne, "Lightning-conductors," *Nature* 116 (15 August 1925): 242.

11. J. M. Thompson, *Robespierre* (New York: Howard Fertig, 1968), 1:34.

12. Jean Matrat, *Robespierre*, trans. Alan Kendall (New York: Charles Scribner's Sons, 1975), 25.

13. G. J. Symonds, ed., *Lightning Rod Conference: Report of the Delegates* (London: 1882), 122.

14. Alexander McAdie, *Protection from Lightning*, Weather Bureau Bulletin no. 15 (Washington: 1895), 7.

15. Killingworth Hedges, *Modern Lightning Conductors: An Illustrated Supplement to the Report of the Lightning Research Committee of 1905* (London:

Crosby Lockwood and Sons, 1905), 88.

16. Symonds, *Lightning Rod Conference*, 195.

17. Hedges, *Modern Lightning Conductors*, 13.

18. Gordon P. McKinnon, ed., *Fire Protection Handbook*, 14th ed. (Boston: National Fire Protection Association, 1976), 15.56–15.61.

19. Symonds, *Lightning Rod Conference*, 61, 65.

20. Mark Twain, *Political Economy*, in *Sketches New and Old* (New York: Harper, 1903).

21. Oliver J. Lodge, *Lightning Conductors and Lightning Guards: A Treatise on the Protection of Buildings* (London: 1892), 24.

22. D. Müller-Hillebrand, "The Protection of Houses by Lightning Conductors—An Historical Review," *Journal of the Franklin Institute* 274 (July 1962): 48.

23. Hedges, *Modern Lightning Conductors*, 56.

24. Hedges, *Modern Lightning Conductors*, 88.

25. Alfred J. Henry, *Loss of Life in the United States by Lightning*, Weather Bureau Bulletin no. 30 (Washington: GPO, 1901), 10.

26. Hedges, *Modern Lightning Conductors*, 49.

27. Hedges, *Modern Lightning Conductors*, 39–40.

28. H. Baatz, "Protection of Structures," in Golde, *Lightning*, 2:623.

9 Sanitation

1. Lawrence Wright, *Clean and Decent* (New York: Viking Press, 1960), 14.

2. Lucinda Lambton, *Temples of Convenience* (London: Gorden Fraser, 1978), 5.

3. Anthony à Wood, quoted in Wright, *Clean and Decent*, 76.

4. "Gardy-loo" is believed to come from the French *Guardez l'eau*, meaning "Watch out for the water!"

5. Christopher Hibbert, *London* (London: Longmans, Green, 1969), 108.

6. W. H. Lewis, *The Splendid Century* (Garden City: Doubleday, 1957), 172.

7. Ray Palmer, *The Water Closet* (Newton Abbot: David and Charles, 1973), 21.

8. Paul Veyne, "The Roman Empire," in *A History of Private Life*, vol. 1, ed. Paul Veyne, trans. Arthur Goldhammer (Cambridge: Harvard University Press, 1987), 198–199.

9. Fernand Braudel, *Civilization and Capitalism, 15th–18th Century*, vol. 1, *The Structures of Everyday Life*, trans. Sian Reynolds (New York: Harper and Row, 1981), 330.

10. Wright, *Clean and Decent*, 137.

11. John Harington, *A New Discourse of a Stale Subject, Called the Metamorphosis of Ajax* (1596), ed. Elizabeth Story Donno (New York: Columbia University Press, 1962), 192–194.

12. British Patent no. 1105. Alexander Cumming, 1775. "Waterclosets."

13. British Patent no. 1177. Joseph Bramah, 1778. "Watercloset."

14. Lambton, *Temples of Convenience*, 11.

15. Glenn Brown, *Water Closets* (New York: 1884), 63.

16. Palmer, *Water Closet*, 33.

17. *Industrial Chicago* (Chicago: 1891), 2:51.

18. *Industrial Chicago*, 2:51.

19. T. D. Turner, "Early Plumbing in the Metropolis of the Mississippi," *Domestic Engineering* (May 1896): 45.

20. Wright, *Clean and Decent*, 202.

21. J. F. C. Harrison, *Early Victorians* (London: Weidenfeld and Nicolson, 1971), 66.

22. Terence McLaughlin, *Dirt* (New York: Stein and Day, 1971), 155.

23. Harrison, *Early Victorians*, 82.

24. Palmer, *Water Closet*, 47.

25. George E. Waring, Jr., "A Royal Instance," *American Architect and Building News* 11 (20 May 1882): 232. Quoting *Lancet*, 18 March 1882.

26. "Royal Instane," 232.

27. Charles Cameron, "The Victorian Era, the Age of Sanitation," *Architecture and Building* 17 (8 October 1892): 176.

28. Allen Hazen, quoted in Joel A. Tarr and Francis Clay McMichael, "Historic Turning Points in Municipal Water Supply and Waste Water Disposal, 1850–1932," *Civil Engineering—ASCE* 47 (October 1977): 86.

29. Trade catalog, American Pipe Company (San Francisco: circa 1878).

30. "The Trap in the House Drain," *Boston Journal of Commerce* (4 June 1892): 134.

31. William Paul Gerhard, "Plumbing," *American Architect and Architecture* 73 (3 August 1901): 35.

32. *Industrial Chicago*, 2:61–62.

33. *Industrial Chicago*, 2:51.

34. All quotations from Gerhard, "Plumbing" (10 August 1901): 44–45.

35. "Trap Venting," *Northwestern Architect and Building Budget* 10 (July 1892): 51.

36. "Un cabinet d'aisance," *Construction Moderne* 2 (23 October 1886): 21.

37. Wright, *Clean and Decent*, 208.

38. Archibald M. Maddock II, *The Polished Earth* (Trenton, New Jersey: 1962), 239.

10 Lighting

1. "Lighting" *All the Year Round* 1 (20 February 1869): 271.

2. W. J. Serrill, "A Century of Light in Philadelphia," *Proceedings, Engineers' Club of Philadelphia* 33 (July 1916): 26.

3. "Lighting," 270.

4. "La grande querelle du verre de lampe. Argand, Quinquet et les frères Montgolfier," *Science Progrès Découverte* 65 (1 May 1937): 404–405.

5. Letter of John Clayton to Robert

Boyle. Quoted in Charles Hunt, *A History of the Introduction of Gas Lighting* (London: Walter King, 1907), 10. There was no such word as "gas" until the seventeenth century, when a Dutch chemist invented the word from a Greek root. Before that time, the words "spirit," "breath," "vapor," "halation," and "wind" were used, and the new word did not come into frequent use until the middle of the eighteenth century.

6. "Illuminating Gas," *Van Nostrand's Engineering Magazine* 6 (April 1872): 399. Reprinted from *American Exchange and Review.*

7. "Murdoch's Paper on Gas Lighting," *Gas World* (9 April 1892): 408. Paper prepared by William Murdoch and read by Sir Joseph Banks at a meeting of the Royal Society, 25 February 1808. Because Englishmen seemed incapable of producing the guttural "ch" in his name with a proper Scottish intonation, Murdoch changed the spelling of his name to Murdock.

8. "Murdoch's Paper," 408.

9. Arthur Elton, "Gas for Light and Heat," in Charles Singer and others, eds., *A History of Technology* (Oxford: Clarendon Press, 1958), 4:270.

10. William Paul Gerhard, *Gas-Lighting and Gas-Fitting* (New York: 1894), 67.

11. John Henderson, "The Manufacture of Light," *Cassier's Magazine* 18 (August 1900): 338.

12. Leon Gaster and J. S. Dow, *Modern Illuminants and Illuminating Engineering*, 2d ed. (London: Sir Isaac Pitman and Sons, 1919), 33–34.

13. Vivian B. Lewes, "A Century of Work on the Development of Light from Coal Gas," *Journal of Gas Lighting, Water Supply, &c.* 59 (21 June 1892): 1177.

14. "Thomas Drummond," *Dictionary of National Biography.*

15. George Foster, *New York by Gas Light* (New York: M. J. Ivers, 1850), 7.

16. Lewes, "Century of Work," 1179–1180.

17. George S. Barrows, "The Work of Dr. Carl Auer von Welsbach in the Field of Artificial Illuminants," *Transactions, Illuminating Engineering Society* 4 (October 1909): 570–571.

18. Harold Baron, "Incandescent Gas Mantles," *Cassier's Magazine* 32 (July 1907): 201, 208.

19. Henderson, "Manufacture of Light," 337.

20. Frank H. Mason, "Auer-Welsbach Patents and Monazite in Germany," *Monthly Consular Reports* 51 (June 1896): 243.

21. "Combination to Test the Welsbach Patent of '93," *Journal of Gas Lighting, Water Supply, &c.* 78 (16 April 1901): 989.

22. "The Welsbach Crisis," *Journal of Gas Lighting, Water Supply, &c.* 79 (21 January 1902): 153.

23. Quoted in S. M. Hammill, "Beginnings and Future of the Arc-Lamp," *Engineering Magazine* 7 (August 1894): 700.

24. W. D'A. Ryan, "Development of Arc Lighting Apparatus from 1810 to 1902," *Canadian Electric News* 12 (September 1902): 161–162.

25. I. C. R. Byatt, *The British Electrical Industry 1875–1914* (Oxford: Clarendon Press, 1979), 12.

26. G. F. Barwick, *The Reading Room of the British Museum* (London: Ernest Benn, 1929), 129.

27. Harry J. Eisenman, "The Brush Double-Arc Lamp," *Technology and Culture* 7 (Fall 1966): 512.

28. Charles Brush, "The Development of Electric Street Lighting," *Journal of the Cleveland Engineering Society* 9 (September 1916): 55. Quoted in Mel Gorman, "Charles F. Brush and the First Public Electric Street Lighting System in America," *Ohio Historical Quarterly* 72 (April 1961): 134.

29. William Ganson Rose, *Cleveland: The Making of a City* (Cleveland: World Publishing Company, 1950), 420.

30. From a "standard textbook published in 1881." Quoted in George R. Metcalf, "The Industrial Development of Electric Lighting," *Electricity* 10 (20 May 1896): 290.

31. L. B. Marks, "Invention of the Enclosed Arc Lamp," *Sibley Journal of Engineering* 22 (October 1907): 2.

32. Robert Grimshaw, "The Lighting of Large Halls," *Architect and Engineer of California* 25 (May 1911): 77.

33. Grimshaw, "Large Halls," 80.

34. W. Mattieu Williams, "A Contribution to the History of Electric Lighting," *Journal of Science* 16 (February 1879): 155.

35. Bernard V. Swenson, "The Electric Incandescent Lamp," *Cassier's Magazine* 21 (January 1902): 238.

36. Hippolyte Fontaine, *Electrical Lighting* (London: 1878). Quoted in Edwin W. Hammer, "Incandescent Lamp Development to the Year 1880," Part 1, *Electrical World and Engineer* 36 (1 December 1900): 841.

37. Letter of Moses G. Farmer. Quoted in Edwin W. Hammer, "Incandescent Lamp Development," part 2 (8 December 1900), 880.

38. Robert Conot, *A Streak of Luck* (New York: Seaview Books, 1979), 73.

39. John W. Howell and Henry Schroeder, *History of the Incandescent Lamp* (Schenectady: Maqua Company, 1927), 77.

40. Howell and Schroeder, *History*, 86.

41. Gaster and Dow, *Modern Illuminants*, 77.

42. Arthur A. Bright, Jr., *The Electric-Lamp Industry* (New York: Macmillan, 1949), 317.

43. C. J. Russell Humphreys, "Relative Cost of Gas and Electricity," *Engineering Magazine* 4 (November 1892): 240.

44. John Henderson, "The Manufacture of Light," *Cassier's Magazine* 18 (August 1900): 337.

45. Henderson, "Manufacture," 338.

46. "Incandescent Gas Burner and Electricity," *Monthly Consular Reports*, no. 295 (April 1905): 140.

47. "Incandescent Gas Burner," 140.

48. W. J. Liberty, "The Centenary of Gas Lighting and Its Historical Development," *Illuminating Engineer* 7 (April 1913): 200.

49. Alfred Lief, *Metering for America* (New York: Appleton-Century-Crofts, 1961), 73.

50. C. Mackechnie Jarvis, "The Generation of Electricity," in Charles Singer and others, eds., *A History of Technology* (London: Oxford University Press, 1958), 5:201–202.

51. "The Incandescent Lamp Suit," *Electrical Review* 20 (7 May 1892): 147.

52. "Electric Lighting," *Municipal Affairs* 2 (December 1898): 738.

53. Thomas Parke Hughes, "British Electrical Industry Lag: 1882–1888," *Technology and Culture* 3 (Winter 1962): 32.

54. Bright, *Electric-Lamp Industry*, 110–111.

55. "The Moore Electric Light," *Electric World and Engineer* 39 (28 June 1902): 1153.

56. Gaster and Dow, *Modern Illuminants*, 110–111.

57. "Moore Electric Light," 1154.

58. Gaster and Dow, *Moderne Illuminants*, 109–110.

59. "Artificial Daylight Floods Garden's Lobby," *New York Times*, 22 December 1905, p. 7, col 6.

60. Gaster and Dow, *Modern Illuminants*, 103.

61. H. E. Watson, *The Jubilee of the Neon Glow Lamp 1911–1961* (Woking: n.p., 1961), 3.

62. Georges Claude, "The Development of Neon Tubes," *Engineering Magazine* 46 (November 1913): 272.

63. Watson, *Jubilee*, 3.

64. Eugene Clute, "Luminous Tubes for Lighting," *Architecture* 71 (February 1935): 67.

65. Bright, *Electric-Lamp Industry*, 390.

66. Arthur A. Bright, "Some Broad Economic Implications of the Introduction of Hot-Cathode Fluorescent Lighting," *Transactions, Electrochemical Society* 87 (October 1945): 375.

11 Heating and Ventilation

1. Charles de La Roncière, "Tuscan Notables on the Eve of the Renaissance," in *A History of Private Life*, vol. 2, ed. Georges Duby, trans. Arthur Goldhammer (Cambridge: Harvard University Press, 1988), 192.

2. Gilles le Bouvier, *Livres de la description des pays*. Quoted in Philippe Contamine, "Peasant Hearth to Papal Palace: The Fourteenth and Fifteenth Centuries," in Duby, *Private Life*, 499–500.

3. L. F. Salzman, *Building in England down to 1540* (Oxford: Clarendon Press, 1967; Millwood, New York: Kraus Reprint Co., 1979), 97.

4. Jeremiah Dwyer, "Stoves and Heating Apparatus," in Chauncey M. Depew, ed., *1795–1895: One Hundred Years of American Commerce* (New York: 1895), 357.

5. Louis Savot, *L'architecture française des bâtiments particuliers* (Paris: 1685; Geneva: Minkoff Reprint, 1973), 158–160.

6. Desaguliers translated this book into English. In 1845 Walter Bernan claimed that "Nicolas Gauger" was a *nom de plume* of Cardinal Polignac, a French diplomat, but in 1893 Billings disagreed and pointed out that "Walter Bernan" was itself the *nom de plume* of a civil engineer, Robert Meikleham.

7. J. Pickering Putnam, *The Open Fireplace in All Ages* (Boston: 1882), 36.

8. Benjamin Franklin, *An Account of the Newly Invented Pennsylvania Fire-Place* (Philadelphia: 1744; Boston: G. K. Hall, 1973), 16–17.

9. Lawrence Wright, *Home Fires Burning* (London: Routledge and Kegan Paul, 1964), 88.

10. Sanborn C. Brown, *Benjamin Thompson, Count Rumford* (Cambridge, Mass.: MIT Press, 1979).

11. Wright, *Home Fires*, 116–117.

12. Wright, *Home Fires*, 138.

13. Details of Arnott's installation are provided in Charles Fowler, "On Warming and Ventilating the Long Room of the Custom House, London, upon Dr. Arnott's Principle," *Transactions, Royal Institute of British Architects* 1 (1842): 168–174.

14. Quoted in Peter Collins, *Changing Ideals of Modern Architecture, 1750–1950* (Montreal: McGill University Press, 1967), 236.

15. Dwyer, "Stoves," 358.

16. Quoted in Josephine H. Peirce, *Fire on the Hearth: The Evolution and Romance of the Heating-Stove* (Springfield, Mass.: Pond-Ekberg Co., 1951), 185.

17. Neville S. Billington, "A Historical Review of the Art of Heating and Ventilating," *Architectural Science Review* 2 (November 1959): 122.

18. Walter Bernan, *On the History and Art of Warming and Ventilating Rooms and Buildings* (London: 1845), 2:240.

19. A. F. Dufton, "Early Application of Engineering to the Warming of Buildings," *Transactions, Newcomen Society*, 21 (1940–1941): 103.

20. Bernan, *On the History*, 2:268.

21. Susan Reed Stifler, *The Beginnings of a Century of Steam and Water Heating by the H. B. Smith Company* (Westfield, Mass.: H. B. Smith, 1960), 15.

22. Letter, G. A. Lee to J. Watt, Jr., 17 June 1802. Quoted in Jennifer Tann, *The Development of the Factory* (London: Cornmarket Press, 1970), 111.

23. Dufton, "Early Applications," 104.

24. Greville Bathe and Dorothy Bathe, *Jacob Perkins* (Philadelphia: Historical Society of Pennsylvania, 1943), 143.

25. *Journal of the Franklin Institute* (September 1837). Quoted in William E. Worthen, "Steam Heating," *Transactions, American Society of Civil Engineers* 24 (March 1891): 208.

26. Bernan, *On the History*, 2:269–271.

27. J. L. Saunders, "Heating in Great Britain," *Heating and Ventilation* 7 (15 December 1897): 7.

28. R. T. Crane, "History of Making Wrought-Iron Pipe," *Valve World* 1 (January 1905): 7.

29. Henry G. Morris, "Beginning of the Manufacture of Wrought-Iron Pipe in the U.S.," *Valve World* 1 (October 1905): 5.

30. "The Origin and Growth of the Fittings Business: Joseph Nason," *Valve World* 1 (December 1905): 7.

31. Ellwood A. Clymer, Jr., "Look What Happened on Chestnut Street . . . !" *District Heating* 62 (April–May 1977): 15.

32. Charles E. Emery, "District Steam Systems," *Transactions, American Society of Civil Engineers* 24 (March 1891): 189.

33. Ellwood A. Clymer and Thomas M. Loughery, "The Philadelphia Steam Story . . . Approaching 90, and Still Going Strong," *District Heating* 61 (April-May-June 1976): 24–25.

34. S. Morgan Bushnell and Fred B. Orr, *District Heating* (New York: Heating and Ventilating Magazine, 1915), 13.

35. J. W. Wallington, "District Heating: Yesterday, Today and Tomorrow," *Chartered Mechanical Engineer* 18 (May 1971): 168.

36. Heating and ventilation of the Houses of Parliament are covered in "The Engineering Features of the Houses of Parliament," *Industries and Iron* 13 (4 November 1892): 445–464; Walter Yates, "Die Ventilation des Britischen Abgeordnetinhauses," *Gesundheits-Ingenieur* 30 (29 June 1907): 421–431; "Yesterday in Parliament," *Journal of the Chartered Institution of Building Services*, 2 (March 1980): 21–25; M. H. Port, ed., *The Houses of Parliament* (New Haven: Yale University Press, 1976).

37. Bernan, *On the History*, 2:85.

38. Billington, "A Historical Review," 129.

39. Robert Cooke, *The Palace of Westminster* (New York: Burton Skira, 1987), 48.

40. "Yesterday in Parliament," 21.

41. Cooke, *Westminster*, 404.

42. Cooke, *Westminster*, 164.

43. C. H. Blackall, "The Heating and Ventilation of the Houses of Parliament, London," *American Architect and Building News* 16 (20 September 1884): 135.

44. Henry W. Lucy, "The Lungs of the House of Commons," *Cornhill Magazine* 91 (February 1905): 170.

45. Detailed accounts of St. George's Hall are given in William Mackenzie, "The Mechanical Ventilation and Warming of St. George's Hall, Liverpool," *Civil Engineer and Architects' Journal* 27 (1 May 1864): 136–139; Charles R. Honiball, "The Mechanical Ventilation and Warming of St. George's Hall, Liverpool," *Heating and Ventilating Magazine* 4 (October 1907): 15–23; John Olley, "St. George's Hall, Liverpool," *Architect's Journal* 183 (18, 25 June 1986): 36–57, 36–61.

46. Olley, "St. George's Hall" (25 June 1986), 54.

47. Hermann Vetter, "Heating from Roman Times to 1870," *Metal Worker, Plumber and Steam Fitter* 68 (28 September 1907): 59. Translated from *Gesundheits-Ingenieur*.

48. The term "cockle" comes from a word for earthenware vessels, perhaps referring to the custom of slowing the consumption of charcoal in a brazier by inverting a bowl over it.

49. [Joshua] Jebb, "Description of the System of Ventilation and Warming Adopted at the Model Prison, Pentonville," *Mechanics' Magazine* 49 (8 July 1848): 26.

50. Robert Brucemann and Donald Prowler, "Nineteenth Century Mechanical System Designs," *Journal of Architectural Education* 30 (February 1977): 12.

51. Nikolaus Pevsner, *A History of Building Types* (Princeton: Princeton University Press, 1976), 164.

52. W. P. Trowbridge in "Rival Systems of Heating," *North American Review* 138 (c. 1883): 200.

53. A. R. J. Ramsey, "The Thermostat or Heat Governor, an Outline of Its History," *Transactions, Newcomen Society* 25 (1945–1946 and 1946–1947): 63–66.

54. Thomas Box, *A Practical Treatise on Heat* (London: 1876), 150–157.

55. J. A. S. Ritson, "Metal and Coal Mining, 1750–1875," in Charles Singer and others, eds., *A History of Technology* (Oxford: Clarendon Press, 1958), 4:93.

56. Bernan, *On the History*, 2:45.

57. H. Shirley Smith, "Bridges and Tunnels," in Singer and others, *A History of Technology*, 5:519.

58. R. C. Carpenter, "Relative Efficiency of Ventilation by a Chimney and by a Fan," *Heating and Ventilation* 6 (15 February 1897): 21.

59. Dr. Neil Arnott. Quoted in John S. Billings, *Ventilation and Heating* (New York: 1893), 35.

60. Catherine E. Beecher and Harriet Beecher Stowe, *The American Woman's Home* (Boston: 1869; Watkins Glen, N. Y.: American Life Foundation, 1979), 35.

61. Thomas Colley Grattan, quoted in Page Smith, *A People's History of the Ante-Bellum Years: The Nation Comes of Age* (New York: McGraw Hill, 1981), 4:878.

62. George Truman Palmer, "Changing Concepts of Ventilation since the Eighteenth Century," *Transactions, American Society of Civil Engineers* 92 (1928): 1264.

63. Charles L. Hubbard, "Recent Developments in the Theory of Ventilation," *Architectural Record* 41 (January 1917): 53.

64. Quoted in George T. Palmer, "What Fifty Years Have Done for Ventilation," in Mazÿck P. Ravenal, ed., *A Half Century of Public Health*

(New York: American Public Health Association, 1921), 339.

65. Yandell Henderson, "The Influence of the Chemical Composition of Air on Animal Life," *Yale Scientific Monthly* 3 (June 1897): 343.

66. Quoted in Palmer, "Fifty Years," 341.

67. A. K. Klauss and others, "History of the Changing Concepts in Ventilation Requirements," *ASHRAE Journal* 18 (July 1976): 44.

68. Russell Sturgis, *Dictionary of Architecture and Building* (New York: Macmillan, 1901–1902); see "Ventilation."

69. L. Borne, "Construction des hôpitaux," *Construction Moderne* 5 (28 June 1890): 455.

70. In 1893 Billings published *Ventilation and Heating*. He founded *Index Medicus*, a monthly guide to medical publications, and after retirement from the U.S. Army was chief librarian of the New York Public Library.

71. [Dr.] Oppert, "A Short Description of the Plans of Hospitals at Paris, Munich and St. Petersburgh," *Royal Institute of British Architects, Papers read at Session 1867–1868* (London: 1868), 283.

72. George Hill, "The Economy of the Office Building," *Architectural Record* 15 (April 1904): 318.

73. "Lecture before New York Chapter on 'Heating the Skyscraper and Its Problems,' " *Heating and Ventilating Magazine* 12 (January 1915): 45.

12 Air Conditioning

1. Isador Schlesinger, *Der Eiskellerbau in Massiv- und Holz-Construction* (Berlin: 1864), plate 4.

2. Robert Maclay, "The Ice Industry," in *1795–1895, One Hundred Years of American Commerce*, ed. Chauncey Mitchell Depew (New York: 1895), 467.

3. "Ice Trade," in *Boston. Board of Trade, Third Annual Report of the Government* (Boston: 1857), 81.

4. S. E. Barnett, "The American Ice Harvests," *ASHRAE Journal* 18 (July 1976): 34–35.

5. W. J. Rushton, "Early Days of the Manufacture of Ice," *Ice and Refrigeration* 51 (November 1916): 151.

6. David L. Fiske, "Time I Speak of—," *Refrigerating Engineering* 28 (December 1934): 290.

7. W. R. Woolrich, *The Men Who Created Cold: A History of Refrigeration* (New York: Exposition Press, 1967), 21–23.

8. Greville Bathe and Dorothy Bathe, *Jacob Perkins* (Philadelphia: Historical Society of Pennsylvania, 1943), 148–151.

9. Conflicting information on Gorrie's life and work is considered in Bernard Nagengast, "John Gorrie: Pioneer of Cooling and Ice Making," *ASHRAE Journal* 33 (January 1991): 552–561.

10. W. R. Woolrich, "The History of Refrigeration: 220 Years of Mechanical and Chemical Cold: 1748–1968," *ASHRAE Journal* 11 (July 1969): 34.

11. Alexander C. Twining, *The Fundamental Ice-making Invention* (Washington: 1870), 14.

12. "What the Refrigerants Have Contributed," *Refrigerating Engineering* 28 (December 1934): 305.

13. Woolrich, *Men Who Created Cold*, 45.

14. "What the Refrigerating Machine Companies Have Contributed," *Refrigerating Engineering* 28 (December 1934): 295.

15. A. Jouglet, "On the Various Systems of Cooling the Air," *Practical Magazine* 1, no. 6 (1873): 452. Translated from *Moniteur Scientifique*.

16. Jouglet, "Various Systems," 454–455.

17. G. Richard Ohmes and Arthur K. Ohmes, "Early Comfort Cooling Plants," *Heating, Piping and Air Conditioning* 8 (June 1936): 310.

18. Jouglet, "Various Systems," 455.

19. Ohmes and Ohmes, "Early Comfort Cooling," 310–311.

20. John J. Harris, "Cooling Auditoriums by the Use of Ice," *Heating and Ventilating Magazine* 4 (September 1907): 30.

21. R. O. Doremus, *North American Review* (May 1893). Quoted in Ohmes and Ohmes, "Early Comfort Cooling," 310. Quoting 1902 article in New York *Evening Journal*.

22. "Cooling the New York Stock Exchange," *Metal Worker, Plumber and Steam Fitter* 53 (5 August 1905): 56.

23. Bernard A. Nagengast, "Alfred Wolff—HVAC Pioneer," *ASHRAE Journal* 32 (January 1990): S76.

24. Margaret Ingels, *Willis Haviland Carrier, Father of Air Conditioning* (Garden City: Country Life Press, 1952), 9.

25. Ingels, *Carrier*, 20–21.

26. Herman Worsham, "The Milam Building," *Heating, Piping and Air Conditioning* 1 (July 1929): 182.

27. *Philadelphia Evening Journal*, 1 June 1859. Quoted in Elliott Harrington, "Air Conditioning for Comfort and Health," *Journal of the Franklin Institute* 215 (June 1933): 656.

28. "Progress during 1932 in Comfort Cooling with Ice," *Refrigerating Engineering* 24 (November 1932): 265.

29. "What the Refrigerants Have Contributed," 307.

30. Bernard A. Nagengast, "The Revolution in Small Vapor Compression Refrigeration," *ASHRAE Journal* 18 (July 1976): 38.

31. E. Macleod, "The History of Room Air Conditioners," *ASHRAE Journal* 18 (July 1976): 41.

32. B. A. Nagengast, "Room Coolers prior to 1930 and the Technical Impediments to Their Development," *ASHRAE Transactions* 92 (1986): 379–380.

33. Stuart W. Cramer, *Useful Information for Cotton Manufacturers*, 2d ed. (Charlotte, N.C.: Queen City Printing Company, 1909), 4:1396.

13 Elevators and Escalators

1. R. J. Forbes, "Power," in Charles Singer and others, eds., *A History of Technology* (Oxford: Clarendon Press, 1956), 2:603.
2. John Albury Bryan, *Evolving the Elevator*, Because of Iron, no. 1 (St. Louis: American Lithographing Company, 1947), 2.
3. Lloyd Morris, *Incredible New York* (New York: Random House, 1951), 11.
4. Paul Anderson, ed., *Tell Me about Elevators* (New York: Otis Elevator Company, 1975), 8.
5. *Elevator World* 11 (September 1963): 20.
6. "Steam versus Stairs: The Movable Room in the Fifth-avenue Hotel," *New York Times*, 23 January 1860.
7. Stephen Birmingham, *The Grandees* (New York: Harper and Row, 1971), 275.
8. "Steam versus Stairs."
9. "Notes on Elevators," *American Architect and Building News* 7 (5 June 1880): 246.
10. Robert M. Sheridan, "The Evolution of the Elevator," *Journal of the Association of Engineering Societies* 9 (December 1890): 585.
11. Robert M. Vogel, *Elevator Systems of the Eiffel Tower*, U.S. National Museum, Bulletin 228 (Washington: Smithsonian Institution, 1961), 9–10.
12. Charles E. Emery, "Modern High Speed Elevators," *Sibley Journal of Engineering* 12 (November 1897): 56.
13. Thomas E. Brown, Jr., "The American Passenger Elevator," *Engineering Magazine* 5 (June 1893): 337.
14. Charles H. Kloman, "The Growth and Development of the Elevator Industry," *Cassier's Magazine* 32 (September 1907): 390.
15. Vogel, *Eiffel Tower*, 10–11.
16. Thomas E. Brown, "Passenger Elevators," *American Architect and Building News* 86 (12 November 1904): 51.
17. Sheridan, "Evolution," 587.
18. W. S. Huyette, "A Comparison between Vertical and Horizontal Cylinder Hydraulic Elevator Engines," part 1, *Power* 17 (July 1897): 8.
19. Horace Gale, "Efficiency of Hydraulic Passenger Elevators," *Industry* 5 (September 1892): 733.
20. Henry D. James, "The Selection and Installation of Hydraulic Elevators," *Engineering News* 48 (31 July 1902): 74.
21. George I. Alden, *The Plunger Elevator*," *Transactions, American Society of Mechanical Engineers* 20 (May 1899): 639.
22. Brown, Jr., "American Passenger Elevator," 339–340.
23. *Elevator World* 11 (September 1963): 31.
24. "The Electric Light [sic] of Dr. Siemens," *Builder* 39 (25 December 1880): 771.
25. Brown, "Passenger Elevators," (10 December 1904), 83–84.
26. Charles G. Darrach, "Office-building Elevators," *American Architect and Building News* 74 (5 October 1901): 5.
27. Frank J. Sprague, "Electric Elevators, with Detailed Description of Special Types," *Transactions, American Institute of Electrical Engineers* 13 (January 1896): 12–13.
28. Brown, Jr., "The American Passenger Elevator," 341.
29. Charles R. Pratt, "Elevators," *Scientific American Supplement* (19 August 1899): 19761.
30. Brown, "Passenger Elevators," (10 December 1904), 91.
31. Harrison P. Reed, "Electric Power Application to Passenger and Freight Elevators," *Journal of the American Institute of Electrical Engineers* 41 (January 1922): 61.
32. Percival Robert Moses, "Standards of Practice in Electric-Elevator Installation," *Engineering Magazine* 14 (December 1897): 479.
33. Bayrd Still, *Mirror for Gotham* (New York: New York University Press, 1956), 206.
34. "Vertical Transportation," *Scientific American* 88 (4 April 1903): 238–239.
35. "Electric Elevators," *Electrical World* 29 (3 April 1897): 447.
36. John Howe, "Going Up!" *American Magazine* 98 (August 1924): 20.
37. Reginald Pelham Bolton, "Elevators for Tall Office Buildings," *Cassier's Magazine* 21 (January 1902): 232.
38. H. Marryat, "Electric Passenger Lifts," *Journal of the Institution of Electrical Engineers* 62 (April 1924): 326.
39. Reed, "Electric Power Application," (November 1922), 829. Comment of David Lindquist in discussion.
40. "Elevators in Private Houses," *Architecture and Building* 28 (19 April 1898): 124. Quoted from *New York Evening Post*.
41. F. P. Boone and S. W. Palmer, "The Modern Passenger Elevator," *National Engineer* 19 (April 1915): 188.
42. "Elevators in Private Houses," 124.
43. Marryat, "Passenger Lifts," 339.
44. "Notes on Elevators," 245.
45. "Elevator Shafts, *American Architect and Building News* 7 (7 February 1880): 44.
46. "Notes on Elevators," 245.
47. "Notes on Elevators," 245.
48. Editorial note, *American Architect and Building News* 7 (13 March 1880): 101.
49. Herbert T. Wade, "The Elevator Installation of the Metropolitan Life Tower," *Scientific American* 110 (30 April 1910): 356.
50. *Figaro* (Paris), translated in *American Architect and Building News* (23 March 1878): 104. Quoted in Bryan, *Evolving the Elevator*, 6–7.
51. W. Barnet LeVan, "Fall of a Hydraulic elevator at the Grand Hotel in Paris, France," *Journal of the Franklin Institute* 105 (May 1878): 346.

52. "A New Pneumatic Safety Device for High-speed Passenger Elevators," *Engineering News* 42 (19 October 1899): 263.

53. *Iron Age*, 4 August 1898. Quoted in Pratt, "Elevators," (12 August 1899), 19747.

54. "A High Drop Test of an Elevator Safety Air Cushion," *Engineering News* 48 (9 October 1902): 295.

55. Richard Snowdon, "Parachuting Down an Elevator Shaft," *Technical World Magazine* 18 (November 1912): 358–359.

56. "Six Hundred Foot Drop Tests Woolworth Building Elevators," *Engineering Record* 70 (5 September 1914): 267.

57. Pratt, "Elevators," 19746.

58. Pratt, "Elevators," 19746.

59. Fred C. Floyd, "Modern Elevators," *Architecture and Building* 28 (29 January 1898): 40.

60. Clayton W. Old, "Effectiveness of Mechanical Elevator Interlocks in Prevention of Accidents," *Monthly Labor Review* 16 (April 1923): 1.

61. Trenton, N.J., *Times* and Chicago *Tribune*. Quoted in Old, "Effectiveness," 2–3.

62. "First 'Dual' Elevator is Run Successfully," *New York Times*, 14 January 1931.

63. *Progressive Architecture* 52 (December 1971): 71.

64. "Tandem Elevators Save Space, Improve Service," *Architectural Forum* 133 (September 1970): 25.

65. "The Reno Continuous Passenger Elevator," *Engineering News* 28 (25 August 1892): 188.

66. "The Reno Inclined Elevator in a Department Store," *Electrical Engineer* 26 (7 July 1898): 3.

67. *Vanity Fair* (London). Quoted in Frank Creden, ed., *Tell Me About Escalators* (New York: Otis Elevator Company, 1974), 9.

68. J. Laverchère, "Les chemins élévateurs à l'Exposition de 1900," *Génie Civile* 37 (14 July 1900): 183–186.

69. "The 'Escalator' or Continuous Elevator," *Engineering* 70 (30 November 1900): 699.

70. Charles Jullien, "Installation d'un escalier à marches mobiles dans la Gare du Quai D'Orsay," *Revue Générale des Chemins de Fer* 31 (June 1908): 391.

71. S. P. Ring, "The Escalator for Department Stores," *Architecture and Building* 44 (December 1912): 512.

72. Otto Friedrich, *Before the Deluge* (New York: Harper and Row, 1972), 165.

14 Fire Protection

1. Quoted in William Bell Dinsmoor, *The Architecture of Ancient Greece*, 3d ed. (London: Batsford, 1950), 150.

2. Steven Runciman, *Byzantine Civilization* (New York: Meridian Books, 1956), 149.

3. C. C. Knowles and P. H. Pitt, *The History of Building Regulation in London, 1189–1972* (London: Architectural Press, 1972), 8.

4. Knowles and Pitt, *Building Regulation*, 20.

5. E. L. Jones, "The Reduction of Fire Damage in Southern England: 1660–1850," *Post-Medieval Archaeology* 2 (1968): 146.

6. "Boston Building Ordinances, 1631–1714," *Journal of the Society of Architectural Historians* 20 (May 1961): 90–91.

7. Quoted in Jacob Landy, *The Architecture of Minard Lafever* (New York: Columbia University Press, 1970), 51.

8. *A History of Real Estate, Building and Architecture During the Last Quarter of a Century* (New York: 1898; New York: Arno Press, 1967), 288–289.

9. Quoted in John Summerson, *Georgian London* (New York: Charles Scribner's Sons, 1946), 33.

10. P. G. M. Dickson, *The Sun Insurance Office: 1710–1960* (London: Oxford University Press, 1960), 91.

11. Christopher Hibbert, *London* (London: Longmans Green, 1969), 150.

12. John Bainbridge, *Biography of an Idea: The Story of Mutual Fire and Casualty Insurance* (Garden City: Doubleday, 1952), 35–36.

13. S. B. Hamilton, *A Short History of the Structural Fire Protection of Buildings* (London: HMSO, 1958), 9.

14. John J. Webster, "Fire-proof Construction," *Minutes of Proceedings, Institution of Civil Engineers* 105 (1891): 263.

15. Quoted in Hamilton, *Short History*, 7–8.

16. Jennifer Tann and L. D. W. Smith, "Early Fireproof Housing in a Staffordshire Factory Village," *Post-Medieval Archaeology* 6 (1972): 195.

17. *Fighting Fire: The Great Fires of History* (Hartford, Connecticut: 1873), 210–211.

18. Harry Chase Brearley, *Fifty Years of a Civilizing Force* (New York: Frederick A. Stokes, 1916), 47.

19. Address by Henry Ward Beecher, 10 November 1872. Quoted in *Fighting Fire*, 396, 403.

20. "Ambrose Godfrey's Fire-extinguisher," *American Architect and Building News* 30 (6 December 1890): 149. Reprinted from *Journal of the Society of Arts*.

21. *Fighting Fire*, 610–611.

22. *Fighting Fire*, 617.

23. Edward V. French, *1860—Fifty Years—1910. Arkwright Mutual Fire Insurance Company* (Boston: 1912), 13.

24. Charles A. Daubney, "Fire-escapes in American Commercial Buildings," *American Architect and Building News* 79 (24 January 1903): 27. *Reprinted from Journal, Royal Institute of British Architects.*

25. Thomas Bolas, "Fire Risks Incidental to Electric-lighting," *American Architect and Building News* 11 (27 May 1882): 244. Reprinted from *Journal of the Society of Arts*.

26. Brearley, *Fifty Years*, 81–82.

27. Edward Atkinson, "Fire Risks on Tall Office Buildings," *Engineering Magazine* 3 (May 1892): 155.

28. John Wellborn Root, "Fire-Insurance and Architecture" in Donald Hoffman, *The Meaning of Architecture: Buildings and Writings of John Wellborn Root* (New York: Horizon Press, 1967), 155.

29. Edward Atkinson, "Lessons of the Park Place Disaster," *Engineering Magazine* 2 (November 1891): 148.

30. Webster, "Fire-proof Construction," 251–252.

31. Francis C. Moore, *How to Build 'Fireproof'*, Publications of the British Fire Prevention Committee, no. 10 (London: 1898), 6.

32. Moore, *How to Build*, 24.

33. Moore, *How to Build*, 21.

34. "Burnt Clay Fireproofing and Its Substitutes," *Architectural Record* 8 (July–September 1898): 112.

35. Horace Cubitt, "A Comparison of English and American Building Laws," *American Architect and Building News* 89 (19, 26 May 1906): 180. Reprinted from *Builder's Journal.*

36. Walter Emden, "Theatres and Fireproof Construction," *Journal of the Society of Arts* 36 (27 January 1888): 222.

37. "L'incendie de l'Opéra-Comique," *Construction Moderne* 3 (29 October 1887): 27–28.

38. "Les mesures de sécurité dans les théâtres de Paris," *Construction Moderne* 3 (15 September 1888): 584–585.

39. Emden, "Theatres," 222.

40. Daubney, "Fire-Escapes," 27.

41. Edwin O. Sachs, "Fire Protection in Europe," *American Architect and Building News* 59 (1 January 1898): 6. Reprinted from *Engineering.*

42. Hamilton, *Short History*, 24.

43. Daubney, "Fire-Escapes," 27.

44. Daubney, "Fire-Escapes" (31 January 1903), 35.

45. Webster, "Fire-proof Construction," 249.

15 Structural Engineering

1. Vitruvius, *The Ten Books of Architecture*, trans. Morris Hicky Morgan (1914; New York: Dover Publications, 1960), 2.9.6–8.

2. Vitruvius, *Ten Books*, 3.3.1.

3. John Harvey, *The Medieval Architect* (London: Wayland Publishers, 1972), 103.

4. Robert Mark, *Experiments in Gothic Structure* (Cambridge: MIT Press, 1982), 119.

5. Leon Battista Alberti, *Architectura* (trans. James Leoni, 1755). Quoted in Friedrich Klemm, *A History of Western Technology* (Cambridge: MIT Press, 1964), 121.

6. William Barclay Parsons, *Engineers and Engineering in the Renaissance* (Baltimore: Williams and Wilkins, 1939), 71.

7. J. P. Richter, ed., *The Literary Works of Leonardo da Vinci* (London: Oxford University Press, 1939). Quoted in Klemm, *A History of Western Technology*, 127–128.

8. Galileo Galilei, *Discorsi e dimostrazioni matematiche*, trans. Henry Crew and Alfonso de Salvio (New York: Macmillan, 1914). Quoted in Klemm, *A History of Technology*, 177.

9. S. B. Hamilton, "The Historical Development of Structural Theory," Structural Paper no. 32, *Proceedings, Institution of Civil Engineers, Part 3: Engineering Division* 1 (1952): 417.

10. Stephen P. Timoshenko, *History of the Strength of Materials* (New York: McGraw-Hill Book Co., 1953; New York: Dover Publications, 1983), 21–23.

11. A. Wolf, *A History of Science, Technology, and Philosophy in the 16th and 17th Centuries* (New York: Macmillan, 1950; New York: Harper and Brothers, 1959), 2:477.

12. Robert Hooke, quoted in Timoshenko, *Strength of Materials*, 20.

13. Jacques Heyman, "The Stone Skeleton," *International Journal of Solids and Space* 2 (1966): 251.

14. E. L. Lasier, "Comparison of Column Formulae," *Engineering Record* 68 (12 July 1913): 41–42.

15. "Column Formulas in Relation to the Practical Column," *Engineering News* 57 (3 January 1907): 15–20.

16. Editorial note appended to Robins Fleming, "The Choice of a Column Formula," *Journal of the Western Society of Engineers* 31 (October 1926): 407.

17. Frederick B. Artz, *The Development of Technical Education in France, 1500–1850* (Cambridge: The Society for the History of Technology, MIT Press, 1966), 153.

18. Aubrey F. Burnstall, *A History of Mechanical Engineering* (Cambridge: MIT Press, 1965), 202.

19. T. K. Derry and Trevor I. Williams, *A Short History of Technology* (London: Oxford University Press, 1970), 404. Artz, *Technical Education*, 245.

20. *Engineering News-Record* (16 September, 18 November, 16 December 1920).

21. P. B. McDonald, "Structural Steel Practice in the United States and Europe," *Mining and Scientific Press* 124 (18 February 1922): 221.

22. David P. Billington, *The Tower and the Bridge* (New York: Basic Books, 1983), 152.

23. S. B. Hamilton, "Building and Civil Engineering Construction," in Charles Singer and others, eds., *A History of Technology* (Oxford: Clarendon Press, 1958), 4:461.

24. Denis Smith, "David Kirkaldy (1820–1897) and Engineering Materials Testing," *Transactions, Newcomen Society* 52 (1980–1981): 52.

25. Timoshenko, *Strength of Materials*, 279.

26. "Factors of Safety," *Engineering News* 56 (6 September 1906): 258.

27. "Factors of Safety," 258.

28. Hans Straub, *A History of Civil Engineering*, trans. Erwin Rockwell (Cambridge: MIT Press, 1964), 114.

29. Frank A. Randall, Jr., "Historical Notes on Structural Safety," *Proceedings of the Annual Convention, American Concrete Institute* 70 (1973): 674.

30. Randall, "Historical Notes," 675.

31. Randall, "Historical Notes," 677.

32. C. C. Schneider, "The Structural Design of Buildings," *Transactions, American Society of Civil Engineers* 54 (June 1905): 372.

33. Frank W. Skinner, "The Development of Building Foundations," *Engineering Record* 57 (4 April 1908): 412.

34. Ralph B. Peck, *History of Building Foundations in Chicago*, University of Illinois, Engineering Experiment Station, Bulletin Series, no. 373 (Urbana-Champaign: University of Illinois, 1948), 11.

35. Harriet Monroe, *John Wellborn Root: Architect* (Boston: 1896). Quoted in Peck, *Building Foundations*, 22.

36. Skinner, "Building Foundations," 414.

37. Ralph B. Peck, Walter E. Hanson, and Thomas G. Thornburn, *Foundation Engineering* (New York: John Wiley and Sons, 1953), 164.

38. Peck, *Building Foundations*, 17.

39. Skinner, "Building Foundations," 414.

40. S. Anglin, "Foundations of Buildings," *Journal of the Royal Institute of British Architects* 11 (November 1899): 15.

41. Skinner, "Building Foundations," 413.

42. Charles Sooysmith, "Concerning Foundations for Heavy Buildings in New York City," *Transactions, American Society of Civil Engineers* 35 (July 1896): 463.

43. Charles Sooysmith, "Foundation Construction for Tall Buildings," *Engineering Magazine* 13 (April 1897): 24.

44. Sooysmith, "Foundation Construction," 29.

45. Guy B. Waite, "Wind Bracing in High Buildings," *Transactions, American Society of Civil Engineers* 33 (March 1895): 191.

46. Waite, "Wind Bracing," 193.

47. Henry H. Quimby, "Wind Bracing in High Buildings," *Transactions, American Society of Civil Engineers* 27 (September 1892): 222.

48. R. Fleming, "Wind Bracing Requirements in Municipal Building Codes," *Engineering News* 73 (11 March 1915): 485.

49. A. G. Davenport, "Wind Engineering," in P. A. Bulson, J. B. Caldwell, and R. T. Stevens, eds., *Engineering Structures* (Bristol: University of Bristol Press, 1983), 265–266.

50. T. M. Charlton, *A History of Theory of Structures in the Nineteenth Century* (Cambridge: Cambridge University Press, 1982), 7.

51. Charlton, *Theory of Structures*, 7–8.

52. Charlton, *Theory of Structures*, 28–32. S. B. Hamilton, "Mechanics in Theory and Practice," *Chartered Mechanical Engineer* 14 (April 1967): 164.

53. Henry J. Cowan, *Science and Building* (New York: John Wiley and Sons, 1978), 90–91. Henry J. Cowan, *Design of Reinforced Concrete Structures*, 2d ed. (Englewood Cliffs: Prentice Hall, 1989), 87.

16 Acoustics

1. Vitruvius, *The Ten Books of Architecture*, trans. Morris Hicky Morgan (1914; New York: Dover Publications, 1960), 5.5.

2. Daniel Drocourt, "L'acoustique au moyen-âge: les résonateurs de l'Eglise Saint-Victor à Marseilles," *Archeologia* 40 (May-June 1971): 29.

3. F. Canac, "On the Acoustics of Grecian and Roman Theatres," *Journal of the Royal Institute of British Architects* 56 (July 1949): 412–414.

4. George Sarton, *Introduction to the History of Science* (Baltimore: Williams and Wilkins, 1948), 3:1569–1570.

5. Drocourt, "L'acoustique au moyen-âge," 31.

6. E. G. Richardson, "Notes on the Development of Architectural Acoustics, Particularly in England," *Journal of the Royal Institute of British Architects* 52 (October 1945): 353.

7. Christopher Wren, *Parentalia* (1750; Farnborough: Gregg Press, 1965), 320.

8. J. B. Upham, *Acoustic Architecture* (New Haven: 1853), 29.

9. The voice range given by Vern O. Knudsen, (*Architectural Acoustics* [New York: John Wiley and Sons, 1932]) indicates 90 percent clarity of speech in an area about 10 percent smaller than that stated by Wyatt; however, Knudsen's tests were made out of doors.

10. George W. Williams, "Robert Mills' Contemplated Addition to St. Michael's Church, Charleston and Doctrine of Sounds," *Journal of the Society of Architectural Historians* 12 (March 1953): 27.

11. Benjamin Latrobe, "Acoustics," in *Edinburgh Encyclopedia* (Philadelphia: 1832), 120.

12. Latrobe, "Acoustics," 123.

13. Glenn Brown, *History of the United States Capitol* (Washington: GPO, 1900–1903), 70.

14. Talbot Hamlin, *Benjamin Latrobe* (New York: Oxford University Press, 1955), 558.

15. T. Roger Smith, "On Acoustics," *Royal Institute of British Architects. Papers read at sessions 1860–1861* (London: 1861), 79.

16. Smith, "On Acoustics," 80.

17. Smith, "On Acoustics," 80.

18. "Paraboloid Sound Reflector at the Senate House, Calcutta," *Indian Engineering* 1 (21 March 1896): 186.

19. Henry W. Arey, *The Girard College and Its Founder* (Philadelphia: 1857), 67–72.

20. Quoted in Upham, *Acoustic Architecture*, 16.

21. Upham, *Acoustic Architecture*, 17.

22. Smith, "On Acoustics," 95.

23. *Builder* 215 (20 March 1847): 129.

24. A relationship between musical harmonies and visual proportions had been stated in Vitruvius's *Ten*

Books on Architecture (1st century B.C.), and the notion had persisted into the nineteenth century.

25. "Improvement and Distribution of Sound," *Builder* 39 (14 August 1880): 221.

26. Luc Gallicane, "Les lois fondamentales de l'acoustique déduites de l'expérience," *Revue Générale des Sciences Pures et Appliqués* 44 (August 1933): 457.

27. Charles Garnier, *L'Opéra* (Paris: 1880). Quoted in Leo Beranek, *Music, Acoustics and Architecture* (New York: Wiley, 1962), 237.

28. P. Planat, "L'acoustique des salles publiques," *La Construction Moderne* 18 (11 April 1903): 327.

29. W. W. Jacques, "Effect of the Motion of the Air within an Auditorium upon its Acoustic Qualities," *Journal of the Franklin Institute* 106 (December 1878): 394–395.

30. Wallace C. Sabine, "Architectural Acoustics," *American Architect* 98 (10 August 1910): 43.

31. Sabine, "Architectural Acoustics" (1910), 44.

32. In spite of the failure of Russell's design of the ship *Great Eastern*, the "viewy Scotchman" (as he was called in the *London Standard's* obituary) continued to be respected, was a member of the initial committee to organize the Great Exhibition of 1851, and designed the dome of the Vienna Exhibition, 360 feet in diameter.

33. [John] Scott Russell, "On the Construction of Buildings with Reference to Sound (Summary)," *Building News* 5 (26 Noevember 1858): 1178.

34. Russell, "Construction of Buildings," (3 December 1858), 1195.

35. Russell, "Construction of Buildings," (3 December 1858), 1196.

36. Quoted in Richardson, "Notes," 353.

37. Joseph Henry, "On Acoustics Applied to Public Buildings," in *Annual Report of the Board of Regents of the Smithsonian Institution* (Washington: 1857), 225–226.

38. Knudsen, *Architectural Acoustics*, 515.

39. Dankmar Adler, "Theatres," *American Architect and Building News* 22 (29 October 1887): 207.

40. Dankmar Adler, "The Theater," *Prairie School Review* 2 (2d qtr. 1965): 26–27.

41. Paul R. Baker, *Richard Morris Hunt* (Cambridge, MIT Press, 1980), 378.

42. William Dana Orcutt, *Wallace Clement Sabine: A Study in Achievement* (Norwood, Mass.: Plimpton Press, 1933), 103–104.

43. Charles Moore, *The Life and Times of Charles Follen McKim* (1929; New York: Da Capo Press, 1970), 111.

44. Wallace C. Sabine, "Architectural Acoustics," *American Architect and Building News* 62 (26 November 1898): 73.

45. Wallace C. Sabine, "Architectural Acoustics," *American Architect and Building News* 68 (16 June 1900): 83.

46. Sabine, "Architectural Acoustics" (1900), 84.

47. Hope Bagenal, "The Leipzig Tradition in Concert Hall Design," *Journal of the Royal Institute of British Architects* 36 (21 September 1929): 756.

48. F. Honoré, "La solution française d'un grand problème d'acoustique," *Illustration* 85 (10 September 1927): 228.

49. Albert Mousset, "L'ancien et le nouveau Trocadéro," *Architecture* 49 (15 April 1936): 109–110.

50. Honoré, "Solution française," 229.

51. Le Corbusier, "La Salle Pleyel: une preuve de l'évolution architecturale," *Cahiers d'Art* 3 (1928): 89.

52. Le Corbusier, "Salle Pleyel," 89.

53. Henry Prunieres, "New Paris Concert Hall," *New York Times*, 30 October 1927.

54. Vern O. Knudsen, quoted in Beranek, *Music, Acoustics and Architecture*, 455.

55. Le Corbusier, *Le Corbusier et Pierre Jeanneret: Oeuvre complète de 1910–1929*, 5th ed. (Zurich: Editions d'Architecture Erlenbach, 1948), 166.

56. F. M. Osswald, "The Acoustics of the Large Assembly Hall of the League of Nations, Geneva, Switzerland," *American Architect* 134 (20 December 1928): 840.

57. P. H. Parkin and others, "The Acoustics of the Royal Festival Hall, London," *Acustica* 3 (1953): 9–10.

58. Parkin and others, "Royal Festival Hall," 19.

59. P. H. Parkin and K. Morgan, " 'Assisted Resonance' in the Royal Festival Hall, London," *Journal of Sound and Vibration* 2 (January 1965): 75.

60. Parkin and Morgan, "Assisted Resonance," 77.

61. Harold C. Schonberg, "Assisted Resonance," *New York Times*, 20 May 1964.

62. Andrew Porter, "How It Played in Peoria and Elsewhere," *New Yorker*, 18 October 1982, 168.

Select Bibliography

The sources listed below include historical studies, definitive works on the subjects, and handbooks that clarify the knowledge at their time. Publishers are listed for all monographs of the twentieth century.

Acoustics

Forsyth, Michael. *Buildings for Music.* Cambridge: Cambridge University Press, 1985.

Knudsen, Vern O. *Architectural Acoustics.* New York: John Wiley and Sons, 1932.

Lyon, Gustave. *L'acoustique architecturale.* Paris: Editions Film et Technique, 1932.

Russell, John Scott. "Elementary Considerations of Some Principles in the Construction of Buildings Designed to Accommodate Spectators and Auditors." *Edinburgh New Philosophical Journal* 27 (July 1839): 131–136.

Sabine, Paul. *Acoustics and Architecture.* New York: McGraw-Hill Book Co., 1932.

Sabine, Wallace C. "Architectural Acoustics." *American Architect and Building News* 68 (7, 21 April; 5, 12, 19 May; 9, 16 June 1900): 3–5, 19–22, 35–37, 43–45, 59–61, 75–76, 83–84.

Smith, T[homas] Roger. *A Rudimentary Treatise on the Acoustics of Public Buildings.* London: 1861.

Upham, J. B. *Acoustic Architecture, or the Construction of Buildings with Reference to Sound and the Best Musical Effect.* New Haven: 1853.

Air Conditioning

Anderson, Oscar Edward, Jr., *Refrigeration in America: A History of a New Technology and Its Impact.* Princeton: Princeton University Press (for University of Cincinnati), 1953.

Feldman, A. M. "Cooling Systems of Buildings." *ASHRAE Journal* 8 (March 1922): 377–392.

Fiske, David L. "The Origins of Air Conditioning." *Refrigeration Engineering* 22 (March 1934): 122–126.

Ingels, Margaret. *Willis Haviland Carrier, Father of Air Conditioning.* Garden City: Country Life Press, 1952.

Jouglet, A. "On the Various Systems of Cooling the Air; and Their Application in Factories, Public Buildings, and Private Houses." *Practical Magazine* 1, no. 6 (1873): 450–463. Translated from *Moniteur Scientifique.*

Ohmes, G. Richard, and Arthur K. Ohmes. "Early Comfort Cooling Plants." *Heating, Piping and Air Conditioning* 8 (June 1936): 310–311.

Thévenot, Roger. *A History of Refrigeration throughout the World.* Paris: International Institute of Refrigeration, 1979.

Cements

Clauss, Adolf. "History of Mortars and Cements." *Stone* 4 (May 1892): 508–509.

Davis, Arthur Charles. *A Hundred Years of Portland Cement, 1824–1924.* London: Concrete Publications Ltd., 1924.

Francis, A. J. *The Cement Industry 1796–1914: A History.* London: David and Charles, 1977.

Gooding, P., and P. E. Halstead. "The Early History of Cement in England." *Proceedings, Third International Symposium on the Chemistry of Cement, 1952.* London: Cement and Concrete Association, 1954.

Halstead, P. E. "The Early History of Portland Cement." *Transactions of the Newcomen Society* 34 (1961–1962): 37–53.

Lesley, Robert W. *History of the Portland Cement Industry in the United States.* Chicago: International Trade Press, Inc., 1924.

Redgrave, Gilbert R., and Charles Spackman. *Calcareous Cements: Their Nature, Manufacture and Uses.* London: Charles Griffin and Co., 1905.

Skempton, A. W. "Portland Cements 1843–1887." *Transactions of the Newcomen Society* 35 (1962–1963): 117–152.

Elevators

Baxter, William, Jr. *Hydraulic Elevators.* Chicago: Engineer Publishing Co., 1905.

Brown, Thomas E. "Passenger Elevators." *American Architect and Building News* 86 (12, 19, 26 November; 10, 17 December 1904): 51–54, 67–71, 78–79, 83–86, 91–94.

Bryan, John Albury. *Evolving the Elevator.* Because of Iron, no. 1. St. Louis: American Lithographing Co., 1947.

Jallings, John H. *Elevators.* Chicago: American Technical Society, 1916.

Kloman, Charles H. "The Growth and Development of the Elevator Industry." *Cassier's Magazine* 32 (September 1907): 389–404.

Marryat, H. "Electric Passenger Lifts." *Journal of the Institution of Electrical Engineers* 62 (April 1924): 325–341, 342–349.

Fire Protection

Espie, J. J. *The Manner of Securing All Sorts of Buildings from Fire.* Trans. L. Dutens. London: 1756.

Hamilton, S. B. *A Short History of the Structural Fire Protection of Buildings.* London: HMSO, 1958.

Moore, Francis C. *How to Build 'Fireproof'.* Publications of the British Fire Prevention Committee, no. 10. London: 1898.

Sturgis, Russell, "An Unscientific Enquiry into Fireproof Building." *Architectural Record* 9 (January 1900): 229–253.

Webster, John J. "Fire-proof Construction." *Minutes of the Proceedings, Institution of Civil Engineers* 105 (1890–1891): 249–288.

Wight, Peter B. "Origin and History of Hollow Tile Fire-Proof Floor Construction." *Brickbuilder* 6 (March, April, May, July 1897): 53–55, 73–75, 98–99, 149–150.

Wight, Peter B. "The Use of Burned Clay Products in the Fireproofing of Buildings." *American Architect and Architecture* 90 (1, 8, 15 September 1906): 67–68, 75–78, 84–86.

Glass

Chance, Henry. "On the Manufacture of Crown and Sheet Glass." *Journal of the Society of Arts* 4 (15 February 1856): 222–231.

Davis, Pearce. *The Development of the American Glass Industry.* Cambridge: Harvard University Press, 1949.

Douglas, R. W., and Susan Frank. *A History of Glassmaking.* Henley-on-Thames: G. T. Foulis and Co., 1972.

Hollister, Paul. "The Glazing of the Crystal Palace." *Journal of Glass Studies* 16 (1974): 95–110.

Turner, W. E. S. "The Modern Production of Sheet Glass." *Journal of the Royal Society of Arts* 73 (24 July 1925): 821–837.

Weeks, Joseph D. "Report on the Manufacture of Glass." In U.S. Census Office, *Report on Manufacturing in the U.S.*, vol. 2. 10th Census (1880).

Heating and Ventilation

Arnott, Neil. *On Warming and Ventilating.* London: 1838.

Bernan, Walter. *On the History and Art of Warming and Ventilating Rooms and Buildings.* 2 vols. London: 1845.

Billings, John S. *Ventilation and Heating.* New York: 1893.

Billington, Neville S. "A Historical Review of the Art of Heating and Ventilating." *Architectural Science Review* 2 (November 1959): 118–130.

Carpenter, Rolla C. *Heating and Ventilating Buildings.* New York: John Wiley and Sons, 1902.

Dufton, A. F. "Early Application of Engineering to the Warming of Buildings." *Transactions of the Newcomen Society* 21 (1940–1941): 99–107.

Planat, P. *Chauffage et ventilation des lieux habités.* Paris: 1880.

Reid, David Boswell. *Illustrations of the Theory and Practice of Ventilation.* London: 1844.

Vetter, Hermann. "Heating from Roman Times to 1870." *Metal Worker, Plumber and Steam Fitter* 68 (28 September 1907): 58–60; (5 October 1907): 49–51. Translated from *Gesundheits-Ingenieur* 30 (31 August 1907): 10–25.

Wright, Lawrence, *Home Fires Burning.* London: Routledge and Kegan Paul, 1964.

Iron and Steel

Ashcroft, J. W. "Notes on the Use of Iron and Steel in Building Construction." *Iron* (27 January, 10 February 1893): 74–75, 114–115.

"The Development of Iron and Steel Roof Design." *Builder* 91 (22, 29 September 1906): 337–339, 363–365.

Fleming, Robins. "Evolution of the Steel Skeleton Type of Building." *Proceedings, Engineers' Society of Western Pennsylvania* 43 (February 1927): 1–26.

Gale, W. K. V. *Iron and Steel.* Industrial Archeology Series, no. 2. New York: Humanities Press, 1969.

Hamilton, S. B. "The Use of Cast Iron in Building." *Transactions of the Newcomen Society* 21 (1940–1941): 139–155.

Scott, W. Basil. "Constructional Steelwork: A Short History." *Architects' Journal* 68 (4, 11 July 1928): 19–23, 55–57.

Temin, Peter. *Iron and Steel in Nineteenth-Century America: An Economic Inquiry.* Cambridge: MIT Press, 1964.

Vierendeel, Arthur. *La construction architecturale en fonte, fer et acier.* Louvain: A. Uystpruyst, 1903.

Walmisley, Arthur T. *Iron Roofs.* London: 1888.

Woodworth, R. B. "Historic Notes on Metallic Beams and Girders." *Proceedings, Engineers' Society of Western Pennsylvania* (February 1912): 11–81.

Lighting

Alglave, Em., and J. Boulard. *The Electric Light: Its History, Production and Applications.* New York: 1884.

Bell, Louis. "Elements of Illumination." *Electrical World and Engineer* 36 (10, 24 November, 8 December 1900): 727–728, 806–808, 882–884.

Bright, Arthur A., Jr. *The Electric-Lamp Industry.* New York: Macmillan Co., 1949.

Gerhard, William Paul. *Gas-Lighting and Gas-Fitting.* New York: 1894.

Hammer, Edwin H. "Incandescent Lamp Development to the Year 1880." *Electrical World and Engineer* 36 (1, 8, 15 December 1900): 839–841, 880–882, 918–919.

Liberty, W. J. "The Centenary of Gas Lighting and Its Historical Development." *Illuminating Engineer* 7 (April 1913): 175–231.

Morgan, M. R. "The Early Development of Gas Lighting." *American Gas Journal* 95 (30 October, 6 November 1911): 278–283, 289–295, 298–301.

Scott, Roscoe. "The Evolution of the Lamp." *Transactions of the Illuminating Engineering Society* 9 (February 1914): 138–163.

Williams, W. Mattieu. "A Contribution to the History of Electric Lighting." *Journal of Science* 16 (February 1879): 155–162.

Lightning Protection

Hedges, Killingworth. *Modern Lightning Conductors: An Illustrated Supplement to the Report of the Lightning Research Committee of 1905.* London: Crosby Lockwood and Sons, 1905.

Lodge, Oliver J. *Lightning Conductors and Lightning Guards: A Treatise on the Protection of Buildings.* London: 1892.

Symonds, G. J., ed. *Lightning Rod Conference: Report of the Delegates.* London: 1882.

Masonry

Bowles, Oliver. *The Stone Industries.* 2d ed. New York: McGraw-Hill Book Company, 1939.

Chamberlain, Humphrey. "The Manufacture of Bricks by Machinery." *Journal of the Royal Society of Arts* 4 (6 June 1856): 491–501.

Green, A. T., and Gerald H. Stewart, eds. *Ceramics: A Symposium.* Stoke-on-Trent: British Ceramic Society, 1953.

Hamilton, S. B. "The History of Hollow Brick." *Transactions, British Ceramic Society* 58 (February 1959): 41–61.

"History of the Manufacture of Hollow Bricks." *Building News* 4 (26 March 1858): 317–318.

Hudson, Kenneth. *The Fashionable Stone.* Park Ridge, N.J.: Noyes Press, 1972.

Mars, G. C. "Brick—Its Manufacture." *American Architect—Architectural Review* 123 (11 April 1923): 313–318, 383–387.

Merrill, George P. *Stones for Building and Decorating.* 3d ed. New York: John Wiley and Sons, 1903.

Ries, Heinrich, and Henry Leighton. *History of the Clay-Working Industry in the United States.* New York: John Wiley and Sons, 1909.

Shaw, W. "British Quarry Practice." *Rock Products* 25 (18 November 1922): 23–24.

Reinforced Concrete

Colby, Albert Ladd. *Reinforced Concrete in Europe.* Easton, Pa.: Chemical Publishing Co., 1909.

de Tedesco, N. "The Historical Evolution of Reinforced Concrete in France." *Concrete and Constructional Engineering* 1 (July 1906): 159–170.

Draffin, Jasper O. "A Brief History of Lime, Cement, Concrete and Reinforced Concrete." *Journal of the Western Society of Engineers* 48 (March 1943): 14–47.

Fling, Russell S. *Early Developments in American Concrete.* Preprint 2791. New York: American Society of Civil Engineers, 1976.

Hamilton, S. B. *A Note on the History of Reinforced Concrete in Buildings.* National Building Studies. Special Report no. 24. London: HMSO, 1956.

Hatt, W. K. "Genesis of Reinforced Concrete Construction." *American Concrete Institute, Proceedings of the 12th Convention* 12 (1916): 21–39.

Kohlmorgen, O. "The Historical Evolution of Reinforced Concrete in Germany." *Concrete and Constructional Engineering* 1 (November 1906): 325–329.

Planat, Paul. "Expériences sur les ciments armés." *Construction Moderne* 14 (25 March, 29 April, 27 May, 3 June, 22 July, 5, 12 August 1899): 309–311, 368–371, 416–418, 429–431, 514–515, 538–540, 548–549.

Potter, Thomas. *Concrete: Its Use in Building.* 2 vols. Winchester: 1891–1894.

Ransome, Ernest L., and Alexis Saurbrey. *Reinforced Concrete Buildings.* New York: McGraw-Hill, 1912.

Titford, R. M., ed. *The Golden Age of Concrete.* London: Dorothy Henry Publications, 1964.

Sanitation

Brown, Glen. *Water-Closets. A Historical, Mechanical and Sanitary Treatise.* New York: 1884.

Clark, Theodore Minot. "Modern Plumbing." *American Architect and Building News* 3–4 (23, 30 March, 27 April, 25 May, 13 July, 3, 31 August, 14 September, 26 October, 30 November 1878): 101–102, 109–110, 144–145, 180; 11–13, 38–40, 73–75, 90–92, 140–141, 178–180.

Gerhard, William Paul. *The Water Supply, Sewerage, and Plumbing of Modern City Buildings.* New York: John Wiley and Sons, 1910.

Lamb, H. A. J. "Sanitation: An Historical Survey." *Architects' Journal* 85 (4 March 1937): 385–403.

Palmer, Roy. *The Water Closet: A New History.* Newton Abbot: David and Charles, 1973.

Wright, Lawrence. *Clean and Decent.* New York: Viking Press, 1960.

Structural Engineering

Charlton, T. M. *A History of Theory of Structures in the Nineteenth Century.* Cambridge: Cambridge University

Press, 1982.

Hamilton, Stanley Baines. "The Historical Development of Structural Theory." *Proceedings, Institution of Civil Engineers, Part 3: Engineering Division* 1 (1952): 374–419.

Peck, Ralph B. *History of Building Foundations in Chicago.* University of Illinois, Engineering Experiment Station, Bulletin Series, no. 373. Urbana-Champaign: University of Illinois, 1948.

Randall, Frank A., Jr. "Historical Notes on Structural Safety." *Proceedings of the Annual Convention, American Concrete Institute* 70 (1973): 669–679.

Sooysmith, Charles. "Foundation Construction for Tall Buildings." *Engineering Magazine* 13 (April 1897) 20–33.

Straub, Hans. *A History of Civil Engineering.* Trans. Erwin Rockwell. Cambridge: MIT Press, 1964.

Westergaard, H. M. "One Hundred Fifty Years Advance in Structural Analysis." *Transactions, American Society of Civil Engineers* 94 (1930): 226–246.

Terra-Cotta

Geer, Walter. *The Story of Terra Cotta.* New York: Tobias A. Wright, 1920.

King, Harry Lee. "A History of Architectural Terra-Cotta." *Architecture and Building* 45 (July 1913): 208–211, 311–316.

"The Manufacture of Terra-Cotta in Chicago." *American Architect and Building News* 1 (30 December 1876): 420–421.

Taylor, James. "The History of Terra Cotta in New York." *Architectural Record* 2 (December 1892): 137–148.

Wood

Hindle, Brooke, ed. *America's Wooden Age: Aspects of Technology.* Tarrytown, N.Y.: Sleepy Hollow Restoration, 1975.

Larson, Agnes M. *History of the White Pine Industry in Minnesota.* Minneapolis: University of Minnesota Press, 1949.

Latham, Bryan. *Timber: Its Development and Distribution: A Historical Survey.* London: George G. Harrap, 1957.

One Hundred Years of Engineering Progress with Wood. Proceedings of Wood Symposium. Washington: Timber Engineering Company, circa 1952.

The Saw in History. Philadelphia: Henry Disston and Sons, 1915.

Wood, Andrew Dick, and Thomas Gray Linn. *Plywoods: Their Development, Manufacture and Application.* Brooklyn: Chemical Publishing Company, Inc., 1943.

Index

New York, continued